Riemann and Klein Surfaces, Automorphisms, Symmetries and Moduli Spaces

CONTEMPORARY MATHEMATICS

629

Riemann and Klein Surfaces, Automorphisms, Symmetries and Moduli Spaces

Conference in Honor of Emilio Bujalance
Riemann and Klein Surfaces, Symmetries and Moduli Spaces
June 24–28, 2013
Linköping University, Linköping, Sweden

Milagros Izquierdo
S. Allen Broughton
Antonio F. Costa
Rubí E. Rodríguez
Editors

American Mathematical Society
Providence, Rhode Island

2010 *Mathematics Subject Classification.* Primary 30Fxx, 14Hxx, 20Hxx, 57Mxx, 14Gxx, 14Kxx, 14Jxx.

Library of Congress Cataloging-in-Publication Data

Riemann and Klein surfaces, automorphisms, symmetries and moduli spaces : conference in honor of Emilio Bujalance on Riemann and Klein surfaces, symmetries and moduli spaces, June 24–28, 2013, Linkoping University, Linkoping, Sweden / Milagros Izquierdo, S. Allen Broughton, Antonio F. Costa, Rubi E. Rodriquez, editors.

pages cm. — (Contemporary mathematics ; volume 629)

Includes bibliographical references.

ISBN 978-1-4704-1093-3 (alk. paper)

1. Riemann surfaces—Congresses. 2. Geometry, Algebraic—Congresses. 3. Automorphisms—Congresses. 4. Moduli theory—Congresses. I. Izquierdo, Milagros, 1963– editor of compilation. II. Broughton, S. Allen, 1951– editor of compilation. III. Costa, A. F., 1960– editor of compilation. IV. Rodriguez, Rubi E., 1953– editor of compilation.

QA333.R54 2014
515′.93—dc23 2014012547

Contemporary Mathematics ISSN: 0271-4132 (print); ISSN: 1098-3627 (online)

DOI: `http://dx.doi.org/10.1090/conm/629`

♾ The paper used in this book is acid-free and falls within the guidelines
established to ensure permanence and durability.
Visit the AMS home page at `http://www.ams.org/`

10 9 8 7 6 5 4 3 2 1 19 18 17 16 15 14

To our master and colleague Prof. Bujalance
and specially to our great friend Emilio who united us all in mathematics

Contents

Preface

Group actions on Riemann surfaces, Klein surfaces, and related structures such as hyperbolic manifolds, complex surfaces, and Teichmüller spaces constitute an important nexus for group theory, combinatorics, geometry, topology, Galois theory and complex analysis. Compact Riemann surfaces are complex projective curves and have been studied for well over one hundred years. These surfaces have spawned a rich, interconnected theory in algebraic geometry, complex analysis, hyperbolic geometry, topology, and combinatorial/geometric group theory using tools such as uniformizing Fuchsian groups, Jacobian varieties, and moduli spaces. The use of combinatorial/geometric group theory for studying Riemann surfaces goes all the way back to Klein, Hurwitz and others; has continued through topological methods such as mapping class groups, Teichmüller and moduli spaces; and is a key part of the more recent developments in cryptographic applications, the Grothendieck theory of dessins d'enfants, and the use of powerful computer algebra systems such as Magma and GAP.

This volume is based on the conference, Riemann and Klein Surfaces, Symmetries, and Moduli Spaces, that took place at Linköping University 24–28 June 2013, in honor of Professor Emilio Bujalance. The conference and this volume are devoted to the mathematics that Professor Bujalance and his many colleagues have worked on, with special focus on Riemann and Klein surfaces, i.e., complex curves, automorphisms of surfaces (real and complex), group actions on surfaces, Grothendieck theory of dessins d'enfants, topological properties of moduli spaces of complex curves, and Abelian varieties. The conference had a computational/combinatorial flavour; the presence of a group action—geometric, combinatorial, or topological—was felt in almost every paper presented.

There is considerable current activity in these fields. As noted, these topics range from classical problems of classifying Riemann and Klein surfaces and their automorphism groups to the discovery of the topological structure of moduli spaces and the Grothendieck theory of dessins d'enfants. This volume should be a very useful source for anyone working in the area since the conference covered such a wide spectrum of topics. To guide mathematicians new to the area, several articles of an expository nature have been included. The volume will also serve as a tool in communicating the interplay among the different areas.

S. Allen Broughton, Antonio F. Costa, Milagros Izquierdo and Rubí E. Rodríguez

On Jacobians of curves with superelliptic components

L. Beshaj, T. Shaska, and C. Shor

Dedicated to Professor Emilio Bujalance on the occasion of his 60th birthday

ABSTRACT. We construct a family of non-hyperelliptic curves whose Jacobians decomposes into a product of superelliptic Jacobians. Moreover, we investigate the decomposition of Jacobians of superelliptic curves based on their automorphisms. For a curve given by the equation $y^n = f(x^m)$, we provide a necessary and sufficient condition in terms of m and n for the the Jacobian of the curve to decompose.

1. Introduction

Let $\mathcal{X}$ be a genus $g \geq 2$ smooth, irreducible projective curve, defined over an algebraically closed field k and $\pi_i : \mathcal{X} \to \mathcal{X}_i$, $1 \leq i \leq s$ be coverings to genus g_i curves $\mathcal{X}_i$, respectively. Denote by $\text{Jac}(\mathcal{X})$ and $\text{Jac}(\mathcal{X}_i)$ the Jacobians of these curves respectively. A classical question is to determine when the Jacobian $\text{Jac}(\mathcal{X})$ is isogenous to a product of $\text{Jac}(\mathcal{X}_i)$. The problem has been studied in the XIX century by Legendre, Jacobi, Hermite, Klein, Kovalevskaya, Hecke, et al. In this context, these curves were of interest due to the fact that split Jacobians give relations between abelian and elliptic integrals. Most of these efforts were focused on the case when the coverings $\pi_i : \mathcal{X} \to \mathcal{X}_i$ are Galois coverings.

The case when genus $g(\mathcal{X}_i) = 1$ for some i and $\pi_i : \mathcal{X} \to \mathcal{X}_i$ not necessarily Galois is especially interesting due to its implications in number theory and relations to Lang's conjecture; see [**7**, **8**] and [**10**]. When $g = 2$ such coverings have different ramification structures when $\deg \pi_i$ is odd or even; see [**4**] for a summary of results in genus two. In both cases the moduli space of such covers with fixed ramification structure can be embedded as a subvariety of $\mathcal{M}_2$.

The Jacobian $\text{Jac}(\mathcal{X})$ is said to (completely) split if it is isogenous to the product of elliptic curves. Ekedahl and Serre [**6**] posed the following questions: i) Is it true that for every integer $g > 1$ there exists a curve of genus g with split Jacobian? ii) Is there a bound on the genus of a curve with split Jacobian? Towards answering these questions they used modular curves, and coverings of curves of genus 2 and 3. The maximal genus of their examples was 1297. Many other authors have considered these questions as well, and in recent work, Shaska [**15**, **16**] and Yamauchi [**20**], have produced somewhat unusual families of curves with decomposable Jacobians.

In this paper we will focus on the cases when the coverings $\pi_i : \mathcal{X} \to \mathcal{X}_i$ are cyclic Galois coverings. Hence, these coverings induce automorphisms on $\mathcal{X}$. By

2010 *Mathematics Subject Classification.* Primary 11G10, 14K02; Secondary 11G30, 14Q05.

focusing on such curves we have a more organized way to study the decompositions of such Jacobians starting from the automorphism group $\mathrm{Aut}(\mathcal{X})$ of $\mathcal{X}$. While we can determine the full list of groups that occur as full automorphism groups of genus $g \geq 2$, the corresponding equations of curves for each group are not known; see [**3**]. There is, however, a nice family of curves when such decomposition can be fully explored and the factors of such Jacobians can be fully determined. This is the class of superelliptic curves, namely the curves which can be written as affine varieties with equation $y^n = f(x)$ for some $n \geq 2$ and discriminant $\Delta(f, x) \neq 0$. For such curves we know precisely the structure of the automorphism group G, the signature, the equations of the curves, and their invariants; see [**5**]. Hence, it is possible to determine explicitly the components of such Jacobians and in some cases the moduli space of such curves in the moduli space $\mathcal{M}_g$. A complete description of curves for which such decompositions is based on their automorphisms is intended in [**3**].

The second goal of the paper is to extend the family of curves introduced by Yamauchi in [**20**]. The family of curves introduced in [**20**] are non-hyperelliptic curves with arbitrary large genus such that the Jacobian has hyperelliptic components. We attempt to extend this family to curves whose Jacobians have superelliptic components. Our proof is based on the automorphisms of curves. We show that such decomposition based on the automorphisms of the curve induces some arithmetic condition among the orders of such automorphisms. This condition implies that the family $F_{m,n}$ constructed in [**20**] agrees with our family of curves only for the values $F_{m,1}$ and $F_{m,2}$.

This paper is organized as follows. In section 2 we briefly define the Jacobian of a curve and give two classical results of Accola; see [**1, 2**]. In these results it is described how one can start from a partition of a group of automorphisms of the curve $\mathcal{X}$ and get a decomposition of the $\mathrm{Jac}(\mathcal{X})$. Further in this section we describe results of Kani and Rosen were the previous results were generalized; see [**9**]. We use such results in our decomposition of Jacobians of the superelliptic curves for the family of curves described in section 4.

In section 3 we study the decomposition of Jacobians of superelliptic curves. A superelliptic curve $\mathcal{X}$ is a curve with equation $y^n = f(x)$ defined over a field k, where $(\mathrm{char}\, k, n) = 1$ and k is algebraically closed. Such curves have the *superelliptic* automorphism $\tau : (x, y) \to (x, \zeta_n y)$, where ζ_n is an n-th primitive root of unity and τ is central in $G := \mathrm{Aut}(\mathcal{X})$. Let σ be another automorphism of $\mathcal{X}$ such that its projection $\bar{\sigma} \in \overline{G} := G/\langle \tau \rangle$ has order m. Then σ and $\sigma\tau$ fix two subfields of the function field $k(\mathcal{X})$ and therefore there are two quotient curves $\mathcal{X}_1 := \mathcal{X}/\langle \sigma \rangle$ and $\mathcal{X}_2 := \mathcal{X}/\langle \sigma\tau \rangle$. We determine equations of $\mathcal{X}_1$ and $\mathcal{X}_2$ and a necessary and sufficient condition in terms of m and n such that $\mathrm{Jac}(\mathcal{X})$ is isogenous to $\mathrm{Jac}(\mathcal{X}_1) \times \mathrm{Jac}(\mathcal{X}_2)$.

In section 4 we generalize a construction of Yamauchi of a family $F_{m,n}$ of non-hyperelliptic curves with decomposable Jacobians, where all components are hyperelliptic Jacobians. Instead we construct a family $\mathcal{X}_{r,s}$ of curves whose Jacobians are superelliptic Jacobians. We prove that the automorphism group of these *component curves* are cyclic or dihedral groups. Moreover, we find a necessary and sufficient condition in terms of r and s for this decomposition to occur. Our decomposition is based solely on the decomposition induced by the automorphisms of the curves.

Notation: Throughout this paper by g we denote an integer ≥ 2 and k denotes an algebraically closed field. Unless otherwise noted, by a *curve* we always mean the isomorphism class of an algebraic curve defined over k. The automorphism group of a curve always means the full automorphism group of the curve over k. We denote the cyclic group of order n by C_n and the dihedral group of order $2n$ by D_{2n}. V_4 denotes the Klein 4-group and ζ_n denotes an n-th primitive root of unity.

2. Preliminaries

Let $\mathcal{X}$ be a genus $g \geq 2$ algebraic curve defined over $\mathbb{C}$. We choose a symplectic homology basis for $\mathcal{X}$, say $\{A_1, \dots, A_g, B_1, \dots, B_g\}$, such that the intersection products $A_i \cdot A_j = B_i \cdot B_j = 0$ and $A_i \cdot B_j = \delta_{ij}$, where δ_{ij} is the Kronecker delta. We choose a basis $\{w_i\}$ for the space of holomorphic 1-forms such that $\int_{A_i} w_j = \delta_{ij}$. The matrix $\Omega = \left[\int_{B_i} w_j \right]$ is the *Riemann matrix* of $\mathcal{X}$ and the matrix $[I \mid \Omega]$ is called the *period matrix*. The columns of the matrix $[I \mid \Omega]$ form a lattice L in $\mathbb{C}^g$. The complex torus $\mathbb{C}^g/L$ is called the Jacobian of $\mathcal{X}$ is denoted by $\mathrm{Jac}(\mathcal{X})$.

Let $\mathcal{H}_g$ be the *Siegel upper-half space*. Then $\Omega \in \mathcal{H}_g$ and there is an injection

$$\mathcal{M}_g \hookrightarrow \mathcal{H}_g / Sp_{2g}(\mathbb{Z}) =: \mathcal{A}_g$$

where $Sp_{2g}(\mathbb{Z})$ is the *symplectic group*. A non-constant morphism $f : A \to B$ between two Abelian varieties which is surjective and of finite kernel is called an **isogeny**. An Abelian variety is called *decomposable* if it is isogenous to a product of Abelian varieties, it is *simple* if it has no non-trivial Abelian subvarieties. An Abelian variety is called **completely decomposable** or **completely split** if it is isogenous to a product of elliptic curves.

A map of algebraic curves $f : \mathcal{X} \to \mathcal{Y}$ is called a **maximal covering** if it does not factor over a nontrivial isogeny. A map $f : \mathcal{X} \to \mathcal{Y}$ induces maps between their Jacobians $f^* : \mathrm{Jac}(\mathcal{Y}) \to \mathrm{Jac}(\mathcal{X})$ and $f_* : \mathrm{Jac}(\mathcal{X}) \to \mathrm{Jac}(\mathcal{Y})$. When f is maximal then f^* is injective and $\ker(f_*)$ is connected, see [**14**, p. 158] for details. Hence, $\mathrm{Jac}(\mathcal{X}) \cong \mathrm{Jac}(\mathcal{Y}) \times A$, where A is some Abelian variety. Thus, coverings $f : \mathcal{X} \to \mathcal{Y}$ give factors of the Jacobian $\mathrm{Jac}(\mathcal{X})$. Such methods have been explored for genus 2 curves by Shaska et al. in [**11, 17, 18**]. They are the only examples that we know when explicit computations have been performed and the corresponding locus has been computed for non-Galois coverings.

If the covering $f : \mathcal{X} \to \mathcal{Y}$ is a Galois covering then its monodromy group is isomorphic to a subgroup H of the automorphism group $G = \mathrm{Aut}(\mathcal{X})$. Hence a common procedure to produce decompositions of Jacobians is to explore the automorphism group of the curve.

Fix an integer $g \geq 2$ and a finite group G. Let $C_1, \dots, C_r$ be conjugacy classes $\neq \{1\}$ of G. Let $\mathbf{C} = (C_1, \dots, C_r)$, be an unordered tuple, repetitions are allowed. We allow r to be zero, in which case $\mathbf{C}$ is empty.

Consider pairs $(\mathcal{X}, \mu)$, where $\mathcal{X}$ is a curve and $\mu : G \to \mathrm{Aut}(\mathcal{X})$ is an injective homomorphism. Mostly we will suppress μ and just say $\mathcal{X}$ is a curve with G-action, or a G-curve, for short. Two G-curves $\mathcal{X}$ and $\mathcal{X}'$ are called equivalent if there is a G-equivariant isomorphism $\mathcal{X} \to \mathcal{X}'$.

We say a G-curve $\mathcal{X}$ is **of ramification type** $(g, G, \mathbf{C})$ if g is the genus of $\mathcal{X}$ and the points of the quotient $\mathcal{X}/G$ that are ramified in the cover $\mathcal{X} \to \mathcal{X}/G$ can be labelled as $p_1, \dots, p_r$ such that C_i is the conjugacy class in G of distinguished inertia group generators over p_i (for $i = 1, \dots, r$).

Let $\mathcal{X}$ be a G-curve and $H < G$. Then there is a covering $\mathcal{X} \to \mathcal{X}/H$. Let the genus of $\mathcal{X}/H$ be denoted by g_H. How is g_H determined in terms of g and G? Consider the following problem: let $H_1, \dots, H_r$ be subgroups of G and $g_1, \dots, g_r$ the genera of the $\mathcal{X}/H_1, \dots, \mathcal{X}/H_r$ respectively. Is there any arithmetic relation between $g, g_1, \dots, g_r$?

Accola proved the following results which provide a method for decomposing Jacobians, see [**1**].

THEOREM 2.1 ([**1**]). *Let $\mathcal{X}_g$ be a G-curve, $H_i < G$, g_i the genus of $\mathcal{X}_g/H_i$ for $i = 1, \dots s$, and $g_0 = g(\mathcal{X}_g/G)$. Assume that $H_i \cap H_j = \{e\}$ for all $i \neq j$. Then,*

$$g_0 \, |G| = g - sg + \sum_{i=1}^{s} |H_i| \, g_i.$$

Moreover, we have the following relation among the genera; see [**2**].

THEOREM 2.2 ([**2**]). *Let $\mathcal{X}_g$ be a G-curve, and H_i, for $i = 1, \dots s$, subgroups of G such that $G = \cup_i^s H_i$. Denote by g_i the genus of $\mathcal{X}_g/H_i$ and by $H_{ij\dots k} = H_i \cap H_j \cap \dots \cap H_k$.*

Then,

$$g_0 \, |G| = \sum_{i=1}^{s} |H_i| \cdot g_i - \sum |H_{ij}| \cdot g_{ij} + \sum |H_{ijk}| \cdot g_{ijk} - \dots - (-1)^s \sum |H_{12\dots s}| \cdot g_{12..s}.$$

See [**2**] for details.

2.1. Decomposing the Jacobian by group partitions. Let $\mathcal{X}$ be a genus g algebraic curve with automorphism group $G := \mathrm{Aut}(\mathcal{X})$. Let $H \leq G$ such that $H = H_1 \cup \dots \cup H_t$ where the subgroups $H_i \leq H$ satisfy $H_i \cap H_j = \{1\}$ for all $i \neq j$. Then,

$$\mathrm{Jac}^{t-1}(\mathcal{X}) \times \mathrm{Jac}^{|H|}(\mathcal{X}/H) \cong \mathrm{Jac}^{|H_1|}(\mathcal{X}/H_1) \times \cdots \mathrm{Jac}^{|H_t|}(\mathcal{X}/H_t).$$

The group H satisfying these conditions is called a group with partition. Elementary Abelian p-groups, the projective linear groups $PSL_2(q)$, Frobenius groups, dihedral groups are all groups with partition.

Let $H_1, \dots, H_t \leq G$ be subgroups with $H_i \cdot H_j = H_j \cdot H_i$ for all $i, j \leq t$, and let g_{ij} denote the genus of the quotient curve $\mathcal{X}/(H_i \cdot H_j)$. Then, for $n_1, \dots, n_t \in \mathbb{Z}$ the conditions $\sum n_i n_j g_{ij} = 0$, $\sum_{j=1}^{t} n_j g_{ij} = 0$, imply the isogeny relation

$$\prod_{n_i > 0} \mathrm{Jac}^{n_i}(\mathcal{X}/H_i) \cong \prod_{n_j < 0} \mathrm{Jac}^{n_j}(\mathcal{X}/H_j). \tag{2.1}$$

In particular, if $g_{ij} = 0$ for $2 \leq i < j \leq t$ and if $g = g_{\mathcal{X}/H_2} + \dots + g_{\mathcal{X}/H_t}$, then

$$\mathrm{Jac}(\mathcal{X}) \cong \mathrm{Jac}(\mathcal{X}/H_2) \times \cdots \times \mathrm{Jac}(\mathcal{X}/H_t). \tag{2.2}$$

The proof of the above statements can be found in [**9**].

3. Jacobians of superelliptic curves

A curve $\mathcal{X}$ is called superelliptic if there exist an element $\tau \in G := \mathrm{Aut}(\mathcal{X})$ such that τ is central and $g\left(\mathcal{X}/\langle\tau\rangle\right) = 0$. Let $\mathcal{X}_g$ have affine equation given by some polynomial in terms of x and y and denote by $K = k(x, y)$ the function field of $\mathcal{X}_g$. Let $H = \langle\tau\rangle$ be a cyclic subgroup of G such that $|H| = n$ and $H \triangleleft G$, where $n \geq 2$. Moreover, we assume that the quotient curve $\mathcal{X}_g/H$ has genus zero. The

reduced automorphism group of $\mathcal{X}_g$ with respect to H is called the group $\bar{G} := G/H = \overline{\mathrm{Aut}}(\mathcal{X})$.

Assume $k(x)$ is the genus zero subfield of K fixed by H. Hence, $[K : k(x)] = n$. Then, the group $\bar{G}$ is a subgroup of the group of automorphisms of a genus zero field. Hence, $\bar{G} < PGL_2(k)$ and $\bar{G}$ is finite. It is a classical result that every finite subgroup of $PGL_2(k)$ is isomorphic to one of the following: C_m, D_{2m}, A_4, S_4, A_5.

The group $\bar{G}$ acts on $k(x)$ via the natural way. The fixed field of this action is a genus 0 field, say $k(z)$. Thus, z is a degree $|\bar{G}|$ rational function in x, say $z = \phi(x)$.

It is obvious that G is a degree n central extension of $\bar{G}$ and $\bar{G}$ is a finite subgroup of $PGL_2(k)$. Hence, if we know all the possible groups that occur as $\bar{G}$ then we can determine G and the equation for K; see [**12**]. In the next Lemma we establish some basic properties of superelliptic curves, the proof of the part i) can be found in [**19**] and part ii) in [**5**].

LEMMA 3.1. *Let $\mathcal{X}_g$ be a superelliptic curve with affine equation $y^n = f(x)$ where $\Delta(f,x) \neq 0$ and $\deg f = d > n$. Then the following hold:*

i) $\mathcal{X}_g$ has genus

$$g = 1 + \frac{1}{2}\left(nd - n - d - \gcd(d,n)\right).$$

If d and n are relatively prime then $g = \frac{(n-1)(d-1)}{2}$.

ii) Let $\sigma \in \mathrm{Aut}(\mathcal{X}_g)$ such that its projection $\bar{\sigma} \in \overline{\mathrm{Aut}}(\mathcal{X}_g)$ has order $m \geq 2$. Then its equation is given as $y^n = g(x^m)$ or $y^n = xg(x^m)$ for some $g \in k[x]$.

Let $\mathcal{X}_g$ be a superelliptic curve and $\sigma \in \mathrm{Aut}(\mathcal{X}_g)$ such that its projection $\bar{\sigma} \in \overline{\mathrm{Aut}}(\mathcal{X}_g)$ has order $m \geq 2$. We can choose a coordinate in $\mathbb{P}^1$ such that $\bar{\sigma}(X) = X^m$. Since σ permutes the Weierstrass points of $\mathcal{X}$ and it has two fixed points then the equation of the curve will be $y^n = f(x^m)$ or $y^n = xf(x^m)$ as claimed in ii) above.

Assume that $\mathcal{X}$ has equation

$$Y^n = f(X^m) := X^{\delta m} + a_1 X^{(\delta-1)m} \cdots + a_{\delta-1}X^m + 1. \tag{3.1}$$

We assume that $\bar{\sigma}$ lifts to G to an element of order m. Then, $\sigma(X,Y) \to (\zeta_m X, Y)$. Denote by $\tau : (X,Y) \to (X, \zeta_n Y)$ its superelliptic automorphism. Since τ is central in G then $\tau\sigma = \sigma\tau$. We will denote by $\mathcal{X}_1$ and $\mathcal{X}_2$ the quotient curves $\mathcal{X}/\langle\sigma\rangle$ and $\mathcal{X}/\langle\tau\sigma\rangle$ respectively. The next theorem determines the equations of $\mathcal{X}_1$ and $\mathcal{X}_2$. We denote by K the function field of $\mathcal{X}_g$ and by F and L the function fields of $\mathcal{X}_1$ and $\mathcal{X}_2$ respectively.

THEOREM 3.1. *Let K be a genus $g \geq 2$ level n superelliptic field and F a degree m subfield fixed by $\sigma : (X,Y) \to (\zeta_m X, Y)$.*

i) Then, $K = k(X,Y)$ such that

$$Y^n = f(X^m) := X^{\delta m} + a_1 X^{(\delta-1)m} \cdots + a_{\delta-1}X^m + 1. \tag{3.2}$$

for $\Delta(f,x) \neq 0$.

ii) $F = k(U,V)$ where $U = X^m$, and $V = Y$ and

$$V^n = f(U). \tag{3.3}$$

iii) There is another subfield $L = k(u,v)$ where

$$u = X^m, \quad v = X^i Y,$$

and

$$v^n = u \cdot f(u), \tag{3.4}$$

for $m = \lambda n$ *and* $i = \lambda(n-1)$.

PROOF. The proof of i) follows from the above remarks. To show that the subfield F is generated by X^m and Y it is enough to show that it is fixed by σ. This is obvious.

In iii) we have to show that $u = X^m$ and $v = X^i Y$ are fixed by $\sigma\tau : (X, Y) \to (\zeta_m X, \zeta_n Y)$. We have that $\sigma\tau(v) = \sigma\tau(X^i Y) = \zeta_m^{\lambda(n-1)} \zeta_n \cdot X^i Y = X^i Y$. It is easily checked that $\zeta_m^{\lambda(n-1)} \zeta_n = 1$. □

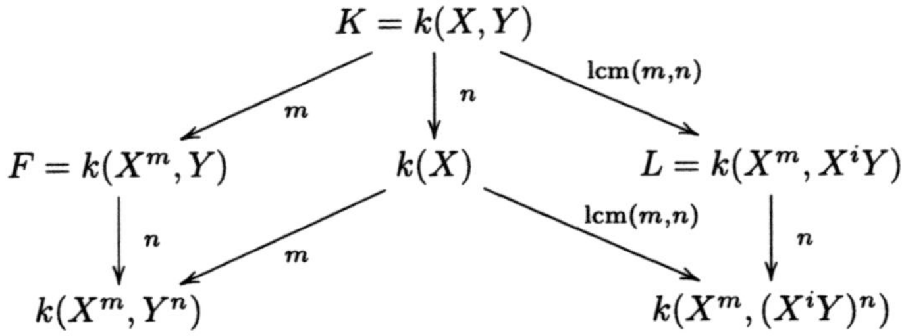

FIGURE 1. The lattice of a genus g superelliptic field with an extra automorphism

In the rest of this section we want to find necessary and sufficient conditions on n and m such that the Jacobian $\text{Jac}(\mathcal{X}_g)$ is isogenous to the product $\text{Jac}(\mathcal{X}_1) \times \text{Jac}(\mathcal{X}_2)$. First we focus on hyperelliptic curves.

3.1. Hyperelliptic curves.

THEOREM 3.2. *Let* $\mathcal{X}_g$ *be a hyperelliptic curve and its reduced automorphism group* $\overline{\text{Aut}}(\mathcal{X}_g) \cong C_m = \langle \sigma \rangle$. *Then* $\mathcal{X}_g$ *is isomorphic to a curve with equation*

$$\mathcal{X}_g : Y^2 = x^{\delta m} + a_1 x^{(\delta-1)m} + \cdots + a_{\delta-1} x^m + 1.$$

There exists subcovers $\pi_i : \mathcal{X}_g \to \mathcal{X}_i$, *for* $i = 1, 2$ *such that*

$$\begin{aligned} \mathcal{X}_1 : \quad & Y^2 = X^\delta + a_1 X^{\delta-1} + \cdots + a_{\delta-1} X + 1, \\ \mathcal{X}_2 : \quad & Y^2 = X(X^\delta + a_1 X^{\delta-1} + \cdots + a_{\delta-1} X + 1). \end{aligned}$$

The Jacobian of $\mathcal{X}$ *is isogenous to the product*

$$\text{Jac}(\mathcal{X}) \cong \text{Jac}(\mathcal{X}_1) \times \text{Jac}(\mathcal{X}_2)$$

if and only if the full automorphism group $\text{Aut}(\mathcal{X})$ *is isomorphic to the Klein 4-group* V_4.

PROOF. Assume $\mathcal{X}_g$ is a hyperelliptic curve and $\sigma \in \bar{G} \cong C_m$. The equation of the hyperelliptic curve $\mathcal{X}_g$ is

$$\mathcal{X}_g : Y^2 = x^{\delta m} + a_1 x^{(\delta-1)m} + \cdots + a_{\delta-1} x^m + 1.$$

Denote with τ the hyperelliptic involution. We have an extra involution $\sigma \in \bar{G} \cong C_m$ and since extra involutions come in pairs we have $\sigma\tau \in \bar{G}$. The two fixed spaces of σ and $\sigma\tau$ have equations respectively

$$\mathcal{X}_1 : \quad Y^2 = X^\delta + a_1 X^{\delta-1} + \cdots + a_{\delta-1} X + 1$$

and

$$\mathcal{X}_2 : \quad Y^2 = X(X^\delta + a_1 X^{\delta-1} + \cdots + a_{\delta-1} X + 1),$$

where $X = x^m$. The genera g_1, g_2 are respectively $[\frac{\delta-1}{2}]$ and $[\frac{\delta}{2}]$. Assume δ is even, say $\delta = 2k$, then we have $g = g_1 + g_2 = \delta - 1$. When δ is odd, say $\delta = 2k+1$ we have $g = g_1 + g_2 = \delta - 1$ and from [**5**, Table 3], δ is as follows $\frac{2(g+1)}{m}$, $\frac{2g+1}{m}$, $\frac{2g}{m} - 1$ and m respectively $\frac{2(g+1)}{\delta}$, $\frac{2g+1}{\delta+1}$, $\frac{2g}{\delta+1}$.

Considering each case we get $m = 2$ in the first case and m not an integer in the other two. Therefore, $G \cong V_4$.

Conversely, assume $G \cong V_4$. The equation of the hyperelliptic curve is

$$\mathcal{X}_g : Y^2 = x^{2g+2} + a_1 x^{2g} + \cdots + a_g x^2 + 1.$$

There are two extra involutions σ and $\sigma\tau$ in G such that $|\sigma| = |\sigma\tau| = 2$. They fix the following curves:

$$\mathcal{X}_1 : \quad Y^2 = X^{g+1} + a_1 X^g + \cdots + a_g X + 1$$

and

$$\mathcal{X}_2 : \quad Y^2 = X(X^{g+1} + a_1 X^g + \cdots + a_g X + 1),$$

where $X = x^2$. If we evaluate the genera we have respectively $g_1 = \left[\frac{(g+1)-1}{2}\right] = \left[\frac{g}{2}\right]$ and $g_2 = \left[\frac{(g+2)-1}{2}\right] = \left[\frac{g+1}{2}\right]$. Thus $g = g_1 + g_2$. Therefore, Jacobian is isogenous to

$$\text{Jac}(\mathcal{X}) \cong \text{Jac}(\mathcal{X}_1) \times \text{Jac}(\mathcal{X}_2).$$

This completes the proof. $\square$

3.2. Non-hyperelliptic curves. Next, we generalize the previous theorem.

THEOREM 3.3. *Let $\mathcal{X}_g$ be a level n superelliptic curve and $C_m = \langle \bar{\sigma} \rangle \hookrightarrow \overline{\text{Aut}}(\mathcal{X}_g)$, where $m \geq 2$ and the equation of $\mathcal{X}_g$ is $y^n = f(x^m)$, with $\deg(f) = d = \delta\, m$, $d > n$. Then there exist degree m coverings $\pi : \mathcal{X}_g \to \mathcal{X}_i$, $i = 1, 2$ where*

$$\mathcal{X}_1 : \quad Y^n = f(X) \quad \textit{and} \quad \mathcal{X}_2 : \quad Y^n = X f(X).$$

Then,

$$\text{Jac}(\mathcal{X}) \cong \text{Jac}(\mathcal{X}_1) \times \text{Jac}(\mathcal{X}_2)$$

if and only if

$$\delta(n-1)(m-2) = 1 - (\gcd(\delta+1, n) + \gcd(\delta, n) - \gcd(\delta m, n)). \tag{3.5}$$

PROOF. Let $\mathcal{X}_g$ be a superelliptic curve with and extra automorphism of order $m \geq 2$ and equation $y^n = f(x^m)$. There is the superelliptic automorphism

$$\tau : \ (x, y) \to (x, \zeta_n y), \quad \textit{and} \quad \bar{\sigma} : \ (x, y) \to (\zeta_m x, y).$$

We denote by σ the lifting of $\bar{\sigma}$ in $\text{Aut}(\mathcal{X})$. Then, $\sigma\tau = \tau\sigma$.

Let $H_1 := \langle \sigma \rangle$ and $H_2 := \langle \sigma\tau \rangle$ be subgroups in G. Then, $|H_1| = n$ and $|H_2| = \text{lcm}(n, m)$. Thus, we have $H := H_1 \times H_2 \hookrightarrow G$. It is easy to check that $g(\mathcal{X}_g/(H_1 H_2)) = 0$.

Moreover, σ and $\sigma\tau$ fix the curves

$$\mathcal{X}_1: \quad Y^n = X^\delta + a_1 X^{\delta-1} + \cdots + a_{\delta-1} X + 1$$

and

$$\mathcal{X}_2: \quad Y^n = X(X^\delta + a_1 X^{\delta-1} + \cdots + a_{\delta-1} X + 1).$$

Let g_1 and g_2 denoted their genera respectively. From Lemma 3.1 we have that

$$g_1 = 1 + \frac{1}{2}\left(n\delta - n - \delta - \gcd(\delta, n)\right)$$

and

$$g_2 = 1 + \frac{1}{2}\left(n(\delta+1) - n - (\delta+1) - \gcd(\delta+1, n)\right).$$

Then we have

$$g_1 + g_2 = \frac{3}{2} + n\delta - \frac{n}{2} - \delta - \frac{1}{2}\left(\gcd(\delta, n) + \gcd(\delta+1, n)\right).$$

The genus of $\mathcal{X}$ is

$$g = 1 + \frac{1}{2}\left(n\delta m - n - \delta m - \gcd(m\delta, n)\right).$$

Then, $g = g_1 + g_2$ implies that

$$\delta(n-1)(m-2) = 1 - \left(\gcd(\delta+1, n) + \gcd(\delta, n) - \gcd(\delta m, n)\right).$$

Thus, from Eq. (2.2) we have that

$$\mathrm{Jac}(X_g) \cong \mathrm{Jac}(\mathcal{X}/H_1) \times \mathrm{Jac}(\mathcal{X}/H_2)$$

which completes the proof. □

COROLLARY 3.2. *Let $\mathcal{X}_g$ be a level n superelliptic curve as in Theorem 3.3. Furthermore, suppose n is prime. Then $m = 2$ or $m = 3$.*

In particular, one of the following situations is true:

- *n is any prime, $m = 2$, and $\delta \equiv 0 \pmod{n}$;*
- *$n = 2$, $m = 2$, and δ is odd;*
- *$n = 3$, $m = 3$, $\delta = 1$;*
- *n is any odd prime, $m = 2$, $\delta \not\equiv 0, -1 \pmod{n}$.*

PROOF. If n is prime, then the gcd's on the right hand side of the Eq. (3.5) are each either 1 or n. We consider cases below. Note that at least one of $\gcd(n, \delta)$ and $\gcd(n, \delta+1)$ equals 1 because n is prime.

If $\gcd(n, \delta) = n$, then $\gcd(n, m\delta) = n$, so

$$1 + \gcd(n, m\delta) - \gcd(n, \delta) - \gcd(n, \delta+1) = 1 + n - n - 1 = 0.$$

Thus, $\delta(m-2) = 0$. But $\delta \neq 0$, so $m = 2$. And since $\gcd(n, \delta) = n$, then $\delta \equiv 0 \pmod{n}$.

If $\gcd(n, \delta+1) = n$, then $\gcd(n, \delta) = 1$, so $\gcd(n, m\delta) = \gcd(n, m)$. Thus,

$$1 + \gcd(n, m\delta) - \gcd(n, \delta) - \gcd(n, \delta+1) = \gcd(n, m) - n.$$

Since we have $m \geq 2$, we know $\delta(n-1)(m-2) \geq 0$, so $\gcd(n, m) - n \geq 0$, so $\gcd(n, m) \geq n$, so $\gcd(n, m) = n$. Thus,

$$1 + \gcd(n, m\delta) - \gcd(n, \delta) - \gcd(n, \delta+1) = 0,$$

so $\delta(n-1)(m-2) = 0$, so we again conclude $m = 2$. Since $\gcd(n, m) = n$ with n prime and $m = 2$, we get $n = 2$. And since $\gcd(n, \delta+1) = n$, δ is odd.

If $\gcd(n,\delta) = \gcd(n,\delta+1) = 1$, then we consider $\gcd(n,m)$. If $\gcd(n,m) = 1$, then again

$$1 + \gcd(n, m\delta) - \gcd(n,\delta) - \gcd(n,\delta+1) = 0,$$

so $m = 2$ and $\delta \not\equiv 0, -1 \pmod n$. If $\gcd(n,m) = n$, then

$$1 + \gcd(n, m\delta) - \gcd(n,\delta) - \gcd(n,\delta+1) = n - 1,$$

so $\delta(m-2) = 1$, so $\delta = 1$ and $m = 3$. Since $\delta = 1$ and $\gcd(n,\delta+1) = 1 \neq 2$, this implies that $2 \nmid n$, so n is an odd prime. □

EXAMPLE 3.3. For triagonal curves, $n = 3$. Then either i) $m = 3$ and $\delta = 1$; or ii) $m = 2$ and $\delta \equiv 0, 1 \pmod 3$.

4. Jacobians with superelliptic components

In this section we study a family of non-hyperelliptic curves introduced in [**20**], whose Jacobians decompose into factors which are hyperelliptic Jacobians. We will extend this family of curves and investigate if we can obtain in this way curves of arbitrary large genus having decomposable Jacobians.

In [**20**] were introduced a family of curves in $\mathbb{P}^{s+2}$ given by the equations

$$\begin{cases} zw = c_0x^2 + c_1xw + c_2w^2 \\ y_1^r = h_1(z,w) := z^r + c_{1,1}z^{r-1}w + \cdots + c_{r-1,1}zw^{r-1} + w^r, \\ \cdots \\ y_s^r = h_s(z,w) := z^r + c_{1,s}z^{r-1}w + \cdots + c_{r-1,s}zw^{r-1} + w^r, \end{cases} \tag{4.1}$$

where $c_i \in k$, $i = 0,1,2$, and $c_{i,j} \in k$ for $i = 1, \cdots, r$, $j = 1, \cdots, s$. The variety $\mathcal{X}_{r,s}$ is an algebraic curve since the function field of $\mathcal{X}_{r,s}$ is a finite extension of $k(z)$. $\mathcal{X}_{r,s}$ is a complete intersection. For a proof of the following facts see [**20**].

REMARK 4.1. Let $\mathcal{X}_{r,s}$ be as above. Assume that $\mathcal{X}_{r,s}$ is smooth and $c_0 \neq 0$. Then the following hold:

i) The genus of $\mathcal{X}_{r,s}$ is

$$g(\mathcal{X}_{r,s}) = (r-1)(rs \cdot 2^{s-1} - 2^s + 1).$$

ii) If $r \geq 3$ and $s \geq 1$, then $\mathcal{X}_{r,s}$ is non-hyperelliptic.

Fix $r \geq 2$. Let λ be an integer such that $1 \leq \lambda \leq s$. Define the superelliptic curve $C_{r,\lambda,m}$ as follows

$$C_{r,\lambda,m}: \qquad Y^r = \prod_{i=1}^{\lambda} h_i(X^m, 1),$$

for some $m \geq 2$. The right side of the above equation has degree $d = rm\lambda$. Using Lemma 3.1 we have that

$$g(C_{r,\lambda,m}) = 1 + \frac{1}{2}\left(r^2m\lambda - r - m\lambda r - \gcd(\lambda rm, r)\right).$$

Hence,

$$g(C_{r,\lambda,m}) = 1 + \frac{r}{2}\left((r-1)\lambda m - 2\right). \tag{4.2}$$

THEOREM 4.1. *Let $C_{r,\lambda,m}$ be a generic curve as above. Then the following hold*

i) $\overline{\mathrm{Aut}}\,(C_{r,\lambda,m}) \cong C_m$.

ii) $\overline{\mathrm{Aut}}\,(C_{2,\lambda,m}) \cong D_{2m}$.

PROOF. i) The equation of $C_{r,\lambda,m}$ is $y^r = g(x^m)$. Then the curve has the following two automorphisms

$$\tau(X,Y) \to (X, \zeta_r Y), \quad \sigma(X,Y) = (\zeta_m X, Y).$$

Since τ commutes with all automorphisms then $\langle\tau\rangle \triangleleft \mathrm{Aut}(C_{r,\lambda,m})$ and $C_{r,\lambda,m}/\langle\tau\rangle$ is a genus 0 curve. Then $\langle\sigma\rangle \hookrightarrow \overline{\mathrm{Aut}}(C_{r,\lambda,m})$. If $C_{r,\lambda,m}$ is a generic curve then $\langle\sigma\rangle \cong \overline{\mathrm{Aut}}(C_{r,\lambda,m})$.

ii) When $r = 2$ then the equation of $C_{2,\lambda,m}$ is given by

$$y^r = \prod_{i=1}^{\lambda} (X^{2m} + a_i X^m + 1).$$

From [5] such curves have reduced automorphism group isomorphic to D_{2m}. This completes the proof. □

The following theorem determines the full automorphism group of such curves.

THEOREM 4.2. *Let $C_{r,\lambda,m}$ be a generic curve as above and G:=*Aut$(C_{r,\lambda,m})$. *Then the following hold*

a) If $\overline{\mathrm{Aut}}\,(C_{r,\lambda,m}) \cong C_m$, then $G \cong C_{mn}$ or G is isomorphic to

$$\langle \gamma, \sigma \,|\, \gamma^n = 1, \sigma^m = 1, \sigma\gamma\sigma^{-1} = \gamma^l \rangle$$

where (l,n)=1 and $l^m \equiv 1 \pmod n$. But if $(m,n) = 1$, then $l = n-1$.
b) If $\overline{\mathrm{Aut}}\,(C_{r,\lambda,m}) \cong D_{2m}$, then

(1) If n is odd then $G \cong D_{2m} \times C_n$.

(2) If n is even and m is odd then $G \cong D_{2m} \times C_n$ or G is isomorphic to the group with presentation

$$\langle \gamma, \sigma, \tau \,|\, \gamma^n = 1, \sigma^2 = \gamma, \tau^2 = \gamma^{n-1}, (\sigma\tau)^m = \gamma^{\frac{n}{2}}, \sigma\gamma\sigma^{-1} = \gamma, \tau\gamma\tau^{-1} = \gamma \rangle.$$

(3) If n is even and m is even then G is isomorphic to one of the following groups $D_{2m} \times C_n$, D_{2mn}, or one of the following

$$G_1 = \langle \gamma, \sigma, \tau \,|\, \gamma^n = 1, \sigma^2 = \gamma, \tau^2 = 1, (\sigma\tau)^m = 1, \sigma\gamma\sigma^{-1} = \gamma, \tau\gamma\tau^{-1} = \gamma^{n-1} \rangle,$$
$$G_2 = \langle \gamma, \sigma, \tau \,|\, \gamma^n = 1, \sigma^2 = \gamma, \tau^2 = \gamma^{n-1}, (\sigma\tau)^m = 1, \sigma\gamma\sigma^{-1} = \gamma, \tau\gamma\tau^{-1} = \gamma \rangle,$$
$$G_3 = \langle \gamma, \sigma, \tau \,|\, \gamma^n = 1, \sigma^2 = \gamma, \tau^2 = 1, (\sigma\tau)^m = \gamma^{\frac{n}{2}}, \sigma\gamma\sigma^{-1} = \gamma, \tau\gamma\tau^{-1} = \gamma^{n-1} \rangle,$$
$$G_4 = \langle \gamma, \sigma, \tau \,|\, \gamma^n = 1, \sigma^2 = \gamma, \tau^2 = \gamma^{n-1}, (\sigma\tau)^m = \gamma^{\frac{n}{2}}, \sigma\gamma\sigma^{-1} = \gamma, \tau\gamma\tau^{-1} = \gamma \rangle.$$

PROOF. The full automorphism group G of such curves is a degree n central extension of C_m or D_{2m}. Such extensions were determined in [**13**, Thm. 3.2] and [**13**, Thm. 3.3]. □

4.1. Decomposition of Jacobians. In [**20**] it is proved that the Jacobians of $C_{2,\lambda,2}$, defined over any algebraic number field k, curves are isogenous to a product of superelliptic Jacobians. A similar theorem was suggested for curves $C_{r,\lambda,2}$ and was remarked that the proof was similar to the case $C_{2,\lambda,2}$ curves. The proof is of arithmetic in nature and is based on Küneth's formula, the Frobenius map on $Gal(\bar{k}/k)$, Chebotarev's theorem, and Faltings theorem.

We will generalize such theorems for curves $C_{r,\lambda,m}$ over an algebraically closed field k of characteristic relatively prime to r. Our proof is based solely on automorphisms of curves and it is much simpler than in [**20**].

Let $\mathcal{X}_{r,s}$ be a generic algebraic curve defined over an algebraically closed field k and $C_{r,\lambda,m}$ as above. Then we have the following.

THEOREM 4.3. *The Jacobian* $\mathrm{Jac}(\mathcal{X}_{r,s})$ *is isogenous to the product of the* $C_{r,\lambda,m}$, *for* $1 \leq \lambda \leq s$, *namely*

$$\mathrm{Jac}(\mathcal{X}_{r,s}) \cong \prod_{1=\lambda}^{s} \mathrm{Jac}(C_{r,\lambda,m}),$$

if and only if

$$r = 4 \cdot \frac{1+s-2^s}{ms(s+1) - s\cdot 2^{s+1}}. \tag{4.3}$$

PROOF. We denote by $\sigma_i(x, y_i, z) \to (x, \zeta_r y_i, z)$, for $i = 1, \ldots, s$. Then the quotient spaces $\mathcal{X}_{r,s}/\langle\sigma_i\rangle$ are the curves $C_{r,i,s}$, for $i = 1, \ldots, s$. Since σ_i is a central element in $G = \mathrm{Aut}(\mathcal{X}_{r,s})$ then $H_i := \langle\sigma_i\rangle \lhd G$, for all $i = 1, \ldots, s$. Obviously, for all $i \neq j$ we have $H_i \cap H_j = \{e\}$. Hence, $H_1, \ldots, H_s$ forms a partition for G.

The genus for every $C_{r,i,s}$, by Lemma 3.1 is given by Eq. (4.2). Then we have

$$\begin{aligned}\sum_{\lambda=1}^{s} g\left(C_{r,\lambda,m}\right) &= \sum_{\lambda=1}^{s}\left(1 + \frac{r}{2}\left((r-1)\lambda m - 2\right)\right)\\ &= s(r-1)\left(\frac{r}{4}m(s+1) - 1\right).\end{aligned}$$

From the results of Eq. (2.2) we have that

$$\frac{r}{4}ms(s+1) - s = rs\cdot 2^{s-1} - 2^s + 1.$$

Hence,

$$r = 4 \cdot \frac{1+s-2^s}{ms(s+1) - s\cdot 2^{s+1}}.$$

This completes the proof. □

REMARK 4.2. For $m = 2$ this result is the case of Theorem 4.2 in [**20**]. We get $r = \frac{2}{s}$. Hence, $s = 1$ or $s = 2$. Therefore, Theorem 4.2 in [**20**] is true only for curves $F_{m,1}$ or $F_{m,2}$.

Next we determine integer combinations of r, m, and s that satisfy Eq. (4.3). First we need the following lemma.

LEMMA 4.3. *For integers* a, n *with* $n > 1$, *suppose* $a^n \equiv 1 \pmod{n}$. *Let* p *be the smallest prime divisor of* n. *Then* $a \equiv 1 \pmod{p}$.

PROOF. Suppose $a^n \equiv 1 \pmod{n}$ for integers a, n with $n > 1$. Let p be the smallest prime divisor of n. Since $a^n \equiv 1 \pmod{n}$, then $a^n \equiv 1 \pmod{p}$. Let d be the order of a modulo p. Since $a^{p-1} \equiv 1 \pmod{p}$, d divides $p-1$, so $d < p$. And since $a^n \equiv 1 \pmod{p}$, d divides n. However, since p is the smallest prime divisor of n, the only divisor of n which is less than p is 1. Hence, $d = 1$, so the order of a modulo p is 1, so $a \equiv 1 \pmod{p}$, as desired. □

PROPOSITION 4.4. *Suppose* $r, m, s \in \mathbb{N}$ *satisfy Eq. 4.3. Then* $mrs = 4k$ *for some odd integer* k. *Moreover,*

i) If $s \equiv 1 \pmod{2}$, *then* $s = 1$.

ii) If $s \equiv 2 \pmod{4}$, *then* $s = 2t$ *for some odd integer* t *which satisfies* $4^t \equiv 1 \pmod{t}$. *Furthermore,* t *is a multiple of 3.*

iii) If $s \equiv 0 \pmod 4$, then $s = 4u$ for some odd integer u which satisfies $16^u \equiv 1 \pmod u$. Furthermore, u is a multiple of 3 or 5.

PROOF. We first determine possible values of s for which r, which is given by the equation

$$r = 4 \cdot \frac{1 + s - 2^s}{ms(s+1) - s \cdot 2^{s+1}},$$

can be an integer. We proceed in cases, considering the highest power of 2 that divides s. In particular, the powers that we consider are 0, 1, 2, and then at least 3.

First, suppose s is an odd integer. Since s divides the denominator, it follows that s divides $4(1 + s - 2^s)$. Since $\gcd(s, 4) = 1$, we conclude that $2^s \equiv 1 \pmod s$. If $s = 1$, then this congruence is satisfied. Now, suppose $s > 1$ and let p be the smallest prime divisor of s. By the Lemma above, one has $2 \equiv 1 \pmod p$, so p divides 1, which is impossible. Thus, if s is odd, then $s = 1$.

Also, note that in this case if $s = 1$, then one finds that $m = 2$ and $r = 2$.

Next, suppose $s = 2t$ for some odd integer t. Substituting in, we get

$$r = 4 \cdot \frac{1 + 2t - 2^{2t}}{m(2t)(2t+1) - 2t \cdot 2^{2t+1}} = 2 \cdot \frac{1 + 2t - 4^t}{mt(2t+1) - 2t \cdot 4^t}.$$

As above, since t divides the denominator, t divides $2(1+2t-4^t)$. Since $\gcd(t, 2) = 1$, we conclude that $4^t \equiv 1 \pmod t$. If $t = 1$, then this congruence is satisfied. Now, suppose $t > 1$ and let p be the smallest prime divisor of t. By the Lemma above, one has $4 \equiv 1 \pmod p$, which implies p divides 3, so $p = 3$. Thus, $s = 2t$ for some integer t which is either 1 or is a multiple of 3 and which satisfies $4^t \equiv 1 \pmod t$.

Next, suppose $s = 4u$ for some odd integer u. Substituting in, we get

$$r = 4 \cdot \frac{1 + 4u - 2^{4u}}{m(4u)(4u+1) - 4u \cdot 2^{4u+1}} = \frac{1 + 4u - 16^u}{mu(4u+1) - 2u \cdot 16^u}.$$

As above, since u divides the denominator, u divides $(1 + 4u - 16^u)$. Thus, $16^u \equiv 1 \pmod u$. If $u = 1$, then this congruence is satisfied. Now, suppose $u > 1$ and let p be the smallest prime divisor of u. By the Lemma above, one has $16 \equiv 1 \pmod p$, which implies p divides 15, so $p = 3$ or $p = 5$. Thus, $s = 4u$ for some integer u which is either 1 or is a multiple of either 3 or 5 and which satisfies $16^u \equiv 1 \pmod u$.

Finally suppose $s = 8v$ for some (even or odd) integer v. Substituting in, we get

$$r = 4 \cdot \frac{1 + 8v - 2^{8v}}{m(8v)(8v+1) - 8v \cdot 2^{8v+1}} = \frac{1 + 8v - 2^{8v}}{2mv(8v+1) - 2v \cdot 2^{8v+1}}.$$

Thus, since 2 divides the denominator, 2 divides $(1+8v-2^{8v})$, so $2^{8v} \equiv 1 \pmod 2$, which occurs only if $v = 0$. Thus, $s = 0$.

To show $mrs = 4k$ for some odd integer k, we have two cases to consider. If s is odd, then $s = 1$, so $m = r = 2$ and thus $mrs = 4$. If s is even, then we clear denominators of Eq. 4.3 and consider the equation modulo 8 to get

$$rms(s+1) - rs2^{s+1} \equiv 4 \cdot (1 + s - 2^s) \pmod 8.$$

Since s is even and thus at least 2, this simplifies to $rms \equiv 4 \pmod 8$, so $mrs = 4k$ for some odd integer k. □

REMARK 4.5. i) When $s = 2t$ as in the proof of the theorem, then the first few integer values of t are $1, 3, 9, 21, 27, 63, 81, 147, 171, 189, 243$. This is sequence A014945 in the Online Encyclopedia of Integer Sequences.

ii) When $s = 4u$ as in the proof of the theorem, then the first few integer values of u are $1, 3, 5, 9, 15, 21, 25, 27, 39, 45, 55, 63, 75, 81, 105, 117$. This is sequence A014957 in the Online Encyclopedia of Integer Sequences.

We then search for values of s that satisfy the above proposition. Searching among $1 \leq s < 500$, we find the following possible values for s.

$$\{1, 2, 4, 6, 12, 18, 20, 36, 42, 54, 60, 84, 100, 108, 126, 156, 162, \\ 180, 220, 252, 294, 300, 324, 342, 378, 420, 468, 486, \dots\}.$$

We now check each of these to find corresponding integer values of m and r.

COROLLARY 4.6. *Suppose* $r, m, s \in \mathbb{N}$ *satisfy Eq.* 4.3, *and suppose* $1 \leq s < 500$ *with* $s \neq 300, 420, 468$. *Then* $s \in \{1, 2, 6, 18, 42, 126, 162, 294, 378, 486\}$. *We display the combinations of integers* s, m, *and* r *in the table below. Note that* m *and* r *grow quickly relative to* s, *hence the scientific notation for the cases where* $s > 100$.

s	m	r
1	2	2
2	2	1
6	18	19
18	27594	29125
42	204560302842	209430786241
126	1.3397×10^{36}	1.3503×10^{36}
162	7.1730×10^{46}	7.2173×10^{46}
294	2.1579×10^{86}	2.8391×10^{62}
294	2.1579×10^{86}	3.3025×10^{31}
294	2.1579×10^{86}	2.2665×10^{27}
378	3.2488×10^{111}	3.2574×10^{111}
486	8.2050×10^{143}	8.2219×10^{143}
486	8.2050×10^{143}	1.4596×10^{20}

Note that there are multiple combinations of integers r and m when $s = 294$ or 486. Also note that we were unable to get results for $s = 300, 420$, or 468. This is because of the time required to factor $(1 + s - 2^s)$ in those cases.

Interestingly, we do not have any cases where $s = 4u$ for some odd integer u.

Acknowledgments

The authors want to thank the anonymous referee for helpful comments and remarks.

References

[1] Robert D. M. Accola, *Two theorems on Riemann surfaces with noncyclic automorphism groups.*, Proc. Amer. Math. Soc. **25** (1970), 598–602. MR0259105 (41 #3747)

[2] Robert D. M. Accola, *Riemann surfaces with automorphism groups admitting partitions*, Proc. Amer. Math. Soc. **21** (1969), 477–482. MR0237764 (38 #6045)

[3] L. Beshaj, K. Magaard, T. Shaska, and S. Shpectorov, *On superelliptic curves and their Jacobians*, NATO Advanced Study Institute, ISO Press (2014), to appear.

[4] T. Shaska and L. Beshaj, *The arithmetic of genus two curves*, Information security, coding theory and related combinatorics, NATO Sci. Peace Secur. Ser. D Inf. Commun. Secur., vol. 29, IOS, Amsterdam, 2011, pp. 59–98. MR2963126

[5] Lubjana Beshaj, Valmira Hoxha, and Tony Shaska, *On superelliptic curves of level n and their quotients, I*, Albanian J. Math. **5** (2011), no. 3, 115–137. MR2846162 (2012i:14036)

[6] Torsten Ekedahl and Jean-Pierre Serre, *Exemples de courbes algébriques à jacobienne complètement décomposable* (French, with English and French summaries), C. R. Acad. Sci. Paris Sér. I Math. **317** (1993), no. 5, 509–513. MR1239039 (94j:14029)

[7] Gerd Faltings, *Diophantine approximation on abelian varieties*, Ann. of Math. (2) **133** (1991), no. 3, 549–576, DOI 10.2307/2944319. MR1109353 (93d:11066)

[8] Gerd Faltings, *The general case of S. Lang's conjecture*, Barsotti Symposium in Algebraic Geometry (Abano Terme, 1991), Perspect. Math., vol. 15, Academic Press, San Diego, CA, 1994, pp. 175–182.

[9] E. Kani and M. Rosen, *Idempotent relations and factors of Jacobians*, Math. Ann. **284** (1989), no. 2, 307–327, DOI 10.1007/BF01442878. MR1000113 (90h:14057)

[10] G. Martens, *On coverings of elliptic curves*, Algebra and number theory (Essen, 1992), de Gruyter, Berlin, 1994, pp. 137–151. MR1285364 (95e:14024)

[11] Kay Magaard, Tanush Shaska, and Helmut Völklein, *Genus 2 curves that admit a degree 5 map to an elliptic curve*, Forum Math. **21** (2009), no. 3, 547–566, DOI 10.1515/FORUM.2009.027. MR2526800 (2010h:14050)

[12] R. Sanjeewa and T. Shaska, *Determining equations of families of cyclic curves*, Albanian J. Math. **2** (2008), no. 3, 199–213. MR2492096 (2010d:14043)

[13] R. Sanjeewa, *Automorphism groups of cyclic curves defined over finite fields of any characteristics*, Albanian J. Math. **3** (2009), no. 4, 131–160. MR2578064 (2011a:14045)

[14] Jean-Pierre Serre, *Groupes algébriques et corps de classes* (French), Publications de l'institut de mathématique de l'université de Nancago, VII. Hermann, Paris, 1959. MR0103191 (21 #1973)

[15] T. Shaska, *Families of genus two curves with many elliptic subcovers*, arXiv:1209.0434.

[16] T. Shaska, *Genus 3 hyperelliptic curves with (2, 4, 4)-split Jacobians*, arXiv:1306.5284.

[17] T. Shaska, *Genus 2 fields with degree 3 elliptic subfields*, Forum Math. **16** (2004), no. 2, 263–280, DOI 10.1515/form.2004.013. MR2039100 (2004m:11097)

[18] T. Shaska, *Curves of genus 2 with (N, N) decomposable Jacobians*, J. Symbolic Comput. **31** (2001), no. 5, 603–617, DOI 10.1006/jsco.2001.0439. MR1828706 (2002m:14023)

[19] Christopher Towse, *Generalized Wronskians and Weierstrass weights*, Pacific J. Math. **193** (2000), no. 2, 501–508, DOI 10.2140/pjm.2000.193.501. MR1755827 (2001c:14055)

[20] Takuya Yamauchi, *On curves with split Jacobians*, Comm. Algebra **36** (2008), no. 4, 1419–1425, DOI 10.1080/00927870701866622. MR2406594 (2009b:14058)

Department of Mathematics, Oakland University, Rochester, Michigan 48309
E-mail address: beshaj@oakland.edu

Department of Mathematics, Oakland University, Rochester, Michigan 48309
E-mail address: shaska@oakland.edu

Department of Mathematics, Western New England University, Springfield, Massachusetts 01119
E-mail address: cshor@wne.edu

Superelliptic surfaces as p-gonal surfaces

S. Allen Broughton

Dedicated to Emilio Bujalance for his sixtieth anniversary

ABSTRACT. In this brief, expository paper, we discuss superelliptic surfaces and p-gonal surfaces, which generalize hyperelliptic surfaces. A superelliptic surface, or more generally, a p-gonal surface, has a conformal automorphism w of prime order such that $S/\langle w\rangle$ has genus zero. Alternatively, the surface has an equation of the form $y^p = f(x)$ for some rational function $f(x)$. We discuss normal forms, automorphism groups, and families of p-gonal surfaces.

1. Introduction

There is a strong interest in superelliptic and p-gonal surfaces. Historically, the surfaces, especially hyperelliptic surfaces, were linked to the study of certain integrals. Most recently, they are of interest in cryptography and fields of moduli. Among all surfaces, p-gonal surfaces are the surfaces with the simplest and most tractable equations. In this brief article we talk about normal forms of p-gonal surfaces, their automorphisms, and very briefly about families of p-gonal surfaces. Space does not allow discussion of applications to cryptography or fields of definition of superelliptic surfaces. See **[19]** and **[20]** for references on these topics.

Acknowledgement

The author thanks the Mathematics department of Linköping University for the hosting the conference for which this paper was written. The author also thanks Aaron Wootton for useful conversations in the preparation of this paper. Our joint paper **[9]** in this conference proceedings has many ideas and development parallel to the current paper. In addition, the author thanks Milagros Izquierdo for pointing out some references.

2. The equation point of view of p-gonal surfaces

2.1. Hyperelliptic surfaces. Hyperelliptic surfaces (curves) were introduced in studying the surface of the function $\sqrt{f(x)}$ and related integrals $\int \sqrt{f(x)}dx$ where $f(x)$ is a rational function, or alternatively the Riemann surface determined by $y^2 = f(x)$. By using elementary algebra we may assume that $f(x)$ is a square

2010 *Mathematics Subject Classification.* Primary 14H37, 30F10, 30F20; Secondary 14E20, 14H50.

Key words and phrases. Riemann surface, automorphisms of Riemann surfaces, p-gonal curve, hyperelliptic surface, superelliptic curve.

free polynomial. For, we may write $f(x) = u^2(x)v(x)$ where $v(x)$ is a square free polynomial and $u(x)$ is a rational function. Setting $y' = \frac{y}{u(x)}, x' = x$ we see that

$$(y')^2 = \frac{y^2}{u^2(x)} = \frac{u^2(x)v(x)}{u^2(x)} = v(x').$$

The transformation $\phi(x,y) = (x',y') = (\frac{y}{u(x)}, x)$ is a birational transformation of $\mathbb{P}^2$ carrying the surface $y^2 = f(x)$ to $(y')^2 = v(x')$. Since we are only interested in surfaces up to birational equivalence, let us assume that $f(x)$ is a square free polynomial and that S is the surface (projective plane curve) defined by

$$y^2 = f(x). \tag{2.1}$$

All of the finite points of S are smooth. At infinity the equation of S has the form $y^2 - x^t h(1/x) = \left(y - x^{t/2}\sqrt{h(1/x)}\right)\left(y + x^{t/2}\sqrt{h(1/x)}\right)$ for some polynomial h with nonzero constant term. If $t = deg(f)$ is even, then the projective completion of S has two branches at infinity otherwise it has a cusp. The normalization $S^\nu \to S$ is the smooth compactification of $S - \{\text{singular points}\}$. Therefore, the normalization $S^\nu \to S$ is 1-1 over the finite points of S and has one or two points lying over the infinite points of S, depending on the parity of t.

The map $\iota : (x,y) \to (x,-y)$ is an involution of S that fixes only the points $(a_i, 0)$ in the finite part of S. The involution ι lifts to the normalization $S^\nu \to S$ and we also denote this map by ι. The action of the involution ι on S^ν fixes all the lifts of the $(a_i, 0)$. In the case S^ν has a single point at infinity lying over the infinite point of S, the lift of ι fixes the infinite point, in case there are two points, the lift interchanges them. The quotient map $\pi : S^\nu \to S^\nu/\langle\iota\rangle$ is given by $(x,y) \to x$ at the finite points and hence $S^\nu/\langle\iota\rangle \simeq \mathbb{P}^1$. So we arrive at another characterization of a hyperelliptic surface namely a smooth surface S with an involutory automorphism ι such that $S/\langle\iota\rangle \simeq \mathbb{P}^1$. Indeed, starting out with such a surface a plane model given by equation 2.1 can be found.

2.2. n-gonal surfaces and equations. Now generalizing, we consider surfaces of the form $y^n = f(x)$ motivated by the study of the function $\sqrt[n]{f(x)}$. These surfaces are called *cyclic n-gonal surfaces.* Using calculations as above, we can show that S has a plane model of the form

$$y^n = f(x) = \prod_{i=1}^{s} (x - a_i)^{t_i} \tag{2.2}$$

where the a_i and t_i satisfy

(1) the a_i are distinct,
(2) $0 < t_i < n$, and
(3) $gcd(n, t_1, \ldots, t_s) = 1$.

The third condition comes from the assumed irreducibility of the surface, for otherwise $y^n - f(x)$ factors. In addition we may wish to impose the following

(4) n divides $t = t_1 + \cdots + t_s = deg(f)$.

The finite singular points of S are the points $(a_i, 0)$ where $t_i > 1$. By writing $y^n - f(x)$ as $y^n - (x - a_i)^{t_i}\phi_i(x)$ with $\phi_i(a_i) \neq 0$ we see that S has $d_i = gcd(t_i, n)$ local branches at $(a_i, 0)$. So the surface has a cusp or single branch at $(a_i, 0)$ if $\gcd(n, t_i) = 1$ and is even smooth if $t_i = 1$. It also follows that the normalization $S^\nu \to S$ has d_i points lying over $(a_i, 0)$. If $t = deg(f) = n$ then S has n smooth points at infinity. Otherwise S has a single point at infinity. Writing, as before, $y^n - f(x) = y^n - x^t h(1/x)$ we see that in every case S has $gcd(t,n)$ local branches at infinity and the normalization $S^\nu \to S$ has $gcd(t,n)$ points lying over the infinite point(s) of S. We call S^ν the *smooth model* and S the *plane model* though we frequently loosely identify the two surfaces.

The projection $\pi : S^\nu \to \mathbb{P}^1$ induced by $(x,y) \to x$ is ramified over all of the points a_i with ramification degree $m_i = \frac{n}{d_i}$ and ramified over ∞ with degree $\frac{n}{\gcd(t,n)}$. If condition 4 holds $\pi : S^\nu \to \mathbb{P}^1$ has d_i points lying over a_i and n points lying over all other points, including the point ∞. By a simple ramified covering argument, the genus σ of S^ν is given by

$$\sigma = \frac{1}{2}\left(2 + (s-2)n - \sum_{i=1}^{s} d_i\right). \tag{2.3}$$

Sometimes it is convenient to shift the branch points away from infinity, i.e., to ensure that t is divisible by n. To this end pick r so that $t+r$ is divisible by n, $0 < r < n$ and pick any linear fractional transformation, L, of the coordinates

$$x' = L(x) = \frac{\alpha x + \beta}{\gamma x + \delta}, x = L^{-1}(x') = \frac{\delta x' - \beta}{-\gamma x' + \alpha}, \text{ and } \alpha\delta - \beta\gamma = 1.$$

It follows that

$$-\gamma x' + \alpha = \frac{1}{\gamma x + \delta}. \tag{2.4}$$

Further, impose the restrictions that

$$\forall i \ L(a_i) \neq \infty, \text{ and } \gamma \neq 0. \tag{2.5}$$

Next set $K = \prod_{i=1}^{s} (\gamma a_i + \delta)^{t_i}$, which is nonzero because of equation 2.5. Then, using $x = \frac{\delta x' - \beta}{-\gamma x' + \alpha}$, we obtain

$$\begin{aligned} y^n &= f\left(\frac{\delta x' - \beta}{-\gamma x' + \alpha}\right) = \prod_{i=1}^{s}\left(\frac{\delta x' - \beta}{-\gamma x' + \alpha} - a_i\right)^{t_i} \\ &= K\frac{(-\gamma x' + \alpha)^r}{(-\gamma x' + \alpha)^{t+r}} \prod_{i=1}^{s}\left(x' - \frac{\alpha a_i + \beta}{\gamma a_i + \delta}\right)^{t_i} \\ &= K(-\gamma)^r (\gamma x + \delta)^{t+r}\left(x' - \frac{\alpha}{\gamma}\right)^r \prod_{i=1}^{s}(x' - L(a_i))^{t_i} \end{aligned}$$

and so

$$\left(\frac{1}{\sqrt[n]{K(-\gamma)^r}} \frac{y}{(\gamma x + \delta)^{\frac{t+r}{n}}}\right)^n = \left(x' - \frac{\alpha}{\gamma}\right)^r \prod_{i=1}^{s}(x' - L(a_i))^{t_i}.$$

Now set

$$\begin{aligned} y' &= \frac{1}{\sqrt[n]{K(-\gamma)^r}} \frac{y}{(\gamma x + \delta)^{\frac{t+r}{n}}}, \\ a_{s+1} &= \infty, \\ b_i &= L(a_i), \ b_{s+1} = \frac{\alpha}{\gamma} = L(\infty), \ t_{s+1} = r, \end{aligned}$$

and we get

$$(2.6) \qquad (y')^n = \prod_{i=1}^{s+1} (x' - b_i)^{t_i}.$$

Because $a_1, \ldots, a_{s+1}$ are distinct, so are $b_1, \ldots, b_{s+1}$ and we arrive at a birationally equivalent surface where $t_1 + \cdots + t_{s+1}$ is divisible by n. There are only a finite number of bad choices for the linear fractional transformation L.

For the remainder of this paper we will assume conditions 1-4 above, unless otherwise noted. We shall call $(a_1, a_2, \ldots, a_s)$ the branch points of S, t_i the local degree at a_i, and $(t_1, t_2, \ldots, t_s)$ the multi-degree of S. If condition 4 holds we call $(t_1, t_2, \ldots, t_s)$ a *complete multi-degree.*

REMARK 2.1. If $deg(f)$ is not divisible by n then we add the additional branch point $a_{s+1} = \infty$, and set $t_{s+1} = r$ defined as above. Then 1-4 all hold. We call t_{s+1} the local degree at ∞ and $(t_1, t_2, \ldots, t_{s+1})$ is a complete multi-degree. It is convenient to also consider $y^n = f(x)$ when $f(x)$ is a rational function. A denominator factor of the form $(x - a_i)^{t_i}$, $0 < t_i < n$, contributes $-t_i$ to the multi-degree. Using the birational equivalence transformations above, the contribution $-t_i$ changes to $n - t_i$ when $f(x)$ is converted to normal polynomial form.

We may determine when two cyclic n-gonal surfaces are birationally equivalent using the following proposition.

PROPOSITION 2.2. *Suppose that two cyclic n-gonal surfaces with the same multi-degree have branch points $(a_1, a_2 \ldots, a_s)$ and $(b_1, b_2 \ldots, b_s)$. Then the surfaces are conformally equivalent if there is an $L \in PSL_2(\mathbb{C})$ and a permutation $\vartheta \in \Sigma_s$, preserving multi-degrees ($t_{\vartheta i} = t_i$), so that*

$$b_i = L(a_{\vartheta i}).$$

for all i.

PROOF. In the proof of equation 2.6 we can assume that $r = 0$ and then observe

$$\prod_{i=1}^{s}(x - L(a_i))^{t_i} = \prod_{i=1}^{s}(x - L(a_{\vartheta i}))^{t_{\vartheta i}} = \prod_{i=1}^{s}(x - b_i)^{t_i}$$

□

2.3. Cyclic n-gonal actions. If ω is a nth root of unity, then $(x, y) \to (x, \omega y)$ is an automorphism of S which fixes the points $(a_i, 0)$ and no others in the finite part of S. Let C be the cyclic group of automorphisms obtained by letting ω range over all n^{th} roots of unity. The action of C on S, and its lift to S^ν, is called a *cyclic n-gonal action.* The map $\pi : S^\nu \to S \to \mathbb{P}^1$, $(x, y) \to x$ is a quotient map for the projection $S^\nu \to S^\nu/C$, and is called the *cyclic n-gonal morphism.* The degree of ramification of π over a_i is $m_i = n/gcd(t_i, n)$. In fact there are $d_i = gcd(t_i, n)$ points lying over a_i, and at each such point P the stabilizer of the C action, C_P, is the unique subgroup of C of order m_i. The quotient group C/C_P transitively permutes the points lying over $\pi(P)$. The map is unramified over ∞ because n divides t and there are n distinct points over ∞. Let w be the generator of C corresponding to $\omega = \exp(2\pi i/n)$. At any point $\widetilde{a}_j$ in S^ν lying over lying over a_j, the rotation number of w, i.e., the differential $dw_{|\widetilde{a}_j}$ is $\exp\left(\frac{2\pi i t_j}{n}\right)$.

Now, if S is any closed Riemann surface with a conformal automorphism w such that $S/\langle w\rangle \simeq \mathbb{P}^1$, then using standard field theory, it can be shown that the function field $\mathbb{C}(S) = \mathbb{C}\left(x, \sqrt[n]{f(x)}\right)$, for some $f(x)$. Hence, S has a plane model of the form given in equation 2.2, satisfying conditions 1-4 above. We state this as a proposition.

PROPOSITION 2.3. *A closed Riemann surface is a cyclic n-gonal surface if and only if either of the following two equivalent conditions hold.*

(1) *The surface S has a conformal automorphism w of order n such that $S/\langle w\rangle$ has genus zero*
(2) *The surface S has a plane model of the form given in equation* 2.2 *and conditions 1-4 in subsection* 2.2 *are satisfied.*

2.4. Cyclic p-gonal and superelliptic surfaces.

DEFINITION 2.4. A *cyclic p-gonal surface* is any closed Riemann surface S with a conformal automorphism w of prime order p such that $S/\langle w\rangle$ has genus zero or alternatively has a plane model of the form

$$y^p = f(x) = \prod_{i=1}^{s}(x - a_i)^{t_i}$$

where the a_i and t_i satisfy

(1) the a_i are distinct,
(2) $0 < t_i < p$, and
(3) p divides $t = t_1 + \cdots + t_s = deg(f)$.

REMARK 2.5. The singularities of the plane model of S are all cusps except possibly a point at infinity with p smooth branches. Moreover, in the genus formula 2.3 all $d_i = 1$ and hence the genus is given by

$$\sigma = \frac{1}{2}(s-2)(p-1). \tag{2.7}$$

Observe that σ only depends on the number of branch points and not on the multi-degree.

Finally, we give the standard definition of a superelliptic surface (see [**18**] for instance) and a generalization of superelliptic introduced discussed in [**8**] and [**9**].

DEFINITION 2.6. A *superelliptic surface* is any p-gonal surface with a plane smooth model of the form $y^p = f(x)$ where $f(x)$ is square free and p does not divide the degree of $f(x)$.

DEFINITION 2.7. A generalized superelliptic surface is any n-gonal surface with a plane model of the form $y^n = f(x)$ where the multi-degree $(t_1, \ldots, t_s)$ of $f(x)$ satisfies

(1) $0 < t_i < n$,
(2) $gcd(n, t_j) = 1$ for all s,
(3) n divides $t = t_1 + \cdots + t_s = deg(f)$.

REMARK 2.8. A superelliptic surface is simply a p-gonal surface given by Definition 2.4 where the multi-degree is $(1, 1, \ldots, 1, t_\infty)$ with $0 < t_\infty < p$. For a generalized superelliptic S the stabilizer C_P of a point $P \in S$ satisfies $C_P = C$ or $C_P = \langle 1\rangle$.

This is guaranteed by condition 2 of the definition. A p-gonal surface is automatically a generalized superelliptic surface but need not be superelliptic. For genus 3 there are two 7-gonal surfaces: the superelliptic surface given by $y^7 = x(x-1)$ with multi-degree $(1,1,5)$ and the non-superelliptic surface $y^7 = x(x-1)^2$ with multi-degree $(1,2,4)$.

3. The action point of view of p-gonal surfaces

We now use the cyclic action point of view to describe p-gonal surfaces. Here we will describe group theoretically a cyclic p-gonal surface as a branch set and a class of generating vectors of C. This allows us to handle equivalence without constructing birational maps. To this end, we introduce Fuchsian groups which captures both the conformal and automorphic structure of the surface.

3.1. Covering actions by Fuchsian groups. A (co-compact) Fuchsian group Γ, a discrete group acting on the hyperbolic plane $\mathbb{H}$, has a presentation by hyperbolic and elliptic generators and relations:

$$\text{generators}: \{\alpha_i, \beta_i, \gamma_j, 1 \leq i \leq \sigma, 1 \leq j \leq s\}$$

$$\text{relations}: \prod_{i=1}^{\sigma}[\alpha_i, \beta_i] \prod_{j=1}^{s} \gamma_j = \gamma_1^{m_1} = \cdots = \gamma_s^{m_s} = 1$$

with $m_i \geq 2$. The signature of Γ is

$$\mathcal{S}(\Gamma) = (\sigma : m_1, \ldots, m_s)$$

which we shorten to $(m_1, \ldots, m_s)$, when the genus is zero. Here are important invariants of and facts about Fuchsian groups.

- The *genus of* Γ: $\sigma(\Gamma) = \sigma$ is the genus of $\mathbb{H}/\Gamma$. If $s = 0$ then Γ is torsion free and usually denoted by Π, as it is isomorphic to $\pi_1(S)$.
- The *area of a fundamental region* of Γ in $\mathbb{H}$ is given by $A(\Gamma) = 2\pi\mu(\Gamma)$ where, $\mu(\Gamma) = 2(\sigma - 1) + \sum_{j=1}^{s}\left(1 - \frac{1}{m_j}\right)$.
- The *Teichmüller dimension* $d(\Gamma)$ of Γ, the dimension of the Teichmüller space of Fuchsian groups with signature $\mathcal{S}(\Gamma)$, is given by

$$d(\Gamma) = 3(\sigma - 1) + s.$$

- If $\gamma \in \Gamma$ fixes a point $z \in \mathbb{H}$ the γ is conjugate to a power of some γ_i.

For any group G acting conformally on a surface S we have pair of Fuchsian groups $\Pi \trianglelefteq \Gamma$ such that Π is torsion free, $S \simeq \mathbb{H}/\Pi$, and the action of G on S is given by the action of Γ/Π on $\mathbb{H}/\Pi$. The isomorphism is given by an exact sequence.

$$\Pi \hookrightarrow \Gamma \overset{\eta}{\twoheadrightarrow} G \tag{3.1}$$

such that the order of elliptic elements is preserved under η. The map η is called a *surface-kernel epimorphism*, and the isomorphism between Γ/Π and G is denoted by $\overline{\eta} : \Gamma/\Pi \longleftrightarrow G$. Also observe that $S/G \simeq \mathbb{H}/\Gamma$. All the proceeding is summarized in the following commutative diagram.

$$\begin{array}{ccc} \mathbb{H} & \overset{\pi_\Pi}{\to} & S \\ \pi_\Gamma \downarrow & & \downarrow \pi_G \\ \mathbb{H}/\Gamma & \simeq & S/G \end{array} \tag{3.2}$$

Now assume Γ is an arbitrary Fuchsian group with $\sigma = 0$, but not necessarily arising from a group action as in the exact sequence 3.1. We have $\mathbb{H}/\Gamma \simeq \mathbb{P}^1$ and our relations then simplify:

$$\prod_{j=1}^{s} \gamma_j = \gamma_1^{m_1} = \cdots = \gamma_s^{m_s} = 1. \tag{3.3}$$

Each γ_j fixes a unique point z_j such that π_Γ is ramified over $a_j = \pi_\Gamma(z_j) \in \mathbb{P}^1$ with degree m_j. In this case Γ is determined up to conjugacy by the branch points $(a_1, a_2, \ldots, a_s)$ and the signature $(m_1, \ldots, m_s)$. If $(a_1', a_2', \ldots, a_s')$ is the branch set of a Γ' with signature $(m_1', \ldots, m_s')$ then Γ and Γ' are conjugate if and only if there is an $L \in PSL_2(\mathbb{C})$ and a permutation $\vartheta \in \Sigma_s$, such that

$$a_i' = L(a_{\vartheta i}),\ m_i' = m_{\vartheta i}.$$

3.2. Generating vectors. With η as in 3.1, set $c_j = \eta(\gamma_j)$, then the s-tuple $(c_1, \ldots, c_s)$ satisfies

$$\begin{aligned} c_1 \cdot c_2 \cdots c_s &= 1 \\ o(c_i) &= m_i > 1 \\ G &= \langle c_1, \ldots, c_s \rangle \end{aligned} \tag{3.4}$$

Such a tuple is called an $(m_1, \ldots, m_s)$-generating vector of G. Given such a vector then $c_j = \eta(\gamma_j)$ defines a surface kernel epimorphism and an action of G on $S = \mathbb{H}/\ker(\eta)$. If $\omega \in \mathrm{Aut}(G)$ then $c_j' = \omega(c_j)$ defines another generating vector such that the epimorphism $\eta' = \omega \circ \eta$ satisfies $c_j' = \eta'(\gamma_j)$ and $\ker(\eta) = \ker(\eta')$. We conclude that $\mathrm{Aut}(G)$-equivalent generating vectors define the same surface with $\mathrm{Aut}(G)$-equivalent G-actions.

REMARK 3.1. If $G = \mathbb{Z}_n$ then generating vectors are in 1-1 correspondence with the s-tuples satisfying

(1) $0 < t_i < n$,
(2) $gcd(n, t_1, \ldots, t_s) = 1$
(3) $t_1 + \cdots + t_s = 0 \bmod n$.

Of course, these are simply the equations satisfied by multi-degrees in the previous section. The two multi-degrees $(t_1, \ldots, t_s)$ and $(et_1, \ldots, et_s) \bmod n$ are $\mathrm{Aut}(\mathbb{Z}_n)$-equivalent if $e \neq 0 \ mod\ n$ and these are the only equivalences. If $n = p$ is prime then $m_i = p$ for all i and 2 automatically holds. The number of generating vectors can be computed using inclusion-exclusion, see [**7**].

3.3. Conformally equivalent actions. We would like to determine when two conformal G-actions on surfaces are conformally equivalent. So assume that $(m_1, \ldots, m_s)$-actions of G on S_1 and S_2 are defined via surface kernel epimorphisms $\eta_1 : \Gamma_1 \to G$ and $\eta_2 : \Gamma_2 \to G$ then the conformal actions of G on S_1 and S_2 are conformally equivalent if and only if there is an $\phi \in \mathrm{Aut}(\mathbb{H})$, and $\omega \in \mathrm{Aut}(G)$ such that the following diagram commutes.

$$\begin{array}{ccccc} \Pi_1 & \hookrightarrow & \Gamma_1 & \overset{\eta_1}{\twoheadrightarrow} & G \\ \downarrow Ad_\phi & & \downarrow Ad_\phi & & \downarrow \omega \\ \Pi_2 & \hookrightarrow & \Gamma_2 & \overset{\eta_2}{\twoheadrightarrow} & G \end{array} \tag{3.5}$$

In the diagram, $\Pi_1 = \ker(\eta_1)$, $\Pi_2 = \ker(\eta_2)$, and $Ad_\phi(\gamma) = \phi\gamma\phi^{-1}$ denotes the adjoint action of ϕ on γ. There is an induced conformal equivalence $h : S_1 \leftrightarrow S_2$ which intertwines the G-actions on S_1 and S_2 :

$$\begin{array}{ccc} G & \overset{\overline{\eta_1}}{\hookrightarrow} & \mathrm{Aut}\,(S_1) \\ \downarrow \omega & & \downarrow Ad_h \\ G & \overset{\overline{\eta_2}}{\twoheadrightarrow} & \mathrm{Aut}\,(S_2) \end{array} \tag{3.6}$$

Next, we want to determine conformal equivalence of p-gonal surfaces simply in terms of the branch set and the multi-degree without producing the element ϕ, i.e., work directly on $\mathbb{P}^1$ and the generating vectors $(c_1, \ldots, c_s)$. To this end, we say that two generating vectors define topologically equivalent actions if there is a homeomorphism $\widetilde{h}$ of $\mathbb{H}$ normalizing Γ such that the following diagram commutes.

$$\begin{array}{ccc} \Gamma & \overset{\eta}{\twoheadrightarrow} & G \\ \downarrow Ad_{\widetilde{h}} & & \downarrow \omega \\ \Gamma & \overset{\eta'}{\twoheadrightarrow} & G \end{array} \tag{3.7}$$

In the case $\ker(\eta) = \ker(\eta') = \Pi$, there is a homeomorphism h of S and $\overline{h}$ of $\mathbb{P}^1$ such that such that the following diagrams are commutative.

$$\begin{array}{ccc} \mathbb{H} & \overset{\widetilde{h}}{\rightarrow} & \mathbb{H} \\ \downarrow \pi_\Pi & & \downarrow \pi_\Pi \\ S & \overset{h}{\rightarrow} & S \end{array} \quad \text{and} \quad \begin{array}{ccc} \mathbb{H} & \overset{\widetilde{h}}{\rightarrow} & \mathbb{H} \\ \downarrow \pi_\Gamma & & \downarrow \pi_\Gamma \\ \mathbb{P}^1 & \overset{\overline{h}}{\rightarrow} & \mathbb{P}^1 \end{array} \tag{3.8}$$

Moreover, given an $\overline{h}$ on $\mathbb{P}^1$ mapping the branch set to itself, in an order-preserving manner, the covering $\widetilde{h}$ may be found. The map $\widetilde{h}$ is conformal if and only if $\overline{h}$ is conformal. A particular homeomorphism, that may be constructed, switches adjacent branch points inducing the switch $a_j \to a_{j+1}$, $a_{j+1} \to a_j$, $m_j \to m_{j+1}$, $m_{j+1} \to m_j$. In the superelliptic case the change in branching orders is always permissible. By lifting the appropriately chosen $\overline{h}$, we get $\gamma_j \to \gamma_j\gamma_{j+1}\gamma_j^{-1}$, $\gamma_{j+1} \to \gamma_j$, and hence $c_j \to c_jc_{j+1}c_j^{-1}$, $c_{j+1} \to c_j$. If G is abelian we simply get $c_j \to c_{j+1}$, $c_{j+1} \to c_j$. It can be shown that the totality of the homeomorphism action is induced by the switch maps and so the action on generating vectors is simply permutation. Thus we arrive at the following.

PROPOSITION 3.2. *Let S be a cyclic p-gonal surface branched over $(a_1, a_2, \ldots, a_s)$ with generating vector (multi-degree) $(t_1, t_2, \ldots, t_s)$. If S' is another such surface with corresponding $(a_1', a_2', \ldots, a_s')$ and $(t_1', t_2', \ldots, t_s')$ then S and S' are conformally equivalent p-gonal surfaces if and only if there is $L \in PSL_2(\mathbb{C})$ and a permutation $\vartheta \in \Sigma_s$, and $e \neq 0 \bmod p$ such that*

$$a_i' = L(a_{\vartheta i}), t_i' = et_{\vartheta i}.$$

REMARK 3.3. We note that the conformal equivalence can be constructed as a birational transformation, and so the Fuchsian group construction is not completely necessary. However, the statement extends to elementary abelian actions but there is no easy construction of birational maps. See [**7**].

4. Automorphism groups of cyclic n-gonal surfaces

There is a great deal of interest in the automorphism group $A = \mathrm{Aut}(S)$ of a cyclic n-gonal surface as these surfaces have tractable automorphism groups. Indeed, the automorphism group can generally be computed directly from the equations but the Fuchsian group methods are easier. Of special interest, is the normal case where $N = \mathrm{Nor}_A(C) = A$, which we discuss next. Automorphisms in $A - N$ are called *exceptional*. See **[8]**, **[9]**, and **[21]** for a discussion of these automorphisms.

4.1. The normal case. In the normal case, $K = N/C = A/C$ is an automorphism group of the sphere, one of five types of Platonic groups $\mathbb{Z}_k$, D_k, A_4, Σ_4, A_5. One "simply" solves an extension problem

$$C \hookrightarrow N \twoheadrightarrow K.$$

For large genus we automatically have $C \trianglelefteq A$ using Accola's theorem on strong branching **[1]**. In the next theorem we state the normality theorem in more general contexts because they all follow from Accola's work in exactly the same way.

THEOREM 4.1. *Let S be a cyclic n-gonal surface of genus σ with cyclic group C. Then the following hold.*

(1) *If $n = p$ is a prime and $\sigma > (p-1)^2$, then C is normal in* $\mathrm{Aut}(S)$ **[1]**.
(2) *Suppose S is generalized superelliptic i.e., $gcd(n, t_i) = 1$ for all. Then, if $\sigma > (n-1)^2$, C is normal in* $\mathrm{Aut}(S)$ **[17]**.
(3) *Suppose the action of C is weakly malnormal i.e., for all $g \in \mathrm{Aut}(S)$ either $gCg^{-1} = C$ or $gCg^{-1} \cap C = \langle 1 \rangle$ Then, if $\sigma > (n-1)^2$, C is normal in* $\mathrm{Aut}(S)$ **[8]**.

REMARK 4.2. Accola's method critically uses results from algebraic geometry and there is no group theoretic proof known to the author. For genus greater than 1 the hyperelliptic involution is always central. A result on centrality for superelliptic surfaces is given in Proposition 4.8. The involution is also unique, namely, for any surface S there can only be one involution ι such that $S/\langle \iota \rangle$ has genus zero. An analogue for p-gonal surfaces is that if $w_1, w_2 \in \mathrm{Aut}(S)$ of the same prime order p are such that $S/\langle w_1 \rangle$ and $S/\langle w_2 \rangle$ both have genus 0 then w_1 and w_2 are power-conjugate **[15]**.

REMARK 4.3. Since $\sigma = \frac{1}{2}(s-2)(p-1)$ then the condition $\sigma > (p-1)^2$ is satisfied when $s > 2p$. So when p is of even moderate size, $N < A$, is possible for high degree polynomials.

Here are some works on the problem of determination of automorphism groups, in increasing generality on the properties of the cyclic action. See also **[9]**.

- The case $n = 2$ (hyperelliptic case) has been studied extensively: Brandt, Bujulance, Etayo, Gamboa, Gromadzki, Martinez, Shaska – **[5]**, **[10]**, **[18]**,
- The case where $n = 3$, (cyclic trigonal surfaces): Accola, Bujalance, Cirre, Costa, Duma, Gromadski, Izquierdo, Martinez, Radtke, Ying – **[2]**, **[11]**, **[14]**, **[13]**, **[23]**.
- The case where $n = p$, for p a prime: Bartolini, Brandt, Costa, Gonzalez-Diez, Harvey, Izquierdo, Wootton – **[4]**, **[5]**, **[15]**, **[16]**, **[21]**.
- General n where the cyclic n-gonal morphism $S \to S/C$ is fully ramified: Kontogeorgis **[17]**.

- General n with weak malnormality conditions: Broughton & Wootton **[8]**.
- The paper **[4]** includes some cases missed in previous works and in **[17]**.

EXAMPLE 4.4. Here are some low genus p-gonal surfaces. The two surfaces with exceptional automorphisms ($N < A$) are the well-known Klein quartic and Bring's curve.

Table 1

genus	A	N	K	$\lvert A/N\rvert$	$\lvert C\rvert$	$(t_1,\ldots,t_s)$
3	$\mathbb{Z}_{14}$	$\mathbb{Z}_{14}$	$\mathbb{Z}_2$	1	7	$(1,1,5)$
3	$PSL_2(7)$	$\mathbb{Z}_3 \ltimes \mathbb{Z}_7$	$\mathbb{Z}_3$	8	7	$(1,2,4)$
4	Σ_5	$\mathbb{Z}_4 \ltimes \mathbb{Z}_5$	$\mathbb{Z}_4$	6	5	$(1,2,3,4)$
4	$\mathbb{Z}_4 \ltimes \mathbb{Z}_5$	$\mathbb{Z}_4 \ltimes \mathbb{Z}_5$	$\mathbb{Z}_4$	1	5	$(1,1,4,4)$
4	$\mathbb{Z}_{15} = \mathbb{Z}_3 \times \mathbb{Z}_5$	$\mathbb{Z}_3 \times \mathbb{Z}_5$	$\mathbb{Z}_3$	1	5	$(1,1,1,2)$

The two normalizers $\mathbb{Z}_4 \ltimes \mathbb{Z}_5$ are not isomorphic.

4.2. Action on multi-degrees and examples. Let us finish this section by giving examples of defining equations and automorphism groups for $K = Z_\ell$ and $K = D_\ell$, where ℓ and p are relatively prime. The method we use is described in detail for all K in **[22]**. Before proceeding we need to know the adjoint action of K on C, and the K-action on the local degrees t_j.

The adjoint action, $Ad_g(x) = gxg^{-1}$ introduced in 3.3, defines a representation $Ad : N \to \mathrm{Aut}(C)$. Since C is abelian, Ad factors through K: $\overline{Ad} : K \to \mathrm{Aut}(C)$. Since $C = \mathbb{Z}_p$ has prime order $\mathrm{Aut}(C) = \mathbb{Z}_p^*$ is cyclic and this severely limits the K-action except in the case where $K = Z_\ell$. Next, let us see how K acts on the local degrees. Let $\widetilde{a}_i$ and $\widetilde{a}_j$ be points lying over branch points a_i and a_j, and suppose that there is an $h \in N$ such that $h(\widetilde{a}_i) = \widetilde{a}_j$ and hence the image $\overline{h} \in K$ satisfies $\overline{h}(a_i) = a_j$. The standard generator w of C fixes $\widetilde{a}_i$ and $\widetilde{a}_j$ and the rotation numbers of w are $dw_{|\widetilde{a}_i} = \exp(2\pi i t_i)$, $dg_{|\widetilde{a}_j} = \exp(2\pi i t_j)$. Since $h^{-1}wh = Ad_{h^{-1}}(w) = w^e$, for some e, then

$$\exp(2\pi i e t_i) = d(w^e)_{|\widetilde{a}_i} = dh^{-1} dw_{|\widetilde{a}_j} dh = dw_{|\widetilde{a}_j} = \exp(2\pi i t_j).$$

Thus

$$t_j = e t_i \bmod p, \text{ i.e., } t_j = \overline{Ad}(\overline{h}^{-1}) t_i \bmod p. \tag{4.1}$$

It follows that if $\overline{h}$ fixes a point then $\overline{Ad}(\overline{h}^{-1})$ acts trivially. The orbits of K on $\mathbb{P}^1$ will be important in what follows. We call an orbit *singular* if the points have non-trivial K-stabilizers, otherwise the orbits are *regular*.

EXAMPLE 4.5. Suppose $K = \mathbb{Z}_\ell$. We use the action of $K = \mathbb{Z}_\ell$ on $\mathbb{P}^1$ generated by $x \to \zeta x$ where $\zeta = exp(\frac{2\pi i}{\ell})$. The group N must be $\mathbb{Z}_\ell \times \mathbb{Z}_p = \mathbb{Z}_{\ell p}$ or $\mathbb{Z}_\ell \ltimes \mathbb{Z}_p$ with the $\mathbb{Z}_\ell$ action is generated by $j \to e'j$ on the additive group $\mathbb{Z}_p$, where $(e')^\ell = 1$ mod p, Pick e so that $ee' = 1 \bmod p$. The action has two fixed points at 0 and ∞. Suppose first, that none of the branch points is 0 or ∞. Then the branch points are made up of regular orbits of K of the form $\{a, a\zeta, \ldots, a\zeta^{\ell-1}\}$. Each such orbit contributes a factor to $f(x)$ of the form $\prod_{i=0}^{\ell-1} \left(x - a\zeta^i\right)^{e^i}$, possibly to some power.

Thus

$$f(x) = \prod_{a\in B}\prod_{i=0}^{\ell-1}\left(x - a\zeta^i\right)^{e^i},$$

where B is some finite list, possibly with repeated elements. Since $1+e+\cdots+e^{\ell-1} = 0 \bmod p$ then $deg(f) = 0 \bmod p$ and so ∞ is not a branch point. If 0 or ∞ is a branch point then equation 4.1 shows that the $\mathbb{Z}_\ell$ action on $\mathbb{Z}_p$ must be trivial so $K = \mathbb{Z}_{\ell p}$ by relative primeness arguments. Note we may also argue that $K = \mathbb{Z}_{\ell p}$ by using ramification arguments whether p and ℓ are relatively prime or not. Then

$$f(x) = z^{t_0}\prod_{a\in B}\prod_{i=0}^{\ell-1}\left(x - a\zeta^i\right),$$

where $0 < t_0 < p$. If $t_\infty < p$ is the local degree at ∞ with then $t_0 + t_\infty + \ell b = 0 \bmod p$, Where $b = |B|$ is the number of repetitions of the factor $\prod_{i=0}^{\ell-1}\left(x - a\zeta^i\right)$.

EXAMPLE 4.6. Suppose $K = D_\ell$. We use the action of $K = D_\ell$ on $\mathbb{P}^1$ generated by $x \to \zeta x$ and $z \to \frac{1}{z}$. The group K must be $D_\ell \times \mathbb{Z}_p$ or $D_\ell \ltimes \mathbb{Z}_p$ with the non-trivial D_ℓ action generated by $j \to -j = (p-1)j$, the only non-trivial involution in $\mathrm{Aut}(C)$. The abelianization of D_ℓ is $\mathbb{Z}_2$ for ℓ odd and $\mathbb{Z}_2 \times \mathbb{Z}_2$ for ℓ even, so there are several non-trivial choices for $\overline{Ad}$. For simplicity let us focus on ℓ odd, so that $\overline{Ad}$ is trivial on the subgroup $\mathbb{Z}_\ell \lhd D_\ell$ The action of D_ℓ has three singular orbits (orbits with fixed points) $\left\{1, \zeta, \ldots, \zeta^{\ell-1}\right\}$, $\left\{-1, -\zeta, \ldots, -\zeta^{\ell-1}\right\}$, and $\{0, \infty\}$. A regular orbit has the form $\left\{a, a\zeta, \ldots, a\zeta^{\ell-1}\right\} \cup \left\{a^{-1}, a^{-1}\zeta, \ldots, a^{-1}\zeta^{\ell-1}\right\}$ for $a \neq \pm 1$. In the trivial action case each regular orbit contributes a factor of the form $\prod_{i=0}^{\ell-1}\left(x - a\zeta^i\right)\prod_{i=0}^{\ell-1}\left(x - a^{-1}\zeta^i\right)$ possibly to some power, the singular orbits can contribute z^{t_0}, and powers of $\prod_{i=0}^{\ell-1}\left(x - \zeta^i\right)$ and $\prod_{i=0}^{\ell-1}\left(x + \zeta^i\right)$. Now the transformation $z \to \frac{1}{z}$ interchanges 0 and ∞ and so the local degrees t_0 and t_∞ must be the same. The sum condition must satisfy

$$t_0 + t_\infty + b_r\ell + b_1\ell + b_{-1}\ell = 0 \bmod p$$

where b_r, b_1, b_{-1} are the number of repetitions of the factors

$$\prod_{i=0}^{\ell-1}\left(x - a\zeta^i\right)\prod_{i=0}^{\ell-1}\left(x - a^{-1}\zeta^i\right),\ \prod_{i=0}^{\ell-1}\left(x - \zeta^i\right),\ \text{and}\ \prod_{i=0}^{\ell-1}\left(x + \zeta^i\right).$$

Now suppose that $\overline{Ad}$ is not trivial. Then the singular orbits $\left\{1, \zeta, \ldots, \zeta^{\ell-1}\right\}$, $\left\{-1, -\zeta, \ldots, -\zeta^{\ell-1}\right\}$ cannot contribute any factors since the stabilizers of these points have non-trivial $\overline{Ad}$ values. Regular orbits contribute factors of the type

$$\frac{\prod_{i=0}^{\ell-1}\left(x - a\zeta^i\right)}{\prod_{i=0}^{\ell-1}\left(x - a^{-1}\zeta^i\right)} \quad\text{or}\quad \prod_{i=0}^{\ell-1}\left(x - a\zeta^i\right)\prod_{i=0}^{\ell-1}\left(x - a^{-1}\zeta^i\right)^{p-1}$$

and we see that $t_\infty = -t_0 \bmod p$. So we have $t_0 + p - t_0 + b_r\ell + (p-1)b_r\ell = 0 \bmod p$, which is no constraint at all.

REMARK 4.7. The general form of factors for any K-action will be

$$\left(\prod_{k\in K}\left(x-\overline{k}\cdot a\right)^{\overline{Ad}(\overline{k})}\right)^{\frac{1}{r}}$$

where r is the order of the stabilizer of the point $a \in \mathbb{P}^1$.

Shaska observed [**19**] that for many automorphism groups of low genus that if a cyclic prime order subgroup had a genus quotient then the subgroup was central. This generalizes the hyperelliptic case. It turns out that many of these surfaces are superelliptic. In fact, we have the following.

PROPOSITION 4.8. *Let S be a superelliptic surface with cyclic p-gonal subgroup C. Then C is central N.*

PROOF. The multi-degree has the form $(1,\ldots,1,t_\infty)$, and there must be at least two ones. But then it is not possible for any automorphism in K to satisfy equation 4.1 for all t_i unless the action is trivial. □

5. Families of p-gonal surfaces

In this section we very briefly discuss families of p-gonal surfaces. Given our cyclic p-gonal equation

$$y^p = f(x) = \prod_{i=1}^{s}(x-a_i)^{t_i} \tag{5.1}$$

we can look at families of surfaces in three different ways.

(1) For each complete multi-degree $T = (t_1, t_2, \ldots, t_s)$ consider the family of surfaces parameterized by the branch points $(a_1, a_2, \ldots, a_s) \in \mathbb{C}^s -$ *diagonals*. The genus of the surfaces so constructed has the constant value $\frac{(s-2)(p-1)}{2}$. If two multi-degrees are equivalent by the equivalence given in Remark 3.4 then a family of equivalent surfaces is determined.
(2) Each family described in (1) for a fixed T and p determines a family of surfaces, or equisymmetric stratum, in the moduli space of surfaces of genus $\frac{(s-2)(p-1)}{2}$. These equisymmetric strata are defined in [**6**]. The complex dimension of this family is $s-3$ as our next proposition shows. The union of these families for all T is called the p-gonal locus of the moduli space. For $p=2$ this is the well known hyperelliptic locus of the moduli space. The trigonal ($p=3$) locus for genus 4 is discussed in [**23**]. Some recent results on the topology of this branch locus for $p>3$ are given in [**3**] and [**12**].
(3) Each polynomial $f(x)$ defines a vector of coefficients in $\mathbb{C}^t$, where $t = deg(f)$. By expanding the right hand side of equation 5.1, we see that each multi-degree defines a locally closed subvariety of $\mathbb{C}^t$. If t is held fixed and s and T are allowed to vary, $\mathbb{C}^t$ is a union of these subvarieties (minus a small piece corresponding to reducible surfaces). The genus of the surface varies over different subvarieties. The set where all t_i equal 1 is open and dense and the genus of the surfaces is largest over this set. In [**19**] Shaska adopts this approach to determine equations of hyperelliptic surfaces with prescribed automorphism group.

PROPOSITION 5.1. *Let* $T = (t_1, t_2, \ldots, t_s)$ *be a complete multi-degree and* Σ_T *be the subgroup of permutations of* Σ_s *that preserve* T. *Let* $PSL_2(\mathbb{C}) \times \Sigma_T$ *act (partially) on* $\mathbb{C}^s$ *– diagonals by* $(L, \vartheta) \cdot (a_1, a_2, \ldots, a_s) = (L(a_{\vartheta 1}), \ldots, L(a_{\vartheta s}))$. *Then*

$$\mathcal{MC}_{n,T} = (\mathbb{C}^s - diagonals)/(PSL_2(\mathbb{C}) \times \Sigma_T)$$

of complex dimension $s - 3$ *is "almost" a moduli space of p-gonal surfaces of multi-degree* T.

Rather than prove the statement, we make some remarks that justify the word almost.

(1) Every cyclic p-gonal action with multi-degree T is accounted for in the quotient space.
(2) The action of $PSL_2(\mathbb{C})$ is only partial and exceptional automorphisms (where $N < A$) need to be taken into account. The partial action can be fixed by looking at $\left(\mathbb{P}^1\right)^s$ but then the cyclic model needs to be fixed.
(3) Each $\mathcal{MC}_{n,T}$ corresponds to a moduli space of the same dimension, of Fuchsian groups determined by the signature $(p, p, \ldots, p)$, specifically there is a finite to one map $\mathcal{MC}_{n,T} \to \mathcal{M}_{(p,p,\ldots,p)}$.

EXAMPLE 5.2. Here is a table of multi-degrees for a small number of branch points, moduli space dimension $\mathfrak{m} = s - 3$, prime p, and genus σ.

	$p = 3$	$p = 5$
$s = 3, \mathfrak{m} = 0$	$\sigma = 1,\ (1,1,1)$	$\sigma = 2,\ (1,1,2)$
$s = 4, \mathfrak{m} = 1$	$\sigma = 2,\ (1,1,2,2)$	$\sigma = 4,\ (1,1,1,2), (1,1,4,4), (1,2,3,4)$

	$p = 7$	$p = 11$
$s = 3, \mathfrak{m} = 0$	$\sigma = 3,\ (1,1,5), (1,2,4)$	$\sigma = 5,\ (1,1,9)$
$s = 4, \mathfrak{m} = 1$	$\sigma = 6,\ (1,1,1,4), (1,1,6,6)$	$\sigma = 10,\ (1,1,1,8), (1,1,10,10)$

(1) The pure superelliptic case is always present, unless p divides s.
(2) All the cases potentially have automorphisms, depending on the position of the branch points.
(3) In the moduli space of surfaces of genus 4 the 5-gonal locus consists of three strata of dimension 1.

References

[1] Robert D. M. Accola, *Strongly branched coverings of closed Riemann surfaces*, Proc. Amer. Math. Soc. **26** (1970), 315–322. MR0262485 (41 #7091)

[2] Robert D. M. Accola, *On cyclic trigonal Riemann surfaces. I*, Trans. Amer. Math. Soc. **283** (1984), no. 2, 423–449, DOI 10.2307/1999139. MR737877 (85j:14052)

[3] Gabriel Bartolini, Antonio F. Costa, and Milagros Izquierdo, *On isolated strata of p-gonal Riemann surfaces in the branch locus of moduli spaces*, Albanian J. Math. **6** (2012), no. 1, 11–19. MR2965666

[4] Gabriel Bartolini, Antonio F. Costa, and Milagros Izquierdo, *On automorphisms groups of cyclic p-gonal Riemann surfaces*, J. Symbolic Comput. **57** (2013), 61–69, DOI 10.1016/j.jsc.2013.05.005. MR3066451

[5] Rolf Brandt and Henning Stichtenoth, *Die Automorphismengruppen hyperelliptischer Kurven* (German, with English summary), Manuscripta Math. **55** (1986), no. 1, 83–92, DOI 10.1007/BF01168614. MR828412 (87m:14033)

[6] S. Allen Broughton, *The equisymmetric stratification of the moduli space and the Krull dimension of mapping class groups*, Topology Appl. **37** (1990), no. 2, 101–113, DOI 10.1016/0166-8641(90)90055-7. MR1080344 (92d:57013)

[7] S. Allen Broughton and Aaron Wootton, *Finite abelian subgroups of the mapping class group*, Algebr. Geom. Topol. **7** (2007), 1651–1697, DOI 10.2140/agt.2007.7.1651. MR2366175 (2009m:20064)

[8] S.A. Broughton and A. Wootton, *Full Automorphism Groups of Generalized Superelliptic Riemann Surfaces*, in preparation.

[9] S.A. Broughton and A. Wootton, *Exceptional Automorphisms of (Generalized) Super Elliptic Surfaces*, this conference proceedings.

[10] E. Bujalance, J. M. Gamboa, and G. Gromadzki, *The full automorphism groups of hyperelliptic Riemann surfaces*, Manuscripta Math. **79** (1993), no. 3-4, 267–282, DOI 10.1007/BF02568345. MR1223022 (94f:20093)

[11] E. Bujalance, J. A. Bujalance, G. Gromadzki, and E. Martinez, *Cyclic trigonal Klein surfaces*, J. Algebra **159** (1993), no. 2, 436–459, DOI 10.1006/jabr.1993.1167. MR1231224 (94g:30040)

[12] A. Costa, M. Izquierdo, H. Parlier, *Connecting p-gonal loci in the compactification of moduli space*, arXiv:1305.0284v2.

[13] Andrei Duma and Wolfgang Radtke, *Automorphismen und Modulraum Galoisscher dreiblättriger Überlagerungen* (German, with English summary), Manuscripta Math. **50** (1985), 215–228, DOI 10.1007/BF01168832. MR784144 (86i:32041)

[14] Antonio F. Costa, Milagros Izquierdo, and Daniel Ying, *On Riemann surfaces with non-unique cyclic trigonal morphism*, Manuscripta Math. **118** (2005), no. 4, 443–453, DOI 10.1007/s00229-005-0593-y. MR2190106 (2006k:30042)

[15] Gabino González-Diez, *On prime Galois coverings of the Riemann sphere*, Ann. Mat. Pura Appl. (4) **168** (1995), 1–15, DOI 10.1007/BF01759251. MR1378235 (97c:14033)

[16] W. J. Harvey, *On branch loci in Teichmüller space*, Trans. Amer. Math. Soc. **153** (1971), 387–399. MR0297994 (45 #7046)

[17] Aristides Kontogeorgis, *The group of automorphisms of cyclic extensions of rational function fields*, J. Algebra **216** (1999), no. 2, 665–706, DOI 10.1006/jabr.1998.7804. MR1692965 (2000f:12005)

[18] Tanush Shaska, *Determining the automorphism group of a hyperelliptic curve*, Proceedings of the 2003 International Symposium on Symbolic and Algebraic Computation, ACM, New York, 2003, pp. 248–254 (electronic), DOI 10.1145/860854.860904. MR2035219 (2005c:14037)

[19] T. Shaska, *On super-elliptic Curves and their Quotients*, Albanian Journal of Math. **5** (2011), no. 3, 131–160.

[20] M. Streit and J. Wolfart, *Galois actions on some series of Riemann surfaces with many automorphisms*, *http://www.math.uni-frankfurt.de/ wolfart/*

[21] A. Wootton, *The full automorphism group of a cyclic p-gonal surface*, J. Algebra **312** (2007), no. 1, 377–396, DOI 10.1016/j.jalgebra.2007.01.018. MR2320463 (2008c:14043)

[22] A. Wootton, *Defining equations for cyclic prime covers of the Riemann sphere*, Israel J. Math. **157** (2007), 103–122, DOI 10.1007/s11856-006-0004-4. MR2342442 (2009b:30084)

[23] D. Ying *On the Moduli Space of Cyclic Trigonal Riemann Surfaces of Genus 4*, PhD thesis, Linköping University Institute of Technology, (2006).

Rose-Hulman Institute of Technology
E-mail address: `brought@rose-hulman.edu`

Exceptional automorphisms of (generalized) super elliptic surfaces

S. Allen Broughton and Aaron Wootton

Dedicated to Emilio Bujalance for his sixtieth anniversary

ABSTRACT. A *super-elliptic* surface is a compact, smooth Riemann surface S with a conformal automorphism w of prime order p such that $S/\langle w \rangle$ has genus zero, extending the hyper-elliptic case $p = 2$. More generally, a cyclic n-gonal surface S has an automorphism w of order n such that $S/\langle w \rangle$ has genus zero. All cyclic $n-$gonal surfaces have tractable defining equations. Let $A = \mathrm{Aut}(S)$ and N be the normalizer of $C = \langle w \rangle$ in A. The structure of N, in principal, can be easily determined from the defining equation. If the genus of S is sufficiently large in comparison to n, and C satisfies a generalized super-elliptic condition, then $A = N$. For small genus $A - N$ may be non-empty and, in this case, any automorphism $h \in A - N$ is called *exceptional*. The exceptional automorphisms of super-elliptic surfaces are known whereas the determination of exceptional automorphisms of all general cyclic n-gonal surfaces seems to be hard. We focus on *generalized super-elliptic* surfaces in which n is composite and the projection of S onto S/C is fully ramified. Generalized super-elliptic surfaces are easily identified by their defining equations. In this paper we discuss an approach to the determination of generalized super-elliptic surfaces with exceptional automorphisms.

1. Cyclic n-gonal surfaces

1.1. Cyclic n-gonal surfaces - introduction. A cyclic n-gonal surface (curve) is a compact, smooth Riemann surface with a plane model of the form

$$y^n = f(x) = \prod_{i=1}^{s} (x - a_i)^{t_i} \tag{1.1}$$

where a_i, t_i and $t = t_1 + \cdots + t_s = deg(f)$ satisfy

- the a_i are distinct,
- $0 < t_i < n$,
- n divides t (this is not the typical requirement), and
- $gcd(n, t_1, \ldots, t_s) = 1$.

If $n = 2$ then the surface is hyperelliptic.

2010 *Mathematics Subject Classification.* Primary 14H37, 30F10, 30F20, Secondary 14E20, 14H50.

Key words and phrases. Riemann surface, automorphisms of Riemann surfaces, p-gonal curve, super-elliptic curve.

The plane model of the surface is smooth except at the points $(a_i, 0)$ where $t_i > 1$, and at a single point at ∞ if $t > n$. Recall that the normalization $S^\nu \to S$ desingularizes S by pulling apart local branches and smoothing out the resulting cusps (if any). The map is an isomorphism away from the singular points. At a singular point of $z \in S$ the local branches are the components of $U - \{z\}$ where U is a small connected neighbourhood of z in S. Thus, for such a z, $\nu^{-1}(z)$ will consist of d_z separate points. Alternatively, we may think of constructing S^ν by removing the singular points of S and them smoothly completing the resulting punctured surface. By considering a local analysis of the defining equation at the singular points, it is easily shown that the normalization $S^\nu \to S$ has $d_i = gcd(t_i, n)$ points lying over $(a_i, 0)$ and $n = gcd(t, n)$ points lying over ∞. We call S^ν the *smooth model* and S the *plane model* though we frequently loosely identify the two surfaces. The genus σ of S^ν is given by

$$\sigma = \frac{1}{2}\left(2 + (s-2)n - \sum_{i=1}^{s} d_i\right). \tag{1.2}$$

If ω is a nth root of unity, then $(x, y) \to (x, \omega y)$ is an automorphism of S which fixes the points $(a_i, 0)$ and no others. Let C be the cyclic group of automorphisms obtained by letting ω range over all n^{th} roots of unity. The action of C on S, and its lift to S^ν, is called a *cyclic n-gonal action.* The map $\pi : S^\nu \to S \to \mathbb{P}^1$, $(x, y) \to x$ is a quotient map for the projection $S^\nu \to S^\nu/C$, and is called the *cyclic n-gonal morphism.* The degree of ramification of π over a_i is $n_i = n/gcd(t_i, n)$. In fact there are $d_i = gcd(t_i, n)$ points lying over a_i, and at each such point P the stabilizer of the C action, C_P, is the unique subgroup of C of order n_i. The quotient group C/C_P transitively permutes the points lying over $\pi(P)$. The map is unramified over ∞ because n divides t and there are n distinct points over ∞. Let w be the generator of C corresponding to $\omega = \exp(2\pi i/n)$. At any point in S^ν lying over lying over a_j, the rotation number of w is $\exp\left(\frac{2\pi i t_j}{n}\right)$.

For the hyperelliptic case, $n = 2$, $w = \iota : (x, y) \to (x, -y)$ is called the hyperelliptic involution. It is well known that the hyperelliptic involution is central in $\mathrm{Aut}(S)$, the conformal automorphism group of S. We can get analogous results for C using Accola's theorem on strong branching [**1**]. In increasing order of generality we have:

(1) Hyperelliptic case: the involution $\iota : (x, y) \to (x, -y)$ is central in $\mathrm{Aut}(S)$.
(2) Prime order or super-elliptic case: $n = p$ is a prime and $\sigma > (p-1)^2$, then C is normal in $\mathrm{Aut}(S)$. If $f(x)$ is square-free then C is central (Accola [**1**]).
(3) Fully ramified or generalized super-elliptic case: $d_i = gcd(n, t_i) = 1$ for all i. If $\sigma > (n-1)^2$, then C is normal in $\mathrm{Aut}(S)$ (Kontogeorgis [**13**]).
(4) Weakly malnormal case: (definition to follow). If C is weakly malnormal in C and $\sigma > (n-1)^2$, then C is normal in $\mathrm{Aut}(S)$. (Broughton-Wootton [**4**])

The rest of this paper concerns our approach to the determination of exceptional automorphisms of groups in case 3 above. We focus on an overview of the problems and some calculations. The detailed proofs and complete results of the classification are beyond the scope of this paper and will appear in the forthcoming paper [**4**].

However, the examples of calculation of explicit automorphisms (Example 1.4) and a description of moduli spaces (Section 2.2) are not discussed in [**4**].

ACKNOWLEDGEMENT 1.1. The authors thank the Mathematics department of Linköping University for the hosting the conference at which this paper was presented. Also, the authors thanks the referee and Milagros Izquierdo for pointing out some references.

1.2. Automorphism groups of cyclic n-gonal surfaces. There is a great deal of interest in the automorphism group $A = \mathrm{Aut}(S)$ of a cyclic n-gonal surface, especially the normal case. In the normal case A/C is an automorphism group of the sphere, one of five types of Platonic groups $\mathbb{Z}_k$, D_k, A_4, Σ_4, A_5. One "simply" solves an extension problem

$$C \hookrightarrow N \twoheadrightarrow K.$$

The automorphisms can be explicitly written down as birational transformations of $\mathbb{P}^2$. See Example 1.4 at the end of this section. Here are some sample works on the problem of determination of automorphism groups, in increasing generality on the properties of the cyclic action.

- The case $n = 2$ (hyperelliptic case) has been studied extensively: Brandt, Stichtenoff, Bujalance, Gamboa, Gromadzki, Shaska ([**5**], [**6**], [**15**]).
- The case where $n = 3$, (cyclic trigonal surfaces): Accola, Bujalance, Bujalance, Costa, Izquierdo, Martinez, Ying, ([**1**],[**7**], [**9**], [**19**]).
- The case where $n = p$, for p a prime: Bartolini-Costa-Izquierdo, Gonzalez-Diez, Wootton, ([**2**], [**12**], [**17**], [**18**]).
- General n where the cyclic n-gonal morphism $S \to S/C$ is fully ramified: Kontogeorgis, Broughton, Wootton ([**13**], [**4**]).
- General n with weak malnormality conditions: Broughton, Wootton, ([**4**]).

Next we set out some notation, definitions, and some facts. Let S be a cyclic n-gonal surface, then:

- σ denotes the genus of S.
- $C = \langle w \rangle$ is a cyclic group of automorphisms of S, of order n, such that S/C has genus zero.
- $A = Aut(S)$ is the group of automorphisms of S.
- $N = N_A(C)$ is the normalizer of C in A.
- The group $K = N/C$ acts on $S/C = \mathbb{P}^1$. If K is not trivial, then it must be one of the five Platonic types noted previously.

Finally, we define exceptional automorphisms.

DEFINITION 1.2. Let S be a cyclic n-gonal surface with cyclic n-gonal group C and other notation be as immediately above. Then an automorphism in $h \in A - N$ is called an *exceptional* automorphism of S.

EXAMPLE 1.3. Here are some low genus cyclic n-gonal surfaces. The two surfaces with exceptional automorphisms are the well known Klein's quartic and

Bring's curve.

Table 1

genus	A	N	K	$\lvert A/N\rvert$	$\lvert C\rvert$	$(t_1,\ldots,t_s)$
3	$\mathbb{Z}_{14}$	$\mathbb{Z}_{14}$	$\mathbb{Z}_2$	1	7	$(1,1,5)$
3	$PSL_2(7)$	$\mathbb{Z}_3 \ltimes \mathbb{Z}_7$	$\mathbb{Z}_3$	8	7	$(1,2,4)$
4	Σ_5	$\mathbb{Z}_4 \ltimes \mathbb{Z}_5$	$\mathbb{Z}_4$	6	5	$(1,2,3,4)$
4	$\mathbb{Z}_4 \ltimes \mathbb{Z}_5$	$\mathbb{Z}_4 \ltimes \mathbb{Z}_5$	$\mathbb{Z}_4$	1	5	$(1,1,4,4)$
4	$\mathbb{Z}_3 \times \mathbb{Z}_5$	$\mathbb{Z}_3 \times \mathbb{Z}_5$	$\mathbb{Z}_3$	1	5	$(1,1,1,2)$

The two normalizers $\mathbb{Z}_4 \ltimes \mathbb{Z}_5$ are not isomorphic.

Ultimately, we want to determine the automorphism group of any cyclic n-gonal surface. We will restrict our attention to generalized super-elliptic surfaces. The normal case $A = N$ is computable using well-known extension methods for the exact sequence

$$C \hookrightarrow N \twoheadrightarrow K.$$

Assuming that S is a generalized super-elliptic surface, $N = A$ if $\sigma > (n-1)^2$. So for fixed n we want to determine the finite number of cases where $N < A$ with exceptional automorphisms. Noting that $d_i = 1$ in equation 1.2 then

$$\sigma = \frac{(n-1)(s-2)}{2}. \tag{1.3}$$

As $\sigma \leq (n-1)^2$ when there are exceptional automorphisms, then the number of branch points satisfies $s \leq 2n$.

The next example shows how the action of C on S and K on S/C can be used to give explicit formulas for the automorphisms of cyclic n-gonal surfaces. The question was posed by Emilio Bujalance and Peter Turbek at the Linköping conference in Emilio Bujalance's honour.

EXAMPLE 1.4. Let us find equations and automorphisms for surfaces where N is non-abelian, $C = \mathbb{Z}_p$, $K = \mathbb{Z}_q$ where p, q are primes satisfying $q|(p-1)$, and there are some additional restrictions. Let $C = \langle w \rangle$. Since p and q are relatively prime then there is an $h \in N$ of order q such that h projects to a generator $\overline{h}$ in K. As conjugation by h is an automorphism of C then $hwh^{-1} = w^r$ where $r^q = 1 \bmod p$. Furthermore, $r^q - 1 = (1 + r + r^2 \cdots + \cdots r^{q-1})(1-r)$, so either $r = 1$ or $p|(1 + r + r^2 \cdots + \cdots r^{q-1})$. As N is non-abelian we conclude that $1 < r < q$ and $1 + r + r^2 \cdots + \cdots r^{q-1} = 0 \mod p$. Let us assume that the h-action on $\mathbb{P}^1$ is given by $h : x \to \zeta x$ where $\zeta = \exp(\frac{2\pi i}{q})$, and let us assume that the a_j constitute a single regular K-orbit, i.e., no a_j equals 0 or ∞. We may then assume that $a_j = \zeta^{j-1}$. Next, the rotation number of w at a point lying over $\overline{h}(a_j)$ is the rotation number of $hwh^{-1} = w^r$ at a point lying over a_j. Therefore $t_{j+1} = rt_j \bmod p$. As we shall see later we may assume that $t_1 = 1$, and hence $t_j = r^{j-1}$, $t = \sum t_j = 1 + r + r^2 \cdots + \cdots r^{q-1}$, divisible by p as required. The corresponding equation is

$$y^p = f(x) = \prod_{j=0}^{q-1} (x - \zeta^j)^{r^j}.$$

Now observe that

$$\begin{aligned} f(\zeta x) &= \prod_{j=0}^{q-1}(\zeta x-\zeta^j)^{r^j} = \left(\prod_{j=0}^{q-1}\zeta^{r^j}\right)(x-\zeta^{-1})\prod_{j=1}^{q-1}(x-\zeta^{j-1})^{r^j} \\ &= \zeta^t\frac{(x-\zeta^{q-1})^{r^q}}{(x-\zeta^{q-1})^{r^q-1}}\left(\prod_{j=1}^{q-1}(x-\zeta^{j-1})^{r^{j-1}}\right)^r \\ &= \left(\frac{\zeta^u}{(x-\zeta^{q-1})^v}\right)^p\left(\prod_{j=0}^{q-1}(x-\zeta^j)^{r^j}\right)^r \\ &= \left(\frac{\zeta^u}{(x-\zeta^{q-1})^v}\right)^p f^r(x) \end{aligned}$$

where $t = pu$ and $r^q - 1 = pv$. Now set

$$\begin{aligned} x' &= \zeta x, \\ y' &= \frac{\zeta^u y^r}{(x-\zeta^{q-1})^v} \end{aligned}$$

and we see that

$$(y')^p = f(x')$$

on the plane model of the curve. Let $\phi(x,y) = (x',y')$ be the corresponding birational transformation of $\mathbb{P}^2$. Then ϕ leaves the plane model invariant, projects to $\overline{h}$ on $\mathbb{P}^1$ and satisfies $\phi\circ w = w^r\circ\phi$, considering w as the map $(x,y)\to(x,\omega y)$ on $\mathbb{P}^2$.

REMARK 1.5. The general method for constructing the above defining equation and others is outlined in [**18**]. We repeated the construction here for completeness and clarity.

2. Super-elliptic surfaces

2.1. What are (generalized) super-elliptic surfaces? A super-elliptic surface is a cyclic n-gonal surface, where

- $n = p$, a prime,
- $f(x)$ is square free, (implies that C will be central in A),
- p need not divide the degree of $f(x)$.

We generalize this definition to non-prime cyclic groups and relax the square-free condition.

DEFINITION 2.1. Let S be a cyclic n-gonal surface, whose plane model satisfies the requirements given in given earlier. If $gcd(n,t_i) = 1$ for all t_i, or alternatively, if the degree of ramification of π over a_i equals n, then S is called a *generalized super-elliptic surface.*

REMARK 2.2. There is much interest – motivated by cryptography – in computing in the Jacobian of super-elliptic surfaces S for fields of prime characteristic. It is a generalization of elliptic cryptography. See the paper of Shaska [**15**] for references.

REMARK 2.3. Alternative conditions for being super elliptic are:

- $S^\nu \to S$ has a single point over each finite singularity $(a_i, 0)$.
- S has a cusp singularity $(a_i, 0)$, i.e., only one local branch.

Generalized super-elliptic surfaces satisfy the *weak malnormality* condition.

Weakly Malnormal Condition: for all $g \in \mathrm{Aut}(S)$ either $gCg^{-1} = C$ or $gCg^{-1} \cap C = \{1\}$.

If the surface satisfies the weak malnormality condition and $\sigma > (n-1)^2$, then C is normal in $\mathrm{Aut}(S)$. Our eventual goal is to determine the automorphism group for weakly malnormal actions, the groundwork is laid out in the paper **[4]**. However, the general weakly malnormal case seems very difficult, so we are initially working on an easier but broadly interesting problem. We address this in the next section by showing that super-elliptic surfaces are fairly ubiquitous.

2.2. How many super-elliptic surfaces are there? Given our cyclic n-gonal equation

$$y^n = f(x) = \prod_{i=1}^{s} (x - a_i)^{t_i},$$

call $(a_1, a_2, \ldots, a_s)$ the branch points of S, $(t_1, t_2, \ldots, t_s)$ the multi-degree of S, and $(n_1, n_2, \ldots, n_s)$ – where $n_i = n/gcd(n, t_i)$ – the branching data or signature of the action of C on S. The following is easily shown

LEMMA 2.4. *Two surfaces with branch points $(a_1, a_2, \ldots, a_s)$ and $(b_1, b_2, \ldots, b_s)$ are conformally equivalent if there is an $L \in PSL_2(\mathbb{C})$ and a permutation $\vartheta \in \Sigma_s$, preserving multi-degree, so that*

$$b_i = L(a_{\vartheta i}).$$

for all i.

Let Σ_T denote the group of permutations preserving the multi-degree T. The lemma leads to the following statement.

Variety of surfaces of given multi-degree: The variety

$$\mathcal{MC}_{n,T} = (\mathbb{C}^s - diagonals)/(PSL_2(\mathbb{C}) \times \Sigma_T)$$

of degree $s-3$ is "almost" a moduli space for the surfaces of multi-degree T.

Rather that make the notion "almost" precise enough to convert the statement into a proposition we note the following.

(1) Every cyclic n-gonal action with multi-degree $(t_1, t_2, \ldots, t_s)$ is accounted for in the quotient space.
(2) The action of $PSL_2(\mathbb{C})$ is only partial and exceptional automorphisms need to be taken into account. Additionally, as in Example 1.4, there may be an automorphism L mapping the branch point set $\{a_i\}$ to itself, for which the induced permutation of the degrees $\{t_i\}$ does not fix the multi-degree, even though the two surfaces are conformally equivalent.
(3) Each $\mathcal{MC}_{n,T}$ corresponds to a moduli space, of the same dimension, of Fuchsian groups determined by the signature $(n_1, n_2, \ldots, n_s)$. See the next section for the notation of signatures of Fuchsian groups.

In Table 1 we give the number of inequivalent multi-degrees for cyclic 35-gonal surfaces with 4 branch points. Only the first line corresponds to generalized super-elliptic surfaces.

(n_1, n_2, n_3, n_4)	# multidegrees	$lcm(n_1, n_2, n_3, n_4)$	genus
$(35, 35, 35, 35)$	26	35	34
$(35, 35, 35, 7)$	18	35	32
$(35, 35, 35, 5)$	13	35	31
$(35, 35, 7, 7)$	12	35	30
$(35, 35, 7, 5)$	8	35	29
$(35, 35, 5, 5)$	6	35	28
$(35, 7, 7, 5)$	2	35	27
$(35, 7, 5, 5)$	3	35	26
$(7, 7, 7, 7)$	4	7	reducible
$(7, 7, 5, 5)$	1	35	24
$(5, 5, 5, 5)$	3	5	reducible

TABLE 1. Multi-degrees for cyclic 35-gonal surfaces with 4 branch points

3. Discovering exceptional automorphisms

Determination of the full automorphism group of a cyclic n-gonal surface requires more work than lifting the action of K on S/C and requires an entirely different method than that suggested in Example 1.4. We first discuss a mechanism that encodes the information in the triple $C \trianglelefteq N < A$. The proofs and more details of the results of this section are given in [**4**].

3.1. Covering actions by Fuchsian groups. A (co-compact) Fuchsian group Γ, a discrete group acting on the hyperbolic plane $\mathbb{H}$, has a presentation by hyperbolic and elliptic generators and relations:

$$\text{generators}: \{\alpha_i, \beta_i, \gamma_j, 1 \leq i \leq \sigma, 1 \leq j \leq s\}$$

$$\text{relations}: \prod_{i=1}^{\sigma}[\alpha_i, \beta_i] \prod_{j=1}^{s} \gamma_j = \gamma_1^{m_1} = \cdots = \gamma_s^{m_s} = 1$$

The signature of Γ is

$$\mathcal{S}(\Gamma) = (\sigma : m_1, \ldots, m_s)$$

($= (m_1, \ldots, m_s)$, when the genus is zero). Here are important invariants of a Fuchsian group, they are all derived from the signature.

- *genus of* Γ: $\sigma(\Gamma) = \sigma$ is the genus of $\mathbb{H}/\Gamma$.
- The area of a fundamental region of Γ is given by $A(\Gamma) = 2\pi\mu(\Gamma)$ where,

$$\mu(\Gamma) = 2(\sigma - 1) + \sum_{j=1}^{s}\left(1 - \frac{1}{m_j}\right).$$

- The *Teichmüller dimension* $d(\Gamma)$ of Γ, the dimension of the Teichmüller space of Fuchsian groups with signature, $\mathcal{S}(\Gamma)$ given by

$$d(\Gamma) = 3(\sigma - 1) + s.$$

For S a cyclic n-gonal surface, we have a covering diagram

$$\begin{array}{ccccc} \Gamma_C & \hookrightarrow & \Gamma_N & \hookrightarrow & \Gamma_A \\ \downarrow \eta & & \downarrow \eta & & \downarrow \eta \\ C & \hookrightarrow & N & \hookrightarrow & A \end{array} \tag{3.1}$$

for an exact sequence

$$\Pi \hookrightarrow \Gamma_A \overset{\eta}{\twoheadrightarrow} A \tag{3.2}$$

such that Π is torsion free and $S = \mathbb{H}/\Pi$. By considering covering triples $\Gamma_C \trianglelefteq \Gamma_N < \Gamma_A$, we may classify $C \trianglelefteq N < A$ using computational group theory methods as follows.

- Determine the Fuchsian group signature triples allowed.
- From the signatures, determine the possible epimorphisms $\eta : \Gamma_C \to C$ and the K-action on $\mathbb{P}^1$. If the K-action is compatible with the epimorphism η then $\Gamma_C \trianglelefteq \Gamma_N$.
- Determine monodromy of Fuchsian group pairs $\Gamma_C \trianglelefteq \Gamma_N$, $\Gamma_N < \Gamma_A$ (definition next section)
- Determine the "word maps" of Fuchsian group pairs $\Gamma_C \trianglelefteq \Gamma_N$, $\Gamma_N < \Gamma_A$ (definition next section)
- Using the monodromy and word maps, the monodromy of the pairs $C \trianglelefteq N$ and $N < A$, may be fused together to produce A.

3.2. Restrictions on structure and signatures. Here we indicate how the weakly malnormal and super-elliptic conditions limit the structure the triple $C < N < A$ and the signatures of the possible triples.

- The weakly malnormal and super-elliptic conditions limit the structure of the signatures of the possible triples.
- There are finitely many cases of parametric families and finitely many exceptional signature pairs to consider.
- The two different conditions (weakly malnormal and super-elliptic) require different computational methods.

The following theorem describes parametric families of surfaces [**4**].

THEOREM 3.1. *If S is generalized super-elliptic (or even if C has a weakly malnormal action), and N is not self-normalizing in A then N contains a copy of $\mathbb{Z}_n \times \mathbb{Z}_n$ and there are three possibilities given in Table 2.*

$\mathcal{S}(\Gamma_A)$	$\mathcal{S}(\Gamma_N)$	$m = \lvert\Gamma_A/\Gamma_N\rvert$	$\lvert C\rvert$	genus	Group
$(2,3,2n)$	$(2,n,2n)$	3	$n \geq 5$	$\frac{(n-1)(n-2)}{2}$	$\Sigma_3 \ltimes (\mathbb{Z}_n \times \mathbb{Z}_n)$
$(2,2,2,n)$	$(2,2,n,n)$	2	$n \geq 3$	$(n-1)^2$	$V_4 \ltimes (\mathbb{Z}_n \times \mathbb{Z}_n)$
$(2,4,2n)$	$(2,2n,2n)$	2	$n \geq 3$	$(n-1)^2$	$D_4 \ltimes (\mathbb{Z}_n \times \mathbb{Z}_n)$

TABLE 2. Generalized super-elliptic surfaces for non-normalizing N

THEOREM 3.2. *If S is weakly malnormal then Γ_N has at most 3 more periods than Γ_A. If Γ_A and Γ_N have the same number of canonical generators, then they appear in Singerman's list. The signatures for Γ_A and Γ_N appear as a pair in Table*

3. *In the table* (a_1, a_2, a_3) *or* (k, k) *is the signature of* $K = \Gamma_N/\Gamma_C$*, as spherical crystallographic group. The signature for* Γ_C *is automatically determined from* Γ_N*. The type is of the form cA or cB, where c is the Teichmüller codimension (difference in the number of generators) and A denotes a non-cyclic* K *and B denotes a cyclic* K*.*

Type	Signature of Γ_N	Signature of Γ_A
0A	$(0; a_1m_1, a_2m_2, a_3m_3, , n_1, \ldots, n_r)$	$(0; b_1, b_2, b_3, n_1, \ldots, n_r)$
0B	$(0; km_1, km_2, , n_1, \ldots, n_r)$	$(0; b_1, b_2, n_1, \ldots, n_r)$
1A	$(0; a_1m_1, a_2m_2, a_3m_3, , n_1, \ldots, n_r)$	$(0; b_1, b_2, n_1, \ldots, n_r)$
1B	$(0; km_1, km_2, , n_1, \ldots, n_r)$	$(0; b_1, n_1, \ldots, n_r)$
2A	$(0; a_1m_1, a_2m_2, a_3m_3, , n_1, \ldots, n_r)$	$(0; b_1, n_1, \ldots, n_r)$
2B	$(0; km_1, km_2, , n_1, \ldots, n_r)$	$(0; n_1, \ldots, n_r)$
3A	$(0; a_1m_1, a_2m_2, a_3m_3, , n_1, \ldots, n_r)$	$(0; n_1, \ldots, n_r)$

TABLE 3. Signatures for weakly malnormal actions

THEOREM 3.3. *If* S *is a generalized super-elliptic surface,* n *is composite, and* N *is self-normalizing in* A *then the signatures of* Γ_N*, and* Γ_A *are as in the Tables* 4 *and* 5*. The examples in Table* 4 *occur in Singerman's list, those in Table* 5 *have positive codimension. The type notation cA or cB is as described in Theorem* 3.2 *and the quantities* m_i *must be* 1 *or* n*.*

REMARK 3.4. Additional constraints on the signatures in Table 5 have not been determined at the time of writing this paper.

Type	$\mathcal{S}(\Gamma_A)$	$\mathcal{S}(\Gamma_N)$	$m = \lvert\Gamma_A/\Gamma_N\rvert$	K	n
0A	$(2,3,8)$	$(3,8,8)$	10	D_3	4
0A	$(2,3,8)$	$(2,8,8)$	6	D_8	4
0A	$(2,3,4q)$	$(q,4q,4q)$	6	D_q	$2q$
0A	$(2,4,2q)$	$(q,2q,2q)$	4	D_q	q
0A	$(2,3,12)$	$(3,4,12)$	4	Σ_4	6
0A	$(2,3,2q)$	$(2,q,2q)$	3	D_q	q
0B	$(2,3,8)$	$(4,8,8)$	12	$\mathbb{Z}_2$	4
0B	$(2,3,9)$	$(9,9,9)$	12	$\mathbb{Z}_9$	9
0B	$(2,3,9)$	$(3,3,9)$	4	$\mathbb{Z}_3$	9

TABLE 4. Co-dimension 0 – from Singerman's list

EXAMPLE 3.5. Table 6 shows some examples of known generalized super-elliptic surfaces. The examples show prime cases, special cases, and families.

The group $G = \mathbb{Z}_4.A_4$ in the third case in table 6, is a central extension of A_4 by $\mathbb{Z}_4$. In the Magma database $G = \text{SmallGroup}(48, 33)$ is group number 33 in the groups of order 48.

Type	$\mathcal{S}(\Gamma_A)$	$\mathcal{S}(\Gamma_N)$	$m = \|\Gamma_A/\Gamma_N\|$
1A	(a_1n, a_2n, n)	(a_1n, a_2n, a_3n_3, n)	$m \leq 6$
1A	(a_1n, b_2, n)	$(a_1n, a_2n_2, a_3n_3, n)$	$m \leq 8$
1B	(b_1, n, n)	(k, k, n, n)	$m \leq 5$
2A	(b_1, b_2, n)	(a_1, a_2, a_3, n)	$m \leq 20$
2A	(b_1, b_2, n, n)	$(a_1m_1, a_2m_2, a_3m_3, n, n)$	$m = 3$
2A	(a_1n, n, n)	(a_1n, a_2n, a_3, n, n)	$m \leq 7$
2A	(a_1n, n, n)	(a_1n, a_2, a_3, n, n)	$m \leq 5$
2A	(b_1, n, n)	(a_1, a_2, a_3, n, n)	$m \leq 10$
2B	$(4, 4, 4)$	$(2, 2, 4, 4, 4)$	$m = 5$
3A	(n, n, n)	(a_1, a_2, a_3, n, n, n)	$m < 3 + \frac{6}{n-3}$

TABLE 5. Co-dimension > 0

$\mathcal{S}(\Gamma_A)$	$\mathcal{S}(\Gamma_N)$	$m = \|\Gamma_A/\Gamma_N\|$	$\|C\|$	genus	Group
$(2, 3, 4n)$	$(2, 2, 3, n)$	4	$n = 2$	2	$GL(2, 3)$
$(2, 3, 7)$	$(3, 3, 7)$	8	7	3	$PSL_2(7)$
$(2, 3, 3n)$	$(3, n, 3n)$	4	$n = 3$	3	$\mathbb{Z}_4.A_4$
$(2, 4, 5)$	$(4, 4, 5)$	6	5	4	Σ_5
$(2, 3, 2n)$	$(2, n, 2n)$	3	$n \geq 5$	$\frac{(n-1)(n-2)}{2}$	$\Sigma_3 \ltimes (\mathbb{Z}_n \times \mathbb{Z}_n)$
$(2, 2, 2, n)$	$(2, 2, n, n)$	2	$n \geq 3$	$(n-1)^2$	$V_4 \ltimes (\mathbb{Z}_n \times \mathbb{Z}_n)$
$(2, 4, 2n)$	$(2, 2n, 2n)$	2	$n \geq 3$	$(n-1)^2$	$D_4 \ltimes (\mathbb{Z}_n \ltimes \mathbb{Z}_n)$

TABLE 6. Known families of generalized super-elliptic surfaces

4. Classification overview

4.1. Fuchsian group pairs. Suppose that $\Gamma < \Delta$ is a Fuchsian group pair. We also suppose that Γ has genus σ and s elliptic generators and that Δ has genus τ and t elliptic generators. For notational convenience, we denote the canonical generating sets of Γ and Δ, respectively, by:

$$\mathcal{G}_1 = \{\theta_1, \ldots, \theta_{2\sigma+s}\}$$

and

$$\mathcal{G}_2 = \{\zeta_1, \ldots, \zeta_{2\tau+t}\},$$

We call the quantity

$$c(\Gamma, \Delta) = d(\Gamma) - d(\Delta)$$

the *Teichmüller codimension* of the pair $\Gamma < \Delta$. In any calculation we will always assume that $\sigma = \tau = 0$ so that $c(\Gamma, \Delta) = s - t$, the difference in the number of generators.

We next discuss two ways of describing the inclusion $\Gamma < \Delta$ namely monodromy groups and word maps.

4.1.1. *Monodromy groups.* The pair $\Gamma < \Delta$ determines a permutation or monodromy representation of Δ on the cosets of Γ

$$\rho : \Delta \to \Sigma_m$$

where m is the index of Γ in Δ. Write

$$\mathcal{P} = (\pi_1, \pi_2, \ldots, \pi_{2\tau+t})$$

for $\pi_i = \rho(\zeta_i) \in \Sigma_m$, to construct the *monodromy vector* of the pair. The cycle types and other properties of $\mathcal{P}$ are determined by signatures $\mathcal{S}(\Gamma)$ and $\mathcal{S}(\Delta)$ and the relations on the generators.

DEFINITION 4.1. Let notation be as above. Then the *monodromy group* of the pair is defined by

$$M(\Delta, \Gamma) = \rho(\Delta) = \langle \pi_1, \pi_2, \ldots, \pi_{2\tau+t} \rangle$$

It is well defined up to conjugacy in Σ_m. We extend the definition of monodromy of a group pair to the sequence $C < N < A$. Then we have these relations because of the diagram 3.1

$$\begin{aligned} M(\Gamma_A, \Gamma_N) &= M(A, N) \\ M(\Gamma_N, \Gamma_C) &= M(N, C) = M(N/C, \langle 1 \rangle) \simeq K \\ M(\Gamma_A, \Gamma_C) &= M(A, C) \simeq A. \end{aligned}$$

The last equation holds because of the super-elliptic or weak malnormality condition both of which implies that $\bigcap_{g \in A} gCg^{-1} = \langle 1 \rangle$ since A strictly contains N.

4.1.2. *Word Maps.* The inclusion $\Gamma < \Delta$ may also be described by a word map which we define now.

DEFINITION 4.2. Let notation be as above. Then the *word map* of the inclusion $\Gamma \hookrightarrow \Delta$ (with respect to the given generating sets) is a set of words $\{w_1, \ldots, w_{2\sigma+s}\}$ in the generators in $\mathcal{G}_2$ such that

$$\theta_i = w_i(\zeta_1 \ldots, \zeta_{2\tau+t}), i = 1, \ldots, 2\tau + t.$$

We note that given a word map for the inclusion $\Gamma \hookrightarrow \Delta$ a monodromy vector $\mathcal{P}$ is easily calculated using the Todd-Coxeter algorithm. Given a monodromy vector $\mathcal{P}$ of a genus zero pair $\Gamma < \Delta$ (both groups), then the word map of the pair may be calculated, in a standard way, using the Reidemeister-Schreier method. See for instance [**10**], [**8**] and [**11**].

EXAMPLE 4.3. Suppose we have the following signatures

$$\mathcal{S}_1 = (2, 2, 2, 5), \mathcal{S}_2 = (2, 4, 5)$$

We want a pair $\Gamma < \Delta$ with

$$\mathcal{S}(\Gamma) = \mathcal{S}_1, \mathcal{S}(\Delta) = \mathcal{S}_2$$

First we find a compatible monodromy vector in Σ_6

$$\pi_1 = (1, 3)(4, 6), \pi_2 = (1, 2)(3, 5, 4, 6), \pi_3 = (1, 2, 3, 4, 5),$$

note that $M(\Delta, \Gamma) = A_6$. Define $\rho\colon \Delta \to \Sigma_6$ by $\rho\colon \zeta_i \to \pi_i, i = 1 \ldots 3$ and set Γ to be the stabilizer of a point for the permutation action of Δ on $\{1, \ldots, 6\}$. From the algorithm, a generating set for Γ is:

$$\begin{aligned} \theta_1 &= (\zeta_1\zeta_2)\zeta_1(\zeta_1\zeta_2)^{-1} \\ \theta_2 &= \zeta_2\zeta_1\zeta_2^{-1} \\ \theta_3 &= \zeta_2^2 \\ \theta_4 &= (\zeta_2^{-1}\zeta_1^{-1}\zeta_2^{-1}\zeta_1\zeta_3\zeta_1)\zeta_3(\zeta_2^{-1}\zeta_1^{-1}\zeta_2^{-1}\zeta_1\zeta_3\zeta_1)^{-1} \end{aligned}$$

4.2. Steps of the classification. The discussion following equation 3.2 may be expanded to classify all super-elliptic surfaces with exceptional automorphisms. Full results will appear in [**4**].

(1) Using Magma or GAP determine all signature pairs $\mathcal{S}(\Gamma_N)$ and $\mathcal{S}(\Gamma_A)$ for codimension 0,1,2,3 treating special pairs and families of pairs separately. In the super-elliptic case the search is considerably cut down because of the restriction on signatures.
(2) The group K and the signature $\mathcal{S}(\Gamma_C)$ is automatically determined. In the super-elliptic case all the periods equal n and only the number of periods needs to be determined.
(3) The action of the group K on $\mathbb{P}^1$ and the possible epimorphisms $\eta : \Gamma_C \to C$ can be determined. The K-action and the epimorphism η must be compatible. If no such compatible action exists then there is no $\Gamma_C \trianglelefteq \Gamma_N$.
(4) For each candidate signature pair, compute all the compatible monodromy vectors up to conjugacy. Use the classification of primitive permutation groups (Magma or GAP).
(5) From monodromy vectors of $\Gamma_N < \Gamma_A$ and $\Gamma_C \trianglelefteq \Gamma_N$ compute the word maps of $\Gamma_C \hookrightarrow \Gamma_N$ and $\Gamma_N \hookrightarrow \Gamma_A$
(6) Compute the word map of $\Gamma_C \hookrightarrow \Gamma_A$ by substitution.
(7) Compute the monodromy group $M(\Gamma_A, \Gamma_C)$ using the Todd-Coxeter algorithm.
(8) If the stabilizer of a point in $M(\Gamma_A, \Gamma_C) \simeq A$ is not cyclic then reject this case.

In the weakly malnormal case there are 202 special pairs and 597 families of pairs $\Gamma_N < \Gamma_A$ that could potentially lead to cyclic n-gonal surfaces. Obviously this cannot be done by hand. In the super-elliptic case computer calculation is still needed but the number of cases is dramatically reduced.

Let us finish with some simple calculations that illustrate steps 1-4. Actual calculation with words maps are cumbersome and beyond the space limitations of this paper. Full details of all calculations will appear in [**4**].

EXAMPLE 4.4. Let us consider the last line of Table 4 with $\mathcal{S}(\Gamma_A) = (2,3,9)$, $\mathcal{S}(\Gamma_N) = (3,3,9)$, $m = |\Gamma_A/\Gamma_N| = 4$, $K = \mathbb{Z}_3$ and $n = |C| = 9$. We see from the table that K has two fixed points of order 3 on S/C and that the branch points of the C-action from a regular K-orbit. The only possibilities for the multi-degree (t_1, t_2, t_3) are $(1,1,7)$, $(1,4,4)$, Now the multi-degree must satisfy $t_{i+1} = rt_i$ for some r satisfying $r^3 = 1 \bmod 9$. This is not possible and we stop here. We do note that the pair $\Gamma_N < \Gamma_A$ does exist. For, we see that the monodromy vector $\mathcal{P} = (\pi_1, \pi_2, \pi_3)$ must have cycle types 2^2, $1 \cdot 3$, and $1 \cdot 3$ respectively. We may choose $\pi_1 = (1,3)(2,4), \pi_2 = (1,2,3), \pi_3 = (2,3,4)$.

EXAMPLE 4.5. Let us consider the third last line of Table 4 with $\mathcal{S}(\Gamma_A) = (2,3,8)$, $\mathcal{S}(\Gamma_N) = (4,8,8)$, $m = |\Gamma_A/\Gamma_N| = 12$, $K = \mathbb{Z}_2$ and $n = |C| = 4$. We see from the table that K has two fixed points of order 2 on S/C and so that the four branch points of order 4 must lie on the K-fixed points and on one regular K-orbit. The possibilities for the multi-degree (t_1, t_2, t_3, t_4) are $(1,1,1,1)$ and $(1,1,3,3)$ permutations if t_1 and t_2 correspond to fixed points. Thus one can proceed to find permutations. The cycle types must be 2^6, $3^4 \cdot 3$, and $1^2 \cdot 2 \cdot 8$. If we are looking for pairs $\Gamma_N < \Gamma_A$ with primitive $M(\Gamma_A, \Gamma_N)$ then there is a monodromy vector $\mathcal{P} = (\pi_1, \pi_2, \pi_3)$ of the prescribed cycle type in the Mathieu group M_{12} of order 95040. Thus there is a pair $\Gamma_N < \Gamma_A$ with $M(\Gamma_A, \Gamma_N) \simeq M_{12}$. However, as $M(\Gamma_A, \Gamma_N)$ is an image of A then, $\frac{|A|}{|M(\Gamma_A,\Gamma_N)|}$ is an integer. But,

$$\frac{|A|}{|M(\Gamma_A,\Gamma_N)|} = \frac{\frac{|A|}{|N|} \times \frac{|N|}{|C|} \times |C|}{|M(\Gamma_A,\Gamma_N)|} = \frac{m \times |K| \times n}{|M_{12}|} = \frac{12 \times 2 \times 4}{95040} = \frac{1}{990}.$$

The other primitive groups can be rejected for other reasons, and so one must look among the 301 transitive groups of degree 12.

References

[1] Robert D. M. Accola, *Strongly branched coverings of closed Riemann surfaces*, Proc. Amer. Math. Soc. **26** (1970), 315–322. MR0262485 (41 #7091)

[2] Gabriel Bartolini, Antonio F. Costa, and Milagros Izquierdo, *On automorphisms groups of cyclic p-gonal Riemann surfaces*, J. Symbolic Comput. **57** (2013), 61–69, DOI 10.1016/j.jsc.2013.05.005. MR3066451

[3] S.A Broughton, presentation at UNED *http://www.rose-hulman.edu/~brought/Epubs/UNED/UNED.html.*

[4] S.A. Broughton and A. Wootton *Full Automorphism Groups of Generalized Superelliptic Riemann Surfaces*, in preparation.

[5] Rolf Brandt and Henning Stichtenoth, *Die Automorphismengruppen hyperelliptischer Kurven* (German, with English summary), Manuscripta Math. **55** (1986), no. 1, 83–92, DOI 10.1007/BF01168614. MR828412 (87m:14033)

[6] E. Bujalance, J. M. Gamboa, and G. Gromadzki, *The full automorphism groups of hyperelliptic Riemann surfaces*, Manuscripta Math. **79** (1993), no. 3-4, 267–282, DOI 10.1007/BF02568345. MR1223022 (94f:20093)

[7] E. Bujalance, J. A. Bujalance, G. Gromadzki, and E. Martinez, *Cyclic trigonal Klein surfaces*, J. Algebra **159** (1993), no. 2, 436–459, DOI 10.1006/jabr.1993.1167. MR1231224 (94g:30040)

[8] Antonio F. Costa and Milagros Izquierdo, *On real trigonal Riemann surfaces*, Math. Scand. **98** (2006), no. 1, 53–68. MR2221544 (2007d:30029)

[9] Antonio F. Costa, Milagros Izquierdo, and Daniel Ying, *On Riemann surfaces with non-unique cyclic trigonal morphism*, Manuscripta Math. **118** (2005), no. 4, 443–453, DOI 10.1007/s00229-005-0593-y. MR2190106 (2006k:30042)

[10] J. Gilman *Computing Adapted Bases for Conformal Automorphism Groups of Riemann Surfaces*, this conference proceedings.

[11] Antonio F. Costa and Peter Turbek, *Lifting involutions to ramified covers of Riemann surfaces*, Arch. Math. (Basel) **81** (2003), no. 2, 161–168, DOI 10.1007/s00013-003-4709-x. MR2009558 (2004k:30097)

[12] Gabino González-Diez, *On prime Galois coverings of the Riemann sphere*, Ann. Mat. Pura Appl. (4) **168** (1995), 1–15, DOI 10.1007/BF01759251. MR1378235 (97c:14033)

[13] Aristides Kontogeorgis, *The group of automorphisms of cyclic extensions of rational function fields*, J. Algebra **216** (1999), no. 2, 665–706, DOI 10.1006/jabr.1998.7804. MR1692965 (2000f:12005)

[14] Tanush Shaska, *Determining the automorphism group of a hyperelliptic curve*, Proceedings of the 2003 International Symposium on Symbolic and Algebraic Computation, ACM, New York, 2003, pp. 248–254 (electronic), DOI 10.1145/860854.860904. MR2035219 (2005c:14037)

[15] T. Shaska, *On super-elliptic Curves and their Quotients*, Albanian Journal of Math. **5** (2011), no. 3, 131–160.

[16] David Singerman, *Finitely maximal Fuchsian groups*, J. London Math. Soc. (2) **6** (1972), 29–38. MR0322165 (48 #529)

[17] A. Wootton, *The full automorphism group of a cyclic p-gonal surface*, J. Algebra **312** (2007), no. 1, 377–396, DOI 10.1016/j.jalgebra.2007.01.018. MR2320463 (2008c:14043)

[18] A. Wootton, *Defining equations for cyclic prime covers of the Riemann sphere*, Israel J. Math. **157** (2007), 103–122, DOI 10.1007/s11856-006-0004-4. MR2342442 (2009b:30084)

[19] D. Ying *On the Moduli Space of Cyclic Trigonal Riemann Surfaces of Genus 4*, PhD thesis, Linkoping University Institute of Technology, (2006).

ROSE-HULMAN INSTITUTE OF TECHNOLOGY
E-mail address: brought@rose-hulman.edu

THE UNIVERSITY OR PORTLAND
E-mail address: wootton@up.edu

Symmetric group actions on Jacobian varieties

Angel Carocca, Rubí E. Rodríguez, and Anita M. Rojas

Dedicated to Professor Emilio Bujalance on the occasion of his 60th birthday.

ABSTRACT. A Jacobian variety that is isogenous to a product of elliptic curves is called a completely decomposable Jacobian. They have been widely studied but there still remain some key questions unsolved. For instance, it is still unknown if there is a completely decomposable Jacobian on each dimension. In this work we study and completely characterize the families of curves with the action of a symmetric group, such that the group algebra decomposition for the corresponding Jacobian is a product of elliptic curves.

1. Introduction

An abelian variety over the complex numbers is said to be *completely decomposable* if it is isogenous to a product of elliptic curves. In particular, the problem of studying completely decomposable Jacobians has received plenty of attention in the last decades because of implications to number theory and also in cryptography.

In [**6**] three questions regarding completely decomposable Jacobians were asked and partially answered. They found, in several dimensions, Jacobians fulfilling such condition. The list is not complete, and one of the questions left opened was if there is a bound for the genus of the curves having completely decomposable Jacobians. Several authors, see for instance [**5**], [**10**] and [**12**], have studied this subject from different viewpoints. In [**5**] it is studied and fully characterized the Riemann matrices of completely decomposable two dimensional Jacobians; in [**12**] the author proposed to study Jacobians isogenous to a selfproduct of an elliptic curve, and look for examples up to genus 10. Recently, in [**9**], the question of finding families of completely decomposable Jacobians is proposed.

In this work, we study the problem of finding (families of) completely decomposable Jacobians using group actions. The main results of this work are Theorems 2.5 and 2.7, which can be summarized as giving a complete classification of the actions of the symmetric group corresponding to completely decomposable Jacobians. Moreover, the respective geometric actions and classification are explicitly given.

It is known (see [**8**] or [**1**, 13.6]) that once there is a group G acting on an abelian variety A, then A is G-equivariantly decomposed into a product of abelian varieties of lower dimension. This is called the *group algebra decomposition* for A.

2010 *Mathematics Subject Classification.* Primary 14H40, 14K10.

Key words and phrases. Group actions, completely decomposable Jacobians, symmetric groups.

The authors were partially supported by Fondecyt grants 1130445, 1141099 and 1140507.

We will not completely enter into the details of this decomposition, since they can be found in the given references; instead, we just recall the facts we will use in this work. We first recall the group algebra decomposition theorem for an abelian variety with group action, giving a simplified version adequate for our purposes.

THEOREM 1.1 (The group algebra decomposition for A). ([**8**]) *Let G be a finite group acting on an abelian variety A. Let $\mathcal{W}_1, \dots, \mathcal{W}_r$ denote the irreducible $\mathbb{Q}$-representations of G up to isomorphism. Assume they are absolutely irreducible.*

Then there are abelian subvarieties $B_1, \dots, B_r$ of A, such that each $B_j^{\dim \mathcal{W}_j}$ admits a G-action with representation (a multiple of) $\mathcal{W}_j$. Moreover, there is a G-equivariant isogeny

$$A \sim B_1^{\dim \mathcal{W}_1} \times \cdots \times B_r^{\dim \mathcal{W}_r}. \tag{1.1}$$

Let X be a smooth projective curve with the action of a finite group G; then there is an induced G-action on the corresponding Jacobian variety JX. The dimensions of the varieties B_j on the group algebra decomposition of JX are captured by the ramification data for the action of G on X. This was proved in [**7**] for groups having all their rational irreducible representations also irreducible over the complex numbers, the so called *absolutely irreducible representations*, and in general in [**13**]. The family of groups we are interested in, the symmetric groups, satisfy this hypothesis.

First we recall that the geometric information about the action of G on X is encoded in the so called *geometric signature* (see [**13**]). It is the tuple

$$\sigma_g = (\gamma; [n_1, C_1], \dots, [n_t, C_t])$$

where γ is the genus of the quotient curve X/G, each C_j is a conjugacy class of cyclic subgroups of G, n_j denotes the number of branch points $y \in X/G$ whose preimages in X are fixed by a subgroup in the class C_j, and $\sum_{j=1}^{t} n_j$ is the number of branch points of $\pi_G : X \to X/G$.

The usual *signature* σ associated to the action of G on X underlying the geometric signature σ_g is defined by

$$\sigma = (\gamma; \underbrace{|G_1|, \dots, |G_1|}_{n_1-\text{times}}, \dots, \underbrace{|G_t|, \dots, |G_t|}_{n_t-\text{times}}),$$

where $|G_j|$ denotes the order of a subgroup G_j in C_j. It is customary to write

$$\sigma = (\gamma; m_1^{b_1}, \dots, m_\ell^{b_\ell}), \quad m_1 < \cdots < m_\ell, \quad b_i \geq 1,$$

and b_i indicates the number of stabilizers of order m_i. See [**4**, Def. 1.1.2] for a detailed discussion about signatures for the Fuchsian equivalent of G.

We will use later a minor variation of the above notation for signature, as it is convenient to simplify some of our proofs, as follows

$$\sigma^* = (\gamma; 2^{a_2}, \dots, j^{a_j}, \dots, \delta^{a_\delta}), \tag{1.2}$$

where γ is the genus of X/G, and a_j is the number of branch points where $\pi_G : X \to X/G$ is locally j to 1; we are allowing the possibility that $a_j = 0$. The tuple stops at the maximum order δ for the elements of G.

The following well known results are going to be used several times along this work; we recall them here for the sake of completeness and to fix the notation. The first one, Theorem 1.2, gives the dimension of each factor B_i in the decomposition

(1.1), in terms of the geometric signature, and Theorem 1.3 is an existence theorem for actions of a group G with a given signature.

THEOREM 1.2 ([**7**]). *Let G be a finite group acting on a curve X with geometric signature $(\gamma; [n_1, C_1], ..., [n_t, C_t])$. Assume that all the rational irreducible representations of G are absolutely irreducible.*

Then the dimension of a subvariety B_i associated to a non-trivial rational irreducible representation $\mathcal{W}_i$, in the group algebra decomposition (1.1) *of the corresponding Jacobian variety JX, is given by*

$$\dim B_i = \dim \mathcal{W}_i(\gamma - 1) + \frac{1}{2}\sum_{k=1}^{t} n_k(\dim \mathcal{W}_i - \dim \operatorname{Fix}_{G_k} \mathcal{W}_i) \ , \tag{1.3}$$

where G_k is a representative of the conjugacy class C_k, and $\operatorname{Fix}_{G_k} \mathcal{W}_r$ *denotes the set of fixed points of G_k in $\mathcal{W}_r$.*

THEOREM 1.3 ([**3**, Prop. 2.1]). *A group G acts on a curve X of genus $\mathbf{g}$ with signature $\sigma = (\gamma; m_1^{b_1}, \ldots, m_\ell^{b_\ell})$ if and only if the Riemann-Hurwitz equation*

$$\mathbf{g} = |G|(\gamma - 1) + 1 + \frac{|G|}{2}\sum_{i=1}^{\ell} b_i\left(1 - \frac{1}{m_i}\right)$$

is satisfied, and there is a tuple

$$\nu = (\alpha_1, \ldots, \alpha_\gamma, \beta_1, \ldots, \beta_\gamma, g_{1,1}, \ldots, g_{1,b_1}, \ldots, g_{\ell,1}, \ldots, g_{\ell,b_\ell})$$

of elements of G that generate G, with $g_{i,k}$ of order m_i, and satisfying

$$\prod_{j=1}^{\gamma}[\alpha_j, \beta_i]\prod_{i=1}^{\ell}\prod_{k=1}^{b_j} g_{i,k} = 1.$$

REMARK 1.4. Given a signature $\sigma = (\gamma; m_1^{b_1}, \ldots, m_\ell^{b_\ell})$, a tuple ν as in Theorem 1.3 is called a *generating vector* for the signature σ (see [**3**, Def. 2.2]).

In what follows, we study curves X with the action of the symmetric group $G = \mathbf{S}_n$ such that the subvarieties B_i of the group algebra decomposition of the corresponding Jacobian variety have dimension less than 2; that is, curves X whose Jacobian variety becomes completely decomposable under the group algebra decomposition. The work is structured in the following way; first in **Hypothesis A**, we capture the conditions under which the classification will be done. In Section 2, we obtain the main results of this work, Theorems 2.7 and 2.5, where a complete classification of the actions of the symmetric group corresponding to completely decomposable Jacobians are given. Finally in Section 3 we present the full classification and the corresponding decomposition for each case.

Although the goal is different, the approach taken in this work is, in some sense, similar to the one used by Bujalance et al. in computing the number of ovals of symmetries of a Riemann surface, [**4**, Section 3.3].

2. Results

Throughout this section let X be a curve of genus $g \geq 2$, with an action of the symmetric group $G = \mathbf{S}_n$. By Theorem 1.1, there are subvarieties $B_1, \ldots, B_r$ of JX such that

$$JX \sim B_1^{\dim \mathcal{W}_1} \times \cdots \times B_r^{\dim \mathcal{W}_r},$$

where $\mathcal{W}_1, \dots, \mathcal{W}_r$ are the rational (absolutely) irreducible representations of G.

We will always assume that $\mathcal{W}_1$ is the trivial representation. Since B_1 is then isogenous to the Jacobian variety of X/G, it follows that $\gamma = \dim B_1$.

We consider the following hypothesis for the action:

Hypothesis A Let $B_1, \dots, B_r$ be the subvarieties in the group algebra decomposition (1.1) of a Jacobian variety JX corresponding to a curve X with the action of the group $\mathbf{S}_n$. Then

$$\dim B_j \leq 1 \text{ for all } j = 1, \dots, r.$$

In particular, we assume that $\gamma = \dim B_1 \leq 1$.

In what follows, we analyze the possibilities for the signature σ and the value of n under Hypothesis A for an action of $G = \mathbf{S}_n$ on a curve X.

LEMMA 2.1. *For $n \geq 6$, let $\mathbf{N}(n)$ be the sum of the degrees of the irreducible representations of the symmetric group $\mathbf{S}_n$. Then $\mathbf{N}(n) < (n-2)!(n-2) - n$.*

PROOF. From [**11**] we know that $\mathbf{N}(n)$ is the number of involutions in $\mathbf{S}_n$, hence it satisfies a recursive relation:

$$\mathbf{N}(n) = \mathbf{N}(n-1) + (n-1)\mathbf{N}(n-2).$$

That is because $\mathbf{N}(n-1)$ counts the number of involutions fixing n and $\mathbf{N}(n-2)$ counts the number of involutions having the cycle $(n-1, n)$, see [**11**] for details. The result follows by induction:

$$\mathbf{N}(n) < (n-3)!(n-3) - (n-1) + (n-1)[(n-4)!(n-4) - (n-2)]$$

Then

$$\mathbf{N}(n) < (n-4)![(n-3)^2 + (n-1)(n-4)] - (n-1)^2$$

Since for all $n > 5$ we have that

$$(n-3)^2 + (n-1)(n-4) < (n-3)(n-2)^2,$$

we obtain

$$\mathbf{N}(n) < (n-4)!(n-3)(n-2)^2 - (n-1)^2 = (n-2)!(n-2) - (n-1)^2 < (n-2)!(n-2) - n.$$

□

LEMMA 2.2. *Let X be a curve of genus $\mathbf{g}$ with an action of $G = \mathbf{S}_n$ with signature $\sigma = (\gamma; m_1^{b_1}, \dots, m_s^{b_s})$. Let $\{\mathcal{W}_1, \dots, \mathcal{W}_r\}$ be a set of non equivalent irreducible representations of G. Set*

$$\mathbf{N}(n) = \sum_{i=1}^{r} \dim \mathcal{W}_i, \quad \textit{and} \quad S := \sum_{i=1}^{s} b_i \left(1 - \frac{1}{m_i}\right).$$

*Then, under **Hypothesis A** we have*

(1) $\mathbf{g} \leq \mathbf{N}(n) - 1 + \gamma$.

(2)
$$S \leq 2\left(\frac{\mathbf{N}(n) - 2 + \gamma}{n!} + 1 - \gamma\right).$$

(3) *If $n \geq 6$, then*
$$S < 2\left(\frac{n-2}{n(n-1)} + \frac{\gamma - 2 - n}{n!} + 1 - \gamma\right).$$

PROOF. First $\mathbf{g}$ is the genus of X, hence the dimension of JX. By comparing dimensions in both sides of (1.1), we have

$$\mathbf{g} = \sum_{j=1}^{r} \dim \mathcal{W}_j \dim B_j \leq \sum_{j=1}^{r} \dim \mathcal{W}_j = \mathbf{N}(n),$$

where the inequality follows from **Hypothesis A**. Item (1) is obtained refining this bound using the fact that if $\gamma = 0$, then the dimension of B_1 is 0.

By the Riemann-Hurwitz formula for the covering $\pi : X \to X/G$, using σ, (1) and $n! = |G|$, we obtain

$$\mathbf{g} = |G|(\gamma - 1) + 1 + \frac{|G|}{2} S \leq \mathbf{N}(n) - 1 + \gamma,$$

and Item (2) follows.

Finally, Item (3) follows from Item (2) and the upper bound of Lemma 2.1. □

Considering that we are under Hypothesis A, we compute the upper bound for S in terms of n in the following corollary.

COROLLARY 2.3. *With the notation of Lemma 2.2 and assuming* ***Hypothesis A****, we have the following bounds for* S.

n	γ	Upper bound for S
3	0	8/3
	1	1
4	0	8/3
	1	3/4
5	0	12/5
	1	5/12
≥ 6	0	$2\left(1 + \frac{n-2}{n(n-1)} - \frac{n+2}{n!}\right)$
	1	$2/n$

REMARK 2.4. The strategy we follow is first to obtain all the possible signatures satisfying the upper bound for S from Corollary 2.3. We use the word *data* to refer to these solutions. Then we examine the following two facts:

(1) Whether such an action exists, using Theorem 1.3, in which case we have an actual signature,
(2) and whether **Hypothesis A** is satisfied for this signature.

We first study the case when the genus γ of the quotient surface is equal to one.

THEOREM 2.5. *Let X be a curve of genus $\mathbf{g}$ with $\mathbf{S}_n$–action satisfying* ***Hypothesis A****, with $X/\mathbf{S}_n$ of genus $\gamma = 1$. Then the only cases which occur are the following.*

(1) *For* $\mathbf{n}=\mathbf{3}$.
- $g=3$ *with signature* $\sigma=(1;\ 3)$. *Generating vectors for* $G=\mathbf{S}_3$ *are*

$$\begin{cases} \alpha_1=(1,2)\ ,\ \beta_1=(1,3)\ ,\ g_1=(1,2,3), \text{ and} \\ \alpha_1=(1,2)\ ,\ \beta_1=(1,2,3)\ ,\ g_1=(1,2,3). \end{cases}$$

- $g=4$ *with* $\sigma=(1;\ 2^2)$. *Generating vectors for* $G=\mathbf{S}_3$ *are*

$$\begin{cases} \alpha_1=(1,2,3)=\beta_1\ ,\ g_1=(1,2)=g_2, \text{ and} \\ \alpha_1=(1,2),\ \beta_1=(1,2,3),\ g_1=(1,2),\ g_2=(1,3). \end{cases}$$

(2) *For* $\mathbf{n}=\mathbf{4}$.
- $g=9$ *with* $\sigma=(1;\ 3)$. *A generating vector for* $G=\mathbf{S}_4$ *is*

$$\alpha_1=(1,2)\ \ ,\ \ \beta_1=(1,2,3,4)\ \ ,\ \ g_1=(1,3,2).$$

PROOF. Consider $\sigma^*=(\gamma;\ldots,j^{a_j},\ldots)$, with the notation as in Equation (1.2). If $n\geq 6$, by Corollary 2.3 we have

$$\sum_{j=1}^{\delta}a_j\left(1-\frac{1}{j}\right)<\frac{2}{n}.$$

Therefore

$$\sum_{j=1}^{\delta}a_j\left(1-\frac{1}{j}\right)=0,$$

and using Theorem 1.3 we obtain the contradiction $\mathbf{S}_n=\langle\alpha_1,\beta_1\ /\ [\alpha_1,\beta_1]=1,R\rangle$. Here and in what follows, with R we denote further relations.

We have then that $n\leq 5$, under our hypothesis.

If $n=5$, from Corollary 2.3 we have

$$\frac{a_2}{2}+\frac{2a_3}{3}+\frac{3a_4}{4}+\frac{4a_5}{5}+\frac{5a_6}{6}\leq\frac{5}{12}.$$

Hence $a_2=a_3=a_4=a_5=a_6=0$ and again using Theorem 1.3 we obtain the contradiction $\mathbf{S}_5=\langle\alpha_1,\beta_1\ /\ [\alpha_1,\beta_1]=1,R\rangle$.

If $n=4$, by Corollary 2.3 we obtain

$$\frac{a_2}{2}+\frac{2a_3}{3}+\frac{3a_4}{4}\leq\frac{3}{4},$$

and therefore $0\leq a_j\leq 1$ for $j\in\{2,3,4\}$.

In the case $a_2=a_3=a_4=0$ we obtain the same contradiction

$$\langle\alpha_1,\ \beta_1\ /\ [\alpha_1,\beta_1]=1\rangle=\mathbf{S}_4.$$

If $a_4=1$ then $a_2=a_3=0$, and $\mathbf{S_4}=\langle\alpha_1,\ \beta_1,g_1\ /\ [\alpha_1,\beta_1]g_1=1,R\rangle$ with $(g_1)^{-1}=[\alpha_1,\beta_1]\in\mathbf{A}_4$ (the alternating group of degree 4), but g_1 is an element of order four, which is impossible.

If $a_4=a_3=0$ and $a_2=1$ then $\mathbf{S}_4=\langle\alpha_1,\beta_1,g_1\ /\ [\alpha_1,\beta_1]g_1=1,R\rangle$ with $[\alpha_1,\beta_1]=g_1$ an even permutation of order 2, which easily seen to be impossible.

If $a_4 = a_2 = 0$ and $a_3 = 1$ then $\mathbf{g} = 9$ and the signature $\sigma = (1;\ 3)$ is realizable with the following generating vector for $\mathbf{S}_4$

$$\alpha_1 = (1,2)\ ;\ \beta_1 = (1,2,3,4)\ ;\ g_1 = (1,3,2).$$

If $n = 3$, by Corollary 2.3

$$\frac{a_2}{2} + \frac{2a_3}{3} \leq 1,$$

and therefore $0 \leq a_2 \leq 2$, $0 \leq a_3 \leq 1$.

If $a_2 = 0 = a_3$, as before we have the contradiction $\mathbf{S}_3 = \langle \alpha_1,\ \beta_1\ /\ [\alpha_1, \beta_1] = 1, R\,\rangle$.

If $a_2 = 1$ and $a_3 = 0$ we have the contradiction $\mathbf{S}_3 = \langle \alpha_1,\ \beta_1, g_1\ /\ [\alpha_1, \beta_1]g_1 = 1, R\rangle$ with $g_1 = [\alpha_1, \beta_1]$ an element of order 2 in the alternating group of degree 3.

If $a_2 = 0$ and $a_3 = 1$ we obtain $\mathbf{g} = 3$ and the signature $\sigma = (1;\ 3)$ is realizable with the following generating vectors for $\mathbf{S}_3$

$$\begin{cases} \alpha_1 = (1,2)\ ;\ \beta_1 = (1,3)\ ;\ g_1 = (1,2,3) \\ \\ \alpha_1 = (1,2)\ ;\ \beta_1 = (1,2,3)\ ;\ g_1 = (1,2,3). \end{cases}$$

If $a_2 = 2$ and $a_3 = 0$ then $\mathbf{g} = 4$ and the signature $\sigma = (1;\ 2^2)$ is realizable with the following generating vectors for $\mathbf{S}_3$

$$\begin{cases} \alpha_1 = (1,2,3) = \beta_1;\ \ g_1 = (1,2) = g_2 \\ \\ \alpha_1 = (1,2);\ \beta_1 = (1,2,3);\ g_1 = (1,2);\ g_2 = (1,3). \end{cases}$$

□

On the other hand, for $\gamma = 0$, the actions of $G = \mathbf{S}_n$ and the values of n corresponding to Jacobians with group algebra decomposition having at most one-dimensional varieties B_j are described in Theorem 2.7. First we need the following Lemma.

LEMMA 2.6. (1) *There is no curve with action of* $\mathbf{S}_n$ *having at most one orbit of ramification points with stabilizer of even order.*

(2) *For* $n \geq 4$ *there is no curve with action of* $\mathbf{S}_n$ *with signature* $\sigma = (0; 2^2, m)$, *for* $m \geq 2$.

PROOF. For case one, since all elements of odd order in $\mathbf{S}_n$ are in $\mathbf{A}_n$, there is no such generating vector of $\mathbf{S}_n$.

As for (2), we observe that two elements of order two in $\mathbf{S}_n$ generate a dihedral group, and therefore a proper subgroup of $\mathbf{S}_n$ for all $n \geq 4$. □

THEOREM 2.7. *Let* X *be a curve of genus* $\mathbf{g}$ *with* $\mathbf{S}_n$*−action satisfying* ***Hypothesis A****, with* $X/\mathbf{S}_n$ *of genus* $\gamma = 0$. *Then the only cases which occur are the following.*

(1) *For* $\mathbf{n} = \mathbf{3}$.
- $\boldsymbol{g} = 2$ *and signature* $\sigma = (0; 2^2, 3^2)$ *which is realizable by choosing*

$$g_1 = (1,2) = g_2;\ g_3 = (1,2,3);\ g_4 = (1,3,2).$$

- $\boldsymbol{g} = 3$ *and* $\sigma = (0; 2^4, 3)$ *which is realizable by choosing*

$$g_1 = (1,2) = g_2 = g_3;\ g_4 = (1,3),\ g_5 = (1,3,2).$$

(2) *For* $\mathbf{n} = \mathbf{4}$.

- $g = 3$ *and* $\sigma = (0; 3, 4^2)$ *which is realizable by choosing*
$$g_1 = (1,2,3);\ g_2 = (1,2,3,4);\ g_3 = (1,4,2,3).$$
- $g = 3$ *and* $\sigma = (0; 2^3, 3)$ *which is realizable by choosing*
$$g_1 = (1,3);\ g_2 = (1,4)(2,3);\ g_3 = (1,4);\ g_4 = (1,3,2).$$
- $g = 4$ *and* $\sigma = (0; 2^3, 4)$ *which is realizable by choosing*
$$g_1 = (3,4);\ g_2 = (2,3);\ g_3 = (1,2);\ g_4 = (1,4,3,2).$$
- $g = 5$ *and* $\sigma = (0; 2^2, 3^2)$ *which is realizable by choosing*
$$g_1 = (2,4);\ g_2 = (1,3);\ g_3 = (2,4,3);\ g_4 = (1,3,2).$$
- $g = 6$ *and* $\sigma = (0; 2^2, 3, 4)$ *which is realizable by choosing*
$$g_1 = (1,4);\ g_2 = (1,2)(3,4);\ g_3 = (1,3,2);\ g_4 = (1,4,3,2).$$
- $g = 7$ *and* $\sigma = (0; 2^2, 4^2)$ *which is realizable by choosing*
$$g_1 = (2,4);\ g_2 = (1,4);\ g_3 = (1,3,4,2);\ g_4 = (1,4,3,2).$$
- $g = 8$ *and* $\sigma = (0; 2, 3^2, 4)$ *which is realizable by choosing*
$$g_1 = (1,2);\ g_2 = (2,4,3);\ g_3 = (1,3,2);\ g_4 = (1,3,2,4).$$

(3) *For* $\mathbf{n} = \mathbf{5}$.

- $g = 4$ *and* $\sigma = (0; 2, 4, 5)$ *which is realizable by choosing*
$$g_1 = (4,5);\ g_2 = (1,2,3,4);\ g_3 = (1,5,4,3,2).$$
- $g = 6$ *and* $\sigma = (0; 2, 4, 6)$ *which is realizable by choosing*
$$g_1 = (1,3)(4,5);\ g_2 = (1,2,3,4);\ g_3 = (2,3)(1,5,4).$$
- $g = 9$ *and* $\sigma = (0; 2, 5, 6)$ *which is realizable by choosing*
$$g_1 = (1,3);\ g_2 = (1,2,3,4,5);\ g_3 = (2,3)(1,5,4).$$
- $g = 11$ *and* $\sigma = (0; 2, 6^2)$ *which is realizable by choosing*
$$g_1 = (1,4)(2,3);\ g_2 = (1,2)(3,4,5);\ g_3 = (2,4)(1,3,5).$$
- $g = 11$ *and* $\sigma = (0; 3, 4^2)$ *which is realizable by choosing*
$$g_1 = (1,4,5);\ g_2 = (2,3,5,4);\ g_3 = (1,5,3,2).$$
- $g = 16$ *and* $\sigma = (0; 3, 4, 6)$ *which is realizable by choosing*
$$g_1 = (1,2,3);\ g_2 = (2,4,3,5);\ g_3 = (1,3)(2,4,5).$$
- $g = 19$ *and* $\sigma = (0; 4^2, 5)$ *which is realizable by choosing*
$$g_1 = (1,2,3,4);\ g_2 = (2,3,4,5);\ g_3 = (1,4,2,5,3).$$
- $g = 24$ *and* $\sigma = (0; 4, 5, 6)$ *which is realizable by choosing*
$$g_1 = (1,2,3,4);\ g_2 = (1,2,3,4,5);\ g_3 = (2,4)(1,5,3).$$

PROOF. Consider $\sigma^* = (\gamma; ..., j^{a_j}, ...)$, with the notation as in Equation (1.2).

(1) Let $n = 3$. Applying Lemma 2.6 we obtain $a_2 > 0$. Moreover, to satisfy the conditions for a generating vector (Theorem 1.3) for $\mathbf{S}_3$, a_2 must be even.

By Corollary 2.3, we have

$$\frac{a_2}{2} + \frac{2a_3}{3} \leq \frac{8}{3} < 3.$$

hence $1 < a_2 < 6$.

Assuming $a_2 = 2$ we have $0 < a_3 < 3$.

If $a_3 = 1$, then $\mathbf{g} = 0$.

If $a_3 = 2$, then $\mathbf{g} = 2$ and the signature $\sigma = (0; 2^2, 3^2)$ is realizable with the following generating vector for $\mathbf{S}_3$: $g_1 = (1,2) = g_2, g_3 = (1,2,3), g_4 = (1,3,2)$.

Assuming $a_2 = 4$ we have $0 \leq a_3 \leq 1$.

If $a_3 = 0$ then $\mathbf{g} = 1$, this genus is not interesting for our case.

If $a_3 = 1$, then $\mathbf{g} = 3$ and the signature $\sigma = (0; 2^4, 3)$ is realizable with the following generating vector for $\mathbf{S}_3$: $g_1 = (1,2) = g_2 = g_3, g_4 = (1,3), g_5 = (1,3,2)$.

(2) Let $n = 4$. Then (see Corollary 2.3) we have

$$\frac{a_2}{2} + \frac{2a_3}{3} + \frac{3a_4}{4} \leq \frac{8}{3};$$

hence $0 \leq a_4 \leq 3$, $0 \leq a_3 \leq 3$ (using the fact that if $a_3 = 4$ the other are 0 and they can not generate $\mathbf{S}_n$), and $0 \leq a_2 \leq 5$.

Assume $a_4 = 3$, then $a_2 = a_3 = 0$. Since the product of two elements of order four in $\mathbf{S}_4$ has order two or three, the data $(0; 4, 4, 4)$ is not realizable (see Theorem 1.3).

Assume $a_4 = 2$, then $0 \leq a_3 \leq 1$ and $0 \leq a_2 \leq 2$. Since the product of two elements $x, y \in \mathbf{S}_4$ of order four has order two if and only if $x = y$, the data $(0; 2, 4, 4)$ is not realizable.

If $a_3 = 0$ and $a_2 = 1$, then $\mathbf{g} = 7$ and the signature $\sigma = (0; 2^2, 4^2)$ is realizable with the following generating vector for $\mathbf{S}_4$: $g_1 = (2,4), g_2 = (1,4), g_3 = (1,3,4,2), g_4 = (1,4,3,2)$.

If $a_3 = 1$ and $a_2 = 0$, then $\mathbf{g} = 3$ and the signature $\sigma = (0; 3, 4^2)$ is realizable with the following generating vector for $\mathbf{S}_4$: $g_1 = (1,2,3), g_2 = (1,2,3,4), g_3 = (1,4,2,3)$.

If $a_3 = 1$ and $a_2 = 1$, let $\mathcal{W}$ be an irreducible module associated to the standard representation of $\mathbf{S}_4$. Then

$$\dim(\mathrm{Fix}_{\langle(1,2,3,4)\rangle}(\mathcal{W})) = 0 \quad \text{and} \quad \dim(\mathrm{Fix}_{\langle(1,2,3)\rangle}(\mathcal{W})) = 1.$$

Hence for all possible realizations of the signature $\sigma = (0; 2, 3, 4^2)$ the subvariety B_i associated to the standard representation has dimension (see Theorem 1.2) greater than one, and **Hypothesis A** does not hold.

Assume $a_4 = 1$.

If $a_3 = 0$ and $a_2 = 3$, then $\mathbf{g} = 4$ and the signature $\sigma = (0; 2^3, 4)$ is realizable with the following generating vector for $\mathbf{S}_4$: $g_1 = (3,4), g_2 = (2,3), g_3 = (1,2), g_4 = (1,4,3,2)$.

If $a_3 = 1$ and $a_2 = 2$, then $\mathbf{g} = 6$ and the signature $\sigma = (0; 2^2, 3, 4)$ is realizable with the following generating vector for $\mathbf{S}_4$: $g_1 = (1,4), g_2 = (1,2)(3,4), g_3 = (1,3,2), g_4 = (1,4,3,2)$.

If $a_3 = 2$ and $a_2 = 1$, then $\mathbf{g} = 8$ and the signature $\sigma = (0; 2, 3^2, 4)$ is realizable with the following generating vector for $\mathbf{S}_4$: $g_1 = (1,2), g_2 = (2,4,3), g_3 = (1,3,2), g_4 = (1,3,2,4)$.

Assume $a_4 = 0$ then $0 \le a_3 \le 3$ and $0 \le a_2 \le 5$.

If $a_3 = 0$ and $a_2 = 4$ then $\mathbf{g} = 1$, we are not interested in this genus.

If $a_3 = 0$ and $a_2 = 5$ we have that $\mathbf{g} = 7$ and the data is $(0; 2^5)$; if $a_3 = 1$ and $a_2 = 4$ we have that $\mathbf{g} = 9$ and the data is $(0; 2^4, 3)$. In both cases, considering the degrees of the representations of $\mathbf{S}_4$ and the dimension of these Jacobians (7 and 9), we have that the dimension of the subvariety B_i associated to the non-trivial representation of degree one must be equal to one. Applying Theorem 1.2, g_1, g_2, g_3 and g_4 are transpositions. But in this situation the subvariety B_j associated to the anti-standard representation has dimension greater than one (by Theorem 1.2).

If $a_3 = 1$ and $a_2 = 3$, then $\mathbf{g} = 3$ and the signature $\sigma = (0; 2^3, 3)$ is realizable with the following generating vector for $\mathbf{S}_4$: $g_1 = (1,3), g_2 = (1,4)(2,3), g_3 = (1,4), g_4 = (1,3,2)$.

If $a_3 = 2$ and $a_2 = 2$, then $\mathbf{g} = 5$ and the signature $\sigma = (0; 2^2, 3^2)$ is realizable with the following generating vector for $\mathbf{S}_4$: $g_1 = (2,4), g_2 = (1,3), g_3 = (2,4,3), g_4 = (1,3,2)$.

(3) Let $n = 5$. Then we have

$$\frac{a_2}{2} + \frac{2a_3}{3} + \frac{3a_4}{4} + \frac{4a_5}{5} + \frac{5a_6}{6} \le \frac{12}{5}.$$

Hence the possible data are

$$\left\{\begin{array}{l}
(2,6^2); (3,6^2) \\
(4,5,6); (3,5,6); (2,5,6) \\
(2,4,6); (3,4,6); (4^2,6) \\
(2,3,6); (3^2,6); (2^2,6); (2^3,6) \\
(4,5^2); (3,5^2); (2,5^2) \\
(4^2,5); (3,4,5); (2,4,5); (3^2,5); (2^2,5); (2^3,5) \\
(4^3); (3,4^2); (2,4^2) \\
(2,3,4); (3^2,4); (2^2,4); (2^3,4) \\
(3^3); (2,3^2); (2^2,3^2) \\
(2^2,3); (2^3,3) \\
(2^3)
\end{array}\right\}$$

Using Lemma 2.6, these data reduce to

$$\left\{\begin{array}{l}
(2,6^2); (3,6^2); (4,5,6); (2,5,6); (2,4,6); (3,4,6) \\
(4^2,6); (2,3,6); (2^3,6); (4^2,5); (2,4,5); (2^3,5) \\
(4^3); (3,4^2); (2,4^2); (2,3,4); (2^3,4); (2^2,3^2) \\
(2^3,3)
\end{array}\right\}$$

Applying the considerations of Remark 2.4, we obtain that the only realizable signatures are as follows, with the generating vectors described in the statement of the Theorem.

$\mathbf{g} = 24$ and signature $(0; 4, 5, 6)$.
$\mathbf{g} = 19$ and signature $(0; 4^2, 5)$.
$\mathbf{g} = 16$ and signature $(0; 3, 4, 6)$.
$\mathbf{g} = 11$ and signature $(0; 3, 4^2)$.
$\mathbf{g} = 11$ and signature $(0; 2, 6^2)$.
$\mathbf{g} = 9$ and signature $(0; 2, 5, 6)$.
$\mathbf{g} = 6$ and signature $(0; 2, 4, 6)$.
$\mathbf{g} = 4$ and signature $(0; 2, 4, 5)$.

(4) Let $n \geq 6$. Then, by Corollary 2.3

$$\sum_{j=2}^{\delta} a_j \left(1 - \frac{1}{j}\right) < \frac{101}{45}.$$

Applying Lemma 2.6, we obtain that the only possible data are

$$\left\{\begin{array}{lll} (2^3, 3) & & \\ (2, 3, m) & \text{with} & m > 7 \ \text{even} \\ (2, 4, m) & \text{with} & 5 \leq m < 179 \\ (2, 5, m) & \text{with} & 6 \leq m < 19 \ \text{even} \\ (2, 6, m) & \text{with} & 6 \leq m < 11 \\ (2, 7, 8) & & \end{array}\right\} \tag{2.1}$$

Since the data $(2, 3, 7)$ is not a possibility, $\mathbf{S}_n$ is not a Hurwitz group for any of these possible signatures, therefore the order of $\mathbf{S}_n$ is less than the maximum possible:

$$|\mathbf{S}_n| = n! < 84(\mathbf{g} - 1) = 84\left(-n! + \frac{n!}{2}\sum_{j=2}^{\delta} a_j \left(1 - \frac{1}{j}\right)\right)$$

Hence,

$$\frac{85}{42} < \sum_{j=2}^{\delta} a_j \left(1 - \frac{1}{j}\right) = S < 2\left(1 + \frac{n-2}{n(n-1)} - \frac{n+3}{n!}\right) < 2\left(1 + \frac{n-2}{n(n-1)}\right)$$

and we obtain $n \leq 82$.

For the tuple $(2^3, 3)$, we use the bound for S from Lemma 2.2

$$S = \frac{3}{2} + \frac{2}{3} \leq 2\left(1 + \frac{\mathbf{N}(n) - 2}{n!}\right).$$

We use the known values for $\mathbf{N}(n)$ with $6 \leq n \leq 82$, (see **[11]**), to obtain computationally that $n = 6$ is the only value for which the inequality holds.

For the triples in the list above (see Equation 2.1), the corresponding generating vector must have two of its elements being odd permutations, in order to generate $\mathbf{S}_n$ (see Theorem 1.3). By Theorem 1.2, it follows that the dimension of the subvariety B_i associated to the non-trivial representation of degree one must be zero.

We can now repeat the argument used to prove (2) in Lemma 2.2, observing that in this case $\mathbf{g} \leq \mathbf{N}(n) - 2$, to conclude that

$$S \leq 2\left(1 + \frac{\mathbf{N}(n) - 3}{n!}\right).$$

Using the known values for $\mathbf{N}(n)$ with $6 \leq n \leq 82$, (see [**11**]), we solve computationally this inequality for the variable m, for all triples in the list above and for each n from 6 to 82, and we find that the only remaining cases are

- For $\mathbf{n = 6}$.

$$\left\{\begin{array}{lll} (2^3, 3) & & \\ (2,3,m) & \text{with} & m > 7; \quad \text{even} \\ (2,4,m) & \text{with} & 5 \leq m \leq 21 \\ (2,5,m) & \text{with} & 6 \leq m \leq 10 \quad \text{even} \\ (2,6,m) & \text{with} & m = 6, 7 \end{array}\right\}$$

- For $\mathbf{n = 7}$.

$$\left\{\begin{array}{lll} (2,3,m) & \text{with} & m > 7; \quad \text{even} \\ (2,4,m) & \text{with} & 5 \leq m \leq 6 \end{array}\right\}$$

As n is now known, m has a natural upper bound in all cases: the maximum order of an element in $\mathbf{S}_n$; that is, 6 for $n = 6$, and 15 for $n = 7$.

To end the proof, we used the software MAGMA [**2**], and verified that none of these data satisfies **Hypothesis A**. For instance, for $G = \mathbf{S}_7$ the signature $(0; 2, 3, 10)$ is realizable in genus 169. In fact, a generating vector for this action is $g_1 = (1,2,3,4,5)(6,7), g_2 = (1,7,2)(3,6,5), g_3 = (2,6)(3,4)(5,7)$. However, we computed the dimensions of the subvarieties in the decomposition, using Theorem 1.2, and we obtained that the subvarieties associated to a representation of degree 14 and to a representation of degree 35 have dimension equal to two.

□

3. Summary.

It is known (see [**4**, Section 1.4.2]) that the dimension of a family $\mathcal{F}$ of curves of a fixed genus where a group G acts with a fixed signature is

$$3\gamma - 3 + k,$$

where γ is the genus of the quotient curve by G and k is the number of branch points for the action.

In this section we summarize the results obtained in Section 2; that is, we give all the actions of $\mathbf{S}_n$, $n \geq 3$, such that the group algebra decomposition for the corresponding Jacobian becomes a product of elliptic curves. We also include this decomposition and the dimensions of the respective families. The last column in the table gives the $G-$equivariant decomposition of the corresponding Jacobian variety as a product of elliptic curves $E_i^{n_i}$, where the subindex in E_i gives the representation acting on $E_i^{n_i}$, as described after the table.

Table 1: S_n actions giving completely decomposable Jacobians.

Group	Genus	Signature	Dimension	Decomposition
S_3	2	$(0; 2^2, 3^2)$	1	E_S^2
	3	$(0; 2^4, 3)$	2	$E_{-1} \times E_S^2$
	3	(1;3)	1	$E_1 \times E_S^2$
	4	$(1; 2^2)$	2	$E_1 \times E_{-1} \times E_S^2$
S_4	3	$(0; 3, 4^2)$	0	E_S^3
	3	$(0; 2^3, 3)$	1	E^3_{-S}
	4	$(0; 2^3, 4)$	1	$E_{-1} \times E^3_{-S}$
	5	$(0; 2^2, 3^2)$	1	$E_{\partial 2}^2 \times E^3_{-S}$
	6	$(0; 2^2, 3, 4)$	1	$E_S^3 \times E^3_{-S}$
	7	$(0; 2^2, 4^2)$	1	$E_{-1} \times E_S^3 \times E^3_{-S}$
	8	$(0; 2, 3^2, 4)$	1	$E_{\partial 2}^2 \times E_S^3 \times E^3_{-S}$
	9	(1;3)	1	$E_1 \times E_{\partial 2}^2 \times E_S^3 \times E^3_{-S}$
S_5	4	(0;2,4,5)	0	E^4_{-S}
	6	(0;2,4,6)	0	$E_{\partial 6}^6$
	9	(0;2,5,6)	0	$E^4_{-S} \times E^5_{-\partial 5}$
	11	$(0; 2, 6^2)$	0	$E_{\partial 5}^5 \times E_{\partial 6}^6$
	11	$(0; 3, 4^2)$	0	$E_{\partial 5}^5 \times E_{\partial 6}^6$
	16	(0;3,4,6)	0	$E_{\partial 5}^5 \times E^5_{-\partial 5} \times E_{\partial 6}^6$
	19	$(0; 4^2, 5)$	0	$E_S^4 \times E^4_{-S} \times E_{\partial 5}^5 \times E_{\partial 6}^6$
	24	(0;4,5,6)	0	$E_S^4 \times E^4_{-S} \times E_{\partial 5}^5 \times E^5_{-\partial 5} \times E_{\partial 6}^6$

Where the representations for each $\mathbf{S}_n$ are as follows

(1) For $\mathbf{S}_3$: The trivial representation χ_1, the alternating representation θ_{-1}, which is of degree one and has the alternating group $\mathbf{A}_3$ as kernel, and the standard representation θ_S which is of degree 2.

(2) For $\mathbf{S}_4$: The trivial representation χ_1, the alternating representation θ_{-1}, which is of degree one and has the alternating group $\mathbf{A}_4$ as kernel, the representation $\theta_{\partial 2}$ of degree 2, the standard representation θ_S which is of degree 3, and the antistandard representation $\theta_{-S} = \theta_S \otimes \theta_{-1}$ which is of degree 3.

(3) For $\mathbf{S}_5$: The trivial representation χ_1, the alternating representation θ_{-1}, which is of degree one and has the alternating group $\mathbf{A}_5$ as kernel, the standard representation θ_S which is of degree 4, and the antistandard representation $\theta_{-S} = \theta_S \otimes \theta_{-1}$ which is of degree 4, the representation $\theta_{\partial 5}$ of degree 5, the other representation $\theta_{-\partial 5} = \theta_{\partial 5} \otimes \theta_{-1}$ of degree 5, the representation $\theta_{\partial 6}$ of degree 6.

Final remarks:

- It is known that the curve of genus 3 where $\mathbf{S}_4$ acts with signature $(0; 3, 4^2)$ belongs to the one dimensional family having the action of $\mathbf{S}_4$ with signature $(0; 2^3, 3)$. Therefore, all the cases for $\mathbf{S}_4$ are one dimensional families of completely decomposable Jacobians, see Table 1.
- To apply our method to other families of groups requires a deep knowledge of the group structure and its complex irreducible representations, for instance a useful bound such as the one found in Lemma 2.1 is not easily obtained in general.

References

[1] Christina Birkenhake and Herbert Lange, *Complex abelian varieties*, 2nd ed., Grundlehren der Mathematischen Wissenschaften [Fundamental Principles of Mathematical Sciences], vol. 302, Springer-Verlag, Berlin, 2004. MR2062673 (2005c:14001)

[2] Wieb Bosma, John Cannon, and Catherine Playoust, *The Magma algebra system. I. The user language*, J. Symbolic Comput. **24** (1997), no. 3-4, 235–265, DOI 10.1006/jsco.1996.0125. Computational algebra and number theory (London, 1993). MR1484478

[3] S. Allen Broughton, *Classifying finite group actions on surfaces of low genus*, J. Pure Appl. Algebra **69** (1991), no. 3, 233–270, DOI 10.1016/0022-4049(91)90021-S. MR1090743 (92b:57021)

[4] Emilio Bujalance, Francisco Javier Cirre, José Manuel Gamboa, and Grzegorz Gromadzki, *Symmetries of compact Riemann surfaces*, Lecture Notes in Mathematics, vol. 2007, Springer-Verlag, Berlin, 2010. MR2683160 (2011h:30062)

[5] Clifford J. Earle, *The genus two Jacobians that are isomorphic to a product of elliptic curves*, The geometry of Riemann surfaces and abelian varieties, Contemp. Math., vol. 397, Amer. Math. Soc., Providence, RI, 2006, pp. 27–36, DOI 10.1090/conm/397/07459. MR2217995 (2007i:14027)

[6] Torsten Ekedahl and Jean-Pierre Serre, *Exemples de courbes algébriques à jacobienne complètement décomposable* (French, with English and French summaries), C. R. Acad. Sci. Paris Sér. I Math. **317** (1993), no. 5, 509–513. MR1239039 (94j:14029)

[7] Amy E. Ksir, *Dimensions of Prym varieties*, Int. J. Math. Math. Sci. **26** (2001), no. 2, 107–116, DOI 10.1155/S016117120101153X. MR1836786 (2002e:14047)

[8] Herbert Lange and Sevin Recillas, *Abelian varieties with group action*, J. Reine Angew. Math. **575** (2004), 135–155, DOI 10.1515/crll.2004.076. MR2097550 (2005h:14107)

[9] Ben Moonen and Frans Oort. The Torelli locus and special subvarieties. *Handbook of Moduli* 2, 549–594, 2011.

[10] Ryo Nakajima, *On splitting of certain Jacobian varieties*, J. Math. Kyoto Univ. **47** (2007), no. 2, 391–415. MR2376963 (2008m:14055)

[11] The OEIS Foundation. *Integer sequences*, `http://oeis.org/A000085`

[12] Jennifer Paulhus, *Decomposing Jacobians of curves with extra automorphisms*, Acta Arith. **132** (2008), no. 3, 231–244, DOI 10.4064/aa132-3-3. MR2403651 (2009c:14049)

[13] Anita M. Rojas, *Group actions on Jacobian varieties*, Rev. Mat. Iberoam. **23** (2007), no. 2, 397–420, DOI 10.4171/RMI/500. MR2371432 (2009e:14047)

Departamento de Matemática, Universidad de La Frontera, Casilla 54-D, Temuco, Chile
E-mail address: angel.carocca@ufrontera.cl

Facultad de Matemáticas, Pontificia Universidad Católica de Chile, Casilla 306-22, Santiago, Chile
E-mail address: rubi@mat.puc.cl

Departamento de Matemáticas, Facultad de Ciencias, Universidad de Chile, Casilla 653, Santiago, Chile
E-mail address: anirojas@uchile.cl

Normal coverings of hyperelliptic real Riemann surfaces

Francisco-Javier Cirre and Rubén A. Hidalgo

Dedicated to Emilio Bujalance on the occasion of his 60th birthday.

ABSTRACT. A real Riemann surface is a pair (R, τ), where R is a compact Riemann surface and $\tau : R \to R$ is an anticonformal involution, called a real structure of R. If R is hyperelliptic then we say that (R, τ) is a hyperelliptic real Riemann surface. In this paper we describe in terms of algebraic equations all normal (possibly branched) coverings $\pi : (R, \tau) \to (S, \eta)$ between hyperelliptic real Riemann surfaces. This extends results due to Bujalance-Cirre-Gamboa, where either the number of ovals fixed by the real structure τ is maximal, or the degree of the covering is two. In this paper we consider any real structure τ and any degree of the covering.

1. Introduction

A compact Riemann surface R of genus $g \geq 2$ is called hyperelliptic if there is a two-fold branched covering $P_R : R \to \widehat{\mathbb{C}}$ over the Riemann sphere $\widehat{\mathbb{C}}$; equivalently, if there exists a conformal involution $j_R : R \to R$ having exactly $2g+2$ fixed points. It is known that j_R, called the hyperelliptic involution of R, is unique, and so it commutes with every (conformal or anticonformal) automorphism of R.

Let us consider a normal (possibly branched) covering $\pi : R \to S$ between compact Riemann surfaces of genus at least two, with $G < \mathrm{Aut}^+(R)$ as its group of covering transformations, where $\mathrm{Aut}^+(R)$ denotes the full group of conformal automorphisms of R. Let us assume first that S is hyperelliptic, say with hyperelliptic involution j_S, and that π is unbranched. If G is an abelian group, then Farkas proved in **[8]** that j_S lifts to a conformal involution j in R, but in general, R need not be hyperelliptic. In fact, if R is hyperelliptic, then Farkas **[7]** and Maclachlan **[15]** proved independently that G is isomorphic to either C_2 or $C_2 \times C_2$, where C_n denotes the cyclic group of order n. The same papers provide examples of unbranched coverings with $G = C_2$ or $G = C_2 \times C_2$ for which R is not hyperelliptic. A description of such lifted involutions j was carried out by Bujalance in **[3]** in the case $G = C_2$. In the case $G = C_2$ or $G = C_2 \times C_2$, Fuertes and González-Diez proved in **[10, 11]** that the quotient orbifold $\mathcal{O} = R/\langle j \rangle$ is a hyperelliptic Riemann surface and the authors provided algebraic equations for $\mathcal{O}$ and for R in terms of

---

2010 *Mathematics Subject Classification*. Primary 30F10, 30F40.

The first author was partially supported by the Spanish Research Project MTM2011-23092.

The second author was partially supported by projects Fondecyt 1110001 and UTFSM 12.13.01.

those for S. Now, let us assume that R is hyperelliptic, say with hyperelliptic involution j_R, and that π may be branched. Then necessarily S is hyperelliptic and j_R induces the hyperelliptic involution on S (see Section 2). A description by algebraic equations of such coverings was done by Horiuchi in **[13]**.

A real Riemann surface is a pair (R, τ), where R is a compact Riemann surface and $\tau : R \to R$ is an anticonformal involution, called a real structure of R. The fixed points of τ (if any) are called the real points of the real structure; each connected component is called an oval or a mirror of τ. Results due to Harnack **[12]** ensures that the number of ovals is bounded above by $g+1$, where g is the genus of R. An automorphism of the real Riemann surface (R, τ) is a conformal automorphism of R which commutes with τ. Let us note that a general Riemann surface may have no real structure and if it has, it may not be unique.

Let us consider a real Riemann surface (R, τ), a compact Riemann surface S and a normal (possibly branched) covering $\pi : R \to S$, whose group of covering transformations G is a group of automorphisms of the real Riemann surface (R, τ) (that is, each element of G commutes with τ). The fact that τ normalizes G ensures the existence of a real structure $\eta : S \to S$ such that $\pi \circ \tau = \eta \circ \pi$. This observation permits to make the following definition. By a normal (possibly branched) covering $\pi : (R, \tau) \to (S, \eta)$ between real Riemann surfaces we mean a normal (possibly branched) covering $\pi : R \to S$ whose group G of covering transformations is a group of automorphisms of the real Riemann surface (R, τ) and such that $\pi \circ \tau = \eta \circ \pi$.

We will assume, from now on, that R is hyperelliptic of genus $g \geq 2$. The study of such (possibly branched) normal coverings between hyperelliptic real Riemann surfaces was carried out in **[4]** if the number of ovals fixed by τ is maximal, and in **[5]** if π has degree two. Algebraic equations were provided for (R, τ), (S, η), the covering group and π; in addition, the topological features and ramification data of such coverings were also described. In this paper we extend these results in all its generality, and consider any real structure τ and any degree of the covering. Some technical lemmas are proved, allowing us to deal with the cases not considered in **[4, 5]**, which otherwise would be unmanageable. Here we focus our attention on the algebraic counterpart, although also the topological description of such coverings can be obtained, see Remark 3.2.

If (R, τ) is a real Riemann surface then the orbit space $X = R/\langle\tau\rangle$ is a compact 2-manifold possibly with boundary and possibly non-orientable with a dianalytic structure, see **[1]**. Such objects are known as Klein surfaces and, in fact, every Klein surface arises in this way. So our results in this paper can be stated in terms of compact Klein surfaces, as it is also done in **[4, 5]**.

2. Normal coverings of hyperelliptic Riemann surfaces

In this section we recall some generalities about hyperelliptic Riemann surfaces and normal (possibly branched) coverings between them that we will need in the rest of this paper.

Let R be a hyperelliptic Riemann surface of genus $g \geq 2$, with hyperelliptic involution j_R. Let G be a (necessarily finite) group of conformal automorphisms of R, and assume that the orbifold $S = R/G$ has genus bigger than one (it has an underlying Riemann surface structure which we still denote by S). Note that the condition for S to have genus at least one ensures that $j_R \notin G$.

Let us denote by $\pi : R \to S$ a normal (possibly branched) covering with G as its group of covering transformations and let $P_R : R \to \widehat{\mathbb{C}}$ be the two-fold branched covering induced by j_R so that $P_R \circ j_R = P_R$. Let $B(P_R)$ be the set of branch values of P_R, that is, the image under P_R of the set $W(R) = \mathrm{Fix}(j_R)$ of Weierstrass points of R, where Fix stands for the fixed point set. If $B(R) = \{a_1, ..., a_{2g+2}\}$, then R can be represented by the hyperelliptic plane curve

$$y^2 = (x - a_1) \cdots (x - a_{2g+2}), \tag{2.1}$$

where we delete the factor $(x - a_j)$ if $a_j = \infty$. In such an algebraic representation, the hyperelliptic involution j_R and the two-fold branched covering $P_R : R \to \widehat{\mathbb{C}}$ are given, respectively, by

$$j_R(x, y) = (x, -y) \quad \text{and} \quad P_R(x, y) = x.$$

As j_R commutes with every element in G (see, for instance [**9**]), there exist a finite group $\widehat{G}$ of conformal automorphisms of $\widehat{\mathbb{C}}$ (that is, a finite group of Möbius transformations) and a surjective homomorphism $\theta : G \to \widehat{G}$ such that $\theta(\gamma) \circ P_R = P_R \circ \gamma$, for every $\gamma \in G$. Since $j_R \notin G$ (as already noted above) it follows that θ is an isomorphism.

It is well known (see, for instance, [**2**]) that $\widehat{G}$ (and hence G itself) is isomorphic to either C_n (a cyclic group of order n) or D_m (a dihedral group of order $2m$) or the alternating groups A_4, A_5 or the symmetric group $\mathfrak{S}_4$. Observe that $B(P_R)$ is necessarily invariant under the action of $\widehat{G}$. Lemma 2.1 below shows that no point in $B(P_R)$ can be fixed by an element of even order of $\widehat{G}$.

LEMMA 2.1. *No fixed point of j_R is a fixed point of a non-trivial element of even order in G. In particular, no point in $B(P_R)$ can be a fixed point of a non-trivial element of even order in $\widehat{G}$.*

PROOF. The stabilizer in $\mathrm{Aut}^+(R)$ of a point is a cyclic group, and we are assuming that $j_R \notin G$; so no fixed point of j_R can be a fixed point of an element of even order in G. □

If $\widehat{\gamma} \in \widehat{G}$, then the corresponding automorphism $\gamma \in G$ is of the form $\gamma(x, y) = (\widehat{\gamma}(x),\ y\, r_{\widehat{\gamma}}(x))$, where $r_{\widehat{\gamma}}(x)$ is a rational function. Its precise form can be found in [**6**, Section 2], but we display it in the following Lemma 2.2 for the reader's convenience.

LEMMA 2.2. *Let us write $\widehat{\gamma}(x) = (ax + b)/(cx + d)$ and $\det \widehat{\gamma} = ad - bc$. Let $Q(x)$ be the polynomial on the right-hand side of formula (2.1). Then, with the above notations, the formula of the rational function $r_{\widehat{\gamma}}(x)$ is*

- *if $\infty \in B(P_R)$ and $\widehat{\gamma}(\infty) = \infty$ then $r_{\widehat{\gamma}}(x) = (-1)^{\varepsilon} \sqrt{\det \widehat{\gamma}}\; a^g/d^{g+1}$;*
- *if $\infty \in B(P_R)$ and $\widehat{\gamma}(\infty) \neq \infty$ then $r_{\widehat{\gamma}}(x) = (-1)^{\varepsilon} \sqrt{-\det \widehat{\gamma}\; Q'(\frac{a}{c})}\; c^g/(cx + d)^{g+1}$;*
- *if $\infty \notin B(P_R)$ and $\widehat{\gamma}(\infty) = \infty$ then $r_{\widehat{\gamma}}(x) = (-1)^{\varepsilon}(a/d)^{g+1}$;*
- *if $\infty \notin B(P_R)$ and $\widehat{\gamma}(\infty) \neq \infty$ then $r_{\widehat{\gamma}}(x) = (-1)^{\varepsilon} \sqrt{Q(\frac{a}{c})}\; c^{g+1}/(cx + d)^{g+1}$,*

where $\varepsilon \in \{0, 1\}$ in all the cases.

The above formulae are also valid for anticonformal automorphisms $\widehat{\sigma}(x) = (a\overline{x} + b)/(c\overline{x} + d)$ of $\widehat{\mathbb{C}}$ which preserve $B(P_R)$. We just have to replace x and y with *their complex conjugates* $\overline{x}$ *and* $\overline{y}$ *respectively to obtain the anticonformal automorphism* $\sigma(x,y) = \big((a\overline{x}+b)/(c\overline{x}+d),\ \overline{y}\, r_{\widehat{\sigma}}(\overline{x})\big)$.

Let $\widehat{\pi} : \widehat{\mathbb{C}} \to \widehat{\mathbb{C}}$ be the branched normal covering induced by $\widehat{G}$. As j_R commutes with every element of G and $j_R \notin G$, it follows that there is a conformal involution $j_S \in \mathrm{Aut}^+(S)$ such that

$$j_S \circ \pi = \pi \circ j_R.$$

This yields a two-fold branched covering $P_S : S \to \widehat{\mathbb{C}}$ such that

$$\widehat{\pi} \circ P_R = P_S \circ \pi. \tag{2.2}$$

Hence, S *is hyperelliptic and* j_S *is its hyperelliptic involution.*

$$\begin{array}{ccc} R & \xrightarrow{\ \pi\ } & S \\ {\scriptstyle P_R}\big\downarrow & & \big\downarrow{\scriptstyle P_S} \\ \widehat{\mathbb{C}} & \xrightarrow{\ \widehat{\pi}\ } & \widehat{\mathbb{C}} \end{array}$$

REMARK 2.3. It follows from equality (2.2) that $\widehat{\pi}(B(P_R)) \cup B(\widehat{\pi}) = P_S(B(\pi)) \cup B(P_S)$. In particular, if π is unbranched, then $\widehat{\pi}(B(P_R)) \cup B(\widehat{\pi}) = B(P_S)$ (that is, all branch values of $\widehat{\pi}$ have branch order 2) and so $\widehat{G}$ cannot have rotations of order greater than two. It follows that $\widehat{G}$ is either C_2 or $\mathrm{C}_2 \times \mathrm{C}_2$, as it should be from the results in [**7**, **15**].

We now study the relation between the sets $W(R) = \mathrm{Fix}(j_R)$ and $W(S) = \mathrm{Fix}(j_S)$ of Weierstrass points of R and S, respectively. This last contains, even properly, the image of $W(R)$ under the covering $\pi : R \to S$. In fact, the following holds, see Lemma 4.1 in [**4**].

LEMMA 2.4. *The set* $W(S)$ *of Weierstrass points of* S *is the union*

$$W(S) = \pi(W(R)) \ \cup\ \pi\Big(\bigcup_{\gamma \in G - \{id\}} \mathrm{Fix}(\gamma \circ j_R)\Big).$$

PROOF. To show the equality, we calculate the fixed points of j_S. Let $p' = \pi(p)$ be a point on S. Then

$$\begin{aligned} j_S(p') = p' \quad &\Longleftrightarrow \quad j_S(\pi(p)) = \pi(p) \Longleftrightarrow \pi(j_R(p)) = \pi(p) \\ &\Longleftrightarrow \quad \text{there exists } \gamma \in G \text{ such that } p = (\gamma \circ j_R)(p). \end{aligned}$$

Therefore $W(S) = \pi\Big(\bigcup_{\gamma \in G} \mathrm{Fix}(\gamma \circ j_R)\Big)$. ☐

REMARK 2.5. In order to get an algebraic equation of the covered surface S we need to know the set $P_S(W(S)) = B(P_S)$ of branch values of P_S and, actually, we just need the finite ones. Let $p = (a,b)$ be a point on R with finite coordinates. We proceed to check when $\pi(a,b) \in W(S)$, that is, when $\widehat{\pi}(a) \in B(P_S)$. We distinguish the following two cases:

(i) If $b = 0$, then $a \in B(P_R)$ and so $\widehat{\pi}(a) \in B(P_S)$ (this corresponds to the $\pi(W(R))$-part in the formula in Lemma 2.4).

(ii) If $b \neq 0$ then $\pi(a,b) \in W(S)$ if and only if (a,b) is fixed by $\gamma \circ j_R$ for some $\gamma \in G - \{id\}$. The automorphism $\gamma \in G$ can be written, in affine coordinates, as $\gamma(x,y) = (\widehat{\gamma}(x),\, y\, r_{\widehat{\gamma}}(x))$ where $r_{\widehat{\gamma}}(x)$ is a suitable rational function (see Lemma 2.2). So,

$$(2.3)\quad \pi(a,b) \in W(S) \quad \text{if and only if} \quad \widehat{\gamma}(a) = a \text{ and } r_{\widehat{\gamma}}(a) = -1 \text{ for some } \widehat{\gamma} \in \widehat{G}.$$

For instance, if $\widehat{G}$ is cyclic generated by a rotation $\widehat{\gamma}$, then we have to check condition (2.3) for the two fixed points of $\widehat{\gamma}$. For a general group $\widehat{G}$, we will see in Lemma 3.4 that condition (2.3) has to be checked just for three points.

3. Normal coverings of hyperelliptic real Riemann surfaces

Let $\pi : (R,\tau) \to (S,\eta)$ be a normal (possibly branched) covering between real Riemann surfaces and let G be its group of covering transformations, say of order n. Recall we are assuming that R is hyperelliptic of genus $g \geq 2$ and that G does not contain the hyperelliptic involution j_R. In this section we provide algebraic equations for R, τ, S, η, π and formulae for a set of generators of G. Since the case $n = 2$ was already considered in [**5**] we will also assume that $n > 2$.

In the same way as G induces the finite group $\widehat{G}$ of Möbius transformations, the real structures τ and η induce anticonformal involutions $\widehat{\tau} : \widehat{\mathbb{C}} \to \widehat{\mathbb{C}}$ and $\widehat{\eta} : \widehat{\mathbb{C}} \to \widehat{\mathbb{C}}$ such that $\widehat{\tau} \circ P_R = P_R \circ \tau$ and $\widehat{\eta} \circ P_S = P_S \circ \eta$. Observe that $\widehat{\tau}$ preserves $B(P_R)$ and $\widehat{\eta}$ preserves $B(P_S)$. As $\pi \circ \tau = \eta \circ \pi$ we also have

$$(3.1)\qquad \widehat{\pi} \circ \widehat{\tau} = \widehat{\eta} \circ \widehat{\pi}.$$

Equality (3.1) ensures that $\widehat{\pi}(a) \in B(P_S)$ if and only if $\widehat{\pi}(\widehat{\tau}(a)) \in B(P_S)$.

Up to conjugation by a suitable Möbius transformation, we may assume that $\widehat{\tau}$ is either

- the reflection $x \mapsto 1/\overline{x}$ if $\widehat{\tau}$ fixes points, or
- the antipodal map $x \mapsto -1/\overline{x}$ if $\widehat{\tau}$ is fixed point free.

Observe that since τ commutes with every element in G, also $\widehat{\tau}$ commutes with every element in $\widehat{G}$. The possible groups $\widehat{G}$, up to conjugation by a suitable Möbius transformation that commutes with $\widehat{\tau}$, are the following:

(1) If $\widehat{\tau}$ is the reflection $x \mapsto 1/\overline{x}$, then all the elements in $\widehat{G}$ must keep invariant the circle of fixed points of $\widehat{\tau}$. It follows that the only possibilities for $\widehat{G}$ are given by either a cyclic group C_n generated by $\widehat{t_n}(x) = xe^{2\pi i/n}$ or a dihedral group D_m generated by $\widehat{t_m}(x) = xe^{2\pi i/m}$ and $\widehat{v}(x) = 1/x$, where $n = 2m$.

(2) If $\widehat{\tau}$ is the antipodal map, then g is odd (see [**6**, Proposition 5.1]) and, in addition to the above groups $\widehat{G} = \mathrm{C}_n$ or D_m we also have to consider the possibilities $\widehat{G} = \mathrm{A}_4$, A_5 or $\mathfrak{S}_4$.

We distinguish cases below according to $\widehat{G}$ being of one of these five types of groups. Observe that when $\widehat{G}$ is A_4, $\mathfrak{S}_4$ or A_5, the involutions τ and $\widehat{\tau}$ have no fixed points. To shorten notation, let us write $\xi_k = e^{2\pi i/k}$ for each $k \geq 2$.

3.1. Cyclic case. As said above, $\widehat{G}$ is generated by $\widehat{t_n}(x) = \xi_n x$, where $n > 2$.

3.1.1. Let us first consider the case when $\widehat{\tau}$ is the reflection $x \mapsto 1/\overline{x}$. The points in $B(P_R)$ are permuted by the elements of $\langle \widehat{t_n}, \widehat{\tau} \rangle$ and so they split up into $\langle \widehat{t_n}, \widehat{\tau} \rangle$-orbits. There are three different types of such orbits corresponding to the $\langle \widehat{t_n}, \widehat{\tau} \rangle$-stabilizer of a point $\alpha \in \widehat{\mathbb{C}}$ being either $\langle \widehat{t_n} \rangle$, or $\langle \widehat{\tau} \rangle$, or trivial. In fact, the $\langle \widehat{t_n}, \widehat{\tau} \rangle$-orbit of a point $\alpha \in \widehat{\mathbb{C}}$ consist of:

- the roots of $(x^n - \alpha^n)(x^n - 1/\overline{\alpha}^n)$ if $|\alpha| \neq 1$ and $\alpha \neq 0, \infty$ (points with trivial stabilizer);
- the roots of $(x^n - \alpha^n)$ if $|\alpha| = 1$ (points with stabilizer $\langle \widehat{\tau} \rangle$);
- $\{0, \infty\}$ if $\alpha \in \{0, \infty\}$ (points with stabilizer $\langle \widehat{t_n} \rangle$).

By Lemma 2.1, if n is even then the fixed points 0 and ∞ of $\widehat{t_n}$ cannot be points in $B(P_R)$. The above yields a defining algebraic equation for R, included in the following theorem.

THEOREM 3.1. *Let $\pi : (R, \tau) \to (S, \eta)$ be a cyclic, branched, n-sheeted covering between real Riemann surfaces where (R, τ) is hyperelliptic of genus $g \geq 2$ and such that $\widehat{\tau}$ fixes points. Assume that $n > 2$ and that the covering transformation group $G = C_n$ does not contain the hyperelliptic involution of R. Then there exist $\delta \in \{0, 1\}$ with $\delta = 0$ if n is even, and non-negative integers L_1 and L_2 with L_2 even, satisfying $2g+2 = 2nL_1 + nL_2 + 2\delta$ such that, up to birational transformation, the following holds.*

(1) *(R, τ) can be represented as follows:*

$$(3.2) \qquad \begin{cases} R : y^2 = x^\delta \prod_{j=1}^{L_1} (x^n - a_j^n)(x^n - 1/\overline{a}_j{}^n) \prod_{k=1}^{L_2} (x^n - b_k^n), \\ \tau(x, y) = \left(\dfrac{1}{\overline{x}}, \dfrac{\overline{y}}{\overline{x}^{g+1}} \right), \end{cases}$$

where $|b_k| = 1$ for all k, the right-hand side polynomial in the defining equation of R has no multiple roots and $\prod_{j=1}^{L_1} a_j^n / \overline{a}_j^n \prod_{k=1}^{L_2} b_k^n = 1$.

(2) *A formula for an automorphism t generating the group G is the following:*
   a) *if n is even (and hence $\delta = 0$) then either $t(x, y) = (x\xi_n, y)$ or $t(x, y) = (x\xi_n, -y)$;*
   b) *if n is odd then $t(x, y) = (x\xi_n, (-\xi_{2n})^\delta y)$.*

(3) *For each R, τ and G, the covered surface (S, η) is given by:*

$$S : w^2 = u^{\delta'} \prod_{j=1}^{L_1} (u - a_j^n)(u - 1/\overline{a}_j{}^n) \prod_{k=1}^{L_2} (u - b_k^n),$$

$$\eta(u, w) = \left(\frac{1}{\overline{u}}, \frac{\overline{w}}{\overline{u}^{\delta' + (g+1-\delta)/n}} \right)$$

where $\delta' = 0$ if $t(x, y) = (x\xi_n, y)$ and $\delta' = 1$ otherwise.

(4) *The covering projection is given by*

$$\pi : (x, y) \mapsto (u, w) = (x^n, x^{(n\delta' - \delta)/2} y).$$

PROOF. The algebraic equation of R in (3.2) and the equality $2g+2 = 2nL_1 + nL_2 + 2\delta$ is a direct consequence of the description of the three different types of $\langle \widehat{t_n}, \widehat{\tau} \rangle$-orbits, where the condition $\delta = 0$ if n is even follows from Lemma 2.1.

Let us show that we may assume the equality $\left(\prod_{j=1}^{L_1} a_j^n/\overline{a}_j^n \prod_{k=1}^{L_2} b_k^n\right)(-1)^{L_2} = 1$ to hold. The product on the left hand side is a unimodular complex number, say $e^{2i\theta_R}$ with $\theta_R \in [0,\pi)$. Rotating the branch values in $B(P_R)$ by an angle $\theta = -\theta_R/(g+1-\delta)$, we see that $\alpha_j = a_j e^{i\theta}$ and $\beta_k = b_k e^{i\theta}$ satisfy

$$\begin{aligned}\prod_{j=1}^{L_1} \frac{\alpha_j^n}{\overline{\alpha}_j^n} \prod_{k=1}^{L_2} \beta_k^n \cdot (-1)^{L_2} &= e^{i\theta n(2L_1+L_2)} \prod_{j=1}^{L_1} \frac{a_j^n}{\overline{a}_j^n} \prod_{k=1}^{L_2} b_k^n \cdot (-1)^{L_2} \\ &= e^{i\theta(2g+2-2\delta)} e^{2i\theta_R} = 1.\end{aligned}$$

Observe that the rotation $x \mapsto xe^{i\theta}$ commutes with $\widehat{t}_n$ and $\widehat{\tau}$ and so the rotated branch values are still permuted by $\langle \widehat{t}_n, \widehat{\tau}\rangle$. In this way, by replacing the values of a_j and b_k by the values of α_j and β_k we will have the desired fact.

In order to prove that L_2 is even, let us compute first the formula for the real structure τ. It is a lifting of $\widehat{\tau} : x \mapsto 1/\overline{x}$ and so, using that $\left(\prod_{j=1}^{L_1} a_j^n/\overline{a}_j^n \prod_{k=1}^{L_2} b_k^n\right) = (-1)^{L_2}$, it is easy to check, see Lemma 2.2, that τ equals τ_0 or $\tau_0 \circ j_R$ where

$$\tau_0(x,y) = (1/\overline{x},\ \overline{y}/\overline{x}^{g+1}).$$

If $\tau = \tau_0$, then the claim concerning the formula of τ is proved. If not, we choose $\alpha_j = a_j e^{i\theta}$ and $\beta_k = b_k e^{i\theta}$ with $\theta = \pi/(g+1-\delta)$ and substitute the real surface $(R, \tau_0 \circ j_R)$ by (T, τ_0) where T has the same equation as R but changing a_j and b_k by α_j and β_k respectively. These two real surfaces are isomorphic via $(x,y) \mapsto (xe^{i\theta},\ ye^{i\theta(g+1-\delta/2)})$. Observe that we still have $\left(\prod_{j=1}^{L_1} a_j^n/\overline{a}_j^n \prod_{k=1}^{L_2} b_k^n\right)(-1)^{L_2} = 1$, with $|\beta_k| = 1$.

We now show that L_2 has to be even. This is clear if n is odd because $nL_2 = 2g+2-2nL_1-2\delta$. If n is even then $\delta = 0$ and a formula for the automorphism t generating G is $t(x,y) = (x\xi_n, \pm y\xi_n^{g+1}) = (x\xi_n, \pm y(-1)^{L_2})$, where in the last equality we have used that $\xi_n^{g+1} = e^{\pi i(2g+2)/n} = e^{\pi i(2L_1+L_2)} = (-1)^{L_2}$. It is straightforward to check, using the above formula of τ, that t commutes with τ if and only if L_2 is even. As a by product, we have also shown claim a) in the statement of the theorem.

In order to show claim b), we assume now that n is odd. If $\delta = 0$ then $t(x,y) = (x\xi_n, \pm y\xi_n^{g+1}) = (x\xi_n, \pm y)$; in fact, $t(x,y) = (x\xi_n, y)$ since the other possibility $t(x,y) = (x\xi_n, -y)$ does not have order n; if $\delta = 1$ then $g/n = L_1 + L_2/2$, which is an integer and so $t(x,y) = (x\xi_n, \pm y\sqrt{\xi_n}\,\xi_n^g) = (x\xi_n, \pm y\sqrt{\xi_n}) = (x\xi_n, \pm y\xi_{2n})$; in fact, $t(x,y) = (x\xi_n, -y\xi_{2n})$, since the other possibility $t(x,y) = (x\xi_n, y\xi_{2n})$ does not have order n. It is straightforward to check that $\tau t\tau(x,y) = (x\xi_n, -y\overline{\xi}_{2n}\xi_n^{g+1}) = (x\xi_n, -y\xi_{2n}^{-1}\xi_n) = (x\xi_n, -y\xi_{2n}) = t(x,y)$, that is, the automorphism t commutes with τ. This proves claim b) in the statement of the theorem.

We now focus our attention on finding algebraic equations of the covered surface (S,η) and the covering projection π. Observe first that the branched normal covering $\widehat{\pi} : \widehat{\mathbb{C}} \to \widehat{\mathbb{C}}$ induced by π is constant on each $\langle \widehat{t}\rangle$-orbit. So we may assume, after composing with a Möbius transformation if necessary, that $x = 0$ is the unique zero and $x = \infty$ is the unique pole of $\widehat{\pi}$, both of order n. Hence, $\widehat{\pi}(x) = \lambda x^n$ for some $\lambda \in \mathbb{C} - \{0\}$ and we may assume $\lambda = 1$ after composing again with a Möbius transformation if necessary.

We now calculate the set $B(P_S)$ of branch values of S and for that we use Lemma 2.4 and Remark 2.5. The points $a_j\xi_n^r$, $1/(\overline{a}_j\xi_n^r)$ and $b_k\xi_n^r$ for $r = 1,\dots,n$, $j = 1,\dots,L_1$ and $k = 1,\dots,L_2$, belong to $B(P_R)$ and so their projections a_j^n,

$1/\overline{a}_j^n$ and b_k^n under $\widehat{\pi}$ belong to $B(P_S)$. If $\delta = 0$ then the above are all the points in $B(P_R)$. If $\delta = 1$ then 0 also belongs to $B(P_R)$, and in this case, $\widehat{\pi}(0) = 0$ also belongs to $B(P_S)$.

In order to find the remaining finite points in $B(P_S)$ we have to consider finite points a fixed by some rotation $\widehat{\gamma} \in \widehat{G}$ and to check condition (2.3) in Remark 2.5. Since $\widehat{G}$ is cyclic generated by $\widehat{t}(x) = x\xi_n$, the only point we have to consider is $a = 0$. If $\delta = 1$ then $\widehat{\pi}(0) = 0$ (and also $\widehat{\pi}(\infty) = \infty$) belongs to $B(P_S)$, as we have just seen. If $\delta = 0$ then either $t(x,y) = (x\xi_n, y)$ or $t(x,y) = (x\xi_n, -y)$; in the first case, with the notations in Remark 2.5, $r_{\widehat{t}}(0) = 1$ and so $\widehat{\pi}(0) = 0$ is not a branch value of S; if $t(x,y) = (x\xi_n, -y)$ then $r_{\widehat{t}}(0) = -1$ and so $\widehat{\pi}(0) = 0$ is a branch value of S.

Summarizing, 0 is not a branch value of S ($\delta' = 0$ with the notations of the theorem) if and only if $t(x,y) = (x\xi_n, y)$. This yields the algebraic equation of S in the statement of the theorem.

We now compute the formula for the covering projection $\pi : R \to S$. It has the form $\pi(x,y) = (\widehat{\pi}(x), f(x,y)) = (x^n, f(x,y)) = (u,w)$ for some rational function f such that $(x^n, f(x,y)) \in S$. Hence,

$$\begin{aligned} w^2 = f(x,y)^2 &= x^{n\delta'} \prod_{j=1}^{L_1} (x^n - a_j^n)(x^n - 1/\overline{a}_j{}^n) \prod_{k=1}^{L_2} (x^n - b_k^n) \\ &= x^{n\delta' - \delta} x^{\delta} \prod_{j=1}^{L_1} (x^n - a_j^n)(x^n - 1/\overline{a}_j{}^n) \prod_{k=1}^{L_2} (x^n - b_k^n) = x^{n\delta' - \delta} y^2. \end{aligned}$$

Therefore $\pi : (x,y) \mapsto (u,w) = (x^n,\ x^{(n\delta' - \delta)/2} y)$, up to an automorphism of S.

Finally, the anticonformal involution $\eta : S \to S$ is determined by the equality $\pi \circ \tau = \eta \circ \pi$. In particular, $\widehat{\pi} \circ \widehat{\tau} = \widehat{\eta} \circ \widehat{\pi}$ which yields $\widehat{\eta}(u) = 1/\overline{u}$. It is easy to see that the second coordinate of η equals $\pm \overline{w}/\overline{u}^{\delta' + (g+1-\delta)/n}$. The equality $\pi \circ \tau = \eta \circ \pi$ yields that the valid sign is "+". So η has the formula in the statement of the theorem and this completes its proof. □

REMARK 3.2. The topology of a real hyperelliptic Riemann surface (R,τ) can be obtained from the number of points in $B(P_R)$ fixed by $\widehat{\tau}$ (see [**6**, Propositions 3.4, 5.1] for a precise statement). Therefore, the above algebraic description of the coverings $\pi : (R,\tau) \to (S,\eta)$ also allows us to study the topology of such coverings. For instance, with the notations in the above Theorem 3.1, if $L_2 > 0$ then the number of ovals of $\mathrm{Fix}(\tau)$ equals $nL_2/2$. The above shows that the maximum number $g+1$ of ovals (Harnack's bound) may be attained by (R,τ) precisely when $L_1 = \delta = 0$. This is the case considered in [**4**]. Observe that the number of points in $B(P_S)$ fixed by $\widehat{\eta}$ equals L_2 and so $\mathrm{Fix}(\eta)$ consists of $L_2/2$ ovals. So Harnack's bound may be attained by the covered surface (S,η) precisely when $L_1 = \delta' = 0$. Similar considerations can be done in the other cases. For instance, if the covering transformation group G is A_4, $\mathfrak{S}_4$ or A_5 then the real structure τ fixes no point (see the beginning of Section 3) and the associated Klein surface R/τ is non-orientable.

3.1.2. Let us consider now the case in which the anticonformal involution $\widehat{\tau}$ is the antipodal map $x \mapsto -1/\overline{x}$. If n is even then $\widehat{t}_n^{n/2} \circ \widehat{\tau}(x) = 1/\overline{x}$ and the results are similar to those included in Theorem 3.1. An analogous analysis as above yields the following.

THEOREM 3.3. *Under the same hypothesis as in Theorem 3.1, let us assume now that* $\widehat{\tau}$ *is fixed point free. Then*

(1) g *is odd.*

(2) *If* n *is even, then there exist* $\varepsilon \in \{0,1\}$ *and non-negative integers* L_1 *and* L_2 *with* L_2 *even, satisfying* $2g+2 = 2nL_1+nL_2$ *such that, up to birational transformation, one has that*

(a) (R, τ) *can be represented as follows:*

$$R: y^2 = \prod_{j=1}^{L_1}(x^n - a_j^n)(x^n - 1/\overline{a}_j{}^n)\prod_{k=1}^{L_2}(x^n - b_k^n)$$

$$\tau(x,y) = \left(\frac{-1}{\overline{x}}, \frac{\overline{y}}{\overline{x}^{g+1}}\right),$$

where $|b_k| = 1$ *for all* k, *the right-hand side polynomial in the defining equation of* R *has no multiple roots and* $\prod_{j=1}^{L_1} a_j^n/\overline{a}_j^n \prod_{k=1}^{L_2} b_k^n = 1$.

(b) *A formula for the automorphism* t *generating the group* G *is*

$$t(x,y) = (x\xi_n, (-1)^{\varepsilon}y).$$

(c) *For each* R, τ *and* G *the covered surface is* (S, η) *where*

$$S: w^2 = u^{\varepsilon}\prod_{j=1}^{L_1}(u - a_j^n)(u - 1/\overline{a}_j{}^n)\prod_{k=1}^{L_2}(u - b_k^n)$$

$$\eta(u,w) = \left(\frac{1}{\overline{u}}, \frac{(-1)^{\varepsilon n/2}\overline{w}}{\overline{u}^{\varepsilon+(g+1)/n}}\right).$$

(d) *The covering projection is given by*

$$\pi: (x,y) \mapsto (u,w) = (x^n, x^{\varepsilon n/2}y).$$

(3) *If* n *is odd then there exist* $\delta \in \{0,1\}$ *and a positive integer* L *satisfying* $g+1 = nL+\delta$ *such that, up to birational transformation, one has that*

(a) (S, η) *can be represented as follows:*

$$R: y^2 = x^{\delta}\prod_{j=1}^{L}(x^n - a_j^n)(x^n + 1/\overline{a}_j{}^n)$$

$$\tau(x,y) = \left(\frac{-1}{\overline{x}}, \frac{\overline{y}}{\overline{x}^{g+1}}\right),$$

where the right-hand side polynomial in the defining equation of R *has no multiple roots and* $\prod_{j=1}^{L} a_j^n/\overline{a}_j^n = 1$.

(b) *A formula for the automorphism* t *generating the group* G *is*

$$t(x,y) = (x\xi_n, (-\xi_{2n})^{\delta}y).$$

(c) *For each* R, τ *and* G *the covered surface is* (S, η) *where*

$$S: w^2 = u^{\delta}\prod_{j=1}^{L}(u - a_j^n)(u + 1/\overline{a}_j{}^n)$$

$$\eta(u,w) = \left(\frac{-1}{\overline{u}}, \frac{(-1)^{\delta(n-1)/2}\overline{w}}{\overline{u}^{\delta+L}}\right)$$

(d) *The covering projection is given by*

$$\pi : (x, y) \mapsto (u, w) = (x^n, x^{\delta(n-1)/2} y).$$

3.2. On the non-cyclic case. The situation is more involved in the remaining cases $\widehat{G} = \mathrm{D}_m$, A_4, $\mathfrak{S}_4$ and A_5 than in the cyclic case. Orbits of points under the action of these groups are cumbersome to describe and, accordingly, the algebraic equations of the surfaces are more difficult to describe. The computation of the branch point set $B(P_S)$ of the covered surface (S, η), however, can be simplified with the use of Lemma 3.4 below.

If $\widehat{G}$ is either D_m, A_4, $\mathfrak{S}_4$ or A_5, then $\widehat{G}$ admits a presentation of the form $\widehat{G} = \langle \widehat{\gamma}_1, \widehat{\gamma}_2, \widehat{\gamma}_3 : \widehat{\gamma}_1^m,\ \widehat{\gamma}_2^2,\ \widehat{\gamma}_3^r,\ \widehat{\gamma}_1 \circ \widehat{\gamma}_2 \circ \widehat{\gamma}_3 \rangle$, where (m, r) is either $(m, 2)$, $(3, 3)$, $(4, 3)$ or $(5, 3)$, respectively. In the case that $\widehat{G} \neq \mathrm{A}_4$, there is a fundamental set (the closure of a fundamental domain) for $\widehat{G}$ given by a spherical triangle, say P, with one vertex having internal angle equal to $2\pi/m$ (this vertex is one of the fixed points of $\widehat{\gamma}_1$), and the other two vertices having internal angles both equal to π/r (one of these two vertices is a fixed point of $\widehat{\gamma}_3$, and the middle point of the boundary arc connecting these two vertices is a fixed point of $\widehat{\gamma}_2$). Inside the triangle P there is another spherical triangle (this being a half of P), say F, which is a fundamental set for the group $\widehat{G} \times \langle \widehat{\tau} \rangle$. The vertices of F are given by a fixed point of $\widehat{\gamma}_1$, a fixed point of $\widehat{\gamma}_2$ and one of $\widehat{\gamma}_3$ (later, in each case, we will describe F explicitly). In the case that $\widehat{G} = \mathrm{A}_4$, we produce a fundamental domain F for $\widehat{G} \times \langle \widehat{\tau} \rangle$, this being a triangle with two internal angles equal to $\pi/4$ and the third angle being equal to $2\pi/3$. The vertex with angle $2\pi/3$ will be a fixed point of $\widehat{\gamma}_1$, one of the vertices of angle $\pi/4$ will be a fixed point of $\widehat{\gamma}_2$.

Let us write $\mathrm{Fix}(\widehat{\gamma}_i) = \{\alpha_i, \alpha_i'\}$ for $i = 1, 2, 3$, where we label the fixed points so that $\{\alpha_1, \alpha_2, \alpha_3\}$ are the vertices of F.

LEMMA 3.4. *With the above notations (together with the definition of $r_{\widehat{\gamma}_j}$ as in Lemma 2.2), the finite branch value set $B(P_S) \cap \mathbb{C}$ consists exactly of the finite points $\widehat{\pi}(a)$ where $a \in B(P_R)$, together with those finite points $\widehat{\pi}(\alpha_i)$ $(i = 1, 2, 3)$ satisfying the condition $r_{\widehat{\gamma}_i}(\alpha_i) = -1$.*

PROOF. For a given point $p = (a, b)$ in R with finite coordinates, we check when $\widehat{\pi}(a) \in B(P_S)$. As we saw in Remark 2.5, this always happens if $b = 0$, whilst if $b \neq 0$ then

$$(3.3) \quad \widehat{\pi}(a) \in B(P_S) \quad \text{if and only if} \quad \widehat{\gamma}(a) = a \text{ and } r_{\widehat{\gamma}}(a) = -1 \text{ for some } \widehat{\gamma} \in \widehat{G}.$$

We first observe that it is not necessary to consider the fixed points of all elements $\widehat{\gamma} \in \widehat{G} - \{id\}$ but just representatives of $\widehat{G}$-orbits. The reason is that all points in the same $\widehat{G}$-orbit are projected by $\widehat{\pi}$ onto the same point. In fact, since each rotation $\widehat{\gamma} \in \widehat{G}$ is conjugate to a power of $\widehat{\gamma}_i$ for some $i \in \{1, 2, 3\}$, we just have to check condition (3.3) for the fixed points of each of these three rotations.

It remains to show that we just have to consider one of the two fixed points of each $\widehat{\gamma}_i$. This is clear if $\widehat{\tau}$ interchanges both fixed points because in this case either the projections under $\widehat{\pi}$ of both points belong to $B(P_S)$ or none of them do (recall that $\widehat{\eta}$ preserves $B(P_S)$). If $\widehat{\tau}$ does not interchange the two fixed points of $\widehat{\gamma}_i$ but fixes them, then $\widehat{\tau}$ is the reflection $x \mapsto 1/\overline{x}$, the rotation $\widehat{\gamma}_i$ is an involution (since it commutes with $\widehat{\tau}$) and $\widehat{G}$ is D_m (since our assumption on $n > 2$ rules out the other possibility $\widehat{G} = \mathrm{C}_2$, see the beginning of Section 3). If m is even then α_2 and

α_2', the two fixed points of $\widehat{\gamma}_2$, are in the same $\widehat{G}$-orbit because $\widehat{\gamma}_1^{m/2}(\alpha_2) = \alpha_2'$. So we just have to consider one of the fixed points of $\widehat{\gamma}_2$. The same holds true for $\widehat{\gamma}_3$. If m is odd then α_2 (respectively α_2') is in the same $\widehat{G}$-orbit as α_3' (respectively α_3) because $\widehat{\gamma}_1^{(m+1)/2}(\alpha_2') = \alpha_3$ (respectively $\widehat{\gamma}_1^{(m-1)/2}(\alpha_3') = \alpha_2$). So also in this case we just have to consider one of the fixed points of $\widehat{\gamma}_2$ and one of the fixed points of $\widehat{\gamma}_3$. □

We omit the lengthy proofs of the theorems corresponding to the remaining cases $\widehat{G} = \mathrm{D}_m$, A_4, $\mathfrak{S}_4$ and A_5, since they follow analogous steps as in the cyclic one, with the help of Lemma 3.4. Generators and invariant functions for $\widehat{G} = \mathrm{A}_4$, $\mathfrak{S}_4$ or A_5 can be found in [**14**]. The ones we use here are adapted to the existence of the antianalytic involution $\widehat{\tau}$.

3.3. Dihedral case. Let us assume that $\widehat{G}$ is the dihedral group generated by $\widehat{t}_m : x \mapsto x\xi_m$ and $\widehat{v} : x \mapsto 1/x$, where $m = n/2$. As before, we consider first the case in which the anticonformal involution $\widehat{\tau}$ is the reflection $x \mapsto 1/\overline{x}$. A fundamental set F for the action of $\widehat{G} \times \langle\widehat{\tau}\rangle$ is

$$F = \{z \in \mathbb{C} : |z| \leq 1, \arg(z) \in [0, \pi/m]\}. \tag{3.4}$$

Let $\mathrm{Int}(F)$ denote the interior of F. The $(\widehat{G} \times \widehat{\tau})$-orbits of the points $\alpha \in F$ can be described in terms of roots of the polynomial $Q_\alpha(x) := (x^m - \alpha^m)(x^m - 1/\alpha^m)$.

THEOREM 3.5. *Let $\pi : (R, \tau) \to (S, \eta)$ be a dihedral, branched, $2m$-sheeted covering between real Riemann surfaces, where (R, τ) is a hyperelliptic real Riemann surface of genus $g \geq 2$ such that $\widehat{\tau}$ fixes points. Assume that the covering transformation group $G = D_m$ does not contain the hyperelliptic involution of R. Let F and $Q_\alpha(x)$ be as above. Then there exist $\delta, \varepsilon_\tau, \varepsilon_t, \varepsilon_v \in \{0, 1\}$ with $\delta = 0$ if m is even, and non-negative integers L_1 and L_2 satisfying $2g + 2 = 4mL_1 + 2mL_2 + 2\delta$ such that, up to birational transformation, one has the following.*

(1) *(R, τ) can be represented as follows:*

$$R : y^2 = x^\delta \prod_{j=1}^{L_1} Q_{\alpha_j}(x) Q_{\overline{\alpha}_j}(x) \prod_{k=1}^{L_2} Q_{\beta_k}(x), \quad \tau(x, y) = \left(\frac{1}{\overline{x}}, (-1)^{\varepsilon_\tau} \frac{\overline{y}}{\overline{x}^{g+1}}\right)$$

where $\alpha_j \in \mathrm{Int}(F)$ for all j, $\beta_k \in F - \mathrm{Int}(F)$ for all k and the right-hand side polynomial in the defining equation of R has no multiple roots.

(2) *Formulae for the automorphisms t and v generating the group G are the following:*

a) if m is even, then

$$t(x, y) = (\xi_m x, (-1)^{\varepsilon_t} y), \qquad v(x, y) = \left(\frac{1}{x}, \frac{(-1)^{\varepsilon_v} y}{x^{g+1}}\right);$$

b) if m is odd, then

$$t(x, y) = (\xi_m x, (-\xi_{2m})^\delta y), \qquad v(x, y) = \left(\frac{1}{x}, \frac{(-1)^{\varepsilon_v} y}{x^{g+1}}\right).$$

(3) *For each R, τ and G the covered surface is (S,η) where*

$$S: w^2=(u-2)^{\varepsilon_v}(u+2)^{\varepsilon_1}\prod_{j=1}^{L_1}(u-a_j)(u-\overline{a}_j)\prod_{k=1}^{L_2}(u-b_k),$$
$$\eta(u,w)=(\overline{u},(-1)^{\varepsilon_v+\varepsilon_\tau}\overline{w})$$

where $a_j:=\alpha_j^m+1/\alpha_j^m$, $b_k:=\beta_k^m+1/\beta_k^m$ and $\varepsilon_1=1$ if $\delta+\varepsilon_t+\varepsilon_v+L_2$ is odd and $\varepsilon_1=0$ otherwise.

(4) *The covering projection is given by*

$$\pi(x,y)=(u,w)=\left(x^m+\frac{1}{x^m},\ \frac{(x^m-1)^{\varepsilon_v}(x^m+1)^{\varepsilon_1}y}{x^{(m\varepsilon_v+m\varepsilon_1+g+1)/2}}\right).$$

In the case in which the anticonformal involution $\widehat{\tau}$ is the antipodal map $x\mapsto -1/\overline{x}$ we see that if m is even then $\widehat{t_m}^{m/2}\circ\widehat{\tau}(x)=1/\overline{x}$ and the results are similar to those included in Theorem 3.5; for instance, a fundamental set F is given by (3.4). If m is odd then a fundamental set D for the action of $\langle\widehat{t}_m,\widehat{v},\widehat{\tau}\rangle$ is

$$D=\{z\in\mathbb{C}:|z|\le 1,\ -\pi/(2m)\le\arg(z)\le\pi/(2m)\}-\{e^{i\theta}:\theta\in[-\pi/(2m),0)\}.$$

THEOREM 3.6. *Under the same hypothesis as in Theorem 3.5 let us assume now that $\widehat{\tau}$ is fixed point free. Then*

(1) *g is odd.*

(2) *If m is even then there exist $\varepsilon_\tau,\varepsilon_t,\varepsilon_v\in\{0,1\}$ and non-negative integers L_1 and L_2 satisfying $2g+2=4mL_1+2mL_2$ such that, up to birational transformation, one has that*

(a) *(R,τ) can be represented as follows:*

$$R: y^2=\prod_{j=1}^{L_1}Q_{\alpha_j}(x)Q_{\overline{\alpha}_j}(x)\prod_{k=1}^{L_2}Q_{\beta_k}(x),\quad \tau(x,y)=\left(\frac{-1}{\overline{x}},\ (-1)^{\varepsilon_\tau}\frac{\overline{y}}{\overline{x}^{g+1}}\right)$$

where $\alpha_j\in\mathrm{Int}(F)$, $\beta_k\in F-\mathrm{Int}(F)$ and the right-hand side polynomial in the defining equation of R has no multiple roots.

(b) *Formulae for the automorphisms t and v generating the dihedral group G are*

$$t(x,y)=(x\xi_m,\ (-1)^{\varepsilon_t}y),\qquad v(x,y)=\left(\frac{1}{x},\ (-1)^{\varepsilon_v}\frac{y}{x^{g+1}}\right).$$

(c) *For each R, τ and G the covered surface is (S,η) where*

$$S: w^2=(u-2)^{\varepsilon_v}(u+2)^{\varepsilon_1}\prod_{j=1}^{L_1}(u-a_j)(u-\overline{a}_j)\prod_{k=1}^{L_2}(u-b_k),$$

$$\eta(u,w)=(\overline{u},(-1)^{\varepsilon_v+\varepsilon_\tau}\overline{w}),$$

where $a_j:=\alpha_j^m+1/\alpha_j^m$, $b_k:=\beta_k^m+1/\beta_k^m$ and $\varepsilon_1=1$ if $\varepsilon_t+\varepsilon_v+L_2$ is odd and $\varepsilon_1=0$ otherwise.

(d) *The covering projection is given by*

$$\pi(x,y)=(u,w)=\left(x^m+\frac{1}{x^m},\ \frac{(x^m-1)^{\varepsilon_v}(x^m+1)^{\varepsilon_1}y}{x^{(m\varepsilon_v+m\varepsilon_1+g+1)/2}}\right).$$

(3) *If m is odd then there exist $\delta, \varepsilon_\tau, \varepsilon_v \in \{0,1\}$ and non-negative integers L_1, L_2 satisfying $2g+2 = 4mL_1 + 2mL_2 + 2\delta$ such that, up to birational transformation, one has that*

(a) *(R,τ) can be represented as follows:*

$$R : y^2 = x^\delta \prod_{j=1}^{L_1} Q_{\alpha_j}(x) Q_{-\overline{\alpha}_j}(x) \prod_{k=1}^{L_2} Q_{\beta_k}(x),$$
$$\tau(x,y) = \left(\frac{-1}{\overline{x}},\, (-1)^{\varepsilon_\tau} \frac{\overline{y}}{\overline{x}^{g+1}}\right),$$

where $\alpha_j, \beta_k \in D - \{1\}$ with $\arg(\alpha_j) \neq \pm\pi/(2m)$, $\arg(\beta_k) \in \{\pm\pi/(2m)\}$ and the right-hand side polynomial in the defining equation of R has no multiple roots.

(b) *Formulae for the automorphisms t and v generating the dihedral group G are*

$$t(x,y) = (\xi_m x, (-\xi_{2m})^\delta y) \qquad v(x,y) = \left(\frac{1}{x}, (-1)^{\varepsilon_v} \frac{y}{x^{g+1}}\right),$$

(c) *For each R, τ and G the covered surface is (S,η) where*

$$S : w^2 = (u^2-4)^{\varepsilon_v} \prod_{j=1}^{L_1} (u - a_j)(u + \overline{a}_j) \prod_{k=1}^{L_2} (u - b_k),$$
$$\eta(u,w) = \left(-\overline{u},\ (-1)^{\varepsilon_\tau + (g+1)/2}\, i^\delta \overline{w}\right),$$

with a_j, b_k as in the even case.

(d) *The covering projection is given by*

$$\pi(x,y) = (u,w) = \left(x^m + \frac{1}{x^m},\ \frac{(x^{2m}-1)^{\varepsilon_v} y}{x^{m\varepsilon_v + (g+1)/2}}\right).$$

3.4. Case $G = \mathrm{A}_4$. The rotations

$$\widehat{t} : x \mapsto \frac{-ix+i}{x+1} \quad \text{and} \quad \widehat{v} : x \mapsto -x$$

satisfy the defining relations $\widehat{v}^2 = \widehat{t}^3 = (\widehat{v} \circ \widehat{t})^3 = 1$ and so they generate a group $\widehat{G}$ isomorphic to the alternating group A_4. Both generators commute with the antipodal map $\widehat{\tau} : x \mapsto -1/\overline{x}$. A fundamental set for the action of $\widehat{G} \times \langle \widehat{\tau} \rangle$ is

$$F = \{z \in \mathbb{C} : |z+i|^2 \leq 2,\ \arg(z) \in [0, \pi/4]\}.$$

This is a curvilinear triangle with vertices 0, 1 and $p = (\sqrt{3}-1)(1+i)/2$ (a fixed point of $\widehat{t}$). The rotation $\widehat{t}$ identifies the side joining p and 1 with the side joining p and 0. For each $\alpha \in F$ let us define $P_\alpha(x) = \prod_t (x - t)$, where t runs over the finite different points in the $(\widehat{G} \times \langle \widehat{\tau} \rangle)$-orbit of α.

If α is fixed by no element in $\widehat{G} \times \langle \widehat{\tau} \rangle$ then $P_\alpha(x) = Q_\alpha(x) \cdot Q_{\overline{\alpha}}(x)$ where

$$Q_\alpha(x) = x^2(x^4-1)^2 \left(\widehat{\pi}(x) - \widehat{\pi}(\alpha)\right) \quad \text{with} \quad \widehat{\pi}(z) = \frac{(z^4+1)(z^8 - 34z^4 + 1)}{z^2(z^4-1)^2}.$$

If $\alpha \in F$ is real then $P_\alpha(x) = Q_\alpha(x)$, and if $\alpha = p$ then $P_\alpha(x) = x^8 + 14x^4 + 1$.

Observe that, by Lemma 2.1, none of the points in the orbit of 0, namely, $\{\infty, 0, 1, -1, i, -i\}$, can be in $B(P_R)$ since each of them is fixed by an involution in $\widehat{G}$. The points in the orbit of p may be branch values. This yields the following.

THEOREM 3.7. *Let $\pi : (R, \tau) \to (S, \eta)$ be a normal branched covering between real Riemann surfaces, where (R, τ) is a hyperelliptic real Riemann surface of genus $g \geq 2$ such that the covering transformation group G is the alternating group A_4. Let F, $P_\alpha(x)$ and $\widehat{\pi}(z)$ be as above. Then there exist $\delta, \varepsilon_\tau \in \{0,1\}$ and non-negative integers L_1 and L_2 satisfying $2g+2 = 24L_1 + 12L_2 + 8\delta$ such that, up to birational transformation,*

(1) *(R, τ) can be represented as follows:*

$$R : y^2 = (x^8 + 14x^4 + 1)^\delta \prod_{j=1}^{L_1} Q_{\alpha_j}(x) Q_{\overline{\alpha}_j}(x) \prod_{k=1}^{L_2} Q_{\beta_k}(x),$$
$$\tau(x, y) = \left(\frac{-1}{\overline{x}},\ (-1)^{\varepsilon_\tau} \frac{\overline{y}}{\overline{x}^{g+1}} \right),$$

where $\alpha_j \in F - (\mathbb{R} \cup \{p\})$, $0 < \beta_k < 1$ and the right-hand side polynomial in the defining equation of R has no multiple roots.

(2) *Formulae for the automorphisms t and v generating the group G are the following:*

$$t(x, y) = \left(\frac{-ix + i}{x + 1},\ \frac{(-2i)^{(g+1)/2} y}{(x+1)^{g+1}} \right),$$

$$v(x, y) = \left(-x,\ (-1)^{(g+1)/2} y \right).$$

(3) *For each R, τ and G, the covered surface (S, η) is*

$$S : w^2 = (u^2 + 108)^\delta \prod_{j=1}^{L_1} \left(u - \widehat{\pi}(\alpha_j)\right)\left(u - \widehat{\pi}(\overline{\alpha}_j)\right) \cdot \prod_{k=1}^{L_2} \left(u - \widehat{\pi}(\beta_k)\right),$$
$$\eta(u, w) = (\overline{u}, (-1)^{\varepsilon_\tau} \overline{w}).$$

(4) *The covering projection is given by*

$$\pi(x, y) = (u, w) = \left(\widehat{\pi}(x),\ \frac{(x^8 + 14x^4 + 1)^\delta}{(x(x^4 - 1))^{(g+1+8\delta)/6}} \cdot y \right).$$

3.5. Case $G = \mathfrak{S}_4$. The rotations

$$\widehat{t} : x \mapsto ix \quad \text{and} \quad \widehat{v} : x \mapsto \frac{-x + i}{x + i}$$

satisfy the defining relations $\widehat{t}^4 = \widehat{v}^3 = (\widehat{t} \circ \widehat{v})^2 = 1$ and so they generate a group $\widehat{G}$ isomorphic to the symmetric group $\mathfrak{S}_4$. Both generators commute with the antipodal map $\widehat{\tau} : x \mapsto -1/\overline{x}$. A fundamental set for the action of $\widehat{G} \times \langle \widehat{\tau} \rangle$ is

$$F = \{z \in \mathbb{C} : |z + 1|^2 \leq 2,\ \arg(z) \in [0, \pi/4]\}.$$

This is a curvilinear triangle with vertices $V_0 = 0$, $V_1 = \sqrt{2} - 1$ and $V_2 = (\sqrt{3} - 1)(1 + i)/2$. For each $\alpha \in F$ let us define $P_\alpha(x) = \prod_t (x - t)$, where t runs over the finite different points in the $(\widehat{G} \times \langle \widehat{\tau} \rangle)$-orbit of α.

If α is fixed by no element in $\widehat{G} \times \langle \widehat{\tau} \rangle$ then $P_\alpha(x) = Q_\alpha(x) \cdot Q_{\overline{\alpha}}(x)$ where

$$Q_\alpha(x) = x^4 (x^4 - 1)^4 \left(\widehat{\pi}(x) - \widehat{\pi}(\alpha) \right) \quad \text{with} \quad \widehat{\pi}(z) = \frac{(z^8 + 14z^4 + 1)^3}{z^4 (z^4 - 1)^4}.$$

If $\alpha \in F - \{V_0, V_1, V_2\}$ lies in the boundary of F then $P_\alpha(x) = Q_\alpha(x)$, and if $\alpha = V_2$ then $P_\alpha(x) = x^8 + 14x^4 + 1$.

Observe that, by Lemma 2.1, none of the points in the orbit of V_0 or V_1 can be in $B(P_R)$. The points in the orbit of V_2 may be branch values. This yields the following.

THEOREM 3.8. *Let $\pi : (R,\tau) \to (S,\eta)$ be a normal branched covering between real Riemann surfaces, where (R,τ) is a hyperelliptic real Riemann surface of genus $g \geq 2$ such that the covering transformation group G is the symmetric group $\mathfrak{S}_4$. Let F, $P_\alpha(x)$ and $\widehat{\pi}(z)$ be as above. Then there exist $\delta, \varepsilon_\tau, \varepsilon_t \in \{0,1\}$ and non-negative integers L_1 and L_2 satisfying $2g+2 = 48L_1 + 24L_2 + 8\delta$ such that, up to birational transformation, one has that*

(1) *(R,τ) can be represented as follows:*

$$R : y^2 = (x^8 + 14x^4 + 1)^\delta \prod_{j=1}^{L_1} Q_{\alpha_j}(x) Q_{\overline{\alpha}_j}(x) \prod_{k=1}^{L_2} Q_{\beta_k}(x)$$
$$\tau(x,y) = \left(\frac{-1}{\overline{x}},\ (-1)^{\varepsilon_\tau} \frac{\overline{y}}{\overline{x}^{g+1}} \right),$$

where $\alpha_j \in \mathrm{Int}(F)$ and $\beta_k \neq V_0, V_1, V_2$ lies in the boundary of F and the right-hand side polynomial in the defining equation of R has no multiple roots.

(2) *Formulae for the automorphisms t and v generating the group G are the following:*

$$t(x,y) = (ix,\ (-1)^{\varepsilon_t} y) \quad \text{and}$$

$$v(x,y) = \left(\frac{-x+i}{x+i},\ \frac{(-4)^{(g+1)/4} y}{(x+i)^{g+1}} \right).$$

(3) *For each R, τ and G, the covered surface (S,η) is*

$$S : w^2 = (u-108)^{\varepsilon_1} u^\delta \prod_{j=1}^{L_1} \left(u - \widehat{\pi}(\alpha_j)\right)\left(u - \widehat{\pi}(\overline{\alpha}_j)\right) \cdot \prod_{k=1}^{L_2} \left(u - \widehat{\pi}(\beta_k)\right),$$
$$\eta(u,w) = (\overline{u}, (-1)^{\varepsilon_\tau} \overline{w}),$$

where $\varepsilon_1 = 1$ if $\varepsilon_t + (g+1)/4$ is odd and $\varepsilon_1 = 0$ otherwise.

(4) *The covering projection is given by $\pi(x,y) = (\widehat{\pi}(x), \pi_2(x,y))$, where*

$$\pi_2(x,y) = \frac{(x^4+1)^{\varepsilon_1}\ (x^8 - 34x^4 + 1)^{\varepsilon_1}\ (x^8 + 14x^4 + 1)^\delta}{\left(x(x^4-1)\right)^{2\varepsilon_1 + (g+1+8\delta)/6}} \cdot y.$$

3.6. Case $G = \mathrm{A}_5$. The rotations

$$\widehat{t} : x \mapsto xe^{2\pi i/5} \quad \text{and} \quad \widehat{v} : x \mapsto \frac{-\varphi x + 1}{x + \varphi}, \quad \text{where } \varphi = 2\cos(\pi/5) \text{ is the golden ratio,}$$

satisfy the defining relations $\widehat{t}^5 = \widehat{v}^2 = (\widehat{t} \circ \widehat{v})^3 = 1$ and so they generate a group $\widehat{G}$ isomorphic to the alternating group A_5. Both generators commute with the antipodal map $\widehat{\tau} : x \mapsto -1/\overline{x}$. A fundamental set for the action of $\widehat{G} \times \langle \widehat{\tau} \rangle$ is

$$F = \{z \in \mathbb{C} : |z + 2\cos(\pi/5)| \leq 2\sin(2\pi/5),\ \arg(z) \in [0, \pi/5]\}.$$

This is a curvilinear triangle with vertices $V_0 = 0$, $V_1 = 2(\sin 2\pi/5 - \cos \pi/5)$ (a point fixed by $\widehat{v}$) and $V_2 =$ the point fixed by $\widehat{t} \circ \widehat{v}$ with positive imaginary part. For each $\alpha \in F$ let us define $P_\alpha(x) = \prod_t (x - t)$, where t runs over the finite different points in the $(\widehat{G} \times \langle \widehat{\tau} \rangle)$-orbit of α.

If α is fixed by no element in $\widehat{G} \times \langle \widehat{\tau} \rangle$ then $P_\alpha(x) = Q_\alpha(x) \cdot Q_{\overline{\alpha}}(x)$ where

$$Q_\alpha(x) = (x^{20} - 228x^{15} + 494x^{10} + 228x^5 + 1)^3 - \widehat{\pi}(\alpha) \cdot x^5(x^{10} + 11x^5 - 1)^5$$

with

$$\widehat{\pi}(\alpha) = \frac{(\alpha^{20} - 228\alpha^{15} + 494\alpha^{10} + 228\alpha^5 + 1)^3}{\alpha^5(\alpha^{10} + 11\alpha^5 - 1)^5}.$$

If $\alpha \in F - \{V_0, V_1, V_2\}$ lies in the boundary of F then $P_\alpha(x) = Q_\alpha(x)$, if $\alpha = V_0$ then $P_\alpha(x) := P_0(x) = x^{11} + 11x^6 - x$, and if $\alpha = V_2$ then $P_\alpha(x) := P_2(x) = x^{20} - 228x^{15} + 494x^{10} + 228x^5 + 1$. Lemma 2.1 yields that none of the points in the orbit of V_1 can be in $B(P_R)$. This yields the following.

THEOREM 3.9. *Let $\pi : (R, \tau) \to (S, \eta)$ be a normal branched covering between real Riemann surfaces, where (R, τ) is a hyperelliptic real Riemann surface of genus $g \geq 2$ such that the covering transformation group G is the alternating group A_5. Let F, $P_\alpha(x)$ and $\widehat{\pi}(\alpha)$ be as above. Then there exist $\delta_0, \delta_1, \varepsilon_\tau \in \{0, 1\}$ and non-negative integers L_1 and L_2 satisfying $2g + 2 = 120L_1 + 60L_2 + 12\delta_0 + 20\delta_1$ such that, up to birational transformation, one has that*

(1) *(R, τ) can be represented as follows:*

$$R : y^2 = P_0(x)^{\delta_0} P_2(x)^{\delta_1} \prod_{j=1}^{L_1} Q_{\alpha_j}(x) Q_{\overline{\alpha}_j}(x) \prod_{k=1}^{L_2} Q_{\beta_k}(x),$$
$$\tau(x, y) = \left(\frac{-1}{\overline{x}},\ (-1)^{\varepsilon_\tau} \frac{\overline{y}}{\overline{x}^{g+1}} \right),$$

where $\alpha_j \in \mathrm{Int}(F)$ and $\beta_k \neq V_0, V_1, V_2$ lies in the boundary of F and the right-hand side polynomial in the defining equation of R has no multiple roots.

(2) *Formulae for the automorphisms t and v generating the group G are the following:*

$$t(x, y) = \left(x e^{2\pi i/5},\ y(-e^{\pi i/5})^{\delta_0} \right),$$

$$v(x, y) = \left(\frac{-\varphi x + 1}{x + \varphi},\ \frac{y\ (-\varphi\sqrt{5})^{(g+1)/2}}{(x + \varphi)^{g+1}} \right),$$

where $\varphi = 2\cos(\pi/5)$ is the golden ratio.

(3) *For each R, τ and G, the covered surface (S, η) is*

$$S : w^2 = (u + 1728)^\varepsilon \cdot u^{\delta_1} \cdot \prod_{j=1}^{L_1} \left(u - \widehat{\pi}(\alpha_j)\right)\left(u - \widehat{\pi}(\overline{\alpha}_j)\right) \cdot \prod_{k=1}^{L_2} \left(u - \widehat{\pi}(\beta_k)\right),$$
$$\eta(u, w) = (\overline{u}, (-1)^{\varepsilon_\tau} \overline{w}),$$

where $\varepsilon = 1$ if $(g + 1)/2$ is odd and $\varepsilon = 0$ otherwise.

(4) *The covering projection is given by $\pi(x, y) = (u, w) = (\widehat{\pi}(x), \pi_2(x, y))$ where*

$$\pi_2(x, y) = \frac{(x^{10} + 1)^\varepsilon (x^{20} + 522x^{15} - 10006x^{10} - 522x^5 + 1)^\varepsilon\, P_2(x)^{\delta_1} \cdot y}{(x(x^{10} + 11x^5 - 1))^{(g+1+20\delta_1+30\varepsilon)/12}}.$$

Acknowledgement

The authors would like to thank the anonymous referee for reading carefully the whole manuscript and for his/her suggestions which have contributed to get a more readable paper.

References

[1] Norman L. Alling and Newcomb Greenleaf, *Foundations of the theory of Klein surfaces*, Lecture Notes in Mathematics, Vol. 219, Springer-Verlag, Berlin-New York, 1971. MR0333163 (48 #11488)

[2] Alan F. Beardon, *The geometry of discrete groups*, Graduate Texts in Mathematics, vol. 91, Springer-Verlag, New York, 1983. MR698777 (85d:22026)

[3] E. Bujalance, *A classification of unramified double coverings of hyperelliptic Riemann surfaces*, Arch. Math. (Basel) **47** (1986), no. 1, 93–96, DOI 10.1007/BF01202505. MR855143 (87k:14030)

[4] E. Bujalance, F. J. Cirre, and J. M. Gamboa, *Normal coverings of hyperelliptic real algebraic curves*, Conform. Geom. Dyn. **11** (2007), 107–127, DOI 10.1090/S1088-4173-07-00163-4. MR2329139 (2010d:30052)

[5] E. Bujalance, F. J. Cirre, and J. M. Gamboa, *Double coverings of hyperelliptic real algebraic curves*, J. Pure Appl. Algebra **212** (2008), no. 9, 2011–2026, DOI 10.1016/j.jpaa.2007.12.008. MR2422189 (2009c:14048)

[6] Francisco Javier Cirre, *Birational classification of hyperelliptic real algebraic curves*, The geometry of Riemann surfaces and abelian varieties, Contemp. Math., vol. 397, Amer. Math. Soc., Providence, RI, 2006, pp. 15–25, DOI 10.1090/conm/397/07458. MR2217994 (2007a:14034)

[7] H. M. Farkas, *Unramified double coverings of hyperelliptic surfaces*, J. Analyse Math. **30** (1976), 150–155. MR0437741 (55 #10664)

[8] H. M. Farkas, *Unramified coverings of hyperelliptic Riemann surfaces*, Complex analysis, I (College Park, Md., 1985), Lecture Notes in Math., vol. 1275, Springer, Berlin, 1987, pp. 113–130, DOI 10.1007/BFb0078347. MR922295 (89c:30112)

[9] Hershel M. Farkas and Irwin Kra, *Riemann surfaces*, Graduate Texts in Mathematics, vol. 71, Springer-Verlag, New York-Berlin, 1980. MR583745 (82c:30067)

[10] Yolanda Fuertes and Gabino González-Diez, *Smooth double coverings of hyperelliptic curves*, The geometry of Riemann surfaces and abelian varieties, Contemp. Math., vol. 397, Amer. Math. Soc., Providence, RI, 2006, pp. 73–77, DOI 10.1090/conm/397/07462. MR2217998 (2006m:14042)

[11] Yolanda Fuertes and Gabino González-Diez, *On unramified normal coverings of hyperelliptic curves*, J. Pure Appl. Algebra **208** (2007), no. 3, 1063–1070, DOI 10.1016/j.jpaa.2006.05.008. MR2283446 (2007i:14026)

[12] Axel Harnack, *Ueber die Vieltheiligkeit der ebenen algebraischen Curven* (German), Math. Ann. **10** (1876), no. 2, 189–198, DOI 10.1007/BF01442458. MR1509883

[13] Ryutaro Horiuchi, *Normal coverings of hyperelliptic Riemann surfaces*, J. Math. Kyoto Univ. **19** (1979), no. 3, 497–523. MR553229 (80k:30044)

[14] Felix Klein, *Lectures on the icosahedron and the solution of equations of the fifth degree*, Second and revised edition, Dover Publications, Inc., New York, N.Y., 1956. Translated into English by George Gavin Morrice. MR0080930 (18,329c)

[15] C. Maclachlan, *Smooth coverings of hyperelliptic surfaces*, Quart. J. Math. Oxford Ser. (2) **22** (1971), 117–123. MR0283194 (44 #427)

Departamento de Matemáticas Fundamentales, Facultad de Ciencias, Universidad Nacional de Educación a Distancia, Madrid, Spain

E-mail address: jcirre@mat.uned.es

Departamento de Matemática, Universidad Técnica Federico Santa María, Casilla 110-V, Valparaiso, Chile

E-mail address: ruben.hidalgo@usm.cl

Large group actions on surfaces

Marston D. E. Conder

Dedicated to Emilio Bujalance on the occasion of his 60th birthday

ABSTRACT. This paper considers the largest numbers of conformal automorphisms and of all automorphisms of a compact Riemann surface of given genus g, and the largest number of automorphisms of an unbordered compact non-orientable Klein surface of given topological genus p. In particular, it reports on some recent computations that provide these numbers exactly for all g in the range $1 < g \leq 301$, and all p in the range $2 < p \leq 302$. Some observations are also made about upper and lower bounds, as well as related matters (namely the strong symmetric genus, the symmetric genus, and the symmetric cross-cap number of a given finite group), and some open questions are stated at the end.

1. Introduction

A well known theorem of Hurwitz (1893) states that if X is a compact Riemann surface of genus $g > 1$, then the number of orientation-preserving automorphisms of X is at most $84(g-1)$, and moreover, when this bound is attained, the group $\mathrm{Aut}^{\mathrm{o}}X$ of all such automorphisms of X is a quotient of the ordinary $(2,3,7)$ triangle group $\langle\, x, y \mid x^2 = y^3 = (xy)^7 = 1\,\rangle$. Finite quotients of the latter group are called *Hurwitz groups*, and are well understood, but not completely known. See [**8**, **9**] for some further details about them.

Hurwitz's theorem is a consequence of the Riemann-Hurwitz formula. Every compact Riemann surface X of genus $g > 1$ may be viewed as quotient space $\mathcal{U}/\Lambda$ of the upper half complex plane $\mathcal{U}$ by some *Fuchsian group* Λ, which is a finitely-generated discrete subgroup of the group $\mathrm{PSL}(2,\mathbb{R})$, and isomorphic to the fundamental group of X. The group $G^{\mathrm{o}} = \mathrm{Aut}^{\mathrm{o}}X$ is then isomorphic to the quotient Γ^{+}/Λ, where Γ^{+} is the normaliser of Λ in $\mathrm{PSL}(2,\mathbb{R})$. Associated with every Fuchsian group is a *signature*, which encodes information about a fundamental domain for its action on $\mathcal{U}$. The Riemann-Hurwitz formula relates the order of

2010 *Mathematics Subject Classification.* Primary 30F10; Secondary 14H37, 20B25, 20H10, 22F50, 30F50.

Key words and phrases. Automorphism group, Riemann surface, Klein surface, symmetric genus, symmetric cross-cap number.

This research was supported in part by the N.Z. Marsden Fund (grant UOA 1015) and a James Cook Fellowship from the Royal Society of New Zealand. It was also made possible by use of the MAGMA system [**2**] and its significant capabilities for finding, constructing and analysing discrete structures and their automorphism groups.

$\mathrm{Aut}^{\circ}X \cong \Gamma^{+}/\Lambda$ to the genus of X via the area of a fundamental domain for Γ^{+}, and for given genus $g > 1$, this order is maximised when the area is minimised.

The smallest possible hyperbolic area comes from Fuchsian groups with signature $(0;\, 2,3,7)$; these are isomorphic to the ordinary $(2,3,7)$ triangle group, and hence we have the Hurwitz theorem. There are infinitely many values of g for which the Hurwitz bound is attained, but these form a sparse subset of the positive integers; for example, there are just 32 such values of g with $1 < g \le 11905$; see [**6**].

The next largest possibilities for $|\mathrm{Aut}^{\circ}X|$ after $84(g-1)$ are $48(g-1)$, $40(g-1)$, $36(g-1)$, $30(g-1)$, $\frac{132}{5}(g-1)$, and $24(g-1)$, which occur when $G^{\circ} = \mathrm{Aut}^{\circ}X$ is a smooth quotient of a Fuchsian group with triangular signature $(0;\, r,s,t)$, for $(r,s,t) = (2,3,8)$, $(2,4,5)$, $(2,3,9)$, $(2,3,10)$, $(2,3,11)$, and either $(2,3,12)$, $(2,4,6)$ or $(3,3,4)$, respectively. This begins a descending sequence of possibilities, with $12(g-1)$ and $8(g-1)$ as the first values coming from non-triangular Fuchsian groups, namely those with signatures $(0;\, 2,2,2,3)$ and $(0;\, 2,2,2,4)$ respectively.

Similar things happen when anti-conformal automorphisms (reversing orientation) are allowed. In that case, the full automorphism group $G = \mathrm{Aut}\,X$ is isomorphic to Γ/Λ, where Γ is the normaliser of the fundamental group Λ in $\mathrm{PGL}(2,\mathbb{R})$. This subgroup Γ is a non-Euclidean crystallographic group (or *NEC-group*), contained in $\mathrm{PGL}(2,\mathbb{R})$ but not $\mathrm{PSL}(2,\mathbb{R})$. The largest possible number of (all) automorphisms of a compact Riemann surface of genus $g > 1$ is then $168(g-1)$, which occurs if and only if the NEC-group Γ has signature $(0;+;[-];\{(2,3,7)\})$, or equivalently, $G = \mathrm{Aut}\,X$ is a smooth quotient of the full $(2,3,7)$ triangle group $\langle\, a,b,c \mid a^2 = b^2 = c^2 = (ab)^2 = (bc)^3 = (ca)^7 = 1\,\rangle$.

Again we have a descending sequence of possibilities, starting from $168(g-1)$. The next is $96(g-1)$, coming from the full $(2,3,8)$ triangle group, but the next after that is $84(g-1)$, occurring when $G = G^{\circ}$ is a smooth quotient of the ordinary $(2,3,7)$ triangle group, acting conformally, and then the next is $80(g-1)$, and so on.

We can also consider automorphisms of unbordered compact *non-orientable Klein surfaces*, in the same way. In that case, the fundamental group Λ of the surface is a particular kind of NEC-group, and the largest possible value for $|G| = |\mathrm{Aut}\,X| = |\Gamma/\Lambda|$ is $84(p-2)$, when X has topological genus p. Again this maximum value occurs when Γ has signature $(0;+;[-];\{(2,3,7)\})$.

Note that there are no such upper bounds on the orders of groups of automorphisms of compact Riemann surfaces of genus 0 and 1, or of compact non-orientable Klein surfaces of genus 1 and 2. In fact, the groups acting on such surfaces have been completely known for some time.

For genus 0, the orientation-preserving automorphism groups are the cyclic groups C_n and dihedral groups D_n (of arbitrary order), and the tetrahedral group A_4, the octahedral group S_4 and the icosahedral group A_5; these are attributed to Maschke [**23**], and all their genus 0 actions admit orientation-reversing automorphisms. For genus 1, the orientation-preserving groups are the subgroups of quotients of the ordinary $(2,3,6)$, $(2,4,4)$ and $(3,3,3)$ triangle groups; see [**27**]. These groups are all solvable, and they include finite 2-generator abelian groups of arbitrary order. Some of the actions admit reflections, while others do not. In the non-orientable case, for genus 1 the groups are again C_n, D_n, A_4, S_4 and A_5, while for genus 2 they are the subgroups of $D_n \times C_2$ for all n; see [**28**].

We note also that *lower bounds* are known for the largest order of a group acting on a compact Riemann surface of genus g or a compact non-orientable Klein surface of genus p. In the orientation-preserving case, it was shown independently by Accola [**1**] and Maclachlan [**22**] that a lower bound is $8(g+1)$, and that this is sharp for infinitely many values of g. The corresponding surfaces are all reflexible, so a lower bound for the largest number of *all* automorphisms of a compact Riemann surface of given genus g is $16(g+1)$, and again this is sharp for infinitely many g. Similarly, in the non-orientable case, it was proved by Conder, Maclachlan, Vasiljevic and Wilson [**16**] that a lower bound is $8(p-2)$ when p is even, and $4p$ when p is odd. It was also conjectured that the latter bound of $4p$ is sharp for infinitely many $p \equiv 3 \bmod 12$.

In this paper, we report on some computations that provide the largest numbers of conformal automorphisms and of all automorphisms of a compact Riemann surface of genus g, for all g in the range $1 < g \leq 301$, and the largest numbers of automorphisms of an unbordered compact non-orientable Klein surface of genus p, for all p in the range $2 < p \leq 302$. In the case of Riemann surfaces, the largest numbers of conformal automorphisms extend the computations undertaken by Thomas Breuer for genus 2 to 48, and published in Table 13 of [**3**], and also correct a mistake for genus 33 in that table. In the non-orientable case, the computations show that the lower bound of $4p$ is sharp for $p = 3, 87$ and 207.

In Section 2 we provide some further background on surface actions, and in Section 3 we describe the computations we carried out and make some observations on the output, which is summarised in Appendix 1 (for the orientable case) and Appendix 2 (for the non-orientable case). In Section 4, we mention related matters, namely the strong symmetric genus, the symmetric genus, and the symmetric cross-cap number of a given group, and finally, we pose some important open questions in Section 5.

2. Further background

Let Σ be any Fuchsian group (that is, a finitely-generated discrete subgroup of $\mathrm{PSL}(2,\mathbb{R})$), acting on the upper half complex plane $\mathcal{U}$ via linear fractional transformations $z \mapsto \frac{az+b}{cz+d}$. The signature of Σ, which encodes information about a fundamental domain for the action on $\mathcal{U}$, is a sequence of the form $(\gamma; m_1, \ldots, m_r)$, where γ is a non-negative integer, and each m_i is either $+\infty$ or an integer greater than 1.

Any Fuchsian group with signature $(\gamma; m_1, \ldots, m_r)$ can be generated by 2γ hyperbolic elements $a_1, b_1, \ldots, a_\gamma, b_\gamma$ together with r elliptic or parabolic elements $x_1, \ldots, x_r$, such that

$$x_1^{m_1} = \cdots = x_r^{m_r} = [a_1, b_1] \ldots [a_\gamma, b_\gamma]\, x_1 \ldots x_r = 1,$$

which are in fact defining relations for Σ (in terms of the given generators). In particular, m_i is the order of x_i, which is elliptic when m_i is finite, and parabolic when $m_i = \infty$. Also the area of the fundamental domain for Σ is

$$\mu(\Sigma) = 2\pi \left(2\gamma - 2 + \sum_{i=1}^{r} \left(1 - \frac{1}{m_i} \right) \right),$$

with $\frac{1}{\infty}$ taken as zero.

The *Riemann-Hurwitz formula* states that if Γ is a Fuchsian group and Λ is a subgroup of finite index in Γ, then $\mu(\Lambda) = |\Gamma:\Lambda|\,\mu(\Gamma)$. In particular, if Λ is the fundamental group of the compact Riemann surface X (isomorphic to the quotient space $\mathcal{U}/\Lambda$), of genus $g>1$, then Λ has signature $(g;\,-)$, so $\mu(\Lambda) = 2\pi(2g-2)$, and if $G^{\circ} = \Gamma/\Lambda$ then it follows that

$$2g-2 = |G^{\circ}|\left(2\gamma - 2 + \sum_{i=1}^{r}\left(1-\frac{1}{m_i}\right)\right).$$

For given genus $g>1$, the order of $G^{\circ} = \Gamma/\Lambda$ is maximised when the large bracketed expression (or equivalently, the area of a fundamental domain for Γ) is minimised. The Hurwitz upper bound of $84(g-1)$ for $|G^{\circ}|$ arises in the case where $\gamma = 0$ and $r = 3$, and $\{m_1, m_2, m_3\} = \{2,3,7\}$, or in other words, when Γ has signature $(0;\,2,3,7)$. The next largest values and corresponding signatures are as described in the Introduction.

When orientation-reversing automorphisms are allowed, or the surface is non-orientable, the role of Fuchsian groups is expanded to *non-Euclidean crystallographic groups* (or *NEC*-groups, for short). These are finitely-generated discrete subgroups of $\mathrm{PGL}(2,\mathbb{R})$, acting on $\mathcal{U}$. Those contained in $\mathrm{PSL}(2,\mathbb{R})$ are Fuchsian groups, while all others are called *proper* NEC-groups.

The signature of an NEC group Σ (introduced by Macbeath [**21**]) has the form

$$(\gamma;\,\pm;\,[m_1,\ldots,m_r];\,\{(n_{11},\ldots,n_{1s_1}),\ldots,(n_{k1},\ldots,n_{ks_k})\}),$$

where γ is a non-negative integer, $\pm$ is called the *sign*, the m_i (called *proper periods*) are as previously (with $m_i<\infty$ for all i), and the n_{ij} are positive integers known as *link periods.* Each bracket $(n_{i1},\ldots,n_{is_i})$ is a *period cycle,* an empty period-cycle (with $s_i = 0$) is denoted by $(-)$, and an empty set of period-cycles (with $k=0$) is denoted by $\{-\}$.

If Σ is an NEC-group with the above signature, then Σ has a natural presentation in terms of the following generators:

- elliptic elements x_i (of finite order m_i), for $1\le i\le r$;
- reflections $c_{i0},\ldots,c_{is_i}$, for $1\le i\le k$;
- orientation-preserving elements e_i, for $1\le i\le k$;
- hyperbolic elements a_i and b_i, for $1\le i\le\gamma$, if the sign is $+$;
- glide reflections d_i, for $1\le i\le\gamma$, if the sign is $-$.

In terms of these generators, the defining relations for Σ are:

- $x_i^{m_i} = 1$ for $1\le i\le r$;
- ${c_{ij-1}}^2 = {c_{ij}}^2 = (c_{ij-1}c_{ij})^{n_{ij}} = 1$ for $1\le j\le s_i$, for $1\le i\le k$;
- $c_{is_i} = e_i c_{i0} {e_i}^{-1}$ for $1\le i\le k$;
- $x_1\ldots x_r\, e_1\ldots e_k\,[a_1,b_1]\ldots[a_\gamma,b_\gamma] = 1$ if the sign is $+$;
- $x_1\ldots x_r\, e_1\ldots e_k\, {d_1}^2\ldots {d_\gamma}^2 = 1$ if the sign is $-$.

The area of a fundamental region for this NEC group Σ is

$$\mu(\Sigma) = 2\pi\left(\alpha\gamma + k - 2 + \sum_{i=1}^{r}\left(1-\frac{1}{m_i}\right) + \frac{1}{2}\sum_{i=1}^{k}\sum_{j=1}^{s_i}\left(1-\frac{1}{n_{ij}}\right)\right),$$

where $\alpha = 2$ if the sign is $+$ and $\alpha = 1$ if the sign is $-$. Also the natural analogue of the Riemann-Hurwitz formula applies for finite-index inclusions of NEC-groups.

In the action of Σ on $\mathcal{U}$, all the reflections c_{ij} and all the glide reflections d_i reverse orientation, while the other generators preserve orientation, so the orientation-preserving elements of Σ are precisely those that are expressible as words (on the generators) in which the total number of occurrences of the c_{ij} and d_i is even. These elements constitute the intersection $\Sigma \cap \mathrm{PSL}(2,\mathbb{R})$, which is known as the *canonical Fuchsian subgroup* Σ^+. This has index 1 or 2 in Σ, with the former occurring only when Σ itself is Fuchsian.

In the orientable case, the surface X is isomorphic to the quotient space of $\mathcal{U}$ by the fundamental group Λ (which is Fuchsian), and the group $G = \mathrm{Aut}\,X$ of all automorphisms of X is isomorphic to Γ/Λ, where Γ is the normaliser of Λ in $\mathrm{PGL}(2,\mathbb{R})$. If the signature of Γ is $(\gamma;\ \pm;\ [m_1,..,m_r];\ \{(n_{11},..,n_{1s_1}),..,(n_{k1},..,n_{ks_k})\})$, and X has genus $g>1$, then the Riemann-Hurwitz formula gives

$$2g-2 = |G|\left(\alpha\gamma + k - 2 + \sum_{i=1}^{r}\left(1-\frac{1}{m_i}\right) + \frac{1}{2}\sum_{i=1}^{k}\sum_{j=1}^{s_i}\left(1-\frac{1}{n_{ij}}\right)\right).$$

Moreover, if Γ is a proper NEC-group, then $G^{\circ} = \mathrm{Aut}^{\circ}X \cong \Gamma^+/\Lambda$, which has index 2 in G, with elements of $G\setminus G^{\circ}$ reversing orientation.

On the other hand, if X is an unbordered compact non-orientable Klein surface of genus $p>2$, then the fundamental group of X is an NEC group Λ with signature $(p;-;[-];\{-\})$, and $\mu(\Lambda)=2\pi(p-2)$. Again the surface X is isomorphic to the quotient space $\mathcal{U}/\Lambda$, and the group $G = \mathrm{Aut}\,X$ of all automorphisms of X is isomorphic to Γ/Λ, where Γ is the normaliser of Λ in $\mathrm{PGL}(2,\mathbb{R})$. In this case the sign of Γ can be $+$ or $-$, and if Γ has signature $(\gamma;\ \pm;\ [m_1,..,m_r];\ \{(n_{11},..,n_{1s_1}),\dots,(n_{k1},..,n_{ks_k})\})$ then the Riemann-Hurwitz formula gives

$$p-2 = |G|\left(\alpha\gamma + k - 2 + \sum_{i=1}^{r}\left(1-\frac{1}{m_i}\right) + \frac{1}{2}\sum_{i=1}^{k}\sum_{j=1}^{s_i}\left(1-\frac{1}{n_{ij}}\right)\right).$$

Here $\Gamma^+\Lambda = \Gamma$ because $\Lambda \not\subseteq \mathrm{PSL}(2,\mathbb{R})$, and so $\Gamma^+\Lambda/\Lambda = \Gamma/\Lambda \cong \mathrm{Aut}\,X$. Hence in this non-orientable case the canonical Fuchsian subgroup Γ^+ is taken to G by the natural epimorphism $\theta : \Gamma \to G$ (rather than to some subgroup of index 2 in G). In particular, G is generated by elements expressible as images of words in the generators of Γ such that the total number of occurrences of the reflections c_{ij} and glide reflections d_i is even.

As noted earlier, upper bounds on the number of automorphisms of a compact Riemann surface of genus $g>1$ are $84(g-1)$ in the orientation-preserving case, and $168(g-1)$ when orientation-reversion automorphisms are allowed; the corresponding signatures are $(0;\ 2,3,7)$ in the orientation-preserving case, and $(0;+;[-];\{(2,3,7)\})$ in the full case. Similarly, for a non-orientable Klein surface of genus $p>2$, the upper bound is $84(p-2)$, again occurring when the signature is $(0;+;[-];\{(2,3,7)\})$. Groups achieving the latter bound were called *H^*-groups* by Singerman in **[26]**; see also **[7]** for more families of examples.

The lower bound of $8(g+1)$ found by Accola **[1]** and Maclachlan **[22]** for the orientation-preserving case occurs for signature $(0;\ 2,4,2(g+1))$, and the corresponding bound of $16(g+1)$ in the full case occurs for signature $(0;+;[-];\{(2,4,2(g+1))\})$. On the other hand, for a non-orientable Klein surface of genus $p>2$, the lower bound of $8(p-2)$ for even p occurs for signature $(0;+;[-];\{(2,2,2,4)\})$, while the lower bound of $4p$ for odd p occurs for signature $(0;+;[-];\{(2,2,2,p)\})$.

Before proceeding, we also need the upper bounds on the order of a single automorphism of a surface of given genus. For a compact Riemann surface of genus $g > 1$, this upper bound is $4g+2$ (first proved by Wiman [**29**]), while for a compact non-orientable Klein surface of genus p, it is $2p+2$ when p is even, or $2p$ when p is odd (proved independently by Wendy Hall [**20**] and Emilio Bujalance [**4**]).

3. Computations and observations

The Riemann-Hurwitz formula helps relate the order of a group $G = \Gamma/\Lambda$ acting on a surface $X = \mathcal{U}/\Lambda$ to the genus of X via the signature of the corresponding Fuchsian group or NEC-group Γ. In each case the natural homomorphism from Γ onto $G = \Gamma/\Lambda$ is 'smooth', in the sense that it preserves the orders of torsion elements given in the standard presentation for Γ, with the kernel Λ (as the fundamental group of X) being torsion-free. Conversely, if θ is any smooth homomorphism from Γ onto a finite group G, with kernel Σ, say, then $X = \mathcal{U}/\Sigma$ is a surface on which G acts faithfully as a group of automorphisms.

Since Γ has a finite presentation, all possibilities for Σ up to a given index (and hence all possibilities for X up to a given genus) can be found with the help of the 'low index normal subgroups' algorithm, developed by Derek Holt and his student David Firth (see [**19**]), and implemented in MAGMA [**2**] as the `LowIndexNormalSubgroups` command.

For example, if Γ is a Fuchsian group with fundamental domain of smallest possible hyperbolic area, then we know Γ has signature $(0;\, 2,3,7)$, and is isomorphic to the ordinary $(2,3,7)$ triangle group $\Delta^+(2,3,7) = \langle\, x_1, x_2, x_3 \mid {x_1}^2 = {x_2}^3 = {x_3}^7 = x_1x_2x_3 = 1\,\rangle$. The surfaces of genus 2 to 301 admitting an action with such signature can be determined by finding all normal subgroups of index up to $84\,(301-1) = 25200$ in $\Delta^+(2,3,7)$. It takes less than two minutes for MAGMA (on a laptop) to find there are 15 such normal subgroups, with the proper ones being torsion-free and giving group actions on surfaces of genus $3, 7, 14, 14, 14, 17, 17, 118, 129, 129, 129, 146, 146$ and 146 — consistent with the determination of all Hurwitz actions for genus 2 to 11905, achieved by hand by the author in [**6**].

The same approach can be taken with all Fuchsian groups Γ for which the area $\mu(\Gamma)$ of a fundamental domain is small (but positive). In fact, since we know that if G is the largest group of orientation-preserving automorphisms of a compact Riemann surface of genus $g > 1$, then $8(g+1) \le |G| \le 84(g-1)$, and so we need only consider those Γ for which $8 < 4\pi/\mu(\Gamma) \le 84$. Equivalently, if Γ has signature $(\gamma;\, m_1, \dots, m_r)$, then

$$\frac{1}{42} \;\le\; 2\gamma - 2 + \sum_{i=1}^{r}\left(1 - \frac{1}{m_i}\right) \;<\; \frac{1}{4}\,.$$

This considerably restricts the possibilities for the signature of Γ: first $\gamma = 0$, and similarly $2 < r < 5$, therefore $r = 3$ or 4, and moreover, up to rearrangement of the indices $1, 2, \dots, r$, if $r = 3$ then $(m_1, m_2) = (2, m)$ with $3 \le m \le 7$, or $(3,3)$ or $(3,4)$, while if $r = 4$ then $(m_1, m_2, m_3, m_4) = (2,2,2,3)$. In fact each m_i is at most $4g+2$ (by the Wiman bound), and so there are only finitely many possibilities for the signature, for given genus $g > 1$.

Here we note that in the case of triangular signature $(0;\, m_1, m_2, m_3)$, we can assume that $2 \le m_1 \le m_2 \le m_3$, since the triangle groups $\Delta^+(m_1, m_2, m_3)$ and $\Delta^+(m_1', m_2', m_3')$ are isomorphic to each other whenever (m_1', m_2', m_3') is a permutation of (m_1, m_2, m_3). More generally, Fuchsian groups with signatures $(0;\ m_1,..,m_r)$ and $(0;\ m_1',..,m_r')$ are isomorphic whenever $(m_1',..,m_r')$ is a permutation of $(m_1,..,m_r)$; see [**21**, Section 2].

For genus 2 to 301, it makes sense to consider the finitely-presented groups corresponding to those Γ with very small $\mu(\Gamma)$ one by one, starting with the ordinary $(2,3,7)$, $(2,3,8)$, $(2,4,5)$, $(2,3,9)$ and $(2,3,10)$ triangle groups, and so on, but some of those with larger values of $\mu(\Gamma)$ can be handled differently. For example, for the signatures $(0;\ 2,3,m)$ with $m \geq 24$, we can find all normal subgroups of index up to 4800 in the ordinary $(2,3,\infty)$ triangle group $\langle\, x_1, x_2, x_3 \mid {x_1}^2 = {x_2}^3 = x_1x_2x_3 = 1\,\rangle$, and consider only those torsion-free normal subgroups for which the image of x_3 in the corresponding quotient has order 24 or more. Similarly, we can handle the Fuchsian signatures $(0;\ 2,4,m)$ with $m \geq 12$ by finding all normal subgroups of index up to 3600 in the ordinary $(2,4,\infty)$ triangle group.

We took this approach in a combined MAGMA computation that ran for almost 7 days in total, and produced the information given in the first column of the table in Appendix 1. Aside from the large number of cases dealt with in a single step via the $(2,4,\infty)$ triangle group, the signatures that required the longest computations were $(0;\ 2,3,12)$, $(0;\ 2,4,8)$, $(0;\ 2,3,18)$, $(0;\ 2,6,6)$, $(0;\ 3,3,6)$ and $(0;\ 2,2,2,3)$.

Understandably, there are very few genera in the range $2 \leq g \leq 301$ for which the Hurwitz bound of $84(g-1)$ is achieved, but many more where the action of a group of order $48(g-1)$ occurs, with signature $(0;\ 2,3,8)$. Also in most cases the largest group acts via a Fuchsian group with triangular signature, but for genus 126, the signature is $(0;\ 2,2,2,3)$.

For the case where orientation-reversing automorphisms are allowed, there are two possible approaches. One is to repeat the above process, using NEC-groups in place of Fuchsian groups, and checking whether each normal subgroup Σ is orientation-preserving. (This can be made in a number of ways, the easiest being to check that the normal subgroup Σ is contained in the canonical Fuchsian subgroup Γ^+ of the relevant NEC-group Γ.) An even easier way, however, is simply to modify the process for the Fuchsian group case, to test whether each normal subgroup Σ of the Fuchsian group Γ is also normal in an NEC-group containing Γ as a subgroup of index 2. For example, we can test whether a normal subgroup of the ordinary triangle group $\Delta^+ = \Delta^+(m_1, m_2, m_3)$ is normal in the full triangle group $\Delta = \Delta(m_1, m_2, m_3) = \langle\, c_0, c_1, c_2 \mid {c_0}^2 = {c_1}^2 = {c_2}^2 = (c_0c_1)^{m_1} = (c_1c_2)^{m_2} = (c_2c_0)^{m_3} = 1\,\rangle$, using the embedding of Δ^+ in Δ given by $(x_1, x_2, x_3) \to (c_0c_1, c_1c_2, c_2c_0)$, together with the observation that conjugation by c_1 inverts the images of each of x_1 and x_2.

We took the latter approach for genus 2 to 301, and the results are given in the second column of the table in Appendix 1. For convenience in this table, we have abbreviated the NEC-group signature $(0;+;[-];\{(m_1,...,m_r)\})$ to $(0;\ m_1,...,m_r)^*$.

In many cases, the largest group of automorphisms for a given genus g has twice the order of the largest orientation-preserving group G^{o}, since at least one of the surfaces admitting G^{o} is 'reflexible', admitting additional automorphisms that reverse orientation. In other cases, the largest group is a quotient of a proper NEC-group Γ with a different signature (and larger $\mu(\Gamma)$ than that for the Fuchsian group giving G^{o}); for example, when $g = 10$ the largest G^{o} has order 432, via Fuchsian signature $(0;\ 2,3,8)$, but no such action/surface is reflexible, and the largest full group of automorphisms has order 720, coming from an NEC-group with signature

$(0;+;[-];\{(2,4,5)\})$. Finally, in some cases the largest order of a full group of automorphisms can be achieved by an orientation-preserving group. This occurs for $g = 27$ in a straightforward fashion, and also for $g = 28$, where there are reflexible actions as well (but having the same group order).

The table in Appendix 1 (for orientable surface actions) corrects and extends a partial and unpublished census for genus 2 to 26, put together by P.R. Hewitt (1989), and reported to the author by David Singerman a few years ago.

Also the first column of this table (for orientation-preserving actions) agrees with the information provided by Thomas Breuer for genus 2 to 48 in Table 13 of [**3**], with one exception, for genus 33. In Breuer's table, the largest order of a group of conformal automorphisms of a compact Riemann surface of genus 33 is given as 768, which is achievable via signature $(0;3,3,4)$. But in fact there exists a larger group, of order 1536, acting with signature $(0;2,3,8)$. This is a solvable group, generated by two elements x and y subject to the defining relations $x^2 = y^3 = (xy)^8 = (xy)^3(xy^2)^2xy(xy^2)^2(xy)^3xy^2xyxy^2xyxy^2 = 1$, and has a faithful permutation representation on 64 points. Moreover, the corresponding surface is reflexible, and so when orientation reversals are allowed, the largest number of automorphisms is 3072, achievable via signature $(0;+;[-];\{(2,3,8)\})$.

The non-orientable case is much more challenging, but also more interesting. In this case, again we have to use NEC-groups in place of Fuchsian groups, but there are many more possibilities for the types of signature that need to be considered, and for the parameters within those signatures.

Of course if $G = \mathrm{Aut}\, X = \Gamma/\Lambda$ where $X = \mathcal{U}/\Lambda$ is non-orientable, then $G = \Gamma^+\Lambda/\Lambda$ is also a smooth quotient of the canonical Fuchsian subgroup Γ^+, but is not possible to simply use the results of our earlier computation for Fuchsian groups, since this is not a two-way association: not every smooth homomorphism θ from Γ^+ onto a finite group G gives rise to a homomorphism from Γ to G with non-orientable kernel containing $\ker\theta$ as a subgroup of index 2. This makes computation more difficult, but the possibilities for the signature of Γ are still limited, with the help of the Hall-Bujalance bound.

For any such proper NEC-group Γ being considered, it is necessary to check each normal subgroup Σ that gives a smooth epimorphism from Γ to $G = \Gamma/\Sigma$, to see if Σ is non-orientable. Again this can be done in a number of ways, but the easiest is to check that $\Gamma^+\Sigma = \Gamma$, or equivalently, that the epimorphism takes Γ^+ to G.

The results are given in the table in Appendix 2. For convenience in this table, we have abbreviated the NEC-group signature $(0;+;[-];\{(m_1,..,m_r)\})$ to $(0;m_1,...,m_r)$, without the '*' used in Appendix 1 (as the '*' is superfluous in the non-orientable case).

In most cases in this table, the largest order occurs for an action with full triangular signature of the form $(0;+;[-];\{(2,m_2,m_3)\})$ with $m_2 \in \{3,4,5,6\}$, or full 'quadrangular' signature $(0;+;[-];\{(2,2,2,n_4)\})$ with $n_4 = 3$ or 4, but in some cases it is achieved when the signature is $(0;+;[2,3];\{(1)\})$ or $(0;+;[2];\{(2,4)\})$. In the exceptional cases of genus $p = 87$ and 207, the largest orders are only 348 and 828, occurring for signatures $(0;+;[-];\{(2,2,2,87)\})$ and $(0;+;[-];\{(2,2,2,207)\})$ respectively, meeting the lower bound of $4p$ (that was conjectured to be sharp for infinitely many values of p in [**16**]).

4. Related matters

Instead of asking for the largest group of automorphisms of a surface of given genus, one can pose the following question: *Given a finite group G, what is the smallest genus of the surfaces on which G has a faithful action?*

More precisely, in the 1980s TomTucker **[27, 28]** defined the *symmetric genus* $\sigma(G)$ as the minimum genus of all closed orientable surfaces on which G has a faithful action as a group of automorphisms, the *strong symmetric genus* $\sigma^{o}(G)$ as the minimum genus of those for which the action *preserves orientation,* and the *symmetric cross-cap number* $\tilde{\sigma}(G)$ as the minimum genus of all closed *non-orientable* surfaces on which G has a faithful action. A related set of questions concerns the set of possible values that each of these parameters can take.

The concept of the strong symmetric genus had been considered previously by other authors (including Burnside), but its formal study was progressed by Tucker and others from the 1980s onwards. The strong symmetric genus is now known for several families of finite groups, including abelian, dihedral, alternating and symmetric groups, the groups $\mathrm{PSL}(2,q)$ for all q, and the 26 sporadic finite simple groups; see for example **[5, 25, 27]** and other references therein.

Also by considering direct products $C_k \times D_n$, May and Zimmerman **[25]** proved that the strong symmetric genus function is surjective; that is, for every natural number $g \in \mathbb{N}$, there exists a finite group G with $\sigma^{o}(G) = g$.

Using a variant of the computations described in the previous section, it is easy to determine the strong symmetric genus of all finite groups of order up to 127 (see **[10]**), and all finite groups with strong symmetric genus 2 to 32 (see **[11]**).

Similarly, the symmetric genus is known for several families of finite groups, including abelian, dihedral and symmetric groups, all simple groups for which the strong symmetric genus is known, and the groups $\mathrm{PSL}(2,q) \times C_2$ and $\mathrm{PGL}(2,q)$ for all q; see for example **[17, 27]** and other references therein. Further, it is easy to determine the symmetric genus of all finite groups of order up to 127 (see **[12]**), and all finite groups with symmetric genus 2 to 32 (see **[13]**). On the other hand, is it not known whether the symmetric genus function is surjective. It was proved in **[17]** (using a small number of families of solvable groups) that the range of values of $\sigma(G)$ covers at least 8/9 of the natural numbers, namely all $g \in \mathbb{N}$ that are odd, or congruent to 0 or 4 mod 6, or congruent to 2 mod 18. The cases where $g \equiv 8$ or 14 mod 18 were left open, but conjectured in **[17]** to be covered as well.

Finally, somewhat less is known about the symmetric cross-cap number.

Of course it is true that if G is an H^*-group (that is, a group of $84(p-2)$ automorphisms of a compact non-orientable Klein surface of genus $p > 2$), then $\tilde{\sigma}(G) = p = 2 + |G|/84$; and although such groups include all but finitely many alternating groups and many other finite simple groups (see **[7]** for example), they are relatively rare. Also Tucker **[28]** classified all finite groups with symmetric cross-cap number 1 or 2, and Coy May **[24]** determined the symmetric cross-cap number of three classes of finite solvable groups, and proved that there is no finite group G with $\tilde{\sigma}(G) = 3$.

More recently, Javier Etayo and Ernesto Martinez **[18]** calculated the symmetric cross-cap number of the groups $C_m \times D_n$, and in doing so, proved that $\tilde{\sigma}(G)$ can be any positive integer p congruent to 0, 1 or 2 mod 4. In addition, they noted that $\tilde{\sigma}(C_3 \times C_{6m}) = 12m-1$ for all m, by a theorem of May **[24]**, and this covers all $p \equiv 11$ mod 12.

We can also cover the class of all $p \equiv 7 \bmod 12$, as follows.

For any positive integer n, let v be a generator for C_{3n}, and consider the semi-direct product $C_{3n} \rtimes S_4$, with each odd element of S_4 conjugating v to its inverse. Next let H_n be the subgroup generated by $w_0 = v(3,4)$, $w_1 = (1,2)(3,4)$ and $w_2 = (1,3)$. These three elements satisfy ${w_0}^2 = {w_1}^2 = {w_2}^2 = (w_0w_1)^2 = (w_1w_2)^4 = (w_2w_0)^{3n} = 1$, with $(w_2w_0)^3 = ((1,3)v(3,4))^3 = (v^{-1}(1,4,3))^3 = v^{-3}$. Also $w_0 = v(3,4)$ conjugates $(w_1w_2)^2 = (1,3)(2,4)$ to $(1,4)(2,3) = w_2w_1w_2$, and so the generators w_0, w_1 and w_2 satisfy $[w_1, (w_2w_0)^3] = w_0w_1w_2w_1w_2w_0w_2w_1w_2 = 1$. The second of these gives a relator of odd length in w_0, w_1 and w_2, and it follows that H_n is also generated by $w_0w_1 = v(1,2)$, $w_1w_2 = (1,2,3,4)$ and $w_2w_0 = v^{-1}(1,4,3)$, and hence that the quotient of H_n by the cyclic normal subgroup $\langle v^3 \rangle$ is isomorphic to the ordinary $(2,4,3)$ triangle group S_4. Thus H_n has order $24n$, and is a smooth quotient of the full $(2,4,3n)$ triangle group, acting faithfully on a non-orientable surface X of genus $p = 3n-2$.

Dropping the relation $w_0w_1w_2w_1w_2w_0w_2w_1w_2 = 1$ gives the group called L in the proof of Corollary 4.2 of [**17**]. This group L of order $48n$ is an extension of C_n by the full $(2,4,3)$ triangle group $S_4 \times C_2$, and has symmetric genus $3n-3$. In particular, L acts faithfully as a group of automorphisms of an orientable surface of genus $3n-3$, which must be the orientable double cover of X, with $L \cong H_n \times C_2$. Then since $\sigma(H_n \times C_2) = \sigma(L) = 3n-3$, we may conclude that H_n has no faithful action on a non-orientable surface of smaller genus, so $\tilde{\sigma}(H_n) = 3n-2$ for all n. Taking $n \equiv 3 \bmod 4$ gives $\tilde{\sigma}(H_n) \equiv 7 \bmod 12$.

This leaves only the class of all $p \equiv 3 \bmod 12$.

Further computations in this non-orientable case give the symmetric cross-cap number of all finite groups of order up to 127 (see [**14**]), and all finite groups with symmetric genus 4 to 65 (see [**15**]). We have also undertaken computations on specific metabelian finite groups, and as a result of these (and the above), we now know that there are groups with symmetric cross-cap number p for every p in the range $4 \le p \le 200$. In particular, we believe we have enough evidence to make the following:

CONJECTURE 4.1. *For every positive integer $p \neq 3$, there is a finite group with symmetric cross-cap number p.*

5. Some open questions

QUESTION 5.1. *Are there infinitely many positive integers $p \equiv 3 \bmod 12$ for which the largest number of automorphisms of a compact non-orientable surface of genus p is $4p$?*

REMARK. The lower bound of $4p$ on the largest number of automorphisms is sharp for $p = 3, 87$ and 203; see [**16**] for additional comments.

QUESTION 5.2. *For every positive integer $g \equiv 8$ or $14 \bmod 18$, is there a finite group with symmetric genus g?*

REMARK. If so, then the symmetric genus spectrum is complete; see [**17**] for additional comments.

QUESTION 5.3. *For every positive integer $p \equiv 3 \bmod 12$ with $p > 3$, is there a finite group with symmetric cross-cap number p?*

REMARK. If so, then Conjecture 4.1 is true; see [**18**] for additional comments.

Appendix 1: LARGEST ACTIONS ON RIEMANN SURFACES OF GENUS 2 TO 301

In the table below, Max° is the largest order of a group of *orientation-preserving* automorphisms of a compact Riemann surface of the given genus, while Max* is the largest order of *any* group of automorphisms of such a surface (allowing orientation reversal).

Genus	Max°	Signature(s)	Max*	Signature(s)
2	48	(0; 2, 3, 8)	96	(0; 2, 3, 8)*
3	168	(0; 2, 3, 7)	336	(0; 2, 3, 7)*
4	120	(0; 2, 4, 5)	240	(0; 2, 4, 5)*
5	192	(0; 2, 3, 8)	384	(0; 2, 3, 8)*
6	150	(0; 2, 3, 10)	300	(0; 2, 3, 10)*
7	504	(0; 2, 3, 7)	1008	(0; 2, 3, 7)*
8	336	(0; 2, 3, 8)	672	(0; 2, 3, 8)*
9	320	(0; 2, 4, 5)	640	(0; 2, 4, 5)*
10	432	(0; 2, 3, 8)	720	(0; 2, 4, 5)*
11	240	(0; 2, 4, 6)	480	(0; 2, 4, 6)*
12	120	(0; 2, 4, 15)	240	(0; 2, 4, 15)*
13	360	(0; 2, 3, 10)	720	(0; 2, 3, 10)*
14	1092	(0; 2, 3, 7)	2184	(0; 2, 3, 7)*
15	504	(0; 2, 3, 9)	1008	(0; 2, 3, 9)*
16	720	(0; 2, 3, 8)	1440	(0; 2, 3, 8)*
17	1344	(0; 2, 3, 7)	1536	(0; 2, 3, 8)*
18	168	(0; 2, 4, 21)	336	(0; 2, 4, 21)*
19	720	(0; 2, 4, 5)	1440	(0; 2, 4, 5)*
20	228	(0; 2, 6, 6)	336	(0; 2, 4, 42)*
21	480	(0; 2, 4, 6)	960	(0; 2, 4, 6)*
22	1008	(0; 2, 3, 8)	2016	(0; 2, 3, 8)*
23	192	(0; 2, 4, 48)	384	(0; 2, 4, 48)*
24	216	(0; 2, 4, 27)	432	(0; 2, 4, 27)*
25	720	(0; 2, 3, 10)	1440	(0; 2, 3, 10)*
26	750	(0; 2, 3, 10)	1500	(0; 2, 3, 10)*
27	624	(0; 2, 3, 12)	624	(0; 2, 3, 12)
28	1296	(0; 2, 3, 8)	1296	(0; 2, 3, 8), (0; 2, 3, 12)*, (0; 2, 4, 6)*
29	672	(0; 2, 4, 6)	1344	(0; 2, 4, 6)*
30	264	(0; 2, 4, 33)	528	(0; 2, 4, 33)*
31	720	(0; 2, 4, 6)	1440	(0; 2, 4, 6)*
32	372	(0; 2, 6, 6)	528	(0; 2, 4, 66)*
33	1536	(0; 2, 3, 8)	3072	(0; 2, 3, 8)*
34	1320	(0; 2, 4, 5)	2640	(0; 2, 4, 5)*
35	544	(0; 2, 4, 8)	576	(0; 2, 4, 72)*
36	672	(0; 2, 3, 16)	1344	(0; 2, 3, 16)*
37	1728	(0; 2, 3, 8)	1728	(0; 2, 3, 8), (0; 2, 3, 12)*, (0; 2, 4, 6)*, (0; 3, 3, 4)*
38	444	(0; 2, 6, 6)	624	(0; 2, 4, 78)*
39	912	(0; 2, 3, 12)	912	(0; 2, 3, 12)
40	936	(0; 2, 3, 12)	936	(0; 2, 3, 12)
41	960	(0; 2, 4, 6)	1920	(0; 2, 4, 6)*
42	410	(0; 2, 5, 10)	720	(0; 2, 4, 45)*
43	1512	(0; 2, 3, 9)	3024	(0; 2, 3, 9)*
44	516	(0; 2, 6, 6)	720	(0; 2, 4, 90)*
45	1320	(0; 2, 3, 10)	2640	(0; 2, 3, 10)*
46	2160	(0; 2, 3, 8)	4320	(0; 2, 3, 8)*
47	384	(0; 2, 4, 96)	768	(0; 2, 4, 96)*

Genus	Max°	Signature(s)	Max*	Signature(s)
48	408	(0; 2, 4, 51)	816	(0; 2, 4, 51)*
49	1920	(0; 2, 4, 5)	3840	(0; 2, 4, 5)*
50	1176	(0; 2, 4, 6), (0; 3, 3, 4)	2184	(0; 2, 3, 13)*
51	1200	(0; 2, 3, 12)	2400	(0; 2, 3, 12)*
52	2448	(0; 2, 3, 8)	4896	(0; 2, 3, 8)*
53	832	(0; 2, 4, 8)	960	(0; 2, 4, 30)*
54	456	(0; 2, 4, 57)	912	(0; 2, 4, 57)*
55	1296	(0; 2, 3, 12), (0; 2, 4, 6)	2592	(0; 2, 3, 12)*, (0; 2, 4, 6)*
56	1320	(0; 2, 3, 12), (0; 2, 4, 6)	2640	(0; 2, 3, 12)*, (0; 2, 4, 6)*
57	1344	(0; 2, 3, 12), (0; 2, 4, 6)	2688	(0; 2, 4, 6)*
58	1368	(0; 2, 3, 12)	1368	(0; 2, 3, 12)
59	928	(0; 2, 4, 8)	960	(0; 2, 4, 120)*
60	504	(0; 2, 4, 63)	1008	(0; 2, 4, 63)*
61	1440	(0; 2, 4, 6)	2880	(0; 2, 4, 6)*
62	732	(0; 2, 6, 6)	1008	(0; 2, 4, 126)*
63	1488	(0; 2, 3, 12)	1488	(0; 2, 3, 12)
64	1512	(0; 2, 3, 12)	2016	(0; 2, 3, 24)*
65	3072	(0; 2, 3, 8)	6144	(0; 2, 3, 8)*
66	1014	(0; 2, 3, 26)	2028	(0; 2, 3, 26)*
67	576	(0; 2, 4, 48)	1152	(0; 2, 4, 48)*
68	804	(0; 2, 6, 6)	1104	(0; 2, 4, 138)*
69	2448	(0; 2, 3, 9)	4896	(0; 2, 3, 9)*
70	660	(0; 2, 5, 11)	1320	(0; 2, 5, 11)*
71	672	(0; 2, 6, 8)	1344	(0; 2, 6, 8)*
72	710	(0; 2, 5, 10)	1200	(0; 2, 4, 75)*
73	2160	(0; 2, 3, 10)	4320	(0; 2, 3, 10)*
74	876	(0; 2, 6, 6)	1200	(0; 2, 4, 150)*
75	1776	(0; 2, 3, 12)	1776	(0; 2, 3, 12)
76	1800	(0; 2, 3, 12)	3600	(0; 2, 3, 12)*
77	912	(0; 2, 6, 6)	1344	(0; 2, 4, 42)*
78	1176	(0; 2, 3, 28)	2352	(0; 2, 3, 28)*
79	1872	(0; 2, 3, 12)	1872	(0; 2, 3, 12)
80	948	(0; 2, 6, 6)	1296	(0; 2, 4, 162)*
81	1536	(0; 2, 3, 16)	3072	(0; 2, 3, 16)*
82	3888	(0; 2, 3, 8)	7776	(0; 2, 3, 8)*
83	1312	(0; 2, 4, 8)	1344	(0; 2, 4, 168)*
84	696	(0; 2, 4, 87)	1392	(0; 2, 4, 87)*
85	4032	(0; 2, 3, 8)	8064	(0; 2, 3, 8)*
86	1200	(0; 2, 3, 40)	2400	(0; 2, 3, 40)*
87	2064	(0; 2, 3, 12)	2064	(0; 2, 3, 12)
88	712	(0; 2, 4, 178)	1424	(0; 2, 4, 178)*
89	2640	(0; 2, 3, 10)	5280	(0; 2, 3, 10)*
90	744	(0; 2, 4, 93)	1488	(0; 2, 4, 93)*
91	3600	(0; 2, 4, 5)	7200	(0; 2, 4, 5)*
92	2184	(0; 2, 3, 12), (0; 2, 4, 6)	4368	(0; 2, 3, 12)*, (0; 2, 4, 6)*
93	768	(0; 2, 4, 96)	1536	(0; 2, 4, 96)*
94	2232	(0; 2, 3, 12)	2232	(0; 2, 3, 12)
95	768	(0; 2, 4, 192)	1536	(0; 2, 4, 192)*
96	3420	(0; 2, 3, 9)	6840	(0; 2, 3, 9)*
97	3840	(0; 2, 4, 5)	7680	(0; 2, 4, 5)*
98	1164	(0; 2, 6, 6)	1584	(0; 2, 4, 198)*
99	2352	(0; 2, 3, 12)	4704	(0; 2, 3, 12)*

Genus	Max°	Signature(s)	Max*	Signature(s)
100	1320	(0; 2, 4, 10)	2640	(0; 2, 4, 10)*
101	2400	(0; 3, 3, 4)	3200	(0; 2, 4, 8)*
102	1010	(0; 2, 5, 10)	1680	(0; 2, 4, 105)*
103	4896	(0; 2, 3, 8)	9792	(0; 2, 3, 8)*
104	1236	(0; 2, 6, 6)	1680	(0; 2, 4, 210)*
105	2496	(0; 2, 3, 12)	4368	(0; 2, 3, 14)*
106	2016	(0; 2, 3, 16)	4032	(0; 2, 3, 16)*
107	1696	(0; 2, 4, 8)	1728	(0; 2, 4, 216)*
108	888	(0; 2, 4, 111)	1776	(0; 2, 4, 111)*
109	5184	(0; 2, 3, 8)	5184	(0; 2, 3, 8), (0; 2, 3, 12)*, (0; 2, 4, 6)*
110	1308	(0; 2, 6, 6)	1776	(0; 2, 4, 222)*
111	2640	(0; 2, 4, 6)	5280	(0; 2, 4, 6)*
112	2664	(0; 2, 3, 12)	2664	(0; 2, 3, 12)
113	2688	(0; 2, 3, 12), (0; 2, 4, 6)	5376	(0; 2, 4, 6)*
114	936	(0; 2, 4, 117)	1872	(0; 2, 4, 117)*
115	3420	(0; 2, 3, 10)	6840	(0; 2, 3, 10)*
116	936	(0; 2, 4, 234)	1872	(0; 2, 4, 234)*
117	1856	(0; 2, 4, 8)	1920	(0; 2, 4, 120)*
118	9828	(0; 2, 3, 7)	19656	(0; 2, 3, 7)*
119	960	(0; 2, 4, 240)	1920	(0; 2, 4, 240)*
120	1734	(0; 2, 3, 34)	3468	(0; 2, 3, 34)*
121	2880	(0; 2, 4, 6)	5760	(0; 2, 4, 6)*
122	5808	(0; 2, 3, 8)	5808	(0; 2, 3, 8), (0; 3, 3, 4)*
123	2928	(0; 2, 3, 12)	2928	(0; 2, 3, 12)
124	1080	(0; 2, 4, 45)	2160	(0; 2, 4, 45)*
125	1488	(0; 2, 6, 6)	2240	(0; 2, 4, 35)*
126	1500	(0; 2, 2, 2, 3)	3000	(0; 2, 2, 2, 3)*
127	3024	(0; 2, 3, 12)	4704	(0; 2, 4, 7)*
128	1524	(0; 2, 6, 6)	2064	(0; 2, 4, 258)*
129	10752	(0; 2, 3, 7)	21504	(0; 2, 3, 7)*
130	3096	(0; 2, 3, 12)	3096	(0; 2, 3, 12)
131	2080	(0; 2, 4, 8)	2400	(0; 2, 4, 30)*, (0; 2, 5, 12)*
132	1310	(0; 2, 5, 10)	2160	(0; 2, 4, 135)*
133	3960	(0; 2, 3, 10)	7920	(0; 2, 3, 10)*
134	1596	(0; 2, 6, 6)	2352	(0; 2, 4, 42)*
135	3216	(0; 2, 3, 12)	3216	(0; 2, 3, 12)
136	6480	(0; 2, 3, 8)	12960	(0; 2, 3, 8)*
137	2176	(0; 2, 4, 8)	2304	(0; 2, 4, 72)*
138	1128	(0; 2, 4, 141)	2256	(0; 2, 4, 141)*
139	1152	(0; 2, 4, 96)	2304	(0; 2, 4, 96)*
140	1668	(0; 2, 6, 6)	2256	(0; 2, 4, 282)*
141	2184	(0; 2, 3, 26)	4368	(0; 2, 3, 26)*
142	1144	(0; 2, 4, 286)	2288	(0; 2, 4, 286)*
143	1152	(0; 2, 4, 288)	2304	(0; 2, 4, 288)*
144	1352	(0; 2, 4, 26)	2704	(0; 2, 4, 26)*
145	6912	(0; 2, 3, 8)	8640	(0; 2, 3, 10)*
146	12180	(0; 2, 3, 7)	24360	(0; 2, 3, 7)*
147	3504	(0; 2, 3, 12)	3504	(0; 2, 3, 12)
148	3528	(0; 2, 3, 12)	7056	(0; 2, 3, 12)*
149	2368	(0; 2, 4, 8)	2496	(0; 2, 4, 78)*
150	1224	(0; 2, 4, 153)	2448	(0; 2, 4, 153)*
151	3600	(0; 2, 3, 12)	7200	(0; 2, 3, 12)*

Genus	Max°	Signature(s)	Max*	Signature(s)
152	1812	(0; 2, 6, 6)	2448	(0; 2, 4, 306)*
153	3648	(0; 2, 3, 12)	4332	(0; 2, 3, 38)*
154	2448	(0; 2, 4, 8)	4896	(0; 2, 4, 8)*
155	1320	(0; 2, 6, 10)	2640	(0; 2, 6, 10)*
156	1550	(0; 2, 5, 10)	2544	(0; 2, 4, 159)*
157	3744	(0; 2, 3, 12)	4032	(0; 2, 3, 84)*
158	1884	(0; 2, 6, 6)	2544	(0; 2, 4, 318)*
159	3792	(0; 2, 3, 12)	3792	(0; 2, 3, 12)
160	1288	(0; 2, 4, 322)	2576	(0; 2, 4, 322)*
161	3840	(0; 2, 4, 6)	7680	(0; 2, 4, 6)*
162	1320	(0; 2, 4, 165)	2640	(0; 2, 4, 165)*
163	3888	(0; 2, 3, 12)	7776	(0; 2, 3, 12)*
164	1956	(0; 2, 6, 6)	2640	(0; 2, 4, 330)*
165	2624	(0; 2, 4, 8)	2880	(0; 2, 4, 45)*
166	6600	(0; 2, 4, 5)	13200	(0; 2, 4, 5)*
167	1344	(0; 2, 4, 336)	2688	(0; 2, 4, 336)*
168	1368	(0; 2, 4, 171)	2736	(0; 2, 4, 171)*
169	8064	(0; 2, 3, 8)	16128	(0; 2, 3, 8)*
170	4056	(0; 2, 4, 6), (0; 3, 3, 4)	4056	(0; 2, 4, 6), (0; 3, 3, 4), (0; 2, 6, 6)*, (0; 3, 4, 4)*, (0; 2, 2, 2, 3)*
171	2720	(0; 2, 4, 8)	4896	(0; 2, 4, 9)*
172	6840	(0; 2, 4, 5)	13680	(0; 2, 4, 5)*
173	2064	(0; 2, 6, 6)	2880	(0; 2, 4, 90)*
174	1416	(0; 2, 4, 177)	2832	(0; 2, 4, 177)*
175	1856	(0; 2, 4, 16)	2880	(0; 2, 4, 120)*
176	3000	(0; 2, 3, 20)	6000	(0; 2, 3, 20)*
177	2640	(0; 2, 5, 6)	5280	(0; 2, 5, 6)*
178	1432	(0; 2, 4, 358)	2864	(0; 2, 4, 358)*
179	2848	(0; 2, 4, 8)	2880	(0; 2, 4, 360)*
180	1464	(0; 2, 4, 183)	2928	(0; 2, 4, 183)*
181	8640	(0; 2, 3, 8)	17280	(0; 2, 3, 8)*
182	2172	(0; 2, 6, 6)	2928	(0; 2, 4, 366)*
183	4368	(0; 2, 3, 12), (0; 2, 4, 6)	8736	(0; 2, 4, 6)*
184	4392	(0; 2, 3, 12)	4392	(0; 2, 3, 12)
185	1600	(0; 2, 4, 50)	3200	(0; 2, 4, 50)*
186	1512	(0; 2, 4, 189)	3024	(0; 2, 4, 189)*
187	4464	(0; 2, 3, 12)	4464	(0; 2, 3, 12)
188	1512	(0; 2, 4, 378)	3024	(0; 2, 4, 378)*
189	1536	(0; 2, 4, 192)	3072	(0; 2, 4, 192)*
190	4536	(0; 2, 3, 12)	6048	(0; 2, 3, 24)*
191	2640	(0; 2, 3, 44)	5280	(0; 2, 3, 44)*
192	1910	(0; 2, 5, 10)	3120	(0; 2, 4, 195)*
193	5760	(0; 2, 3, 10)	11520	(0; 2, 3, 10)*
194	2316	(0; 2, 6, 6)	3120	(0; 2, 4, 390)*
195	4656	(0; 2, 3, 12)	4656	(0; 2, 3, 12)
196	3420	(0; 2, 3, 19)	6840	(0; 2, 3, 19)*
197	4704	(0; 2, 4, 6), (0; 3, 3, 4)	6272	(0; 2, 4, 8)*
198	1608	(0; 2, 4, 201)	3216	(0; 2, 4, 201)*
199	2640	(0; 2, 4, 10)	5280	(0; 2, 4, 10)*
200	2388	(0; 2, 6, 6)	3216	(0; 2, 4, 402)*
201	4800	(0; 2, 3, 12)	9600	(0; 2, 3, 12)*
202	4824	(0; 2, 3, 12)	4824	(0; 2, 3, 12)

Genus	Max°	Signature(s)	Max*	Signature(s)
203	3232	(0; 2, 4, 8)	3264	(0; 2, 4, 408)*
204	1656	(0; 2, 4, 207)	3312	(0; 2, 4, 207)*
205	7344	(0; 2, 3, 9)	14688	(0; 2, 3, 9)*
206	2050	(0; 2, 5, 10)	3312	(0; 2, 4, 414)*
207	4944	(0; 2, 3, 12)	4944	(0; 2, 3, 12)
208	1672	(0; 2, 4, 418)	3344	(0; 2, 4, 418)*
209	4368	(0; 2, 3, 14)	8736	(0; 2, 3, 14)*
210	2904	(0; 2, 3, 44)	5808	(0; 2, 3, 44)*
211	5040	(0; 2, 3, 12), (0; 2, 4, 6)	10080	(0; 2, 3, 12)*
212	2532	(0; 2, 6, 6)	3408	(0; 2, 4, 426)*
213	3392	(0; 2, 4, 8)	3456	(0; 2, 4, 216)*
214	1720	(0; 2, 4, 430)	3440	(0; 2, 4, 430)*
215	1728	(0; 2, 4, 432)	3456	(0; 2, 4, 432)*
216	1800	(0; 2, 4, 90)	3600	(0; 2, 4, 90)*
217	6480	(0; 2, 3, 10)	12960	(0; 2, 3, 10)*
218	2604	(0; 2, 6, 6)	3504	(0; 2, 4, 438)*
219	5232	(0; 2, 3, 12)	5232	(0; 2, 3, 12)
220	5256	(0; 2, 3, 12)	5256	(0; 2, 3, 12)
221	5280	(0; 2, 4, 6)	10560	(0; 2, 4, 6)*
222	1800	(0; 2, 4, 225)	3600	(0; 2, 4, 225)*
223	5328	(0; 2, 3, 12)	5328	(0; 2, 3, 12)
224	2676	(0; 2, 6, 6)	3600	(0; 2, 4, 450)*
225	5376	(0; 2, 3, 12)	7680	(0; 2, 3, 20)*
226	5400	(0; 2, 3, 12), (0; 3, 3, 4)	10800	(0; 2, 3, 12)*
227	3616	(0; 2, 4, 8)	3648	(0; 2, 4, 456)*
228	1848	(0; 2, 4, 231)	3696	(0; 2, 4, 231)*
229	6840	(0; 2, 3, 10)	13680	(0; 2, 3, 10)*
230	2748	(0; 2, 6, 6)	3696	(0; 2, 4, 462)*
231	6072	(0; 2, 3, 11)	12144	(0; 2, 3, 11)*
232	3276	(0; 2, 3, 39)	6552	(0; 2, 3, 39)*
233	3712	(0; 2, 4, 8)	3840	(0; 2, 4, 120)*
234	1896	(0; 2, 4, 237)	3792	(0; 2, 4, 237)*
235	11232	(0; 2, 3, 8)	22464	(0; 2, 3, 8)*
236	1896	(0; 2, 4, 474)	3792	(0; 2, 4, 474)*
237	1920	(0; 2, 4, 240)	3840	(0; 2, 4, 240)*
238	5688	(0; 2, 3, 12)	5688	(0; 2, 3, 12)
239	2176	(0; 2, 4, 32)	3840	(0; 2, 4, 480)*
240	1944	(0; 2, 4, 243)	3888	(0; 2, 4, 243)*
241	9600	(0; 2, 4, 5)	19200	(0; 2, 4, 5)*
242	2892	(0; 2, 6, 6)	3888	(0; 2, 4, 486)*
243	5808	(0; 2, 3, 12)	11616	(0; 2, 3, 12)*
244	11664	(0; 2, 3, 8)	23328	(0; 2, 3, 8)*
245	3904	(0; 2, 4, 8)	4160	(0; 2, 4, 65)*
246	4116	(0; 2, 3, 21)	8232	(0; 2, 3, 21)*
247	2624	(0; 2, 4, 16)	4032	(0; 2, 4, 168)*
248	2964	(0; 2, 6, 6)	3984	(0; 2, 4, 498)*
249	5952	(0; 2, 3, 12)	5952	(0; 2, 3, 12)
250	2008	(0; 2, 4, 502)	4016	(0; 2, 4, 502)*
251	12000	(0; 2, 3, 8)	12000	(0; 2, 3, 8), (0; 2, 4, 6)*
252	2510	(0; 2, 5, 10)	4080	(0; 2, 4, 255)*
253	6048	(0; 2, 3, 12)	11232	(0; 2, 3, 13)*
254	12144	(0; 2, 3, 8)	24288	(0; 2, 3, 8)*

Genus	Max°	Signature(s)	Max*	Signature(s)
255	6096	$(0;2,3,12)$	6096	$(0;2,3,12)$
256	4896	$(0;2,3,16)$	9792	$(0;2,3,16)^*$
257	12288	$(0;2,3,8)$	24576	$(0;2,3,8)^*$
258	2088	$(0;2,4,261)$	4176	$(0;2,4,261)^*$
259	6192	$(0;2,3,12)$	6192	$(0;2,3,12)$
260	3108	$(0;2,6,6)$	4176	$(0;2,4,522)^*$
261	4160	$(0;2,4,8)$	4800	$(0;2,4,30)^*$
262	3132	$(0;2,4,12)$	4208	$(0;2,4,526)^*$
263	2112	$(0;2,4,528)$	4224	$(0;2,4,528)^*$
264	2136	$(0;2,4,267)$	4272	$(0;2,4,267)^*$
265	7920	$(0;2,3,10)$	15840	$(0;2,3,10)^*$
266	2200	$(0;2,4,110)$	4400	$(0;2,4,110)^*$
267	6384	$(0;2,3,12)$	6384	$(0;2,3,12)$
268	2152	$(0;2,4,538)$	4304	$(0;2,4,538)^*$
269	3216	$(0;2,6,6)$	4416	$(0;2,4,138)^*$
270	2184	$(0;2,4,273)$	4368	$(0;2,4,273)^*$
271	5040	$(0;2,4,7)$	10080	$(0;2,4,7)^*$
272	3252	$(0;2,6,6)$	4368	$(0;2,4,546)^*$
273	4896	$(0;2,3,18)$	9792	$(0;2,3,18)^*$
274	6552	$(0;2,3,12)$, $(0;2,4,6)$	13104	$(0;2,3,12)^*$, $(0;2,4,6)^*$
275	4384	$(0;2,4,8)$	4416	$(0;2,4,552)^*$
276	3750	$(0;2,3,50)$	7500	$(0;2,3,50)^*$
277	2304	$(0;2,4,96)$	4608	$(0;2,4,96)^*$
278	3324	$(0;2,6,6)$	4464	$(0;2,4,558)^*$
279	6672	$(0;2,3,12)$	6672	$(0;2,3,12)$
280	6696	$(0;2,3,12)$	6696	$(0;2,3,12)$
281	2688	$(0;2,4,24)$, $(0;2,6,8)$, $(0;3,3,8)$	5376	$(0;2,4,24)^*$, $(0;2,6,8)^*$, $(0;3,3,8)^*$
282	2810	$(0;2,5,10)$	4560	$(0;2,4,285)^*$
283	2304	$(0;2,4,192)$	4608	$(0;2,4,192)^*$
284	3396	$(0;2,6,6)$	4560	$(0;2,4,570)^*$
285	2400	$(0;2,4,75)$	4800	$(0;2,4,75)^*$
286	10260	$(0;2,3,9)$	20520	$(0;2,3,9)^*$
287	4368	$(0;2,3,28)$	8736	$(0;2,3,28)^*$
288	2328	$(0;2,4,291)$	4656	$(0;2,4,291)^*$
289	13824	$(0;2,3,8)$	13824	$(0;2,3,8)$, $(0;2,3,12)^*$, $(0;2,4,6)^*$, $(0;3,3,4)^*$
290	13872	$(0;2,3,8)$	13872	$(0;2,3,8)$, $(0;3,3,4)^*$
291	4640	$(0;2,4,8)$	4800	$(0;2,4,120)^*$
292	6984	$(0;2,3,12)$	6984	$(0;2,3,12)$
293	4672	$(0;2,4,8)$	4800	$(0;2,4,150)^*$
294	2376	$(0;2,4,297)$	4752	$(0;2,4,297)^*$
295	7056	$(0;2,3,12)$	14112	$(0;2,3,12)^*$
296	2376	$(0;2,4,594)$	4752	$(0;2,4,594)^*$
297	7104	$(0;2,3,12)$	7104	$(0;2,3,12)$
298	3888	$(0;2,3,72)$	7776	$(0;2,3,72)^*$
299	4768	$(0;2,4,8)$	4800	$(0;2,4,600)^*$
300	4056	$(0;2,3,52)$	8112	$(0;2,3,52)^*$
301	9000	$(0;2,3,10)$	18000	$(0;2,3,10)^*$

Appendix 2: LARGEST ACTIONS ON NONORIENTABLE SURFACES OF GENUS 3 TO 302

Genus	Max	Signature(s)	Genus	Max	Signature(s)
3	12	$(0;2,2,2,3)$	4	48	$(0;2,4,6)$
5	120	$(0;2,4,5)$	6	160	$(0;2,4,5)$
7	120	$(0;2,4,6)$	8	504	$(0;2,3,7)$
9	336	$(0;2,3,8)$	10	192	$(0;2,4,6)$
11	216	$(0;2,4,6)$	12	240	$(0;2,4,6)$
13	120	$(0;2,4,15)$	14	360	$(0;2,3,10)$
15	1092	$(0;2,3,7)$	16	504	$(0;2,3,9)$
17	720	$(0;2,3,8)$	18	192	$(0;2,2,2,3)$
19	168	$(0;2,4,21)$	20	240	$(0;2,4,10)$
21	132	$(0;2,6,22)$	22	240	$(0;2,6,6)$, $(0;2,2,2,3)$
23	1008	$(0;2,3,8)$	24	176	$(0;2,2,2,4)$
25	216	$(0;2,4,27)$	26	720	$(0;2,3,10)$
27	300	$(0;2,2,2,3)$	28	240	$(0;2,4,30)$
29	648	$(0;2,3,12)$	30	672	$(0;2,4,6)$
31	264	$(0;2,4,33)$	32	720	$(0;2,4,6)$
33	204	$(0;2,6,34)$	34	384	$(0;2,4,12)$, $(0;2,2,2,3)$
35	1320	$(0;2,4,5)$	36	272	$(0;2,2,2,4)$
37	392	$(0;2,4,14)$	38	480	$(0;2,4,10)$
39	222	$(0;+;[2,3];\{(1)\})$	40	336	$(0;2,4,42)$
41	360	$(0;3,4,5)$	42	600	$(0;2,5,6)$
43	360	$(0;2,4,45)$	44	672	$(0;2,4,8)$
45	276	$(0;2,6,46)$	46	1320	$(0;2,3,10)$
47	2160	$(0;2,3,8)$	48	368	$(0;2,2,2,4)$
49	408	$(0;2,4,51)$	50	1920	$(0;2,4,5)$
51	1092	$(0;2,3,13)$	52	480	$(0;2,6,8)$
53	408	$(0;+;[2];\{(2,4)\})$	54	416	$(0;2,2,2,4)$
55	456	$(0;2,4,57)$	56	1296	$(0;2,4,6)$
57	1320	$(0;2,3,12)$, $(0;2,4,6)$	58	1008	$(0;2,3,18)$
59	456	$(0;+;[2];\{(2,4)\})$	60	464	$(0;2,2,2,4)$
61	504	$(0;2,4,63)$	62	1440	$(0;2,4,6)$
63	366	$(0;+;[2,3];\{(1)\})$	64	528	$(0;2,4,66)$
65	720	$(0;2,5,8)$	66	960	$(0;2,5,6)$
67	552	$(0;2,4,69)$	68	528	$(0;2,2,2,4)$, $(0;+;[2];\{(2,4)\})$
69	420	$(0;2,6,70)$	70	2448	$(0;2,3,9)$
71	552	$(0;+;[2];\{(2,4)\})$	72	672	$(0;2,6,8)$
73	600	$(0;2,4,75)$	74	2160	$(0;2,3,10)$
75	438	$(0;+;[2,3];\{(1)\})$	76	624	$(0;2,4,78)$
77	1000	$(0;2,4,10)$	78	608	$(0;2,2,2,4)$
79	648	$(0;2,4,81)$	80	624	$(0;2,2,2,4)$, $(0;+;[2];\{(2,4)\})$
81	492	$(0;2,6,82)$	82	1152	$(0;2,4,9)$
83	1944	$(0;2,4,6)$	84	656	$(0;2,2,2,4)$
85	696	$(0;2,4,87)$	86	4032	$(0;2,3,8)$
87	348	$(0;2,2,2,87)$	88	720	$(0;2,4,90)$
89	696	$(0;+;[2];\{(2,4)\})$	90	1320	$(0;2,5,6)$
91	744	$(0;2,4,93)$	92	1440	$(0;2,4,8)$
93	2184	$(0;2,3,12)$, $(0;2,4,6)$	94	768	$(0;2,4,96)$
95	744	$(0;+;[2];\{(2,4)\})$	96	752	$(0;2,2,2,4)$
97	3420	$(0;2,3,9)$	98	1920	$(0;2,5,5)$
99	582	$(0;+;[2,3];\{(1)\})$	100	816	$(0;2,4,102)$
101	1320	$(0;2,4,10)$	102	1200	$(0;2,2,2,3)$

Genus	Max	Signature(s)	Genus	Max	Signature(s)
103	840	$(0;2,4,105)$	104	816	$(0;2,2,2,4)$, $(0;+;[2];\{(2,4)\})$
105	636	$(0;2,6,106)$	106	2184	$(0;2,3,14)$
107	1320	$(0;2,4,11)$	108	848	$(0;2,2,2,4)$
109	888	$(0;2,4,111)$	110	1440	$(0;2,4,10)$
111	654	$(0;+;[2,3];\{(1)\})$	112	2640	$(0;2,4,6)$
113	888	$(0;+;[2];\{(2,4)\})$	114	1344	$(0;2,4,12)$, $(0;3,4,4)$, $(0;2,2,2,3)$
115	936	$(0;2,4,117)$	116	3420	$(0;2,3,10)$
117	708	$(0;2,6,118)$	118	960	$(0;2,4,120)$
119	2184	$(0;2,4,7)$	120	944	$(0;2,2,2,4)$
121	984	$(0;2,4,123)$	122	2880	$(0;2,4,6)$
123	1452	$(0;2,2,2,3)$	124	1008	$(0;2,4,126)$
125	984	$(0;+;[2];\{(2,4)\})$	126	1120	$(0;2,4,35)$
127	1500	$(0;2,2,2,3)$	128	1296	$(0;2,4,18)$
129	780	$(0;2,6,130)$	130	1920	$(0;2,5,6)$
131	1032	$(0;+;[2];\{(2,4)\})$	132	1200	$(0;2,4,30)$
133	1080	$(0;2,4,135)$	134	3960	$(0;2,3,10)$
135	798	$(0;+;[2,3];\{(1)\})$	136	1104	$(0;2,4,138)$
137	2160	$(0;2,4,8)$	138	1088	$(0;2,2,2,4)$
139	1128	$(0;2,4,141)$	140	1104	$(0;2,2,2,4)$, $(0;+;[2];\{(2,4)\})$
141	852	$(0;2,6,142)$	142	2184	$(0;2,3,26)$
143	1128	$(0;+;[2];\{(2,4)\})$	144	1136	$(0;2,2,2,4)$
145	1352	$(0;2,4,26)$	146	3456	$(0;2,4,6)$
147	12180	$(0;2,3,7)$	148	1200	$(0;2,4,150)$
149	1764	$(0;2,2,2,3)$	150	1184	$(0;2,2,2,4)$
151	1224	$(0;2,4,153)$	152	1440	$(0;2,6,8)$
153	924	$(0;2,6,154)$	154	1344	$(0;2,4,42)$
155	1224	$(0;+;[2];\{(2,4)\})$	156	1320	$(0;2,6,10)$
157	1272	$(0;2,4,159)$	158	1248	$(0;2,2,2,4)$, $(0;+;[2];\{(2,4)\})$
159	942	$(0;+;[2,3];\{(1)\})$	160	1296	$(0;2,4,162)$
161	1272	$(0;+;[2];\{(2,4)\})$	162	3840	$(0;2,4,6)$
163	1320	$(0;2,4,165)$	164	2160	$(0;2,4,10)$
165	996	$(0;2,6,166)$	166	1440	$(0;2,4,45)$
167	6600	$(0;2,4,5)$	168	1328	$(0;2,2,2,4)$
169	1368	$(0;2,4,171)$	170	5040	$(0;2,3,10)$
171	2028	$(0;2,2,2,3)$	172	2448	$(0;2,4,9)$
173	6840	$(0;2,4,5)$	174	1376	$(0;2,2,2,4)$
175	1416	$(0;2,4,177)$	176	1392	$(0;2,2,2,4)$, $(0;+;[2];\{(2,4)\})$
177	3000	$(0;2,3,20)$	178	1920	$(0;2,4,15)$
179	1416	$(0;+;[2];\{(2,4)\})$	180	1424	$(0;2,2,2,4)$
181	1464	$(0;2,4,183)$	182	4320	$(0;2,4,6)$
183	1086	$(0;+;[2,3];\{(1)\})$	184	4368	$(0;2,4,6)$
185	1464	$(0;+;[2];\{(2,4)\})$	186	1600	$(0;2,4,50)$
187	1512	$(0;2,4,189)$	188	1488	$(0;2,2,2,4)$, $(0;+;[2];\{(2,4)\})$
189	1140	$(0;2,6,190)$	190	1536	$(0;2,4,192)$
191	2184	$(0;2,4,13)$	192	1680	$(0;2,4,42)$
193	1560	$(0;2,4,195)$	194	2304	$(0;2,6,6)$, $(0;2,2,2,3)$
195	1158	$(0;+;[2,3];\{(1)\})$	196	1584	$(0;2,4,198)$
197	2184	$(0;2,4,14)$	198	2352	$(0;2,2,2,3)$
199	1608	$(0;2,4,201)$	200	2640	$(0;2,4,10)$
201	1212	$(0;2,6,202)$	202	2400	$(0;2,2,2,3)$
203	1608	$(0;+;[2];\{(2,4)\})$	204	1616	$(0;2,2,2,4)$
205	1656	$(0;2,4,207)$	206	4896	$(0;2,4,6)$

Genus	Max	Signature(s)	Genus	Max	Signature(s)
207	828	$(0;2,2,2,207)$	208	1680	$(0;2,4,210)$
209	1656	$(0;+;[2];\{(2,4)\})$	210	2184	$(0;2,6,7)$
211	1704	$(0;2,4,213)$	212	5040	$(0;2,3,12)$
213	1284	$(0;2,6,214)$	214	1728	$(0;2,4,216)$
215	1704	$(0;+;[2];\{(2,4)\})$	216	1712	$(0;2,2,2,4)$
217	1752	$(0;2,4,219)$	218	5184	$(0;2,4,6)$
219	1302	$(0;+;[2,3];\{(1)\})$	220	1776	$(0;2,4,222)$
221	1752	$(0;+;[2];\{(2,4)\})$	222	5280	$(0;2,4,6)$
223	1800	$(0;2,4,225)$	224	1776	$(0;2,2,2,4)$, $(0;+;[2];\{(2,4)\})$
225	1356	$(0;2,6,226)$	226	2688	$(0;3,4,4)$
227	2700	$(0;2,2,2,3)$	228	1808	$(0;2,2,2,4)$
229	1848	$(0;2,4,231)$	230	6840	$(0;2,3,10)$
231	1374	$(0;+;[2,3];\{(1)\})$	232	6072	$(0;2,3,11)$
233	1848	$(0;+;[2];\{(2,4)\})$	234	1856	$(0;2,2,2,4)$
235	1896	$(0;2,4,237)$	236	11232	$(0;2,3,8)$
237	1428	$(0;2,6,238)$	238	1920	$(0;2,4,240)$
239	1896	$(0;+;[2];\{(2,4)\})$	240	1904	$(0;2,2,2,4)$
241	1944	$(0;2,4,243)$	242	2880	$(0;2,2,2,3)$, $(0;+;[2];\{(2,3)\})$
243	1446	$(0;+;[2,3];\{(1)\})$	244	1968	$(0;2,4,246)$
245	9720	$(0;2,4,5)$	246	2080	$(0;2,4,65)$
247	2744	$(0;2,4,14)$	248	1968	$(0;2,2,2,4)$, $(0;+;[2];\{(2,4)\})$
249	1500	$(0;2,6,250)$	250	2112	$(0;2,4,66)$
251	1992	$(0;+;[2];\{(2,4)\})$	252	6000	$(0;2,4,6)$
253	2040	$(0;2,4,255)$	254	3024	$(0;2,2,2,3)$, $(0;+;[2];\{(2,3)\})$
255	12144	$(0;2,3,8)$	256	2064	$(0;2,4,258)$
257	4896	$(0;2,3,16)$	258	3072	$(0;2,4,12)$, $(0;2,2,2,3)$
259	2088	$(0;2,4,261)$	260	2064	$(0;2,2,2,4)$, $(0;+;[2];\{(2,4)\})$
261	1680	$(0;2,8,15)$	262	2112	$(0;2,4,264)$
263	2088	$(0;+;[2];\{(2,4)\})$	264	2096	$(0;2,2,2,4)$
265	2136	$(0;2,4,267)$	266	2240	$(0;2,4,70)$
267	1100	$(0;2,2,2,55)$	268	2160	$(0;2,4,270)$
269	2136	$(0;+;[2];\{(2,4)\})$	270	2144	$(0;2,2,2,4)$
271	2184	$(0;2,4,273)$	272	5040	$(0;2,4,7)$
273	1644	$(0;2,6,274)$	274	4896	$(0;2,3,18)$
275	6552	$(0;2,3,12)$, $(0;2,4,6)$	276	2192	$(0;2,2,2,4)$
277	3000	$(0;2,4,15)$	278	2208	$(0;2,2,2,4)$, $(0;+;[2];\{(2,4)\})$
279	1662	$(0;+;[2,3];\{(1)\})$	280	2256	$(0;2,4,282)$
281	2232	$(0;+;[2];\{(2,4)\})$	282	2688	$(0;2,4,24)$, $(0;2,6,8)$
283	2280	$(0;2,4,285)$	284	2256	$(0;2,2,2,4)$, $(0;+;[2];\{(2,4)\})$
285	1716	$(0;2,6,286)$	286	2400	$(0;2,4,75)$
287	6840	$(0;2,4,6)$	288	2288	$(0;2,2,2,4)$
289	2328	$(0;2,4,291)$	290	3840	$(0;2,4,10)$
291	3468	$(0;2,2,2,3)$	292	2352	$(0;2,4,294)$
293	2328	$(0;+;[2];\{(2,4)\})$	294	2336	$(0;2,2,2,4)$
295	2376	$(0;2,4,297)$	296	2352	$(0;2,2,2,4)$, $(0;+;[2];\{(2,4)\})$
297	1788	$(0;2,6,298)$	298	2496	$(0;2,4,78)$
299	2376	$(0;+;[2];\{(2,4)\})$	300	2384	$(0;2,2,2,4)$
301	2424	$(0;2,4,303)$	302	3600	$(0;2,2,2,3)$

References

[1] Robert D. M. Accola, *On the number of automorphisms of a closed Riemann surface*, Trans. Amer. Math. Soc. **131** (1968), 398–408. MR0222281 (36 #5333)

[2] Wieb Bosma, John Cannon, and Catherine Playoust, *The Magma algebra system. I. The user language*, J. Symbolic Comput. **24** (1997), no. 3-4, 235–265, DOI 10.1006/jsco.1996.0125. Computational algebra and number theory (London, 1993). MR1484478

[3] Thomas Breuer, *Characters and automorphism groups of compact Riemann surfaces*, London Mathematical Society Lecture Note Series, vol. 280, Cambridge University Press, Cambridge, 2000. MR1796706 (2002i:14034)

[4] Emilio Bujalance, *Cyclic groups of automorphisms of compact nonorientable Klein surfaces without boundary*, Pacific J. Math. **109** (1983), no. 2, 279–289. MR721920 (85f:30081)

[5] Marston D. E. Conder, *Some results on quotients of triangle groups*, Bull. Austral. Math. Soc. **30** (1984), no. 1, 73–90, DOI 10.1017/S0004972700001738. MR753563 (85j:20029)

[6] Marston Conder, *The genus of compact Riemann surfaces with maximal automorphism group*, J. Algebra **108** (1987), no. 1, 204–247, DOI 10.1016/0021-8693(87)90135-9. MR887205 (88f:20063)

[7] Marston Conder, *Groups of minimal genus including C_2 extensions of* PSL(2, q) *for certain q*, Quart. J. Math. Oxford Ser. (2) **38** (1987), no. 152, 449–460, DOI 10.1093/qmath/38.4.449. MR916227 (89a:20043)

[8] Marston Conder, *Hurwitz groups: a brief survey*, Bull. Amer. Math. Soc. (N.S.) **23** (1990), no. 2, 359–370, DOI 10.1090/S0273-0979-1990-15933-6. MR1041434 (91d:20032)

[9] Marston Conder, *An update on Hurwitz groups*, Groups Complex. Cryptol. **2** (2010), no. 1, 35–49, DOI 10.1515/GCC.2010.002. MR2672551 (2011f:30081)

[10] M.D.E. Conder, *The strong symmetric genus of all finite groups of order 2 to 127*, listed at www.math.auckland.ac.nz/~conder/StrongSymmGenusSmallGroups127.txt.

[11] M.D.E. Conder, *The finite groups with strong symmetric genus 2 to 32*, listed at www.math.auckland.ac.nz/~conder/GroupsWithStrongSymmGenus2to32.txt.

[12] M.D.E. Conder, *The symmetric genus of all finite groups of order 2 to 127*, listed at www.math.auckland.ac.nz/~conder/SymmGenusSmallGroups127.txt.

[13] M.D.E. Conder, *The finite groups with symmetric genus 2 to 32*, listed at www.math.auckland.ac.nz/~conder/GroupsWithSymmGenus2to32.txt.

[14] M.D.E. Conder, *The symmetric cross-cap number of all finite groups of order 2 to 127*, listed at www.math.auckland.ac.nz/~conder/CrosscapNumberSmallGroups127.txt.

[15] M.D.E. Conder, *The finite groups with symmetric cross-cap number 3 to 65*, listed at www.math.auckland.ac.nz/~conder/GroupsWithCrosscapNumber3to65.txt.

[16] Marston Conder, Colin Maclachlan, Sanja Todorovic Vasiljevic, and Steve Wilson, *Bounds for the number of automorphisms of a compact non-orientable surface*, J. London Math. Soc. (2) **68** (2003), no. 1, 65–82, DOI 10.1112/S0024610703004277. MR1980244 (2004b:57025)

[17] Marston D. E. Conder and Thomas W. Tucker, *The symmetric genus spectrum of finite groups*, Ars Math. Contemp. **4** (2011), no. 2, 271–289. MR2825649

[18] J. J. Etayo and E. Martínez, *The symmetric crosscap number of the groups of small-order*, J. Algebra Appl. **12** (2013), no. 2, 1250164, 16, DOI 10.1142/S0219498812501642. MR3005612

[19] D. Firth, *An algorithm to find normal subgroups of a finitely presented group up to a given index*, PhD Thesis, University of Warwick, 2005.

[20] W. Hall: *Automorphisms and coverings of Klein surfaces*, Ph.D. Thesis, University of Southampton, UK, 1978.

[21] A. M. Macbeath, *The classification of non-euclidean plane crystallographic groups*, Canad. J. Math. **19** (1967), 1192–1205. MR0220838 (36 #3890)

[22] C. Maclachlan, *A bound for the number of automorphisms of a compact Riemann surface.*, J. London Math. Soc. **44** (1969), 265–272. MR0236378 (38 #4674)

[23] H. Maschke, *The Representation of Finite Groups, Especially of the Rotation Groups of the Regular Bodies of Three-and Four-Dimensional Space, by Cayley's Color Diagrams*, Amer. J. Math. **18** (1896), no. 2, 156–194, DOI 10.2307/2369680. MR1505708

[24] Coy L. May, *The symmetric crosscap number of a group*, Glasg. Math. J. **43** (2001), no. 3, 399–410, DOI 10.1017/S0017089501030038. MR1878584 (2002k:57047)

[25] Coy L. May and Jay Zimmerman, *There is a group of every strong symmetric genus*, Bull. London Math. Soc. **35** (2003), no. 4, 433–439, DOI 10.1112/S0024609303001954. MR1978995 (2004b:57026)

[26] D. Singerman, *Automorphisms of compact non-orientable Riemann surfaces*, Glasgow Math. J. **12** (1971), 50–59. MR0296286 (45 #5347)

[27] Thomas W. Tucker, *Finite groups acting on surfaces and the genus of a group*, J. Combin. Theory Ser. B **34** (1983), no. 1, 82–98, DOI 10.1016/0095-8956(83)90009-6. MR701174 (85b:20055)

[28] Thomas W. Tucker, *Symmetric embeddings of Cayley graphs in nonorientable surfaces*, Graph theory, combinatorics, and applications. Vol. 2 (Kalamazoo, MI, 1988), Wiley-Intersci. Publ., Wiley, New York, 1991, pp. 1105–1120. MR1170849 (93b:05053)

[29] A. Wiman, *Über die hyperelliptischen Curven und diejenigen vom Geschlechte $p = 3$, welche eindeutigen Transformationen in sich zulassen*, Bihang Kongl. Svenska Vetenskaps-Akademiens Handlingar (Stockholm 1895-96).

Department of Mathematics, University of Auckland, Private Bag 92019, Auckland 1142, New Zealand

E-mail address: m.conder@auckland.ac.nz

Maximal bordered surface groups versus maximal handlebody groups

Marston D. E. Conder and Bruno P. Zimmermann

Dedicated to Emilio Bujalance on the occasion of his 60th birthday

ABSTRACT. The order of a finite group of diffeomorphisms of a compact bordered surface of algebraic genus $g > 1$ is at most $12(g-1)$, and this bound is sharp for infinitely many values of g. The same upper bound holds for the maximum possible order of a finite group of orientation-preserving diffeomorphisms of a handlebody of genus $g > 1$. Every finite group which acts on a compact bordered surface acts also on a handlebody of the same genus. In this paper, we show that there are infinitely many values of g for which the upper bound is attained for handlebodies but not for bordered surfaces. Quite remarkably, for $2 \leq g \leq 2000$ the only such value is $g = 161$.

1. Introduction

By a classical result of Hurwitz (1893), the order of a finite group of orientation-preserving diffeomorphisms (or homeomorphisms) of a closed orientable surface of genus $g > 1$ is bounded above by $84(g-1)$, and the finite groups of this maximal order are called *Hurwitz groups*; these are exactly the finite quotients of the $(2,3,7)$ triangle group $\langle\, x, y \mid x^2 = y^3 = (xy)^7 = 1 \,\rangle$. There is a long history and literature on the construction and classification of Hurwitz groups (see [**3**–**5**]), but their complete determination, or even just the determination of the corresponding genera, remains an open problem.

Analogously, the order of a finite group of diffeomorphisms (which can preserve or reverse orientation) of a compact bordered surface of algebraic genus $g > 1$ is bounded above by $12(g-1)$; see [**11**]. (The algebraic genus of such a bordered surface is defined as the rank of its free fundamental group.) The finite groups achieving this maximum possible order are often called M^*-groups, and we will call them *maximal bordered surface groups* in what follows. Again, a classification remains open. Maximal bordered surface groups are precisely the finite smooth quotients of the extended modular group

$$\mathrm{PGL}_2(\mathbb{Z}) \cong \mathbb{D}_2 *_{\mathbb{Z}_2} \mathbb{D}_3$$

2010 *Mathematics Subject Classification.* Primary 57M60; Secondary 20H10, 30F40, 57S25.
Key words and phrases. Handlebody, bordered surface, automorphism group.
The first author was supported by a James Cook Fellowship from the Royal Society of New Zealand, and the second author by an FRA grant at the Università degli Studi di Trieste.

(the free product of two dihedral groups of orders 4 and 6, amalgamated via a common cyclic subgroup of order 2), with 'smooth' meaning that the kernel is torsion-free, so the orders of the elements of $\mathbb{D}_2$ and $\mathbb{D}_3$ are preserved in the quotient. The extended modular group acts by isometries on the hyperbolic plane, and the torsion-free normal subgroups of finite index uniformise punctured surfaces, which may be compactified to compact bordered surfaces, maintaining the induced finite group actions.

The product of a bordered surface with an interval (the twisted product if the surface is non-orientable) is an orientable handlebody of genus g, and a finite group action on the surface induces an orientation-preserving action of the group on the handlebody. Again, the order of a finite group of orientation-preserving diffeomorphisms of a handlebody of genus $g > 1$ is bounded above by $12(g-1)$ (see [**14**] and [**12**, Theorem 7.2]), and the finite groups occurring are the smooth finite quotients of any one (or more) of the following four products with amalgamation:

$$G_2 := \mathbb{D}_2 *_{\mathbb{Z}_2} \mathbb{S}_3, \quad G_3 := \mathbb{D}_3 *_{\mathbb{Z}_3} \mathbb{A}_4, \quad G_4 := \mathbb{D}_4 *_{\mathbb{Z}_4} \mathbb{S}_4, \quad G_5 := \mathbb{D}_5 *_{\mathbb{Z}_5} \mathbb{A}_5,$$

where $\mathbb{D}_k$ is the dihedral group of order $2k$, and $\mathbb{A}_k$ and $\mathbb{S}_k$ are the alternating and symmetric groups of degree k; see [**12, 15**].

In what follows, for $2 \le i \le 5$ any finite smooth quotient of G_i will be called a *maximal handlebody group of type* G_i. In particular, the maximal bordered surface groups are precisely the maximal handlebody groups of type G_2. Each of these groups G_i acts by orientation-preserving isometries on hyperbolic 3-space, and the torsion-free normal subgroups of finite index in G_i are free Schottky groups uniformising 3-dimensional handlebodies with an induced finite group action.

It is easy to find examples of maximal handlebody groups which are not maximal bordered surface groups (see [**16**]); an interesting example is the Mathieu group M_{12}, of order 95040. Hence, naturally, one would expect that there exist genera for which there exists a maximal handlebody action but no maximal bordered surface action, but this has been an open question for some time – see [**13**, section 3]. (Note that there are other groups of order 95040 which are maximal bordered surface groups, namely extensions of solvable groups by PGL(2, 11).)

We now have a solution. Quite surprisingly, the following hold:

PROPOSITION 1.1. *The only integer g in the range $2 \le g \le 2000$ for which there exists a maximal handlebody group but no maximal bordered surface group of genus g is* 161. *Furthermore, there is just one maximal handlebody group of genus* 161 *(and this group occurs for all three types G_3, G_4 and G_5).*

COROLLARY 1.2. *There are infinitely many g for which there exists a maximal handlebody group but no maximal bordered surface group of genus g.*

Our starting point is a recent paper by Bujalance, Cirre and Conder [**2**] which provides a classification (obtained by computational methods) of all maximal bordered surface groups of order up to 1200, or equivalently, of genus 2 to 101.

In the next section, we generalise this to find for each type G_i, all values of g in the range $2 \le g \le 300$ for which there exists a handlebody of genus g admitting a maximal handlebody group of type G_i. The only value which occurs for G_3, G_4 or G_5 but not G_2 is $g = 161$, and this genus occurs for all three of G_3, G_4 and G_5.

2. The genera of maximal handlebody groups

For small values of g, all maximal handlebody groups of genus g may be found by determining all smooth quotients of the four groups G_i, of order $12(g-1)$. This can be achieved with the help of an algorithm developed by Derek Holt and his PhD student David Firth [**9**] for finding all normal subgroups of up to a given index in a given finitely-presented group, and available as the `LowIndexNormalSubgroups` command in MAGMA [**1**]. In contrast to the standard algorithm for finding all subgroups of up to a given index, which uses coset table methods, this algorithm systematically finds all possibilities for a composition series of the quotient. It has been used in many contexts, including the determination of all regular maps of Euler characteristic -1 to -600 (see [**6**] and the website given there).

We can take the following as presentations for the groups G_2 to G_5:

$G_2 = \langle\, a, b, c \mid a^2 = b^2 = c^2 = (ab)^2 = (bc)^3 = 1 \,\rangle$, with $\langle a, b\rangle \cong \mathbb{D}_2$ and $\langle b, c\rangle \cong \mathbb{S}_3$,

$G_3 = \langle\, a, b, c \mid a^2 = b^3 = c^2 = (ab)^2 = (bc)^3 = 1 \,\rangle$, with $\langle a, b\rangle \cong \mathbb{D}_3$ and $\langle b, c\rangle \cong \mathbb{A}_4$,

$G_4 = \langle\, a, b, c \mid a^2 = b^4 = c^2 = (ab)^2 = (bc)^3 = 1 \,\rangle$, with $\langle a, b\rangle \cong \mathbb{D}_4$ and $\langle b, c\rangle \cong \mathbb{S}_4$,

$G_5 = \langle\, a, b, c \mid a^2 = b^5 = c^2 = (ab)^2 = (bc)^3 = 1 \,\rangle$, with $\langle a, b\rangle \cong \mathbb{D}_5$ and $\langle b, c\rangle \cong \mathbb{A}_5$.

Then for genus 2 to 300 we have the following, as a result of a 10-minute computation in MAGMA [**1**] on an Apple laptop:

PROPOSITION 2.1. *In each case below, the integers listed are the values of g with $2 \le g \le 300$ for which there exists a maximal handlebody group of type G_i:*

• *Type G_2 (61 genera)*: 2, 3, 4, 5, 6, 9, 10, 11, 13, 17, 21, 25, 26, 28, 29, 31, 33, 37, 41, 43, 49, 50, 55, 57, 61, 65, 73, 76, 82, 85, 92, 97, 101, 109, 111, 113, 121, 122, 126, 129, 145, 148, 169, 170, 181, 183, 193, 197, 201, 205, 217, 221, 226, 241, 244, 251, 253, 257, 286, 289, 290.

• *Type G_3 (23 genera)*: 3, 6, 9, 11, 17, 29, 31, 33, 55, 65, 85, 92, 111, 121, 129, 161, 183, 205, 217, 241, 251, 257, 286.

• *Type G_4 (33 genera)*: 3, 5, 9, 13, 17, 29, 33, 37, 49, 55, 57, 61, 65, 97, 101, 109, 111, 121, 129, 145, 161, 169, 181, 183, 193, 197, 205, 217, 221, 241, 251, 257, 289.

• *Type G_5 (7 genera)*: 6, 11, 61, 111, 161, 181, 286.

In particular, the only value for g which does not occur for type G_2 but does occur for some of the other types (in fact for all of them) is 161.

The MAGMA computation (using the `LowIndexNormalSubgroups` algorithm) for genus 2 to 2000 which resulted in the first part of Proposition 1.1 took eight days and used a little under 10Gb memory, on a 3.0GHz computer. The next smallest value of g for which there exists a maximal handlebody group but no maximal bordered surface group of genus g is 3781; this required a significantly bigger computation: six weeks of time and 60.5Gb memory, on a bigger machine.

For genus 161, each of the groups G_3, G_4 and G_5 has just one quotient of order 1920 ($= 12 \cdot 160$), and the quotient is the same group H in all three cases. (This is amazing, since there are 241004 different groups of order 1920.) The group H itself is a semi-direct product $(\mathbb{Z}_2)^5 \rtimes \mathbb{A}_5$, but is not the wreath product $\mathbb{Z}_2 \wr \mathbb{A}_5$, since the centre of H is trivial while the centre of $\mathbb{Z}_2 \wr \mathbb{A}_5$ has order 2. The group

H is isomorphic to the quotient of the wreath product $\mathbb{Z}_2 \wr \mathrm{PSL}(2,5) \cong (\mathbb{Z}_2)^6 \rtimes \mathbb{A}_5$ by its centre of order 2.

For genus 3781, again there is just one group, namely a non-split extension of $(\mathbb{Z}_3)^2$ by $\mathbb{S}_7$, of order 45360. This is a quotient of G_3 and G_4 (but not G_2 or G_5).

We will not give a computer-free proof of Proposition 1.1 or Proposition 2.1, but we can give an explanation of some of what happens for genus 161.

PROOF OF THE SECOND PART OF PROPOSITION 1.1. First, suppose H is any group of order 1920 ($= 2^7 \cdot 3 \cdot 5$). Then since the only non-abelian simple group whose order divides 1920 is $\mathbb{A}_5$, we find that either H is soluble, or its composition factors are $\mathbb{A}_5$ and fives copies of $\mathbb{Z}_2$. Also a Sylow 2-subgroup Q of H has index 15 in H. This subgroup Q cannot be maximal, for otherwise the natural permutation representation of H on cosets of Q would give a primitive permutation group of degree 15, but the order of each of the six primitive groups of degree 15 is divisible by 9 (see [**8**]). Hence H must have a subgroup S containing Q, with $|H:S| = 3$ and $|S:Q| = 5$, or $|H:S| = 5$ and $|S:Q| = 3$.

If $|H:S| = 3$ then the core of S in H is a normal subgroup of index 3 or 6, and so H is soluble. Moreover, the permutation group induced by S on the 5 cosets of Q must be isomorphic to a subgroup of $\mathbb{S}_5$ of order divisible by 5 but not 3, and hence must be isomorphic to $\mathbb{Z}_5$, $\mathbb{D}_5$ or $\mathbb{Z}_5 \rtimes \mathbb{Z}_4$, of order 5, 10 or 20 respectively.

Now suppose that H is a smooth quotient of G_2 (and $|H:S| = 3$). Then the pre-images of S and Q in G_2 are subgroups of index 3 and 15, and the natural actions of G_2 on the cosets of those two subgroups give the same permutation representations as H on cosets of S and Q. It is not difficult to see that the group G_2 has two conjugacy classes of subgroups of index 3, with representatives $\langle a, b, cac\rangle$ and $\langle a, b, cabc\rangle$, and both are isomorphic to the free product $\mathbb{D}_2 * \mathbb{Z}_2$. The latter group has precisely three distinct transitive permutation representations of degree 5 with image of order divisible by 5 but not 3. In each case, one of the three elements of the $\mathbb{D}_2$ factor is taken to the identity, and the resulting permutation group is $\mathbb{D}_5$. In all three cases, however, the permutation group induced by G_2 on the cosets of the point-stabiliser is a group of order 300, 300 or 3000 (respectively); in particular, its order does not divide 1920, which gives a contradiction.

The same argument holds for the group G_3, except in this case there is just one conugacy class of subgroups of index 3 (with representative $\langle a, c, bcb^{-1}\rangle$, which is again isomorphic to $\mathbb{D}_2 * \mathbb{Z}_2$). Similarly, the group G_4 has two classes of subgroups of index 3, with representatives $\langle a, b, cac, cb^2c\rangle$ and $\langle a, b, cabc, cab^{-1}c\rangle$, each isomorphic to $\mathbb{D}_4 *_{\mathbb{Z}_2} \mathbb{D}_4 *_{\mathbb{Z}_2} \mathbb{D}_2$. Again, the latter group has precisely three distinct transitive permutation representations of degree 5 with image of order divisible by 5 but not 3, and in each case, the resulting permutation group is $\mathbb{D}_5$, but then the permutation group induced by G_4 has order 300 or 3000. On the other hand, the group G_5 has no subgroups of index 3 at all.

Hence we can rule out the possibility that $|H:S| = 3$ completely, and deduce that $|H:S| = 5$.

It is easy to show that each of the groups G_2, G_3, G_4 and G_5 has at least one subgroup of index 5. Indeed both G_2 and G_4 have exactly one conjugacy class of such subgroups, while G_3 has three classes, and G_5 has five. Also in all ten cases, the permutation group induced by G_i on the cosets of every subgroup of index 5 is isomorphic to $\mathbb{A}_5$. Thus H is an extension by A_5 of a 2-group of order $32 = 2^5$.

Next, we show that the wreath product $W = \mathbb{Z}_2 \wr \mathrm{PSL}(2,5) \cong (\mathbb{Z}_2)^6 \rtimes \mathbb{A}_5$ is a smooth quotient of each of the groups G_3, G_4 and G_5. Every element of W can be written in the form (v, π) where π is an element of the permutation group P induced by $\mathrm{PSL}(2,5)$ on the projective line over $\mathbb{Z}_5$, which we shall write as $\{0,1,2,3,4,\infty\}$, and v is a word on a natural basis $\{e_0, e_1, \ldots, e_4, e_\infty\}$ for the base group $V = (\mathbb{Z}_2)^6$.

There exist smooth epimorphisms as follows:

- $\psi_3 \colon G_3 \to W$, taking
  $a \mapsto (e_2e_3e_4, (1,\infty)(3,4))$, $b \mapsto (1_V, (0,\infty,1)(2,3,4))$, $c \mapsto (1_V, (0,3)(1,2))$;
- $\psi_4 \colon G_4 \to W$, taking
  $a \mapsto (e_0e_2, (1,\infty)(3,4))$, $b \mapsto (e_\infty, (0,2)(1,\infty))$, $c \mapsto (e_\infty, (1,4)(2,3))$;
- $\psi_5 \colon G_5 \to W$, taking
  $a \mapsto (e_\infty, (0,1)(2,4))$, $b \mapsto (1_V, (0,1,2,3,4))$, $c \mapsto (1_V, (1,3)(2,\infty))$.

Factoring out the subgroup $Z(W)$ generated by the central involution $e_0e_1e_2e_3e_4e_\infty$ gives smooth epimorphisms from each of G_3, G_4 and G_5 onto $W/Z(W) \cong (\mathbb{Z}_2)^5 \rtimes \mathbb{A}_5$, of order 1920.

Note: by conjugacy there are other possibilities for the epimorphism ψ_i in each case, all giving the same image. Remarkably, our MAGMA computations show that $H \cong W/Z(W)$ is the only quotient of G_i of order 1920, for each $i \in \{3,4,5\}$.

Finally, we prove that G_2 has no such quotient. To do this, we assume the contrary, and suppose that one exists, say H. Then we know H is an extension by A_5 of a 2-group J of order $32 = 2^5$. There is a unique epimorphism from G_2 to $\mathbb{A}_5$, the kernel K of which is free of rank 6. Hence any quotient of G_2 that is an extension by $\mathbb{A}_5$ of a 2-group is of the form G_2/L where $L \subseteq K$ and $|K:L|$ is a power of 2. We will consider possibilities for H by looking at other normal subgroups of G_2 contained in K, and their images in H. An immediate observation we can make is that every chief factor of H (that is, every characteristically simple factor of a normal series for H) must be $\mathbb{A}_5$ or an elementary abelian 2-group.

The abelianisation $G_2/(G_2)'$ of G_2 is $(\mathbb{Z}_2)^2$, generated by the images of a and b. Also $(G_2)'$ is the index 4 subgroup of G_2 generated by bc and $a(bc)a$, and since both of these have order 3, we find that $(G_2)'$ has no non-trivial 2-quotient. It does have $\mathbb{A}_5$ as a quotient, via a unique normal subgroup N, with G_2/N isomorphic to $\mathbb{A}_5 \times (\mathbb{Z}_2)^2$. In turn, this has three distinct quotients isomorphic to $\mathbb{A}_5 \times \mathbb{Z}_2$, of order 120, as well as a unique quotient isomorphic to $\mathbb{A}_5$ (namely G_2/K). Accordingly, the derived subgroup H' of H has index 1, 2 or 4 in H, and is perfect, and is therefore an extension by $\mathbb{A}_5$ of a 2-subgroup of order 8, 16 or 32.

Next, G_2 is isomorphic to the extended modular group $\mathrm{PGL}(2,\mathbb{Z})$, which is the universal group for 2-arc-regular automorphism group actions on finite trivalent graphs — that is, actions in which the group is sharply-transitive on paths of length 2. It follows that the quotient H of G_2 is a 2-arc-regular group of automorphisms of a regular cover of the Petersen graph, with covering group $J = K/L$ (see [**7**]). Similarly, $H/J' = H/[J,J]$ is a 2-arc-regular group of automorphisms of a regular cover of the Petersen graph, with abelian covering group J/J'. By Theorem 7.1 of [**7**], we find that $J/J' \cong (\mathbb{Z}_m)^6$ or $(\mathbb{Z}_m)^2 \times (\mathbb{Z}_{m/2})^4$ or $\mathbb{Z}_m \times (\mathbb{Z}_{m/2})^5$ where $m = 2^e$ for some $e > 0$. In particular, since $|J/J'| \leq |J| = 2^5$, it follows that $m = 2$ and $J/J' \cong Z_2$ or $(\mathbb{Z}_2)^2$, and $|H/J'| = 120$ or 240. Moreover, by the proof of Theorem 7.1 of [**7**], there are precisely three candidates for the subgroup J' when $J/J' \cong Z_2$, and just one when $J/J' \cong (Z_2)^2$. By our earlier observations (two paragraphs above), we conclude that $H/J' \cong \mathbb{A}_5 \times \mathbb{Z}_2$ or $\mathbb{A}_5 \times (\mathbb{Z}_2)^2$. In particular, $|J'| = 16$ or 8, and $H'/J' \cong \mathbb{A}_5$ (since H' is perfect).

Now consider the action of H' on J' by conjugation. This gives a homomorphism from H' to a subgroup of $\mathrm{Aut}(J')$ with kernel $C_{H'}(J')$, so $H'/C_{H'}(J')$ is isomorphic to a subgroup of $\mathrm{Aut}(J')$. If J' is not central in H', then $H'/C_{H'}(J')$ is non-trivial and so must have $\mathbb{A}_5$ as a quotient, in which case $|\mathrm{Aut}(J')|$ has a subgroup of order divisible by 60. The only group of order 8 or 16 with this property is $(\mathbb{Z}_4)^2$, but if $J' \cong (\mathbb{Z}_4)^2$ then $\mathrm{Aut}(J') \cong \mathrm{SL}(4,2) \cong \mathbb{A}_8$, which has no subgroup isomorphic to $\mathbb{A}_5 \times \mathbb{Z}_2$, so J' must be central in J, and then since $J/J' \cong \mathbb{Z}_2$ is cyclic, it follows that J is abelian, contradiction. Thus J' is central in H'. But now we have H' perfect, $J' \subseteq Z(H')$, and $J'/H' \cong \mathbb{A}_5$, so J' must be a subgroup of the Schur multiplier of $\mathbb{A}_5$, which is $\mathbb{Z}_2$. This is another contradiction, and shows that no such quotient H exists. □

Proof of Corollary 1.2. We know that the group $W/Z(W)$ of order 1920 we encountered above is a maximal handlebody group of genus 161, but is not a maximal bordered surface group. Now for $i \in \{3,4,5\}$, let ψ_i be any smooth epimorphism from G_i onto this finite group H. Then in each case the kernel of ψ_i is a free group K of rank 161.

Taking a similar approach to the one pioneered by Macbeath in [**10**] to construct infinite families of covers of a given Hurwitz surface, we may abelianise this free group K and reduce mod p for any positive integer p (by factoring out the characteristic subgroup of K generated by all commutators $[x,y]$ and all pth powers x^p of elements of K), and thereby obtain a smooth epimorphism from G_i onto an extension of $(\mathbb{Z}_p)^{161}$ by H.

Hence there exists a maximal handlebody group of order $1920p^{161}$, for every positive integer p.

On the other hand, for infinitely many values of p, there is no such maximal bordered surface group. For suppose the group G_2 has a smooth quotient of order $1920p^{161}$, where $p > 1920$ and p is prime. Then by Sylow theory, the quotient has a normal p-subgroup of order p^{161}, and factoring it out gives a smooth quotient of G_2 of order 1920. But no such quotient exists, contradiction. □

3. Concluding remarks

The four groups G_2, G_3, G_4 and G_5 in this paper belong to an infinite series

$$G_n := \mathbb{D}_n *_{\mathbb{Z}_n} \Delta(2,3,n)$$

where $\Delta(2,3,n)$ is the ordinary $(2,3,n)$ triangle group $\langle\, x, y \mid x^2 = y^3 = (xy)^n = 1 \,\rangle$.

Smooth finite quotients of these groups are precisely the largest finite groups of diffeomorphisms of a product with handles (or compression body); equivalently, these are the largest finite groups acting faithfully on a closed surface that is reducible along a finite equivariant system of disjoint simple closed curves. Here the genus g must be greater than 1, and again the maximum possible order is $12(g-1)$.

The cases $n \le 5$ correspond to handlebodies, and clearly the case $n = 7$ is related to Hurwitz groups. The class of groups obtained for arbitrary n consists precisely of the smooth quotients of the ordinary quadrangle group of type $(2,2,2,3)$.

A related question of interest is whether there exists some integer $g > 1$ such that there exists a maximal group action of the more general sort, but no maximal handlebody action of genus g? Equivalently, does there exist $g > 1$ such that G_n

has a smooth quotient of order $12(g-1)$ for some $n > 5$ but for no $n \in \{2,3,4,5\}$? We undertook another extensive computation (using MAGMA) and found that the answer is 'Yes'. The smallest genus for which this happens is $g = 820$: the group $\mathrm{PSL}(2,27)$ of order 9828 occurs as a quotient of the group $G_n = \mathbb{D}_n *_{\mathbb{Z}_n} \Delta(2,3,n)$ for $n = 7$, 13 and 14, but for no other value of n.

References

[1] Wieb Bosma, John Cannon, and Catherine Playoust, *The Magma algebra system. I. The user language*, J. Symbolic Comput. **24** (1997), no. 3-4, 235–265, DOI 10.1006/jsco.1996.0125. Computational algebra and number theory (London, 1993). MR1484478

[2] E. Bujalance, F. J. Cirre, and M. D. E. Conder, *Finite group actions of large order on compact bordered surfaces*, Computational algebraic and analytic geometry, Contemp. Math., vol. 572, Amer. Math. Soc., Providence, RI, 2012, pp. 25–56, DOI 10.1090/conm/572/11357. MR2953422

[3] Marston Conder, *The genus of compact Riemann surfaces with maximal automorphism group*, J. Algebra **108** (1987), no. 1, 204–247, DOI 10.1016/0021-8693(87)90135-9. MR887205 (88f:20063)

[4] Marston Conder, *Hurwitz groups: a brief survey*, Bull. Amer. Math. Soc. (N.S.) **23** (1990), no. 2, 359–370, DOI 10.1090/S0273-0979-1990-15933-6. MR1041434 (91d:20032)

[5] Marston Conder, *An update on Hurwitz groups*, Groups Complex. Cryptol. **2** (2010), no. 1, 35–49, DOI 10.1515/GCC.2010.002. MR2672551 (2011f:30081)

[6] Marston D. E. Conder, *Regular maps and hypermaps of Euler characteristic −1 to −200*, J. Combin. Theory Ser. B **99** (2009), no. 2, 455–459, DOI 10.1016/j.jctb.2008.09.003. MR2482963 (2010b:05084)

[7] Marston D. E. Conder and Jicheng Ma, *Arc-transitive abelian regular covers of cubic graphs*, J. Algebra **387** (2013), 215–242, DOI 10.1016/j.jalgebra.2013.02.035. MR3056695

[8] John D. Dixon and Brian Mortimer, *Permutation groups*, Graduate Texts in Mathematics, vol. 163, Springer-Verlag, New York, 1996. MR1409812 (98m:20003)

[9] D. Firth, *An algorithm to find normal subgroups of a finitely presented group up to a given index*, PhD Thesis, University of Warwick, 2005.

[10] A. M. Macbeath, *On a theorem of Hurwitz*, Proc. Glasgow Math. Assoc. **5** (1961), 90–96 (1961). MR0146724 (26 #4244)

[11] Coy L. May, *A bound for the number of automorphisms of a compact Klein surface with boundary*, Proc. Amer. Math. Soc. **63** (1977), no. 2, 273–280. MR0435385 (55 #8345)

[12] Darryl McCullough, Andy Miller, and Bruno Zimmermann, *Group actions on handlebodies*, Proc. London Math. Soc. (3) **59** (1989), no. 2, 373–416, DOI 10.1112/plms/s3-59.2.373. MR1004434 (90h:57014)

[13] A. Miller and B. Zimmermann, *Large groups of symmetries of handlebodies*, Proc. Amer. Math. Soc. **106** (1989), no. 3, 829–838, DOI 10.2307/2047442. MR962246 (90a:57015)

[14] Bruno Zimmermann, *Über Abbildungsklassen von Henkelkörpern* (German), Arch. Math. (Basel) **33** (1979/80), no. 4, 379–382, DOI 10.1007/BF01222772. MR564296 (82k:57002)

[15] Bruno Zimmermann, *Über Homöomorphismen n-dimensionaler Henkelkörper und endliche Erweiterungen von Schottky-Gruppen* (German), Comment. Math. Helv. **56** (1981), no. 3, 474–486, DOI 10.1007/BF02566224. MR639363 (83f:57025)

[16] Bruno Zimmermann, *Finite group actions on handlebodies and equivariant Heegaard genus for 3-manifolds*, Topology Appl. **43** (1992), no. 3, 263–274, DOI 10.1016/0166-8641(92)90161-R. MR1158872 (93c:57012)

DEPARTMENT OF MATHEMATICS, UNIVERSITY OF AUCKLAND, PRIVATE BAG 92019, AUCKLAND 1142, NEW ZEALAND

E-mail address: `m.conder@auckland.ac.nz`

DIPARTIMENTO DI MATEMATICA E GEOSCIENZE, UNIVERSITÀ DEGLI STUDI DI TRIESTE, 34127 TRIESTE, ITALIA

E-mail address: `zimmer@units.it`

Maximal and Non-maximal NEC and Fuchsian groups uniformizing Klein and Riemann surfaces

Antonio F. Costa, Milagros Izquierdo, and Ana M. Porto

Dedicated to our friend Emilio.

ABSTRACT. The aim of this article is to expose a topic where professor Emilio Bujalance has made very important contributions: the theory of maximal Fuchsian and NEC groups and their use as uniformization groups of Riemann and Klein surfaces.

1. Introduction

An n-manifold M has a spherical (resp. Euclidean, resp. hyperbolic) structure if M is the quotient space by the action of a discrete group of $\mathrm{Isom}(\mathbb{S}^n)$ (resp. $\mathrm{Isom}(\mathbb{E}^n)$, resp. $\mathrm{Isom}(\mathbb{H}^n)$) acting freely on $\mathbb{S}^n$(resp. $\mathbb{E}^n$ and $\mathbb{H}^n$). These are the three isotropic geometrical structures on manifolds, i. e. the geometries of constant sectional curvature. Geometric structures play an important rôle in understanding the topology of 3-dimensional manifolds. The celebrated Geometrization Theorem (conjectured by W. Thurston in the 70's and completed by G. Perelman) states that any compact 3 -manifold can be decomposed, cutting along a finite collection of disjoint embedded surfaces, in 3-manifolds with geometrical structure. The 3 - dimensional possible geometries are eight: the three isotropic and five non-isotropic. The manifolds with hyperbolic geometry form the richest family and, by the Mostow Rigidity Theorem [**43**], the topology can be studied using hyperbolic geometry:

THEOREM 1.1. *Let Γ_1 and Γ_2 be Kleinian groups such that $\mathbb{H}^3/\Gamma_1$ and $\mathbb{H}^3/\Gamma_2$ have finite volume. Every isomorphism between Γ_1 and Γ_2 is realized by conjugation in* $\mathrm{Isom}(\mathbb{H}^3)$.

Note that the above theorem applies to hyperbolic 3-dimensional orbifolds as well; i.e. when the elements of the groups Γ_1 and Γ_2 have fixed points. This theorem reduces the study of the topology of hyperbolic 3-manifolds and orbifolds to the study of the algebraic structure of Kleinian groups (see [**9**]). Hyperbolic orbifolds provide an important tool for the study of automorphisms of algebraic curves and Riemann and Klein surfaces.

2010 *Mathematics Subject Classification.* Primary 30F10, 30F50, 30F60; Secondary 14H55, 20H15, 57M60.

Key words and phrases. Riemann surface, Klein surface, Fuchsian group, Non-euclidean crystallographic group, Teichmüller space, Moduli space.

The authors were supported in part by MTM2011-23092.

For the hyperbolic plane and Fuchsian or Non-Euclidean Crystallographic (NEC) groups there is no an analogous to Mostow's rigidity theorem. Also in contrast with dimension three, the topological classification of compact surfaces and orbifolds can be considered as an easy subject. But discrete groups of the isometry group of the hyperbolic plane produce objects that are intrinsically interesting: hyperbolic surfaces and orbifolds.

Hyperbolic surfaces are important objects in mathematics; for instance, after the works of Riemann, Poincaré, Klein and others, every complex (real) algebraic curve can be uniformized by a class of Fuchsian and NEC groups. This provides a well defined hyperbolic structure on the surface underlying the algebraic curve; unlike the case of 3-manifolds, these hyperbolic structures in surfaces are not rigid, making it necessary to consider Teichmüller space: the space of marked hyperbolic structures on surfaces. For the 2-dimensional orbifolds it is necessary also to consider Teichmüller and Moduli spaces of Fuchsian an NEC groups following Fenchel & Nielsen [**32**], Bers [**8**], Ahlfors [**1**], Macbeath & Singerman [**44**].

There are some phenomena in dimension two that can be considered as rigidity aspects, since the isomorphy class of the fundamental group of an orbifold provides some geometrical (not only topological) information: these phenomena are Teichmüller spaces of dimension zero and non-maximal signatures. The former corresponds to triangular Fuchsian and NEC groups and explains the existence of isolated points in the branch loci of the moduli space of Riemann surfaces (see [**41**]); in this case the algebraic structure of the orbifold fundamental group determines completely the geometrical structure of the hyperbolic 2-orbifold. The non-maximal signatures phenomenon explains why the isomorphic class of some 2-orbifold fundamental groups provides information on the existence of orbifold automorphisms or of orbifolds coverings. These last situations have been studied by Greenberg [**34**], [**35**], Singerman [**49**], E. Bujalance [**14**], Estévez & Izquierdo [**28**], [**29**] (following the suggestion of E. Bujalance). It has have been applied extensively by many authors in the theory of Riemann and Klein surfaces. This article is an exposition of the last topic by selecting some examples arising as consequences of this phenomenon.

2. Non-euclidean crystallographic and Fuchsian groups

A non-Euclidean crystallographic group or NEC group Γ is a discrete subgroup of the group $Aut^{\pm}(\mathbb{H}^2)$ of conformal and anticonformal automorphisms of $\mathbb{H}^2$ and we shall assume that the orbit space $\mathbb{H}^2/\Gamma$ is compact. If the NEC group Γ does not contain any orientation-reversing automorphism of $\mathbb{H}^2$, then we say that Γ is a Fuchsian group.

The so called canonical presentation for NEC groups first appeared in [**51**] and their structure was clarified by the introduction of signatures in [**42**] (see also [**50**]).

Given an NEC group Γ, the subgroup of Γ consisting of the orientation-preserving elements is called the *canonical Fuchsian subgroup of* Γ. The algebraic structure of Γ and the geometric and topological structure of the quotient orbifold $\mathbb{H}^2/\Gamma$ are given by the signature:

$$\sigma(\Gamma) = (h; \pm; [m_1, ..., m_r]; \{(n_{1,1}, ..., n_{1,r_1}), ..., (n_{k,1}, ..., n_{k,r_k})\}). \tag{2.1}$$

The orbit space $\mathbb{H}^2/\Gamma$ is an orbifold with underlying surface of genus h, having $r \geq 0$ cone points and k boundary components of the underlying surface, each with

$r_i \geq 0$ corner points, $i = 1, ..., k$. The signs "$+$" and "$-$" correspond to orientable and non-orientable orbifolds respectively. The integers m_i are called the proper periods of Γ and they are the orders of the cone points of $\mathbb{H}^2/\Gamma$. The brackets $(n_{i,1}, ..., n_{i,r_i})$ are the period cycles of Γ. The integers $n_{i,j}$ are the link periods of Γ and the orders of the corner points of $\mathbb{H}^2/\Gamma$. The group Γ is isomorphic to the *fundamental group* of the orbifold $\mathbb{H}^2/\Gamma$.

The hyperbolic area of the orbifold $\mathbb{H}^2/\Gamma$ coincides with the hyperbolic area of an arbitrary fundamental region of Γ and depends only on the signature of σ of Γ:

$$\mu(\Gamma) = 2\pi(\varepsilon h - 2 + k + \sum_{i=1}^{r}(1 - \frac{1}{m_i}) + \frac{1}{2}\sum_{i=1}^{k}\sum_{j=1}^{r_i}(1 - \frac{1}{n_{i,j}})), \tag{2.2}$$

where $\varepsilon = 2$ if there is a $+$ sign and $\varepsilon = 1$ otherwise. If Γ' is a subgroup of Γ of finite index then Γ' is an NEC group and the following Riemann-Hurwitz formula holds:

$$[\Gamma : \Gamma'] = \mu(\Gamma')/\mu(\Gamma). \tag{2.3}$$

How the structure of finite index subgroups of NEC and Fuchsian groups depends on the structure of the NEC/Fuchsian groups and the embedding is studied in [**12**], [**13**], [**39**], [**48**] and [**50**].

A signature as (2.1) is the signature of an NEC group if and only if the corresponding area is > 0. An NEC or Fuchsian group Γ without elliptic elements is called an NEC (resp. Fuchsian) *surface group* and it has signature $(h; \pm; [-], \{(-), .\overset{k}{.}., (-)\})$ (resp. $(h; +; [-], \{-\})$).

Given a Klein surface X (see [**2**]) with topological type $(h, \pm, k)$, i. e. X has genus h, is orientable if the sign is $+$ and non-orientable if the sign is $-$ and has k boundary components, the algebraic genus of X is the integer $\varepsilon h + k \geq 2$, where $\varepsilon = 1$ if there is a sign $-$ in t and 2 otherwise. By the uniformization theorem, X is isomorphic to $\mathbb{H}^2/\Gamma$, where Γ is a surface NEC group with signature $(h; \pm; [-], \{(-), .\overset{k}{.}., (-)\})$, see [**15**]. Given a Riemann surface Y of genus g, there is a Fuchsian group Γ of signature $(g; +; [-], \{-\})$ such that $Y = \mathbb{H}^2/\Gamma$. If a finite group G is isomorphic to a group of automorphisms of X then there exist an NEC group Δ and an epimorphism $\theta : \Delta \to G$ with $\ker(\theta) = \Gamma$. The NEC group Δ is the lifting of G to the universal covering $\pi : \mathbb{H}^2 \to \mathbb{H}^2/\Gamma$. Analogously for the automorphism groups of a Riemann surface X.

3. Teichmüller and Moduli spaces of Fuchsian and NEC groups

Let σ be a signature of NEC groups and $\mathcal{G}$ be an abstract group isomorphic to the NEC groups with signature σ. We denote by $\mathbf{R}(\sigma)$ the set of monomorphisms $r : \mathcal{G} \to Aut^{\pm}(\mathbb{H}^2)$ such that $r(\mathcal{G})$ is an NEC group with signature σ. The set $\mathbf{R}(\sigma)$ has a natural topology given by the topology of $Aut^{\pm}(\mathbb{H}^2)$. Two elements r_1 and $r_2 \in \mathbf{R}(s)$ are said to be equivalent, $r_1 \sim r_2$, if there exists $g \in Aut^{\pm}(\mathbb{H}^2)$ such that for each $\gamma \in \mathcal{G}$, $r_1(\gamma) = gr_2(\gamma)g^{-1}$. The space of classes $\mathbf{T}(\sigma) = \mathbf{R}(\sigma)/\sim$ is called the *Teichmüller space* of NEC groups with signature σ (see [**44**]). If the signature σ is given in (1), the Teichmüller space $\mathbf{T}(\sigma)$ is homeomorphic to $\mathbb{R}^{d(\sigma)}$ where

$$d(\sigma) = 3(\varepsilon h - 1 + k) - 3 + (2r + \sum_{i=1}^{k} r_i).$$

If σ is the signature of a Fuchsian group then $\mathbf{T}^+(\sigma) = \mathbf{R}(\sigma)/\sim^+$, where $r_1 \sim^+ r_2$, if there exists $g \in Aut^+(\mathbb{H}^2)$ such that for each $\gamma \in \mathcal{G}$, $r_1(\gamma) = gr_2(\gamma)g^{-1}$.

The modular group $\text{Mod}(\mathcal{G})$ is the quotient $\text{Mod}(\mathcal{G}) = \text{Aut}(\mathcal{G})/\text{Inn}(\mathcal{G})$, where $\text{Inn}(\mathcal{G})$ denotes the inner automorphisms of $\mathcal{G}$. The *moduli space* for NEC groups with signature σ is the quotient $\mathcal{M}_\sigma = \mathbf{T}(\sigma)/\text{Mod}(\mathcal{G})$ endowed with the quotient topology. Hence $\mathcal{M}_\sigma$ is an orbifold with orbifold fundamental group $\text{Mod}(\mathcal{G})$.

For signatures corresponding to Fuchsian groups the spaces to be considered are $\mathcal{M}_\sigma^+ = \mathbf{T}^+(\sigma)/\text{Mod}^+(\mathcal{G})$, where $\text{Mod}^+(\mathcal{G}) = \text{Aut}^+(\mathcal{G})/\text{Inn}(\mathcal{G})$ and $\text{Aut}^+(\mathcal{G})$ are the automorphisms given by orientation preserving transformations.

If $\sigma = (h; \pm; [-], \{(-), \overset{k}{\ldots}, (-)\})$ then $t = (h, \pm, k)$ and we denote by $\mathbf{T}(\sigma) = \mathbf{T}_t$, $\mathcal{M}_\sigma = \mathcal{M}_t$ the Teichmüller and the moduli space of Klein surfaces of topological type t. For Riemann surfaces, $t = (g, +, 0)$, $\mathbf{T}^+(\sigma) = \mathbf{T}_g$ and $\mathcal{M}_\sigma^+ = \mathcal{M}_g$.

Let $\mathcal{G}$ and $\mathcal{G}'$ be abstract groups isomorphic to NEC groups with signatures σ and σ' respectively. The inclusion mapping $\alpha : \mathcal{G} \to \mathcal{G}'$ induces an embedding $\mathbf{T}(\alpha) : \mathbf{T}(\sigma') \to \mathbf{T}(\sigma)$ defined by $[r] \mapsto [r \circ \alpha]$ (see [**44**] and [**35**]).

If a finite group G is isomorphic to a group of automorphisms of Klein surfaces with topological type $t = (g, \pm, k)$, then the action of G is determined by an epimorphism $\omega_\theta : \mathcal{D} \to G$, where $\mathcal{D}$ is an abstract group isomorphic to NEC groups with a given signature σ and $\ker(\omega_\theta) = \mathcal{G}$ is a group isomorphic to surface NEC groups uniformizing Klein surfaces of topological type t. Then there is an inclusion $\alpha : \mathcal{G} \to \mathcal{D}$ and an embedding $\mathbf{T}(\alpha) : \mathbf{T}(\sigma) \to \mathbf{T}_t$. The continuous map $\mathbf{T}(\alpha)$ induces a continuous map $\mathcal{M}_\sigma \to \mathcal{M}_t$. The image of this continuous map is the set of Klein surfaces where G acts in a given topological way as an automorphism group.

The singular set of the orbifold $\mathcal{M}_t$, the branch locus $\mathcal{B}_t$, is formed by surfaces with automorphisms, with the exception of the moduli space of Riemann surfaces of algebraic genus 2, where $\mathcal{B}_t$ is the set of surfaces with automorphisms different from the hyperelliptic one, see [**44**], [**38**].

The automorfism groups of hyperbolic surfaces of genus $g > 1$ are given by the conjugacy classes of finite groups of $\text{Mod}(\mathcal{G})$, where $\mathcal{G}$ is a surface NEC or Fuchsian group.

4. Maximal NEC groups

Definition 4.1. An NEC group Γ is called a maximal NEC group if there is no other NEC group Δ containing Γ and $\Gamma \neq \Delta$.

Definition 4.2. An NEC signature σ is said to be non-maximal if NEC groups with signature σ are not maximal. The pair $(\sigma, \delta), \sigma \neq \delta$, is said to form a pair of signatures if σ is a non-maximal signature and each group with signature σ is a subgroup of some group with signature δ.

Equivalently, the pair (σ, δ) is a pair of signatures if and only if there exist NEC groups Γ and Δ with signatures σ and δ respectively such that $\Gamma \leq \Delta$ and $d(\sigma) = d(\delta)$.

In terms of orbifolds: if (σ, δ) is a pair of signatures and $\mathbb{H}^2/\Gamma$ is an orbifold with $\sigma(\Gamma) = \sigma$, then there is a finite orbifold covering $\mathbb{H}^2/\Gamma \to \mathbb{H}^2/\Delta$, where Δ has signature δ.

A pair (σ, δ) is said a normal pair of signatures if each group with signature σ is a normal subgroup of some group with signature δ. In term of orbifolds the orbifold covering $\mathbb{H}^2/\Gamma \to \mathbb{H}^2/\Delta$ is regular.

The lists of pairs of signatures and normal pairs of signatures of Fuchsian groups have been obtained by Singerman in [**49**] (see also Greenberg [**34**]). The following is the list of pairs of normal signatures for Fuchsian groups:

TABLE 1. Normal pairs of Fuchsian signatures

σ	δ
$(1;+;[t])$	$(0;+;[2,2,2,2t])$
$(1;+;[t,t])$	$(0;+;[2,2,2,2,t])$
$(2;+;[-])$	$(0;+;[2,2,2,2,2,2])$
$(0;+;[t,t,t,t])$	$(0;+;[2,2,2,t]), t \geq 3$
$(0;+;[t,t,u,u])$	$(0;+;[2,2,t,u]), t+u \geq 5$
$(0;+;[t,t,t])$	$(0;+;[3,3,t]), t \geq 4$
$(0;+;[t,t,t])$	$(0;+;[2,3,2t]), t \geq 4$
$(0;+;[t,t,u])$	$(0;+;[2,t,2u]), t \geq 4, t+u \geq 7$

For instance, the pair $((2;+;[-]),(0;+;[2,2,2,2,2,2]))$ in Table 1 tells us that every Riemann surface of genus two is hyperelliptic. The non-normal pairs consist of pairs of signatures of Fuchsian triangular groups that are given by irregular coverings between orbifold structures on the sphere with three conic points (Table 2).

TABLE 2. Non-normal pairs of Fuchsian signatures

σ	δ
$(0,+,[7,7,7])$	$(0,+,[2,3,7])$
$(0,+,[2,7,7])$	$(0,+,[2,3,7])$
$(0,+,[3,3,7])$	$(0,+,[2,3,7])$
$(0,+,[4,8,8])$	$(0,+,[2,3,8])$
$(0,+,[3,8,8])$	$(0,+,[2,3,8])$
$(0,+,[9,9,9])$	$(0,+,[2,3,9])$
$(0,+,[4,4,5])$	$(0,+,[2,4,5])$
$(0,+,[n,4n,4n])$	$(0,+,[2,3,4n]), n \geq 2$
$(0,+,[n,2n,2n])$	$(0,+,[2,4,2n]), n \geq 3$
$(0,+,[n,3,3n])$	$(0,+,[2,3,3n]), n \geq 3$
$(0,+,[n,2,2n])$	$(0,+,[2,3,2n]), n \geq 4$

Emilio Bujalance obtained the list of normal pairs for NEC groups in [**14**] (Table 3). To obtain the list he used his results on signatures of normal subgroups of NEC groups obtained in [**12**] and [**13**].

Each pair of signatures comes from a phenomenon in hyperbolic geometry, for instance, in the list of normal NEC groups (Table 3) appear the three following pairs:

$((0;+;[-];\{(t,t,u,u)\}),(0;+;[-];\{(2,2,t,u)\}))$, $\max(t,u)\geq 3$
$((0;+;[-];\{(t,u,t,u)\}),(0;+;[2];\{(t,u)\}))$, $\max(t,u)\geq 3$
$((0;+;[-];\{(t,t,t,t)\}),(0;+;[-];\{(2,2,2,t)\}))$, $t\geq 3$

The first pair tells us that every hyperbolic quadrilateral with angles $(\pi/t,\pi/t,\pi/u,\pi/u)$ (the angle mesures are written following the perimeter of the quadrilateral) has axial symmetry.

The second one tells us that every hyperbolic quadrilateral with angles $(\pi/t,\pi/u,\pi/t,\pi/u)$ has an order two rotational symmetry.

Finally the third one says that every hyperbolic quadrilateral with angles $(\pi/t,\pi/t,\pi/t,\pi/t)$ has dihedral symmetry.

A pair that informs us about a property of involutions of Klein surfaces is $((3;-;[-];\{\}),(0;+;[2,2,2];\{\}))$, this pair tells us that every Klein surface homeomorphic to a closed non-orientable surface of topological genus 3 admits an involution with exactly three fixed points (see [**21**]).

Later Hoare (in [**39**]) gave a method to determine signatures of non-normal subgroups of NEC groups and using this method and the suggestion of E. Bujalance, Estévez & Izquierdo obtained the list of pairs of NEC groups (see [**29**]).

In the list of non-normal pairs of signatures, Table 4, appears the case:

$$((0;+;[-];\{(2,n,2n)\}),(0;+;[-];\{(2,3,2n)\})),n\geq 4$$

that is a consequence of the geometrical fact that considering three triangles with angles $(\pi/2,\pi/3,\pi/2n)$, two of then with an orientation and the third one with the oposite orientation, we can place them with a common vertex (the one corresponding to the angle measuring $\pi/3$) in such a way that the union of the three triangles becomes a triangle with angles $(\pi/2,\pi/n,\pi/2n)$.

In Table 4 we have added at the end some lines to the original Table appeared in [**29**]. These lines correspond to pairs of signatures (σ,δ) where σ is the signature of a Fuchsian group, in [**29**] the authors obtain the list of pairs of proper NEC signatures of non-normal subgroups.

We explain quickly how we obtain such added lines. If δ is a signature of NEC groups we denote by δ^+ the signature of the canonical Fuchsian groups of NEC groups with signature δ. Now, if (σ,δ) is a non-normal pair with σ the signature of a Fuchsian group, then (σ,δ^+) must be in Table 1 or Table 2 and (δ^+,δ) must be in Table 3. When (σ,δ^+) appears in Table 1 and (δ^+,δ) in Table 3, we have verified that the pair (σ,δ) is in Table 3 and there is no a non-normal inclusion of groups with signature σ in groups with signature δ. Hence the cases that we must add to Table 4 are those when (σ,δ^+) appears in Table 2 and the pair (σ,δ) is not in Table 3.

5. Applications

The lists of pairs of signatures have been largely used in the study of Riemann and Klein surfaces. We will only present some examples without the intention of been exhaustive. The most obvious application of maximal NEC and Fuchsian groups is their use in determining the full automorphism groups of surfaces, see for instance [**11**], [**22**], [**23**] [**30**], [**33**] among a long list of articles. An example of a general result in this area is the following result in [**18**]:

THEOREM 5.1. *Let S be a proper (bordered or non-orientable) Klein surface and let G be any nontrival finite group. Then there exists a normal covering* $\Phi : T \longrightarrow S$ *whose covering group is isomorphic to G and it is the full automorphism group of T.*

The above result has a nice reformulation in terms of function fields:

THEOREM 5.2. *Let K be a function field in one variable (over* $\mathbb{R}$*) and let G be a finite group. Then there is a Galois extension L of K, such that*

$$Aut(L) = Aut_{\mathbb{R}}(L) = Aut_K(L) = G.$$

The possible full groups of automorphisms of Riemann surfaces of given genus produce the equisymmetric stratification of the moduli space (see [**10**], [**25**], [**5**]). The use of such stratification, and then the use of maximal Fuchsian groups, has been of essential importance in the study of several properties of moduli and Teichmüller space, for instance in the study of the connectedness and unconnectedness of the branch loci (see [**6**], [**4**], [**7**], [**25**], [**26**]).

Other subject where the maximal groups are important is in the study of M*-groups (the groups of automorphisms of compact bordered Klein surfaces of maximal possible order for a given genus) (see [**31**], [**19**], [**20**]).

Let $X = \mathbb{H}^2/\Gamma$ be a Riemann surface, where Γ is a surface Fuchsian group. Let us further consider that X has an automorphism group G in such a way that the orbifold X/G is isomorphic to $\mathbb{H}^2/\Gamma'$ where Γ' is a Fuchsian group such that $\Gamma \lhd \Gamma'$ and Γ' has signature σ appearing as the first coordinate of one of the pairs of signatures in the list of non-maximal signatures of Fuchsian groups. Then, it is possible to establish algebraic conditions to be satisfied by G so that the automorphism group of X is strictly greater than G, i.e., the surface X is more symmetric that we are supposing (see [**16**] and independently [**27**]). See [**17**] for the analogous case for Klein surfaces. In these cases, it is possible to obtain analytic information on S from topological and algebraic conditions, this may be considered as rigidity phenomena in Riemann and Klein surfaces. As an example where there is no algebraic condition on the group G (in this case G is a cyclic group) we have ([**16**] and [**27**]):

THEOREM 5.3. *If X is a cyclic n-fold covering of an elliptic curve having exactly two branched points. Then the group of automorphisms of X contains the dihedral group* D_n.

The origin of Klein surfaces is the study of real algebraic curves ([**40**], [**2**], [**47**]). A real form of an algebraic curve is a pair (X, τ), where X is a Riemann surface and τ is an anticonformal involution of X. Two real forms (X, τ) and (X, τ') are birationally equivalent if τ and τ' are conjugate by an automorphism of X. To determine the possible real forms of an algebraic curve it is necessary the knowledge of the full automorphism groups of Riemann surfaces and then maximal Fuchsian and NEC groups play an important rôle. In [**3**], [**24**], [**36**], [**37**], [**45**], [**46**] one can see applications of the lists of non-maximal signatures in the study of real algebraic curves. For instance the possible topological types of real algebraic forms of cyclic p-gonal curves are obtained in [**24**].

TABLE 3. Normal pairs of NEC signatures

σ	δ
$(3,-,[-],\{-\})$	$(0,+,[2,2,2],\{(-)\})$
$(2,-,[t],\{-\})$	$(0,+,[2,2],\{(t)\})$
$(2,-,[-],\{(-)\})$	$(0,+,[2,2],\{(2,2)\})$
$(1,+,[-],\{(-)\})$	$(0,+,[2,2,2],\{(-)\})$
$(1,-,[-],\{(t)\})$	$(0,+,[2],\{(2,2t)\})$
$(1,-,[t],\{(-)\})$	$(0,+,[2],\{(2,2,t)\})$
$(1,-,[-],\{(t,t)\})$	$(0,+,[2],\{(2,2,t)\})$
$(1,-,[-],\{(-),(-)\})$	$(0,+,[2],\{(2,2,2,2)\})$
$(1,-,[t,t],\{-\})$	$(0,+,[2,t],\{(-)\}), t \geq 3$
$(1,-,[t,u],\{-\})$	$(0,+,[2],\{(t,u)\}), max(t,u) \geq 3$
$(0,+,[-],\{(-),(-),(-)\})$	$(0,+,[-],\{(2,2,2,2,2,2)\})$
$(0,+,[-],\{(t),(-)\})$	$(0,+,[-],\{(2,2,2,2t)\})$
$(0,+,[-],\{(t),(t)\})$	$(0,+,[2,2],\{(t)\})$
$(0,+,[t],\{(-),(-)\})$	$(0,+,[-],\{(2,2,2,2,t)\})$
$(0,+,[-],\{(t,t),(-)\})$	$(0,+,[-],\{(2,2,2,2,t)\})$
$(0,+,[-],\{(t,u,t,u)\})$	$(0,+,[2],\{(t,u)\}), max(t,u) \geq 3$
$(0,+,[-],\{(t,t,u,u)\})$	$(0,+,[-],\{(2,2,t,u)\}), max(t,u) \geq 3$
$(0,+,[t],\{(u,u)\})$	$(0,+,[-],\{(2,t,2,u)\}), max(t,u) \geq 3$
$(0,+,[t,u],\{(-)\})$	$(0,+,[-],\{(2,2,t,u)\}), max(t,u) \geq 3$
$(0,+,[t,t],\{(-)\})$	$(0,+,[t],\{(2,2)\}), t \geq 3$
$(0,+,[t,t],\{(-)\})$	$(0,+,[t,2],\{(-)\}), t \geq 3$
$(0,+,[t],\{(u)\})$	$(0,+,[-],\{(2,t,2u)\}), t \geq 3, t+u \geq 7$
$(0,+,[-],\{(t,t,u)\})$	$(0,+,[-],\{(2,t,2u)\}), t \geq 3, t+u \geq 7$
$(0,+,[t,t,u],\{-\})$	$(0,+,[t],\{(u)\}), t \geq 3, t+u \geq 7$
$(0,+,[t,t,u],\{-\})$	$(0,+,[-],\{(t,t,u)\}), t \geq 3, t+u \geq 7$
$(0,+,[m,t,u],\{-\})$	$(0,+,[-],\{(m,t,u)\}), t \geq 7, t,u \geq 4, m \neq t \neq u \neq m$
$(0,+,[-],\{(t,t,t)\})$	$(0,+,[3],\{(t)\}), t \geq 4$
$(1,-,[t,t],\{-\})$	$(0,+,[-],\{(2,2,2,t)\}), t \geq 3$
$(0,+,[t,t],\{(-)\})$	$(0,+,[-],\{(2,2,2,t)\}), t \geq 3)$
$(0,+,[-],\{(t,t,t,t)\})$	$(0,+,[-],\{(2,2,2,t)\}), t \geq 3$
$(0,+,[t,t,u],\{-\})$	$(0,+,[-],\{(2,t,2u)\}), t \geq 3, t+u \geq 7$
$(0,+,[-],\{(t,t,t)\})$	$(0,+,[-],\{(2,3,2t)\}), t \geq 4$
$(0,+,[t,t,t],\{-\})$	$(0,+,[3],\{(t)\}), t \geq 4$
$(0,+,[t,t,t],\{-\})$	$(0,+,[-],\{(3,3,t)\}), t \geq 4$
$(0,+,[t,t,t],\{-\})$	$(0,+,[-],\{(2,3,2t)\}), t \geq 4$

TABLE 4. Non-normal pairs of NEC signatures

σ	δ
$(0,+,[-],\{(7,7,7)\})$	$(0,+,[-],\{(2,3,7)\})$
$(0,+,[3],\{(7)\})$	$(0,+,[-],\{(2,3,7)\})$
$(0,+,[7],\{(2)\})$	$(0,+,[-],\{(2,3,7)\})$
$(0,+,[8],\{(4)\})$	$(0,+,[-],\{(2,3,8)\})$
$(0,+,[-],\{(4,8,8)\})$	$(0,+,[-],\{(2,3,8)\})$
$(0,+,[8],\{(3)\})$	$(0,+,[-],\{(2,3,8)\})$
$(0,+,[9],\{(9)\})$	$(0,+,[-],\{(2,3,9)\})$
$(0,+,[4],\{(5)\})$	$(0,+,[-],\{(2,4,5)\})$
$(0,+,[4n],\{(n)\})$	$(0,+,[-],\{(2,3,4n)\})$
$(0,+,[-],\{(4n,4n,n)\})$	$(0,+,[-],\{(2,3,4n)\})$
$(0,+,[2n],\{(n)\})$	$(0,+,[-],\{(2,4,2n)\})$
$(0,+,[-],\{(2n,2n,n)\})$	$(0,+,[-],\{(2,4,2n)\})$
$(0,+,[-],\{(3n,3,n)\})$	$(0,+,[-],\{(2,3,3n)\})$
$(0,+,[-],\{(2,n,2n)\})$	$(0,+,[-],\{(2,3,2n)\}), n \geq 4$
$(0,+,[n],\{(n)\})$	$(0,+,[-],\{(2,3,2n)\}), n \geq 4$
$(0,+,[n],\{(n)\})$	$(0,+,[-],\{(3,3,n)\}), n \geq 4$
$(0,+,[n],\{(n,n)\})$	$(0,+,[2],\{(2,n)\}), n \geq 3$
$(0,+,[7,7,7])$	$(0,+,[-],\{(2,3,7)\})$
$(0,+,[2,7,7])$	$(0,+,[-],\{(2,3,7)\})$
$(0,+,[3,3,7])$	$(0,+,[-],\{(2,3,7)\})$
$(0,+,[4,8,8])$	$(0,+,[-],\{(2,3,8)\})$
$(0,+,[3,8,8])$	$(0,+,[-],\{(2,3,8)\})$
$(0,+,[9,9,9])$	$(0,+,[-],\{(2,3,9)\})$
$(0,+,[4,4,5])$	$(0,+,[-],\{(2,4,5)\})$
$(0,+,[n,4n,4n])$	$(0,+,[-],\{(2,3,4n)\}), n \geq 2$
$(0,+,[n,2n,2n])$	$(0,+,[-],\{(2,4,2n)\}), n \geq 3$
$(0,+,[n,3,3n])$	$(0,+,[-],\{(2,3,3n)\}), n \geq 3$
$(0,+,[n,2,2n])$	$(0,+,[-],\{(2,3,2n)\}), n \geq 4$

References

[1] Lars V. Ahlfors, *Teichmüller spaces*, Proc. Internat. Congr. Mathematicians (Stockholm, 1962), Inst. Mittag-Leffler, Djursholm, 1963, pp. 3–9. MR0176055 (31 #330)

[2] Norman L. Alling and Newcomb Greenleaf, *Foundations of the Theory of Klein surfaces*, Lecture Notes in Mathematics, Vol. 219, Springer-Verlag, Berlin-New York, 1971. MR0333163 (48 #11488)

[3] Czesław Bagiński and Grzegorz Gromadzki, *Minimal genus problem for pseudo-real Riemann surfaces*, Arch. Math. (Basel) **95** (2010), no. 5, 481–492, DOI 10.1007/s00013-010-0186-1. MR2738867 (2012a:30101)

[4] Gabriel Bartolini, Antonio F. Costa, and Milagros Izquierdo, *On the connectivity of branch loci of moduli spaces*, Ann. Acad. Sci. Fenn. Math. **38** (2013), no. 1, 245–258, DOI 10.5186/aasfm.2013.3820. MR3076808

[5] Gabriel Bartolini, Antonio F. Costa, and Milagros Izquierdo, *On isolated strata of pentagonal Riemann surfaces in the branch locus of moduli spaces*, Computational algebraic and analytic geometry, Contemp. Math., vol. 572, Amer. Math. Soc., Providence, RI, 2012, pp. 19–24, DOI 10.1090/conm/572/11372. MR2953421

[6] Gabriel Bartolini, Antonio F. Costa, Milagros Izquierdo, and Ana M. Porto, *On the connectedness of the branch locus of the moduli space of Riemann surfaces* (English, with English and Spanish summaries), Rev. R. Acad. Cienc. Exactas Fís. Nat. Ser. A Math. RACSAM **104** (2010), no. 1, 81–86, DOI 10.5052/RACSAM.2010.08. MR2666443 (2011e:32013)

[7] Gabriel Bartolini and Milagros Izquierdo, *On the connectedness of the branch locus of the moduli space of Riemann surfaces of low genus*, Proc. Amer. Math. Soc. **140** (2012), no. 1, 35–45, DOI 10.1090/S0002-9939-2011-10881-5. MR2833515 (2012f:14059)

[8] Lipman Bers, *Spaces of Riemann surfaces*, Proc. Internat. Congress Math. 1958, Cambridge Univ. Press, New York, 1960, pp. 349–361. MR0124484 (23 #A1796)

[9] Michel Boileau, Sylvain Maillot, and Joan Porti, *Three-dimensional orbifolds and their geometric structures* (English, with English and French summaries), Panoramas et Synthèses [Panoramas and Syntheses], vol. 15, Société Mathématique de France, Paris, 2003. MR2060653 (2005b:57030)

[10] S. Allen Broughton, *The equisymmetric stratification of the moduli space and the Krull dimension of mapping class groups*, Topology Appl. **37** (1990), no. 2, 101–113, DOI 10.1016/0166-8641(90)90055-7. MR1080344 (92d:57013)

[11] Broughton, S.A.; Wootton, A., Exceptional automorphisms of (generalized) superelliptic surfaces. This conference proceedings.

[12] E. Bujalance, *Proper periods of normal N.E.C. subgroups with even index*, Proceedings of the Eighth Portuguese-Spanish Conference on Mathematics, Vol. I (Coimbra, 1981), Univ. Coimbra, Coimbra, 1981, pp. 81–83. MR672966 (84a:20054)

[13] Emilio Bujalance, *Normal subgroups of N.E.C. groups*, Math. Z. **178** (1981), no. 3, 331–341, DOI 10.1007/BF01214870. MR635202 (83e:20047)

[14] E. Bujalance, *Normal N.E.C. signatures*, Illinois J. Math. **26** (1982), no. 3, 519–530. MR658461 (84i:20049)

[15] E. Bujalance, *Klein surfaces as orbit spaces of NEC groups*, Bol. Soc. Mat. Mexicana (2) **27** (1982), no. 2, 45–49. MR745315 (85g:30079)

[16] Emilio Bujalance, F. J. Cirre, and Marston Conder, *On extendability of group actions on compact Riemann surfaces*, Trans. Amer. Math. Soc. **355** (2003), no. 4, 1537–1557 (electronic), DOI 10.1090/S0002-9947-02-03184-7. MR1946404 (2003k:20079)

[17] E. Bujalance, F. J. Cirre, and M. D. E. Conder, *Extensions of finite cyclic group actions on non-orientable surfaces*, Trans. Amer. Math. Soc. **365** (2013), no. 8, 4209–4227, DOI 10.1090/S0002-9947-2013-05776-5. MR3055694

[18] E. Bujalance, A. Costa, and A. Fernandez, *Uniformization of Klein surfaces by maximal NEC groups*, Rev. Roumaine Math. Pures Appl. **40** (1995), no. 1, 39–54. Travaux de la Conférence Internationale d'Analyse Complexe et du 7e Séminaire Roumano-Finlandais (1993). MR1406119 (97g:30043)

[19] Emilio Bujalance, Francisco-Javier Cirre, and Peter Turbek, *Subgroups of M^*-groups*, Q. J. Math. **54** (2003), no. 1, 49–60, DOI 10.1093/qjmath/54.1.49. MR1967069 (2004b:20073)

[20] Emilio Bujalance, Francisco-Javier Cirre, and Peter Turbek, *Riemann surfaces with real forms which have maximal cyclic symmetry*, J. Algebra **283** (2005), no. 2, 447–456, DOI 10.1016/j.jalgebra.2004.09.010. MR2111204 (2006d:30065)

[21] E. Bujalance, A. F. Costa, S. M. Natanzon, and D. Singerman, *Involutions of compact Klein surfaces*, Math. Z. **211** (1992), no. 3, 461–478, DOI 10.1007/BF02571439. MR1190222 (94i:30039)

[22] E. Bujalance, J. J. Etayo, and E. Martínez, *The full group of automorphisms of non-orientable unbordered Klein surfaces of topological genus 3, 4 and 5*, Rev. Mat. Complut. **27** (2014), no. 1, 305–326, DOI 10.1007/s13163-013-0121-7. MR3149189

[23] Cirre, F.J.: Hidalgo, R.A., Normal coverings of hyperelliptic real Riemann surfaces. This conference proceedings

[24] Antonio F. Costa and Milagros Izquierdo, *Symmetries of real cyclic p-gonal Riemann surfaces*, Pacific J. Math. **213** (2004), no. 2, 231–243, DOI 10.2140/pjm.2004.213.231. MR2036917 (2004k:30098)

[25] Antonio F. Costa and Milagros Izquierdo, *Equisymmetric strata of the singular locus of the moduli space of Riemann surfaces of genus 4*, Geometry of Riemann surfaces, London Math. Soc. Lecture Note Ser., vol. 368, Cambridge Univ. Press, Cambridge, 2010, pp. 120–138. MR2665007 (2011g:30095)

[26] Antonio F. Costa and Milagros Izquierdo, *On the existence of connected components of dimension one in the branch locus of moduli spaces of Riemann surfaces*, Math. Scand. **111** (2012), no. 1, 53–64. MR3001358

[27] Antonio F. Costa and Hugo Parlier, *Applications of a theorem of Singerman about Fuchsian groups*, Arch. Math. (Basel) **91** (2008), no. 6, 536–543, DOI 10.1007/s00013-008-2817-3. MR2465872 (2009j:30098)

[28] José L. Estévez, *Pairs of nonnormal NEC signatures* (Spanish, with English summary), Proceedings of the XVth Portuguese-Spanish Conference on Mathematics, Vol. III (Portuguese) (Évora, 1990), Univ. Évora, Évora, 1991, pp. 113–117. MR1161800

[29] José Luis Estévez and Milagros Izquierdo, *Non-normal pairs of non-Euclidean crystallographic groups*, Bull. London Math. Soc. **38** (2006), no. 1, 113–123, DOI 10.1112/S0024609305017984. MR2201610 (2006k:20103)

[30] Beatriz Estrada, *Automorphism groups of orientable elliptic-hyperelliptic Klein surfaces*, Ann. Acad. Sci. Fenn. Math. **25** (2000), no. 2, 439–456. MR1762428 (2001i:30047)

[31] J. J. Etayo Gordejuela, *Klein surfaces with maximal symmetry and their groups of automorphisms*, Math. Ann. **268** (1984), no. 4, 533–538, DOI 10.1007/BF01451857. MR753412 (86g:30058)

[32] Werner Fenchel and Jakob Nielsen, *Discontinuous Groups of isometries in the hyperbolic plane*, de Gruyter Studies in Mathematics, vol. 29, Walter de Gruyter & Co., Berlin, 2003. Edited and with a preface by Asmus L. Schmidt; Biography of the authors by Bent Fuglede. MR1958350 (2003k:30066)

[33] Ernesto Girondo and Gou Nakamura, *Compact non-orientable hyperbolic surfaces with an extremal metric disc*, Conform. Geom. Dyn. **11** (2007), 29–43 (electronic), DOI 10.1090/S1088-4173-07-00157-9. MR2295996 (2007k:30087)

[34] Leon Greenberg, *Maximal groups and signatures*, Discontinuous groups and Riemann surfaces (Proc. Conf., Univ. Maryland, College Park, Md., 1973), Princeton Univ. Press, Princeton, N.J., 1974, pp. 207–226. Ann. of Math. Studies, No. 79. MR0379835 (52 #740)

[35] Leon Greenberg, *Maximal Fuchsian groups*, Bull. Amer. Math. Soc. **69** (1963), 569–573. MR0148620 (26 #6127)

[36] Grzegorz Gromadzki, *On ovals on Riemann surfaces*, Rev. Mat. Iberoamericana **16** (2000), no. 3, 515–527, DOI 10.4171/RMI/282. MR1813327 (2002a:14032)

[37] Grzegorz Gromadzki and Ewa Kozłowska-Walania, *On ovals of non-conjugate symmetries of Riemann surfaces*, Internat. J. Math. **20** (2009), no. 1, 1–13, DOI 10.1142/S0129167X09005145. MR2488712 (2010g:30056)

[38] W. J. Harvey, *On branch loci in Teichmüller space*, Trans. Amer. Math. Soc. **153** (1971), 387–399. MR0297994 (45 #7046)

[39] A. H. M. Hoare, *Subgroups of N.E.C. groups and finite permutation groups*, Quart. J. Math. Oxford Ser. (2) **41** (1990), no. 161, 45–59, DOI 10.1093/qmath/41.1.45. MR1044754 (91d:20055)

[40] Felix Klein, *On Riemann's theory of algebraic functions and their integrals. A supplement to the usual treatises*, Translated from the German by Frances Hardcastle, Dover Publications, Inc., New York, 1963. MR0158068 (28 #1295)
[41] Ravi S. Kulkarni, *Isolated points in the branch locus of the moduli space of compact Riemann surfaces*, Ann. Acad. Sci. Fenn. Ser. A I Math. **16** (1991), no. 1, 71–81. MR1127697 (93a:30048)
[42] A. M. Macbeath, *The classification of non-euclidean plane crystallographic groups*, Canad. J. Math. **19** (1967), 1192–1205. MR0220838 (36 #3890)
[43] G. D. Mostow, *Strong rigidity of locally symmetric spaces*, Princeton University Press, Princeton, N.J.; University of Tokyo Press, Tokyo, 1973. Annals of Mathematics Studies, No. 78. MR0385004 (52 #5874)
[44] A. M. Macbeath and D. Singerman, *Spaces of subgroups and Teichmüller space*, Proc. London Math. Soc. (3) **31** (1975), no. 2, 211–256. MR0397022 (53 #882)
[45] Adnan Melekoğlu, *A family of M-surfaces whose automorphism groups act transitively on the mirrors*, Rev. Mat. Complut. **13** (2000), no. 1, 163–181. MR1794908 (2002e:30038)
[46] Adnan Melekoğlu, *Automorphism groups of* (M − 1)*-surfaces with the M-property*, Turkish J. Math. **27** (2003), no. 3, 349–367. MR1994796 (2004k:30101)
[47] S. M. Natanzon, *Moduli of Riemann surfaces, real algebraic curves, and their superanalogs*, Translations of Mathematical Monographs, vol. 225, American Mathematical Society, Providence, RI, 2004. Translated from the 2003 Russian edition by Sergei Lando. MR2075914 (2005d:32020)
[48] David Singerman, *Subgroups of Fuschian groups and finite permutation groups*, Bull. London Math. Soc. **2** (1970), 319–323. MR0281805 (43 #7519)
[49] David Singerman, *Finitely maximal Fuchsian groups*, J. London Math. Soc. (2) **6** (1972), 29–38. MR0322165 (48 #529)
[50] David Singerman, *On the structure of non-Euclidean crystallographic groups*, Proc. Cambridge Philos. Soc. **76** (1974), 233–240. MR0348003 (50 #501)
[51] H. C. Wilkie, *On non-Euclidean crystallographic groups*, Math. Z. **91** (1966), 87–102. MR0185013 (32 #2483)

Departamento de Matemáticas Fundamentales, Facultad de Ciencias, UNED, Senda del rey, 9, 28040 Madrid, Spain
E-mail address: acosta@mat.uned.es

Matematiska institutionen, Linköpings Universitet, 581 83 Linköping, Sweden
E-mail address: miizq@mai.liu.se

Departamento de Matemáticas Fundamentales, Facultad de Ciencias, UNED, Senda del rey, 9, 28040 Madrid, Spain
E-mail address: asilva@mat.uned.es

On the minimum genus problem on bordered Klein surfaces for automorphisms of even order

J. J. Etayo and E. Martínez

This paper is dedicated to Professor Emilio Bujalance on occasion of his sixtieth birthday.

ABSTRACT. The minimum genus problem consists on determining the minimum algebraic genus of a surface on which a given group G acts. For cyclic groups G this problem on bordered Klein surfaces was solved in 1989. The next step is to fix the number of boundary components of the surface and to obtain the minimum algebraic genus, and so the minimum topological genus. It was achieved for cyclic groups of prime and prime-power order in the nineties.

In this work the corresponding results for cyclic groups of order $N = 2q$, where q is an odd prime, are obtained. There appear different results depending on the orientability of the surface.

Finally, using the above mentioned results and those of this paper, we state explicitly the general values for arbitrary number of boundary components, which are valid for each $N < 12$, and show how to deal with $N = 12$.

1. Introduction

A bordered Klein surface of algebraic genus $p \geq 2$ has at most $12(p-1)$ automorphisms ([**M1**], [**M2**]). This upper bound is a particular case of the problem of finding the minimum algebraic genus of a surface whose group of automorphisms is a given finite group G. For cyclic groups this value was obtained in [**BEGM**]. Earlier, Harvey had obtained the latter result for Riemann surfaces [**H**].

A further step is to minimize the topological genus when the number of connected components of the boundary is fixed, say k. The interest in this problem comes from the field of real algebraic geometry. In fact, under the equivalence between compact bordered Klein surfaces and real algebraic curves, the boundary components of the surfaces correspond to the ovals of the curve. Therefore, in terms of real algebraic geometry, the above problem consists of computing the minimum genus of the projective, smooth, irreducible, real algebraic curves with k connected components admitting G as a group of birational automorphisms. If G is a cyclic group of order N, this problem was solved in [**BEGG**] for N prime, in [**BGM**] for N a prime-power and in [**EM**] for N a product of two odd primes. For this and

2010 *Mathematics Subject Classification.* Primary 57M60, 20H10; Secondary 30F50.
Key words and phrases. Klein surfaces, algebraic genus, boundary components.
The first author was partially supported by UCM910444 and MTM2011-22435.
The second author was partially supported by MTM2011-23092.

other related questions see [**BCEGM**] where in particular (page 166) the pending question of solving the problem for an arbitrary N is recalled.

In this work we solve this question for $N = 2q$, where q is an odd prime. Although it is quite impossible to obtain an explicit general solution for arbitrary N, we obtain information enough to have it totally solved for small values of N, and show how to deal with a given arbitrary N.

We shall use the notation of [**BEGG**] which we now describe. Let $\mathcal{K}$ be the class of compact bordered Klein surfaces. Given $S \in \mathcal{K}$ we call g its topological genus, k the number of connected components of its boundary, and take $\alpha = 2$ if S is orientable, and $\alpha = 1$ if S is non-orientable. Then the algebraic genus of S is $p = \alpha g + k - 1$. So for a given k, minimizing g is equivalent to minimizing p. Now fix $N \geq 2$, $k \geq 1$. We denote by $\mathcal{K}_+^+(N,k)$ (resp. $\mathcal{K}_+^-(N,k)$) the set of orientable surfaces $S \in \mathcal{K}$ with $p \geq 2$, having k connected boundary components, which admit an orientation-preserving (resp. orientation-reversing) automorphism of order N. In the same way $\mathcal{K}_-(N,k)$ is the set of non-orientable surfaces $S \in \mathcal{K}$ with $p \geq 2$ and k boundary components which admit an automorphism of order N. Then define

$$p_+^+(N,k) = \min\{p \mid \text{there exists } S \in \mathcal{K}_+^+(N,k) \text{ with algebraic genus } p\}$$
$$p_+^-(N,k) = \min\{p \mid \text{there exists } S \in \mathcal{K}_+^-(N,k) \text{ with algebraic genus } p\}$$
$$p_-(N,k) = \min\{p \mid \text{there exists } S \in \mathcal{K}_-(N,k) \text{ with algebraic genus } p\}$$

Of course, these results determine the corresponding values of the topological genus, namely

$$\begin{aligned} g_+^+(N,k) &= \frac{1}{2}(p_+^+(N,k) + 1 - k) \\ g_+^-(N,k) &= \frac{1}{2}(p_+^-(N,k) + 1 - k) \\ g_-(N,k) &= p_-(N,k) + 1 - k \end{aligned}$$

From now on we fix an even number $N = 2q$ where q is an odd prime, and $k \geq 1$. We shall devote Section 2 to computing $p_+^+(N,k)$, Section 3 to computing $p_+^-(N,k)$ and Section 4 to computing $p_-(N,k)$. In the Section 5, we use the results of previous Sections and those in [**BEGG**], [**BGM**], for arbitrary k and each $N < 12$. Section 6 is devoted to the case $N = 12$ which is the first that is not a consequence of the general previous results and it gives an indication of how to deal with arbitrary N, which is described in Section 7.

2. The computation of $p_+^+(N,k)$

Chapter 3 in [**BEGG**] is essentially devoted to determine necessary and sufficient conditions for the existence of a surface $S \in \mathcal{K}_+^+(N,k)$ (and also in $\mathcal{K}_+^-(N,k)$ and $\mathcal{K}_-(N,k)$) with topological genus g.

More precisely, from Theorem 3.1.5 and Corollary 3.2.3 of [**BEGG**], the existence of $S \in \mathcal{K}_+^+(N,k)$ with topological genus g is equivalent to the existence of non-negative integers $g', m_1, \ldots, m_r, t_1, \ldots, t_{k'}$, with $m_i \geq 2$, m_i and t_j divisors of N, such that

$$(2.1)\ \mu = 2g' + k' - 2 + \sum_{i=1}^{r}\left(1 - \frac{1}{m_i}\right) > 0.$$

(2.2) $k = \sum_{j=1}^{k'} \frac{N}{t_j}$.

(2.3) If $g' = 0$, then $\mathrm{lcm}(m_1, \dots, m_r, t_1, \dots, t_{k'}) = N$.

(2.4) The set $X = \{m_1, \dots, m_r, t_1, \dots, t_{k'}\}$ has the *elimination property*, that is to say, the lcm does not change if we delete one of the numbers in X.

If the conditions (2.1)-(2.4) are satisfied, then $N\mu = 2(g-1)+k = p-1$. Hence in order to minimize g and so p, we shall minimize μ constrained to conditions (2.1)-(2.4).

In our strategy to minimize μ, observe that the conditions (2.3) and (2.4) are satisfied, for any choice of k', of the t_j and of the other m_i, if there exist two values $m_1 = m_2 = N$. Also, the addition of a term m_i adds less than a unity to μ. On the other hand, from condition (2.2), k must be obtained as a sum of divisors of N. Dividing a number by a non-trivial divisor of N one has the Euclidean division $x = ay+b$; for any other non-negative integers a', b', such that $x = a'y+b'$, we have $a+b \leq a'+b'-2$, with just one exception. When $y = 2$, taking $a' = a-1$, $b' = b+2$, then $a + b = a' + b' - 1$. With this exception, the best is to take the Euclidean division, k' as small as possible, and add as much (at most two) m_i as necessary. Hence, in all what follows any division will be the Euclidean division, excepting when we divide by 2 and if choosing Euclidean division two m_i are necessary. In that case the result obtained for $a' = a - 1$, $b' = b + 2$, will be better if no m_1 is necessary. This will be the situation in 2.1.2, and the discussion there makes clear the procedure.

Finally observe that g' is only useful in order to assure that $\mu > 0$. For each unity in g' the value of μ increases in 2. Since the same goal can be obtained by adding one or two $m_i = N$, which increase μ in $1 - 1/N$ each, the minimum is always attained for $g' = 0$. Hence throughout this section we take $g' = 0$ without indicating it explicitly.

We distinguish four cases according to the relationship between k and $N = 2q$, namely i) $k < q$; ii) $q \leq k < N$; iii) $k = aN$; and iv) $k > N$, k not a multiple of N.

2.1. $\boxed{k < q}$

2.1.1. $k = 1$. By condition (2.2) $k' = 1$, $t_1 = N$. Then condition (2.3) is satisfied and in order to satisfy condition (2.4) either $m_1 = N$ or $r \geq 2$. Hence for having $\mu > 0$ (condition (2.1)), the minimum value is obtained for $r = 2$, $m_1 = 2$, $m_2 = q$. Then $\mu = \frac{1}{2} - \frac{1}{q}$, and so $p_+^+(N, 1) = 1 + N(\frac{1}{2} - \frac{1}{q}) = q - 1$.

2.1.2. k even. If we take $k' = \frac{k}{2}$, $t_1 = \cdots = t_{k'} = q$, we need $r = 2$, and the minimum μ is attained for $m_1 = 2$ and $m_2 = N$, what gives $\mu = \frac{k-1}{2} - \frac{1}{N}$. However, taking $k' = \frac{k}{2} + 1$, $t_1 = \cdots = t_{k'-2} = q$, $t_{k'-1} = t_{k'} = N$, we have $r = 0$ and $\mu = \frac{k}{2} - 1$, which is lower than the former value. So that $\mu = \frac{k}{2} - 1$, and $p_+^+(N, k) = 1 + N(k/2 - 1)$. Observe that for $k = 2$ this value is not admissible because μ would be 0. Then we still have $\mu = \frac{1}{2} - \frac{1}{N}$, and $p_+^+(N, 2) = q$.

2.1.3. $k \geq 3$, odd. Then $k' = [k/2]+1$, $t_1 = \cdots = t_{[k/2]} = q$, $t_{k'} = N$. The minimum μ is given by $r=1$, $m_1 = 2$. Now $\mu = k/2-1$ and $p_+^+(N,k) = 1+N(k/2-1)$.

2.2. $\boxed{q \leq k < N}$

2.2.1. $k = q$. Then $k' = 1$, $t_1 = 2$. Hence $r = 2$, $m_1 = N$, $m_2 = q$. Thus $\mu = 1-1/N-1/q$ and $p_+^+(N,k) = N-2$.

2.2.2. $k = q+1$. Then $k' = 2$, $t_1 = 2$ and $t_2 = N$. Hence $r = 1$, $m_1 = q$. Thus $\mu = 1-1/q$ and $p_+^+(N,k) = N-1$.

2.2.3. $k = q+2$. Then $k' = 2$, $t_1 = 2$ and $t_2 = q$. Hence $r = 1$, $m_1 = N$. Thus $\mu = 1-1/N$ and $p_+^+(N,k) = N$.

2.2.4. $k = q+b$, $1 < b$ odd. Then $k' = [b/2]+2$, $t_1 = 2$, $t_2 = \cdots = t_{[b/2]+1} = q$, $t_{k'} = N$. Hence $r = 0$. Thus $\mu = (k-q-1)/2$ and $p_+^+(N,k) = 1+q(k-q-1)$.

2.2.5. $k = q+b$, $2 < b$ even. Then $k' = b/2+1$, $t_1 = 2$, $t_2 = \cdots = t_{k'} = q$. Hence $r = 1$, $m_1 = 2$. $\mu = (k-q-1)/2$ and $p_+^+(N,k) = 1+q(k-q-1)$.

2.3. $\boxed{k = aN}$

Then $k' = a$, $t_1 = \cdots = t_a = 1$ and so $r = 2$, $m_1 = m_2 = N$. So that $\mu = (k-2)/N$ and $p_+^+(N,k) = k-1$.

2.4. $\boxed{k > N}$ k not a multiple of N.

It is also clear that in this case the value of $p_+^+(N,k)$ depends on $\left[\frac{k}{N}\right]$ and $k - N\left[\frac{k}{N}\right]$. Let us divide k by N and write $k = aN+N'$. First, we suppose that $N' \neq 2$. Since $N' < N$, we had already obtained the minimum μ corresponding to $k = N'$. Let us call it μ_0 corresponding to certain k'_0, $t_1, \ldots, t_{k'_0}$, r_0, $m_1, \ldots, m_{r_0}$. Then in order to get the minimum μ for N it suffices to take $k' = k'_0 + a$, $t_1, \ldots, t_{k'_0}$, and $t_{k'_0+1} = \cdots = t_{k'_0+a} = 1$, r_0, $m_1, \ldots, m_{r_0}$. So the minimum μ for $k = N$ is $\mu_0 + a$, and hence

$$p_+^+(N,k) = 1+N(\mu_0+a) = 1+N\mu_0+Na.$$

Since $p_+^+(N,N') = 1+N\mu_0$, we have $p_+^+(N,k) = p_+^+(N,N')+Na$.

Now, let $N' = 2$, and recall that $k = 2$ had an exceptional treatment in 2.1.2. So take $k' = 2+a$, $t_1 = t_2 = N$, $t_3 = \cdots = t_{k'} = 1$, $r = 0$. So $\mu = a$, and $p_+^+(N,k) = Na+1$.

We have proved the following Theorem,

THEOREM 2.1. *Let $N = 2q$, q an odd prime, and $1 \leq k$. Then*

$$p_+^+(N,k) = \begin{cases} q-1 & k=1 \\ q & k=2 \\ N\left(\frac{k}{2}-1\right)+1 & 2<k<q \\ q+k-2 & q\le k\le q+2 \\ q(k-q-1)+1 & q+2<k<N \\ k-1 & k>N,\ k\equiv 0,2 \mod (N) \\ p_+^+\left(N, k-N\left[\frac{k}{N}\right]\right)+N\left[\frac{k}{N}\right] & k>N,\ k\not\equiv 0,2 \mod (N) \end{cases}$$

3. The computation of $p_+^-(N,k)$

The biggest difference with the case N odd, is that when N is even the set $\mathcal{K}_+^-(N,k)$ is not void, and so $p_+^-(N,k)$ must be calculated. Besides it is much more involved than the computation of $p_+^+(N,k)$ because the surfaces in $\mathcal{K}_+^-(N,k)$ can come from two ways given by Theorems 3.1.5 and 3.1.9 in [**BEGG**].

Since $N=2q$, this number is even but not a multiple of 4. So Corollary 3.2.3 (3) in the same book gives the first kind of surfaces. For this one by Theorem 3.1.5, there exists $S\in\mathcal{K}_+^-(N,k)$ with topological genus g if and only if there exist non-negative integers $g', m_1,\ldots,m_r, t_1,\ldots,t_{p'}, r'_{s'+1},\ldots,r'_{k'}$ with $m_i\ge 2$, m_i and t_j divisors of $\dfrac{N}{2}$, r'_l even, $0\le p'\le s'\le k'$, $p'<k'$ such that

(3.1.1) $\mu = 2g'+k'-2+\displaystyle\sum_{i=1}^{r}\left(1-\frac{1}{m_i}\right)+\frac{1}{2}\sum_{l=s'+1}^{k'}\frac{r'_l}{2}>0.$

(3.1.2) $k=\displaystyle\sum_{j=1}^{p'}\frac{N}{t_j}+\frac{N}{2}\sum_{l=s'+1}^{k'}\frac{r'_l}{2}.$

(3.1.3) If $g'=0$ and $k'=p'+1$ then $\operatorname{lcm}(m_1,\ldots,m_r,t_1,\ldots,t_{p'})=\dfrac{N}{2}$.

If the conditions (3.1.1)-(3.1.3) are satisfied, then $N\mu=2(g-1)+k=p-1$.

The second kind of surfaces is given by Corollary 3.2.3 in [**BEGG**]. In this case by Theorem 3.1.9, there exists $S\in\mathcal{K}_+^-(N,k)$ with topological genus g if and only if there exist non-negative integers $g'\ge 1$, $m_1,\ldots,m_r$, $t_1,\ldots,t_{p'}$, $r'_{s'+1},\ldots,r'_{k'}$ with $m_i\ge 2$, m_i and t_j divisors of N, r'_l even, $0\le p'\le s'\le k'$, such that

(3.2.1) $\mu = g'+k'-2+\displaystyle\sum_{i=1}^{r}\left(1-\frac{1}{m_i}\right)+\frac{1}{2}\sum_{l=s'+1}^{k'}\frac{r'_l}{2}>0.$

(3.2.2) $k=\displaystyle\sum_{j=1}^{p'}\frac{N}{t_j}+\frac{N}{2}\sum_{l=s'+1}^{k'}\frac{r'_l}{2}.$

(3.2.3) $M=\operatorname{lcm}(m_1,\ldots,m_r,t_1,\ldots,t_{p'})\ne N$.

(3.2.4) The set of even m_i and t_j has even cardinality.

(3.2.5) There exist α_i, β_j such that $\gcd(\alpha_i,m_i)=\gcd(\beta_j,t_j)=1$, and all $\alpha_i\dfrac{N}{m_i}$ and $\beta_j\dfrac{N}{t_j}$ are even.

(3.2.6) If $p' < k'$, then all m_i and t_j are odd.

(3.2.7) If $g' = 1$ and $p' = k'$, then $M = \frac{N}{2}$.

If the conditions (3.2.1)-(3.2.7) are satisfied, then $N\mu = 2(g-1) + k = p - 1$.

We start by minimizing μ on conditions (3.1.1)-(3.1.3). As in the previous Section we can always state $g' = 0$ and k' as small as possible.

For each unity in p' the value of μ increases by one, and the same happens for each unity in $k' - s'$, because both parameters increase k'. Hence p' must be as small as possible, and the same happens with $k' - s'$. Since r'_l has a unique restriction, to be even, and its value provides $r'_l/4$ to μ, we can always take $k' - s'$ to be 0 or 1, and in the latter case, r'_l as big as necessary.

Hence we make the successive Euclidean divisions of k, first by N, the remainder by q, and the new remainder by 2, and we express $k = aN + bq + c2 + d$, with $0 \le bq + 2c + d < N$, $0 \le 2c + d < q$, $0 \le d \le 1$.

If either $a \ge 1$ or $b = 1$, that is to say, $k \ge q$, then we take $k' = s' + 1$, and $r'_{s'+1} = 4a + 2b$. Otherwise, for $k < q$, we take $k' = s'$.

Now if $d = 0$ then $p' = c$, $t_1 = \cdots = t_c = q$. Since condition (3.1.3) is satisfied, we can take $k' = p' + 1$.

Finally, if $d = 1$ in this way it is necessary to take $p' = c + 1$, $t_1 = \cdots = t_c = q$ and $t_{c+1} = N$. But this choice violates the condition that all t_j must divide $N/2 = q$. We shall consider this case separately afterwards.

In this way we get conditions (3.1.2) and (3.1.3) and we have minimized μ, which is $a + b/2 + c - 1$, always with $d = 0$. We must now check the cases in which μ appears to be non-positive. They are

(i) $a = 0$, $b = 1$, $c = 0$, $d = 0$. Then $k = q$ and we have put $p' = 0$, $k' = 1$, $r'_1 = 4a + 2b = 2$. Then $\mu = -1/2$ and so we must add $r = 1$, $m_1 = q$, and so μ becomes $1/2 - 1/q$.

(ii) $a = 1$, $b = 0$, $c = 0$, $d = 0$. Then $k = N$. We have put $p' = 0$, $k' = 1$, $r'_1 = 4a = 4$. Then $\mu = 0$ and we add again $r = 1$, $m_1 = q$ and μ becomes $\mu = 1 - 1/q$.

(iii) $a = 0$, $b = 0$, $c = 1$, $d = 0$. Then $k = 2$, $p' = 1$, $k' = 2$ and $\mu = 0$. We add $r = 1$, $m_1 = q$ and μ becomes $\mu = 1 - 1/q$.

Now we deal with the pending case $d = 1$, that is to say, the remainder of the division of k by q is odd. Since the value $t_j = N$ is not admissible, the best solution is to change $r'_{s'+1}$ into $4a + 2b - 2$, and take $p' = c + (q+1)/2$, $t_1 = \cdots = t'_p = q$, $s' = p'$, and then have $\mu = a + b/2 + c + q/2 - 1$.

Let us now check the exceptional possibilities in this special case. First, if $a = b = 0$, this solution and any other one satisfying (3.1.2) is impossible, and so for all odd values $k < q$, there is no solution under conditions (3.1.1)-(3.1.3). And second, if $a = 0$, $b = 1$, then $r'_{s'+1}$ should be 0, and so $s' = k$. In order to have $p' < k'$, we must put $s' = p' + 1$, and so $\mu = c + (q-1)/2$.

We now compare these results for μ with those corresponding to the second kind of surfaces, satisfying conditions (3.2.1)-(3.2.7).

Since $g' \geq 1$, $k' \geq 1$, the value of μ is always at least 0. Since condition (3.2.2) is just the same as condition (3.1.2), it is easy to check that all numbers satisfying conditions (3.2.1)-(3.2.7) also satisfy conditions (3.1.1)-(3.1-3), excepting maybe the fact that m_i and t_j must divide $N/2$ in the former case. However, in the minimizing process we had always m_i and t_j equal to q if they exist. Hence the second type of surfaces does not give values of μ minor than those of the first kind, excepting possibly the four special cases, namely (i) $a = 0$, $b = 1$, $c = 0$, $d = 0$; (ii) $a = 1$, $b = 0$, $c = 0$, $d = 0$; (iii) $a = 0$, $b = 0$, $c = 1$, $d = 0$ and (iv) $d = 1$. We study now these four values of k. Observe that the condition (3.2.1) together with $g' \geq 1$, $k' \geq 1$, implies that if $\mu > 0$, then $\mu \geq 1/2$.

(i) $a = 0$, $b = 1$, $c = 0$, $d = 0$. Then $k = q$. The previous value $1/2 - 1/q$ cannot be minimized.

(ii) $a = 1$, $b = 0$, $c = 0$, $d = 0$. Then $k = N$. The unique value of $\mu \leq 1 - 1/q$ is $1/2$, which is only attained for $g' = s' = k' = 1$, $r = 1$, $m_1 = 2$. But then (condition (3.2.2)) $p' = 1$, $t_1 = 1$, and neither condition (3.2.4) nor (3.2.7) are satisfied. Again the previous value is not minimized.

(iii) $a = 0$, $b = 0$, $c = 1$, $d = 0$. Then $k = 2$. This implies $p' = 1$, $t_1 = q$. If we put $k' = 1$, then (condition (3.2.1)) we need $r \geq 1$ for having μ positive. But (condition (3.2.7)) then $m_1 = q$. So the minimum μ is again $1 - 1/q$ as in the first kind of surfaces.

(iv) $a = 0$, $b = 0$, $d = 1$. Then we have an odd $k < q$. From condition (3.2.2) there must be some $t_j = N$. But condition (3.2.3) implies that this is not possible. Hence again there are not surfaces with this number of boundary components. For arbitrary a and b, also because of condition (3.2.3) the previous value of μ is not bettered.

We have obtained the minimum value of μ for each k and hence using that $p_{+}^{-}(N,k) = N\mu + 1$ we state the following result, where we express $k = aN + bq + 2c + d$, a, b and c being the quotients of the successive Euclidean divisions.

PROPOSITION 3.1. *Let $N = 2q$, q an odd prime, and $1 \leq k = aN + bq + 2c + d$, excluding odd $k < q$. Then*

$$p_{+}^{-}(N,k) = \begin{array}{ll} N-1 & k=2 \\ q-1 & k=q \\ N-1 & k=N \\ N(a+b/2+c+q/2-1)+1 & k>q,\ k-q[\frac{k}{q}] \text{ odd} \\ N(a+b/2+c-1)+1 & \text{otherwise} \end{array}$$

If $k < q$ is odd, the set $\mathcal{K}_{+}^{-}(N,k)$ is void and hence $p_{+}^{-}(N,k)$ does not exist.

We now express this result directly in terms of $N = 2q$ and k. Observe that $a + b/2 = \frac{1}{2}[\frac{k}{q}]$, $c = [\frac{k-q[\frac{k}{q}]}{2}]$. We state this result as follows

THEOREM 3.2. *Let $N = 2q$, q an odd prime, and $1 \leq k$, excluding odd $k < q$. Then*

$$p_+^-(N,k) = \begin{cases} 2q-1 & k=2 \\ q(k-2)+1 & 2<k<q,\ k \text{ even} \\ q-1 & k=q \\ 2q-1 & k=2q \\ q\left([\frac{k}{q}]+2[\frac{k-q[\frac{k}{q}]}{2}]+q-2\right)+1 & k>q,\ k-q[\frac{k}{q}] \text{ odd} \\ q\left([\frac{k}{q}]+2[\frac{k-q[\frac{k}{q}]}{2}]-2\right)+1 & \text{otherwise} \end{cases}$$

REMARK 3.3. It looks convenient to highlight some aspects of this Section. First, the origin of the two kinds of surfaces comes from the fact that the quotient of the surfaces under the automorphism can be orientable or not. This distinction leads to each of Theorems 3.1.5 and 3.1.9 in [**BEGG**].

And second, consider the case of odd $k < q$. In both kinds of surfaces, respectively conditions (3.1.2) and (3.2.2) are equal. If $k < q$ they reduce to $k = \sum_{j=1}^{p'} \frac{N}{t_j}$. Since $N = 2q$, q prime, all the summands are 2 or 1.

If k is odd, there must be some $N/t_j = 1$, and hence $t_j = N$. But in the first set of conditions, all t_j divide $N/2$, and in the second one their lcm is not N. Hence the conditions cannot be fulfilled, and there exists no reversing-orientation automorphism of order $N = 2q$ for odd $k < q$. Observe that this situation also happens in $\mathcal{K}_+^-(N,k)$ for every n multiple of 4 and arbitrary k odd, [**BGM**, Prop. 3.1].

4. The computation of $p_-(N,k)$

The study of this problem is very similar to the previous Section. We have again two kinds of surfaces given by Theorems 3.1.6 and 3.1.8 in [**BEGG**].

For the first one there exists $S \in \mathcal{K}_-(N,k)$ with topological genus g if and only if there exist non-negative integers g', $m_1,\ldots,m_r$, $t_1,\ldots,t_{p'}$, $r'_{s'+1},\ldots,r'_{k'}$ with $m_i \geq 2$, m_i and t_j divisors of N, r'_l even, $0 \leq p' \leq s' \leq k'$, $p' < k'$ such that

(4.1.1) $\mu = 2g' + k' - 2 + \sum_{i=1}^{r}\left(1 - \frac{1}{m_i}\right) + \frac{1}{2}\sum_{l=s'+1}^{k'} \frac{r'_l}{2} > 0.$

(4.1.2) $k = \sum_{j=1}^{p'} \frac{N}{t_j} + \frac{N}{2}\sum_{l=s'+1}^{k'} \frac{r'_l}{2}.$

(4.1.3) If $g' = 0$ and $k' = p' + 1$ then $\operatorname{lcm}(m_1,\ldots,m_r,t_1,\ldots,t_{p'}) = N$.

If the conditions (4.1.1)-(4.1.3) are satisfied, then $N\mu = g - 1 + k = p - 1$.

The second kind of surfaces is given by Theorem 3.1.8 in [**BEGG**]. There exists $S \in \mathcal{K}_-(N,k)$ with topological genus g if and only if there exist non-negative integers $g' \geq 1, m_1, \dots, m_r, t_1, \dots, t_{p'}, r'_{s'+1}, \dots, r'_{k'}$ with $m_i \geq 2$, m_i and t_j divisors of N, r'_l even, $0 \leq p' \leq s' \leq k'$, such that

(4.2.1) $\mu = g' + k' - 2 + \sum_{i=1}^{r} \left(1 - \frac{1}{m_i}\right) + \frac{1}{2} \sum_{l=s'+1}^{k'} \frac{r'_l}{2} > 0.$

(4.2.2) $k = \sum_{j=1}^{p'} \frac{N}{t_j} + \frac{N}{2} \sum_{l=s'+1}^{k'} \frac{r'_l}{2}.$

(4.2.3) If $p' = k'$, there exist α_i, β_j such that $\gcd(\alpha_i, m_i) = \gcd(\beta_j, t_j) = 1$, and $\sum_{i=1}^{r} \alpha_i \frac{N}{m_i} + \sum_{j=1}^{p'} \beta_j \frac{N}{t_j}$ is even.

(4.2.4) If $g' = 1$ and $p' = k'$, then $\operatorname{lcm}(m_1, \dots, m_r, t_1, \dots, t_{p'}) = N$.

If the conditions (4.2.1)-(4.2.4) are satisfied, then $N\mu = g - 2 + k = p - 1$.

We start by minimizing μ by conditions (4.1.1)-(4.1.3). They are exactly the same as in Section 3, with just one exception which is m_i and t_j are divisors of N (instead of $N/2$), and their lcm in condition (4.1.3) is also N instead of $N/2$. So the unique important difference arises when $d = 0$. Then $p' = c$, with $t_1 = \cdots = t_c = q$. Thus in order to satisfy condition (4.1.3) it is necessary to take $r = 1$, and $m_1 = 2$ if $c \neq 0$, and $m_1 = N$ if $c = 0$. Conversely there exists no problem when $d = 1$ because the value $t_j = N$ is now admissible (see Remark 3.3), and besides it guarantees condition (4.1.3).

Observe that this addition of a term m_1 means that μ is always strictly positive, even in the three exceptional cases appearing in Section 3. Only the case $a = b = c = 0$, $d = 1$, that is to say $k = 1$, must be considered. In that case $p' = 1$, $t_1 = N$, $k' = 2$ and thus we need $r = 1$ and $m_1 = 2$. The values of μ are respectively

$\mu = c + a + b/2 - 1/2$	if $a \geq 1$ or $b = 1$, $c \neq 0$, $d = 0$
$\mu = a + b/2 - 1/N$	if $a \geq 1$ or $b = 1$, $c = 0$, $d = 0$
$\mu = c + a + b/2$	if $a \geq 1$ or $b = 1$, $d = 1$
$\mu = c - 1/2$	if $a = b = d = 0$
$\mu = c$	if $a = b = 0$, $c \neq 0$, $d = 1$
$\mu = 1/2$	if $a = b = c = 0$, $d = 1$

In order to compare these results with those corresponding to surfaces satisfying conditions (4.2.1)-(4.2.4), observe that (4.1.2) and (4.2.2) are the same, and so we need only to check the case $d = 0$, when condition (4.1.3) implies $r = 1$.

If $k \geq q$, then $k' > p'$, and so conditions (4.2.3) and (4.2.4) do not apply. Then from (4.2.1) and (4.2.2) we have $k' = s'+1$, $r'_{s'+1} = 4a+2b$, $p' = c$, $t_1 = \cdots = t_c = q$, and then $\mu = c + a + b/2$. By comparing with the previous result, we need only to check the cases $k = q$ and $k = N$. In the first one we have $\mu = 1/2$, and in the second one we have $\mu = 1$. In both of them the value of μ under conditions (4.1.1)-(4.1.3) was lesser than this one.

Now we suppose $k < q$. Then $k' = p'$, $t_1 = \dots t_c = q$. Because of that, condition (4.2.4) implies $r = 1$, $m_1 = 2$, and again the value of μ is bigger than the value obtained under conditions (4.1.1)-(4.1.3).

From the above values of μ we obtain $p_-(N,k)$ in the following

PROPOSITION 4.1. Let $N = 2q$, q an odd prime and $1 \leq k = aN + bq + 2c + d$. Then

$$p_-(N,k) = \begin{array}{ll} N(c+a+b/2-1/2)+1 & \text{if } a \geq 1 \text{ or } b=1,\ c\neq 0,\ d=0 \\ N(a+b/2-1/N)+1 & \text{if } a \geq 1 \text{ or } b=1,\ c=0,\ d=0 \\ N(c+a+b/2)+1 & \text{if } a \geq 1 \text{ or } b=1,\ d=1 \\ N(c-1/2)+1 & \text{if } a=b=d=0 \\ Nc+1 & \text{if } a=b=0,\ c\neq 0,\ d=1 \\ q+1 & \text{if } a=b=c=0,\ d=1. \end{array}$$

Analogously to Section 3, we express this result on terms of $N = 2q$ and k in the following

THEOREM 4.2. *Let $N = 2q$, q an odd prime and $1 \leq k$. Then*

$$p_-(N,k) = \begin{array}{ll} q+1 & k=1 \\ q(k-1)+1 & 1<k<q \\ k & k \textit{ multiple of } q \\ q[\frac{k}{q}] + 2q[\frac{k-q[\frac{k}{q}]}{2}]+1 & k>q,\ k \textit{ not multiple of } q,\ k-q[\frac{k}{q}] \textit{ odd} \\ q[\frac{k}{q}] + 2q[\frac{k-q[\frac{k}{q}]}{2}]-q+1 & k>q,\ k \textit{ not multiple of } q,\ k-q[\frac{k}{q}] \textit{ even} \end{array}$$

5. The values for $N < 12$

As we saw in [**EM**], there is no hope for an explicit general formula giving $p_+^+(N,k)$, $p_+^-(N,k)$ or $p_-(N,k)$ for an arbitrary N. However, given a pair (N,k), the respective genera can be obtained because all numbers appearing in the relevant sets of conditions are bounded above, see also Example (3.1.10) in [**BEGG**].

We obtained in [**EM**] general formulae when N is odd and k 'small', but the arguments used there only apply in case N even for $k < 3$. Hence we replace those considerations on given values of k by the following ones which involve small values of N. All numbers $N < 12$ are prime, or prime-power, or twice an odd prime. Hence the results obtained in [**BEGG**], [**BGM**] and this paper can be used in order to obtain explicitly $p_+^+(N,k)$, $p_+^-(N,k)$ and $p_-(N,k)$ for each $N < 12$ and arbitrary k. We give them in the following lists.

$\boxed{N=2}$

$$\begin{array}{lll} p_+^+(2,k) = & k+1 & k=1,2 \\ & k-1 & k\geq 3 \\ p_+^-(2,k) = & k+1 & k=1,2 \\ & k-1 & k\geq 3 \\ p_-(2,k) = & 2 & k=1 \\ & k & k>1 \end{array}$$

$\boxed{N=3}$ $k = 3a + b, \quad 0 \le b \le 2.$

$p_+^+(3,k) =$	$3(a+b-2)+1$	$k \neq 2$, $k \equiv 2 \mod (3)$
	$3(a+b-1)$	$k \neq 1$, $k \equiv 1 \mod (3)$, or $k=2$
	$3(a+b)-1$	$k \equiv 0 \mod (3)$, or $k=1$
$p_+^-(3,k)$	does not exist	
$p_-(3,k) =$	$3(a+b-1)+1$	$k \not\equiv 0 \mod (3)$, $k \neq 1$
	$3(a+b)$	$k \equiv 0 \mod (3)$, or $k=1$

$\boxed{N=4}$ $k = 4a + 2b + c, \quad 0 \le b, c \le 1.$

$p_+^+(4,k) =$	$k+1$	$k = 1, 2$
	$4(a+b)$	$k > 2$, odd
	$4(a+b-1)+1$	$k > 2$, $k \equiv 2 \mod (4)$
	$4(a+b)-1$	$k \equiv 0 \mod (4)$
$p_+^-(4,k) =$	5	$k=2$
	$4(a+b)-1$	$k \equiv 0 \mod (4)$
	$4(a+b-2)+1$	$k \equiv 2 \mod (4)$, $k > 2$
	does not exist	k odd
$p_-(4,k) =$	3	$k=1$
	$2(k-2)+1$	$k > 1$, odd
	$2k$	k even

$\boxed{N=5}$ $k = 5a + b, \quad 0 \le b \le 4.$

$p_+^+(5,k) =$	$5(a+b-2)+1$	$k \neq 2$, $k \not\equiv 0, 1 \mod (5)$
	$5(a+b-1)$	$k \neq 1$, $k \equiv 1 \mod (5)$, or $k=2$
	$5(a+b)-1$	$k \equiv 0 \mod (5)$, or $k=1$
$p_+^-(5,k)$	does not exist	
$p_-(5,k) =$	$5(a+b-1)+1$	$k \not\equiv 0 \mod (5)$, $k \neq 1$
	$5(a+b)$	$k \equiv 0 \mod (5)$, or $k=1$

$\boxed{N=6}$ $k = 6a + 3b + 2c + d, \quad 0 \le b, c, d \le 1.$

$p_+^+(6,k) =$	$k-1$	$k \equiv 0, 2 \mod (6), k \neq 2$
	$k+1$	$k \not\equiv 0, 2 \mod (6)$, or $k=2$
$p_+^-(6,k) =$	5	$k=2$
	$k-1$	$k = 3, 6$
	$3(2a+b+2c)+4$	$k \equiv 1 \mod (3)$, $k \neq 1$
	$3(2a+b+2c)-5$	$k \not\equiv 1 \mod (3)$, $k \neq 2, 3, 6$
	does not exist	$k=1$
$p_-(6,k) =$	4	$k = 1, 2$
	k	$k \equiv 0 \mod (3)$
	$3(2a+b+2c)+1$	$k \equiv 1 \mod (3)$, $k \neq 1$
	$3(2a+b+2c)-2$	$k \equiv 2 \mod (3)$, $k \neq 2$

$\boxed{N=7}$ $k = 7a + b, \quad 0 \le b \le 6.$

$$
\begin{array}{lll}
p_+^+(7,k) = & 7(a+b-2)+1 & k \neq 2,\ k \not\equiv 0,1 \mod (7) \\
 & 7(a+b-1) & k \neq 1,\ k \equiv 1 \mod (7), \text{ or } k=2 \\
 & 7(a+b)-1 & k \equiv 0 \mod (7), \text{ or } k=1 \\
p_+^-(7,k) & \text{does not exist} & \\
p_-(7,k) = & 7(a+b-1)+1 & k \not\equiv 0 \mod (7),\ k \neq 1 \\
 & 7(a+b) & k \equiv 0 \mod (7), \text{ or } k=1
\end{array}
$$

$\boxed{N=8}$ $k = 8a + 4b + 2c + d, \quad 0 \le b, c, d \le 1,$

$$
\begin{array}{lll}
p_+^+(8,k) = & k+3 & k=1,2 \\
 & 8(a+b+c) & k>1, \text{ odd} \\
 & 8(a+b+c-1)+1 & k>2,\ k \equiv 2 \mod (4) \\
 & 8(a+b+c)-1 & k \equiv 0 \mod (4) \\
p_+^-(8,k) = & 5 & k=2 \\
 & 8(a+b+c)-1 & k \equiv 0 \mod (4) \\
 & 8(a+b+c-1)+1 & k>2,\ k \equiv 2 \mod (4) \\
 & \text{does not exist} & k \text{ odd} \\
p_-(8,k) = & 5 & k=1 \\
 & 4(2a+b+2c)+1 & k>1, \text{ odd} \\
 & 4(2a+b+2c) & k \text{ even}
\end{array}
$$

$\boxed{N=9}$ $k = 9a + 3b + c, \quad 0 \le b, c \le 2.$

$$
\begin{array}{lll}
p_+^+(9,k) = & k+5 & k=1,2 \\
 & 9(a+b)+1 & k \neq 2,\ k \equiv 2 \mod (3) \\
 & 9(a+b) & k \neq 1,\ k \equiv 1 \mod (3) \\
 & 9(a+b)-1 & k \equiv 0 \mod (3) \\
p_+^-(9,k) & \text{does not exist} & \\
p_-(9,k) = & 7 & k=1 \\
 & 9(a+b+c-1)+1 & k>1,\ k \not\equiv 0 \mod (3) \\
 & 9(a+b) & k \equiv 0 \mod (3)
\end{array}
$$

$\boxed{N=10}$ $k = 10a + 5b + 2c + d, \quad 0 \le b \le 1, \quad 0 \le c \le 2, \quad 0 \le d \le 1.$

$$
\begin{array}{lll}
p_+^+(10,k) = & k-1 & k \equiv 0,2 \mod (10), k \neq 2 \\
 & k+7 & k \equiv 4,9 \mod (10) \\
 & k+3 & \text{otherwise} \\
p_+^-(10,k) = & k+7 & k=2,4 \\
 & k-1 & k=5,10 \\
 & 5(2a+b+2c)+16 & k \equiv 1,3 \mod (5),\ k \neq 1,3 \\
 & 5(2a+b+2c)-9 & k \equiv 0,2,4 \mod (5),\ k \neq 2,4,5,10 \\
 & \text{does not exist} & k=1,3 \\
p_-(10,k) = & 6 & k=1 \\
 & 5k-4 & k=2,3,4 \\
 & k & k \equiv 0 \mod (5) \\
 & 5(2a+b+2c)+1 & k \equiv 1,3 \mod (5),\ k \neq 1,3 \\
 & 5(2a+b+2c)-4 & k \equiv 2,4 \mod (5),\ k \neq 2,4
\end{array}
$$

$\boxed{N = 11}$ $k = 11a + b, \quad 0 \le b \le 10.$

$p_+^+(11,k) =$	$11(a+b-2)+1$	$k \neq 2,\ k \not\equiv 0,1 \mod (11)$
	$11(a+b-1)$	$k \neq 1,\ k \equiv 1 \mod (11)$, or $k = 2$
	$11(a+b)-1$	$k \equiv 0 \mod (11)$, or $k = 1$
$p_+^-(11,k)$	does not exist	
$p_-(11,k) =$	$11(a+b-1)+1$	$k \not\equiv 0 \mod (11),\ k \neq 1$
	$11(a+b)$	$k \equiv 0 \mod (11)$, or $k = 1$

6. The case $N = 12$

Since 12 is not a prime nor a prime-power, nor twice a prime, the general results quoted above cannot be applied. However, still the respective sets of conditions can be used for N arbitrary, and we make it here for the value $N = 12$. Since 12 has a lot a divisors, the analysis is much more involved, but it gives an idea of how to rule out cases for a specific value of N.

In order to obtain $p_+^+(12,k)$, from Theorem 3.1.5 and Corollary 3.2.3(2) in [**BEGG**], the conditions (2.1)-(2.4) in Section 2 must be satisfied. We obtain first for each $k \le 12$ the value of k' and the corresponding t_j satisfying the condition (2.2), and then the values of m_i such that the conditions (2.1), (2.3) and (2.4) are fulfilled with minimal μ. It is very important to note that for $k = 2$ and $k = 8$, the minimal value of μ is not obtained making minimal k' because of the necessity of the elimination property.

In the next Table we list the values of k, k', t_j and m_i:

$k=1$	$k'=1$	$t_1=12$	$m_1=12$	$m_2=2$	
$k=2$	$k'=2$	$t_1=12$	$t_2=12$	$m_1=2$	
$k=3$	$k'=1$	$t_1=4$	$m_1=3$	$m_2=12$	
$k=4$	$k'=1$	$t_1=3$	$m_1=4$	$m_2=12$	
$k=5$	$k'=2$	$t_1=3$	$t_2=12$	$m_1=4$	
$k=6$	$k'=1$	$t_1=2$	$m_1=12$	$m_2=12$	
$k=7$	$k'=2$	$t_1=2$	$t_2=12$	$m_1=12$	
$k=8$	$k'=3$	$t_1=2$	$t_2=12$	$t_3=12$	
$k=9$	$k'=2$	$t_1=2$	$t_2=4$	$m_1=3$	$m_2=12$
$k=10$	$k'=2$	$t_1=2$	$t_2=3$	$m_1=4$	$m_2=12$
$k=11$	$k'=3$	$t_1=3$	$t_2=3$	$t_3=4$	$m_1=4$
$k=12$	$k'=1$	$t_1=1$	$m_1=12$	$m_2=12$	

Now we suppose $k > 12$, $k = 12a + b$, with $0 \le b \le 11$. Then the minimal value for μ is obtained by adding a to the value of k' corresponding to b, and all the new t_j equal to 1. The m_i added to satisfy conditions (2.3) and (2.4) must be kept, and in just two cases an m_i for making $\mu > 0$. Hence we arrive now to the values of k', t_j and m_i for each b, and the two exceptional cases.

b	k'	t_j	m_i
1	$a+1$	$t_1 = 12,\ t_2 = \cdots = t_{k'} = 1$	$m_1=12$; if $k=1$, $m_2=2$
2	$a+2$	$t_1 = t_2 = 12,\ t_3 = \cdots = t_{k'} = 1$	if $k = 2$ then $m_1 = 2$
3	$a+1$	$t_1 = 4,\ t_2 = \cdots = t_{k'} = 1$	$m_1 = 3,\ m_2 = 12$
4	$a+1$	$t_1 = 3,\ t_2 = \cdots = t_{k'} = 1$	$m_1 = 4,\ m_2 = 12$
5	$a+2$	$t_1 = 3,\ t_2 = 12,\ t_3 = \cdots = t_{k'} = 1$	$m_1 = 4$
6	$a+1$	$t_1 = 2,\ t_2 = \cdots = t_{k'} = 1$	$m_1 = m_2 = 12$
7	$a+2$	$t_1 = 2,\ t_2 = 12,\ t_3 = \cdots = t_{k'} = 1$	$m_1 = 12$
8	$a+3$	$t_1 = 2,\ t_2 = t_3 = 12,\ t_4 = \cdots = t_{k'} = 1$	
9	$a+2$	$t_1 = 2,\ t_2 = 4,\ t_3 = \cdots = t_{k'} = 1$	$m_1 = 3,\ m_2 = 12$
10	$a+2$	$t_1 = 2,\ t_2 = 3,\ t_3 = \cdots = t_{k'} = 1$	$m_1 = 4,\ m_2 = 12$
11	$a+3$	$t_1 = t_2 = 3,\ t_3 = 4,\ t_4 = \cdots = t_{k'} = 1$	$m_1 = 4$
0	a	$t_1 = \cdots = t_{k'} = 1$	$m_1 = m_2 = 12$

Now we calculate μ for each value of b and using that $12\mu + 1 = p$, we have

$$p_+^+(12,k) = \begin{array}{ll} k-1 & k \equiv 0,1,2 \mod (12),\ k \neq 1,2 \\ k+5 & k \equiv 3,\dots,8 \mod (12), \text{ or } k = 1,2 \\ k+11 & k \equiv 9,10,11 \mod (12) \end{array}$$

Considering $p_+^-(12,k)$, first observe that if k is odd, then $\mathcal{K}_+^-(12,k)$ is void, see Remark 3.3 and [**BGM**, Prop 3.1].

Since 12 is a multiple of 4, surfaces of only one kind appear in this case, by Corollary 3.2.3(4) of [**BEGG**]. They are those defined by Corollary 3.2.3(2) and Theorem 3.1.9 there. We explicitly describe the conditions. There exists $S \in \mathcal{K}_+^-(12,k)$ with topological genus g if and only if there exist non-negative integers $g' \geq 1$, $k' \geq 1$, $m_1, \dots, m_r$, $t_1, \dots, t_{k'}$ with $m_i \geq 2$, m_i and t_j divisors of 12 such that

(6.1) $\mu = g' + k' - 2 + \sum_{i=1}^{r} \left(1 - \dfrac{1}{m_i}\right) > 0.$

(6.2) $k = \sum_{j=1}^{k'} \dfrac{12}{t_j}.$

(6.3) $M = \operatorname{lcm}(m_1, \dots, m_r, t_1, \dots, t_{k'}) \neq 12.$

(6.4) The set of those m_i and t_j which are a multiple of 4 has even cardinality.

(6.5) There exist α_i, β_j such that $\gcd(\alpha_i, m_i) = \gcd(\beta_j, t_j) = 1$, and all $\alpha_i \dfrac{12}{m_i}$ and $\beta_j \dfrac{12}{t_j}$ are even. Moreover the sum of all $\alpha_i \dfrac{12}{m_i}$ and $\beta_j \dfrac{12}{t_j}$ is a multiple of 4 if and only if g' is even.

(6.6) If $g' = 1$, then $M = 6$.

As we made before we take $g' = 1$ and k' as small as possible. We begin with $k \leq 12$, always even.

$k = 2$	$k' = 1$	$t_1 = 6$		$m_1 = 3$
$k = 4$	$k' = 1$	$t_1 = 3$		$m_1 = 2$
$k = 6$	$k' = 1$	$t_1 = 2$		$m_1 = 3$
$k = 8$	$k' = 2$	$t_1 = 2$	$t_2 = 6$	$m_1 = 2$
$k = 10$	$k' = 2$	$t_1 = 2$	$t_2 = 3$	
$k = 12$	$k' = 1$	$t_1 = 1$		$m_1 = 6$

Observe that the choice of m_1 when $k = 2, 8$ is made in order to satisfy the second part of condition (6.5).

Now if $k > 12$, $k = 12a + b$, it suffices to add a to the value of k' corresponding to b. It is necessary to keep m_1 which guarantees the condition (6.6). Hence we have

b	k'	t_j	m_i
2	$a+1$	$t_1 = 6,\ t_2 = \cdots = t_{k'} = 1$	if $k = 2$ then $m_1 = 3$
4	$a+1$	$t_1 = 3,\ t_2 = \cdots = t_{k'} = 1$	$m_1 = 2$
6	$a+1$	$t_1 = 2,\ t_2 = \cdots = t_{k'} = 1$	$m_1 = 3$
8	$a+2$	$t_1 = 2,\ t_2 = 6,\ t_3 = \cdots = t_{k'} = 1$	$m_1 = 2$
10	$a+2$	$t_1 = 2,\ t_2 = 3,\ t_3 = \cdots = t_{k'} = 1$	
0	a	$t_1 = \cdots = t_{k'} = 1$	$m_1 = 6$

Now we calculate μ for these values of b and using $12\mu + 1 = p$ we obtain

$$p_+^-(12,k) = \begin{array}{ll} k-1 & k \equiv 0,2 \mod (12),\ k \neq 2 \\ k+3 & k \equiv 4,6,10 \mod (12) \\ k+11 & k \equiv 8 \mod (12) \\ 9 & k = 2 \end{array}$$

Finally for $p_-(12, k)$ we have two kinds of surfaces. Those given by Theorem 3.1.6 in **[BEGG]** satisfying conditions (4.1.1)-(4.1.3) in Section 4, and those give by Theorem 3.1.8 satisfying conditions (4.2.1)-(4.2.4) plus an additional condition in case $g' = 2$. Since this assumption never holds when minimizing μ we do not need to check it. We act as above, with conditions (4.1.1)-(4.1.3) firstly for $k \leq 6$.

$k = 1$	$p' = 1$	$t_1 = 12$	$s' = k' = 2$	$m_1 = 2$
$k = 2$	$p' = 1$	$t_1 = 6$	$s' = k' = 2$	$m_1 = 4$
$k = 3$	$p' = 1$	$t_1 = 4$	$s' = k' = 2$	$m_1 = 3$
$k = 4$	$p' = 1$	$t_1 = 3$	$s' = k' = 2$	$m_1 = 4$
$k = 5$	$p' = 2$	$t_1 = 3,\ t_2 = 12$	$s' = k' = 3$	
$k = 6$	$p' = s' = 0$		$k' = 1,\ r_1' = 2$	$m_1 = 12$

Now if $k > 6$, $k = 6a + b$, we take $k' = s' + 1$, and $r'_{s'+1} = 2a$ if $b \neq 0$, $r'_{s'+1} = 2a + 2$ if $b = 0$. Hence we have

b	k'	p'	t_j	$r'_{k'}$	m_i
1	2	1	$t_1 = 12$	$r_2' = 2a$	if $k = 1$, $m_1 = 2$
2	2	1	$t_1 = 6$	$r_2' = 2a$	$m_1 = 4$
3	2	1	$t_1 = 4$	$r_2' = 2a$	$m_1 = 3$
4	2	1	$t_1 = 3$	$r_2' = 2a$	$m_1 = 4$
5	3	2	$t_1 = 3, t_2 = 12$	$r_3' = 2a$	
0	1	0		$r_1' = 2a + 2$	$m_1 = 12$

As it happens in Section 4, the second set of conditions does not give minor values for μ. Hence we calculate μ according to the above numbers and we obtain

$$p_-(12,k) = \begin{array}{ll} k & k \equiv 1 \mod (6),\ k \neq 1 \\ k+8 & k \equiv 2,5 \mod (6) \\ k+6 & k \equiv 3,4 \mod (6) \text{ or } k=1 \\ k+5 & k \equiv 0 \mod (6) \end{array}$$

7. Final remarks

The analysis of the case $N = 12$ recalls that for arbitrary N the determination of $p_+^+(N,k)$, $p_+^-(N,k)$ and $p_-(N,k)$ is much more involved as the complexity of the prime decomposition of N increases. Anyway, the respective values can be obtained for a given pair (N,k) applying directly Theorems 3.1.2 to 3.1.9 in [**BEGG**], and keeping in mind that Corollary 3.2.3 of the same book determines whether an automorphism of an orientable surface is or not orientation preserving. The key point is that k must be expressed as a sum of divisors of N. So there is a finite set of candidate numbers, which determine the value of μ that must be minimized, and it is always feasible. On the contrary, there is no explicit expression for general (N,k).

The authors wish to thank the referee for the very careful reading of the paper which spotted a mistake in Section 2.

References

[BCEGM] E. Bujalance, F. J. Cirre, J. J. Etayo, G. Gromadzki, and E. Martínez, *A survey on the minimum genus and maximum order problems for bordered Klein surfaces*, Groups St Andrews 2009 in Bath. Volume 1, London Math. Soc. Lecture Note Ser., vol. 387, Cambridge Univ. Press, Cambridge, 2011, pp. 161–182. MR2858856 (2012j:30120)

[BEGG] E. Bujalance, J. J. Etayo, J. M. Gamboa, and G. Gromadzki, *Automorphism groups of compact bordered Klein surfaces. A combinatorial approach*, Lecture Notes in Mathematics, vol. 1439, Springer-Verlag, Berlin, 1990. MR1075411 (92a:14018)

[BEGM] E. Bujalance, J. J. Etayo, J. M. Gamboa, and G. Martens, *Minimal genus of Klein surfaces admitting an automorphism of a given order*, Arch. Math. (Basel) **52** (1989), no. 2, 191–202, DOI 10.1007/BF01191274. MR985604 (90b:30052)

[BGM] E. Bujalance, J. M. Gamboa, and C. Maclachlan, *Minimum topological genus of compact bordered Klein surfaces admitting a prime-power automorphism*, Glasgow Math. J. **37** (1995), no. 2, 221–232, DOI 10.1017/S0017089500031128. MR1333741 (96c:57026)

[EM] J. J. Etayo and E. Martínez, *On the minimum genus problem on bordered Klein surfaces*, Mathematical contributions in honor of Juan Tarrés (Spanish), Univ. Complut. Madrid, Fac. Mat., Madrid, 2012, pp. 149–158. MR3060529

[H] W. J. Harvey, *Cyclic groups of automorphisms of a compact Riemann surface*, Quart. J. Math. Oxford Ser. (2) **17** (1966), 86–97. MR0201629 (34 #1511)

[M1] C. L. May, *Automorphisms of compact Klein surfaces with boundary*, Pacific J. Math. **59** (1975), no. 1, 199–210. MR0399451 (53 #3295)

[M2] C. L. May, *Cyclic automorphism groups of compact bordered Klein surfaces*, Houston J. Math. **3** (1977), no. 3, 395–405. MR0457710 (56 #15914)

Departamento de Álgebra, Facultad de Matemáticas, Universidad Complutense. 28040-Madrid, Spain
E-mail address: jetayo@mat.ucm.es

Departamento de Matemáticas Fundamentales. UNED. Paseo Senda del Rey 9. 28040-Madrid, SPAIN
E-mail address: emartinez@mat.uned.es

Computing adapted bases for conformal automorphism groups of Riemann surfaces

Jane Gilman

This paper is dedicated to Emilio Bujalance on his 60th birthday

ABSTRACT. The concept of an adapted homology basis for a prime order conformal automorphism of a compact Riemann surface originated in a series of papers by the author in the 1970s and is currently extended to arbitrary finite groups of conformal automorphisms. Here we compute some examples of adapted homology bases for some groups of automorphisms. The method is to begin by apply the Schreier-Reidemeister rewriting process and the Schreier-Reidemeister Theorem as formulated by Magnus, Karrass and Solitar and then to eliminate generators and relations until there is one single large defining relation for the fundamental group in which every generator and its inverse occurs. We are then able to compute the action of the group on the homology image of these generators in the first homology group. The matrix of the action is in a simple form.

1. Introduction

This paper is about adapted homology bases for conformal automorphism groups of compact Riemann surfaces. The concept of an adapted homology basis can be extended from a cyclic group of prime order to an arbitrary conformal automorphism group [**12**]. Our emphasis here is on the method. Adapted bases can be considered from two different equivalent approaches: one is from the geometric/topological perspective of coverings and curve lifting and the other is from the algebraic perspective of groups and their subgroups using the Schreier-Reidemeister theory. In particular, although in the original concept and initial results [**6**, **7**] used curve lifting, it became clear that using Shreier-Reidemeister theory [**25**] gave a much less ad hoc method and more robust, complete results [**9**, **10**]. Here we give three explicit examples (sections 6, 7, and 8) demonstrating how the Schreier-Reidemeister method is used to find adapted homology bases.

2010 *Mathematics Subject Classification.* Primary 20,32,30,52.

Some of this work was carried out while the author was a supported visitor at the Institute for Computational and Experimental Research in Mathematics (ICERM) and other parts under an NSF IPA grant as well as a grant from the NSA.

2. The Set-Up and Notation

Let S, a compact Riemann surface of genus $g \geq 2$ and G, a group of conformal automorphisms of S, with order, $o(G) = n$. G acts on S with quotient $S_0 = S/G$, of genus g_0.

G induces an action on the free homotopy gropup of S and on the first homology of S. Here by abuse of notation we do not distinguish between the action of G on S, on the free homotopy group, or on homology.

2.1. Well-known facts. If $y_1, ..., y_m$ are the fixed points on S of elements of G, then G_{y_i}, the stabilizer of y_i is cyclic for each i. Set $n_i = o(G_{y_i})$, the order of G_{y_i}. The covering $\pi : S \to S_0$ is branched over t points $p_1, ..., p_t$. If $\pi(y_i) = p_j$, then there are $\frac{n}{n_i}$ distinct points projecting onto p_j. As this is a regular covering, the points over each given p_j have the same branch order. The Riemann-Hurwitz Theorem implies

$$2g = 2n(g_0) - 2n + 2 + n\Sigma_{i=1}^{t}(1 - \frac{1}{n_i}).$$

We assume that S and S_0 are uniformized by the Fuchsian groups Γ and Γ_0 and that $U = \{x + iy | x, y \in \mathbb{R} y > 0\}$ is the upper half-plane in $\mathbb{C}$ so that $S = U/\Gamma$ and $S_0 = U/\Gamma_0$. While Γ is a surface group, Γ_0 is either a surface group or an orbifold group depending upon whether it contain elliptics. We note that Γ_0/Γ is isomorphic to G and that $\exists \phi : \Gamma_0 \to G$ with $\Gamma = Ker\phi$.

The map ϕ is known as surface kernel map or *the surface kernel homomorphism*. A vector that gives the same information as the surface kernel map is known as *the generating vector*. It is the vector

$$(\phi(a_1), ..., \phi(a_{g_0}), \phi(b_1), ..., \phi(b_{g_0}), \phi(x_1), ..., \phi(x_t))$$

where

$$\Gamma_0 = \langle a_1, ..., a_{g_0}, b_1, ..., b_{g_0}, x_1, ..., x_t \mid R = (\Pi_{j=1}^{g_0}[a_j, b_j]) \cdot x_1 \cdots x_t = 1;\ x_i^{n_i} = 1, i = 1, ..., t \rangle.$$

Γ_0/Γ is isomorphic to G and acts as outer automorphisms on Γ. Both G and Γ_0/Γ induce an action on the first homology group of S. By abuse of notation we do not distinguish between the action of G on the surface and the action induced on the first homology group. We write the fundamental group multiplicatively and the first homology group additively. We use $=$ for equality both in Γ and in the homology group, but sometimes emphasize equality in homology by using $\approx_h$.

2.2. Original motivation and potential applications. The original motivation for considering adapted bases was to establish an algebraic structure on the moduli space of compact surfaces with punctures. Moduli space is the quotient of the Teichmüller Space under the action of the Mapping-class group, but it can also be obtained by factoring through the Siegel Upper Half plane, that is by taking the images of the period matrices in the Siegel Upper half Plane factored out by the action of the Siegel Modular group. It was known by the work of Bailey using the Satake Compactification of the quotient of the Siegel Upper half Plane under the Siegel Modular group, that the moduli space of compact surfaces was a quasi-projective variety.

If one begins with a punctured surface, one can form a two sheeted covering that is branched over the punctures. The covering will have a conformal involution that fixes the points lying over the punctures. The matrix of the involution with respect to the adapted basis will then have an image in the Siegel modular group and as such will act on the period matrices of all surfaces. The fixed points then of the involution will be a subvariety corresponding to punctured surfaces and its image in moduli space an algebraic variety. There is some work to extend the action of the image of the adapted automorphism in the Siegel Modular group to the compactification of moduli space.

We mention a number of potential applications of adapted bases which will be addressed more fully in [**12**]. One area of applications is to the representation variety [**14, 15, 27**]. We consider the representation variety of the fundamental group of a compact surface of genus $g \geq 2$, $Hom(\pi_1(S), PSL(2,\mathbb{R})/PSL(2,\mathbb{R}))$ where the action of $PSL(2,\mathbb{R})$ is given by conjugation. If S is a surface with a conformal automorphism group G, there is a representation of the fundamental group of S where the images of the curves in $PSL(2,\mathbb{R})$ have equal traces. That is, there is a set of generators for the fundamental group of S whose images give an adapted homology basis and, therefore, whose traces satisfy certain relations. Conversely, if $\pi_1(S)$ has a representation whose traces satisfy these relations, S has a conformal automorphism group isomorphic to G. That is, the length function will have certain symmetries. A corollary of the definition of an adapted basis, definition 3.1 below, and the existence of such a basis is:

COROLLARY 2.1. Representations of discrete groups with conformal automorphisms can be identified by the traces of elements in the representation.

The existence of adapted bases will give further results about strata in the Teichmuller space and their algebraic structure. It would be nice if it these together with the results for all finite groups could be used to say something new about the disparity in dimensions count for $\mathcal{H}_g$, the Siegel Upper-half plane (i.e. $(g(g+1)/2)$) and that of the period matrices for compact Riemann surfaces that it contains since Teichmuller space or moduli space has dimension $(3g-3)$ for compact surfaces.

3. Adapted homology basis

Heuristically, an adapted basis is basis in which the action of a group is in a particularly nice simple form. The original definition for prime order automorphisms [**6, 7, 9**] can be extended to any finite group G of conformal automorphisms.

DEFINITION 3.1. (*Adapted Homology Basis*) [**12**] The homology basis of S, $\mathcal{B}$, is *adapted* to a finite group G if for each $\gamma \in \mathcal{B}$ one of the following occurs:

(1) γ and $g(\gamma)$ are in the basis for all $g \in G$ and $g(\gamma) \neq \gamma$.
(2) γ and $h^j(\gamma)$ are in the basis for all $j = 0, 1, ..., m_{i-2}$ where $h \in G$ is of order m_i, and

$$h^{m_{i-1}}(\gamma) = -(\gamma + h(\gamma) + \cdots + h^{m_{i-2}}(\gamma)).$$

Further for each right coset representative, g_h, for G modulo $\langle h \rangle$, we have that $g_h(\gamma)$ and $(g_h \circ h^j)(\gamma)$ are in the basis for all $j = 0, 1, ..., m_{i-2}$ and

$$(g_h \circ h^{m_{i-1}})(\gamma) = -(g_h(\gamma) + (g_h \circ h)(\gamma) + \cdots (g_h \circ h^{m_{i-2}})(\gamma)).$$

(3) $\gamma = h^r(\gamma_0)$, $r \neq m_{i-1}$ where γ_0 is one of the curves in item 2 above.
(4) $g(\gamma) \in \mathcal{B}$ for all $g \in G_0$, G_0 a subgroup of G of order m. All of the other n/m images of γ under G are permuted appropriately by conjugate elements. Further if h is a coset representative of G/G_0, then either $h(\gamma)$ or $-h(\gamma)$ is in $\mathcal{B}$.

Here by abuse of notation we do not distinguish between the action of G on the surface and the action induced on the first homology group or between a curve on the surface and its image in homology. We do not use different notation for equality in the group G, the fundamental group, or the homology group except we use $\approx_h$ when it seems important to emphasize that this is equality of curves in homology.

When we construct adapted homology bases we refer to curves of type (1), (2), (3), and (4) as in definition 3.1 above.

We note that they can be more than one homology basis that is an adapted basis for any given group.

The question of the existence of an adapted homology basis for any G of finite order is addressed in [**12**]. Further, the numbers of the different types of elements in the basis depends upon the fixed point structure of the group G which is determined by the surface kernel homomorphism or equivalently by the generating vector in a manner that can be made precise again using definition 3.1 and the existence to give

COROLLARY 3.2. [**12**] If G is a group of conformal automorphisms of S of order n, then the matrix of the action of an element of G on an appropriately ordered adapted homology basis breaks up into blocks: an element of type (1) will give a block that is an $n \times n$ permutation matrix; elements of type (2) will gives blocks that are $(m_i - 1) \times (m_i - 1)$ matrices with 1's along the super diagonal, all elements of the last row -1 and all other entries 0; elements of type (4) will give $n \times n$ square matrices that contain permuted blocks of n/m_i blocks of $(m_i) \times (m_i)$ permutation matrices.

Note that every element of G either fixes this basis (but not the order basis) or maps some elements into their inverse. That is, elements of G will permute the blocks and/or the entries within the blocks, but may map some generators into their inverses.

REMARK 3.3. The matrix will, of course, not necessarily be a symplectic matrix, but the intersection numbers of the curves can be computed from the surface kernel map ϕ using the methods of [**13**]. The method there is applied to the prime order case and is due to Patterson, but it also applies to an arbitrary finite order group and has been used by subsequent authors. Further, and there are methods for finding the corresponding symplectic matrix and the action of the group on the corresponding homology basis [**10, 13**].

REMARK 3.4. We note that if the group has no fixed points, then all elements are of type 1 or 4. If the quotient has genus 0, all elements are of type 2 or 3, so a large automorphism group in the sense of Kulkarni with quotient of genus 0, will have all elements γ of type 2 or 3 and every curve on the surface will satisfy $\Sigma_{g\in G} g(\gamma) \approx_h 1$. Note that Kulkarni defines a large automorphism groups one where the order of the group is greater than $4(g-1)$ so that the genus of the quotient mist be 0 or 1. The latter occurs only for triangle groups. Here we speak of *large automorphism groups in the sense of Kulkarni of genus* 0.

4. Outline of the Method

In order to find the adapted basis pick a set of right coset representatives for Γ_0/Γ. One must choose a Schreier system of cosets, that is one where the initial segment of any representative is itself a representative. Use the theorem from [25] Theorem 5.1 page and given below. This gives a presentation for the kernel, that is generators and relations for the kernel. This uses the coset representatives to obtain generators for the kernel. There is a rewriting system τ rewriting any word in the kernel as a word in the chosen generators. One then eliminates of generators and relations via an algorithm [12] using Teitze transformations until there is only one main relation of length $4g$. The relation contains each word and its inverse exactly once. One can compute the action of elements of the group by conjugating the word by a coset representative on a word using and computing the image of the conjugate under τ.

The full algorithm for the elimination of generators and relations is given in [12]. We give examples of the computations and the full elimination process in the case of three groups.

Treatments of the Schreier-Reidemeister methods can be found in Lyndon-Schupp [22] and the text of D. Johnson [18]. See also works of Marston Condor, Gareth Jones [20] and David Singerman as well as that of Costa and Turbek [5].

5. Schreier-Reidemeister Theorem

We remind the reader of the Schreier-Reidemeister Theorem.

THEOREM 5.1. (Schreier-Reidemeister [25] as stated for our situation) *Let* Γ_0 *have generators*

$$a_1, ..., a_{g_0}, b_1, ..., b_{g_0}, x_1, ...x_t$$

and relations

$$R = \Pi_{i=1}^{g_0}[a_i, b_i] \cdot x_1 \cdots x_t = 1, \quad x_j^{n_j} = 1, j = 1, ..., t.$$

and let Γ *be the subgroup of* Γ_0 *with* G *as above* G *isomorphic* Γ_0/Γ.

If τ *is a Reidemeister-Schreier rewriting process, then* Γ *can be presented as a group with generators*

$$S_{K,a_i}, \quad S_{K,b_i}, i = 1, ..., g_0 \text{ and, } \quad S_{K,x_j}, \ j = 1, ..., t$$

and relations

$$\tau(KRK^{-1}) = 1, \ \tau(Kx_j^{n_j}K^{-1}) = 1, \ j = 1, ..., t, \ S_{M,a} = 1$$

where K *runs over a complete set of coset representative and* M *is a coset representative and* a *any element of* Γ_0 *with* Ma *freely equal to* $\overline{Ma}$.

Here K is a system of Schreier right coset representatives for Γ_0/Γ (that is, a systems of coset representatives where any initial segment of a representative is also a representative). We let $\overline{w}$ denote the coset representative for a word $w \in \Gamma_0$. $S_{K,y} = Ky\overline{Ky}^{-1}$ for any generator y. The rewriting process τ expresses a word in the generators for Γ_0 as a word in the generators for the kernel, if the word is in the kernel and is defined as follows: let $c_i, i = 1, ..., r$ be generators of Γ_0 and $\epsilon_i = \pm 1$.

$$\tau(c_1^{\epsilon_1} c_2^{\epsilon_2} c_3^{\epsilon_3} \cdots c_r^{\epsilon_r}) = S_{K_1,c_1}^{\epsilon_1} S_{K_2,c_2}^{\epsilon_2} \cdots S_{K_r,c_r}^{\epsilon_r}$$

where $K_i = \overline{c_1 c_2 \cdots c_{i-1}}$ if $\epsilon_i = 1$ and $K_i = \overline{c_1 c_2 \cdots c_{i-1} c_i}$ if $\epsilon_i = -1$.

DEFINITION 5.2. For ease of reference, when q is in the generating set for Γ_0, we refer to the $S_{K,q}$ as the q-generators for Γ.

One can speak of relators for the groups Γ_0 and Γ or of relations. Relators omit the equal sign, but relations do not. We also use the equal sign to denote equality in homology but use $\approx_h$ to denote equality in the homology group for emphasis at times. That is, we use $\approx_h$ to denote homology when the distinction between $=$ and $\approx_h$ is significant.

Reminder: As noted in section 3 we use the same notation for the conjugation action of an element of Γ on Γ_0 as for its induced action on curves in the corresponding homology basis and/or homotopy basis.

6. Example #1: A group with no fixed points: $G = \mathbb{Z}_6$

The Fuchsian group Γ_0*:* $\Gamma_0 = \langle a, b, c, d \mid [a, b][c, d]| \rangle$

The finite group: $G = \mathbb{Z}_6 = \langle h \rangle$.

The homomorphism, ϕ*:* $\phi(q) = \phi(b) = \phi(c) = 1; \phi(d) = h$. *The generating vector:* $(\phi(a), \phi(b), \phi(c), \phi(d)) = (1, 1, 1, h)$

The Coset representatives: $\{1, , d, d^2, d^3, d^4, d^5\}$

A Riemann-Hurwitz calculation gives $g = 7$ as $g_0 = 2$.

6.1. Subgroup generators.

For a. Write the subgroup generators for a, $S_{1,a}, S_{b,a}, S_{d,a}, S_{d^2,a}, S_{bd,a}, S_{bd^2,a}$, as:

- $S_{1,a} = a\overline{a}^{-1} =: A$
- $S_{d,a} = da\overline{da}^{-1} = h(A)$
- $S_{d^d,a} = d^2 a\overline{d^2}^{-1} = h^2(A)$
- $S_{d^3,a} = d^3 a\overline{d^3}^{-1} = h^3(A)$
- $S_{d^4,a} = d^4 a\overline{d^4 a}^{-1} = h^4(A)$
- $S_{d^5,a} = d^5 a\overline{d^5 a}^{-1} = h^5(A)$

Compute $\tau(d^j a d^{-j}) = h^j(A)$.

Computation for images of b and c are similar. Set $S_{1,b} = B$ and $S_{1,c} = C$ (i.e. $S_{1,b} := B$ and $S_{1,c} := C$) so that we have

$$B, h(B), h^2(B), h^3(B), h^4(B), h^5(B), h^6(B) = B$$

and

$$C, h(C), h^2(C), h^3(C), h^4(C), h^5(C), h^6(C) = C$$

Images of d. Here, by contrast, we will set $S_{1,d^6} = D$ (i.e. $S_{1,d^6} =: D$).

- $S_{1,d} = d\overline{d}^{-1} = 1$
- $S_{d,d} = dd\overline{dd}^{-1} = 1$
- $S_{d^d,d} = d^2 d\overline{d^3}^{-1} = 1$
- $S_{d^3,a} = d^3 d\overline{d^4}^{-1} = 1$
- $S_{d^4,d} = d^4 d\overline{d^4 d}^{-1} = 1$
- $S_{d^5,d} = d^5 d\overline{d^5 d}^{-1} = d^5 d\overline{d^6 d}^{-1} = d^6 \cdot 1 = d^6 := D$

We compute $\tau(d^j d^6 d^{-j}) = \tau(d^j D d^{-j})$ for all j to see that, $h^j(D) = D$.

$$\tau(R) = \tau([a,b][c,d]).$$

$$\tau(aba^{-1}b^{-1}cdc^{-1}d^{-1}) = S_{1,a} \cdot S_{\overline{a},b} \cdot S^{-1}_{\overline{aba^{-1}},a} \cdot S^{-1}_{\overline{aba^{-1}b^{-1}},b} \cdot S_{\overline{aba^{-1}b^{-1}},c} \cdot S_{\overline{aba^{-1}b^{-1}c},d} \cdot S^{-1}_{\overline{aba^{-1}b^{-1}cdc^{-1}},c} \cdot S^{-1}_{\overline{aba^{-1}b^{-1}cdc^{-1}d^{-1}},d}$$

Set $X = [A,B]$ and note that the $S_{*,d}$ words are all 1. We obtain

$$XCh(C) = 1 \tag{6.1}$$

Compute $\tau(d^jRd^{-j}) = \tau(d^j[a,b][c,d]d^{-j})$. For $j = 1,2,3,4$ we obtain

$$h(X)h(C)h^2(C) = 1 \tag{6.2}$$

$$h^2(X)h^2(C)h^3(C) = 1 \tag{6.3}$$

$$h^3(X)h^3(C)h^4(C) = 1 \tag{6.4}$$

$$h^4(X)h^4(C)h^5(C) = 1 \tag{6.5}$$

But when we compute

$$\tau(d^5Rd^{-5}) = \tau(d^5[a,b][c,d]d^{-5}), \tag{6.6}$$

we obtain D and D^{-1} in the relation:

$$h^5(X)h^5(C)DC^{-1}D^{-1} = 1 \tag{6.7}$$

That is, let

$$Q = S_{1,d} \cdot S_{d,d} \cdot S_{d^2,d} \cdot S_{d^3,d} \cdot S_{d^4}$$

We obtain

$$Q \cdot S_{d^5,a} \cdot S_{\overline{d^5a},b} \cdot {}^{-1}_{\overline{d^5aba^{-1}},a} \cdot S^{-1}_{\overline{d^5aba^{-1}b^{-1}},b} \cdot S_{\overline{d^5aba^{-1}b^{-1}},c} \cdot S_{\overline{d^5aba^{-1}b^{-1}c},d} \cdot S^{-1}_{\overline{d^5aba^{-1}b^{-1}cdc^{-1}},c} \cdot S^{-1}_{\overline{d^5aba^{-1}b^{-1}cdc^{-1}d^{-1}},d} \cdot Q^{-1}.$$

Since $Q = 1$, we obtain equation 6.7.

6.1.1. **Eliminations.** Now we could eliminate images of C from the six relations, but instead we note that $XCh(C)^{-1} = 1$ gives $h(C) = XC$ since X is a commutator we have $h(C) \approx_h C$, where $\approx_h$ denotes homologous.

We end up with a single long relation the product of X and its five images under h and the $CDC^{-1}D^{-1}$.

Since $g_0 = 2$, the order of G is six, and there is no branching $2g-2 = g(2g_0-2)$, we have $2g = 14$. Thus we have the correct number of generators and their inverses in the long defining relation.

Now $X = [A,B]$ so that $h^j(A)$ and $h^j(B)$ are in the basis for all $j = 1,...,6$. We note that $h^j(D) \approx_h D$ and $h^j(C) \approx_h C$ for all j. That is, we have twelve generators of type (1) and two generators of type (4). Types (1) and (4) refer to the types in definition 3.1.

Wirth respect to this basis, the matrix of the action of any element of G will have two 6×6 blocks that are permutation matrices and one 2×2 identity matrix.

7. The $G = \mathbb{Z}_2 \times \mathbb{Z}_2$ example, example #2

The Fuchsian group Γ_0; $\Gamma_0 = \langle a, b, c, d \mid abcd; a^2 = b^2 = c^2 = d^2 \rangle$

The finite group: $G = \mathbb{Z}_2 \times \mathbb{Z}_2 = \langle g, h | g^2 = h^2 = 1, gh = hg \rangle$

The homomorphism ϕ: $\phi(a) = \phi(b) = g$, $\phi(c) = \phi(d) = h$,

The generating vector: $(\phi(a), \phi(b), \phi(c), \phi(d) = (g, g, h, h)$

Coset representatives: $1, b, d, bd$

Riemann Hurwitz: $2 - 2g = 4(2 - 2g_0) - 4[(1 - \frac{1}{2}) + (1 - \frac{1}{2} + (1 - \frac{1}{2} + (1 - \frac{1}{2}]$ since $g_0 = 0$ Thus $g = 1$. The homology basis will have two curves.

7.1. a-generators.

$S_{1,a} = ab^{-1} =: A$
$S_{b,a} = ba$
$g(A) = \tau(bab^{-1}a^{-1}) = S_{1,b}S_{b,a}S_{b,b}^{-1}S_{1,b}^{-1} = S_{b,a} = g(S_{1,a})$
$\tau(aa) = S_{1,a}S_{b,a}$
$\implies S_{b,a} = S_{1,a}^{-1}$
$S_{d,a} = da\overline{bd}^{-1} = dad^{-1}b^{-1}$
$(\tau(dab^{-1}d^{-1}) = S_{1,d}S_{d,a}S_{d,b}^{-1}S_{1,d}^{-1} = S_{d,a}S_{d,b}^{-1}$ -because all $S_{*,d} = 1$, as we will see below.
$h(S_{1,a}) = S_{d,a}S_{d,b}^{-1}$
$h(S_{d,a}) = \tau(ddad^{-1}b^{-1}d^{-1}) = S_{1,a}S_{b,d}^{-1}S_{d,b}^{-1}S_{1,d}^{-1} = S_{1,a}S_{b,d}^{-1}$.
$S_{bd,a} = bdad^{-1}$
$h(S_{bd,a}) = \tau(dbdad^{-1}) = S_{1,d}S_{d,b}S_{bd,d}S_{b,a}S_{1,d} = S_{d,b}S_{b,a}$
$S_{bd,a} = S_{d,a}^{-1}$

7.2. b-generators.

$S_{1,b} = bb^{-1} = 1$
$S_{b,b} = b^2 \times 1 = 1$
$S_{d,b} = db(bd)^{-1} = dbd^{-1}b^{-1} = [d, b] \approx_h 1$
$S_{bd,b} = bdbd^{-1} = bdb^{-1}d^{-1} = [b, d] \approx_h = 1$
$S_{bd,b} = S_{d,b}^{-1}$
Note $h(S_{d,b}) = \tau(ddbd^{-1}b^{-1}d^{-1})\tau(bd^{-1}b^{-1}d^{-1} = S_{1,b}S_{bd,d}^{-1}S_{d,b}^{-1}S_{1,d}^{-1} = S_{d,b}^{-1} = S_{bd,b}$.
$g(S_{d,b}) = \tau(bdbd^{-1}b^{-1}b^{-1}) = \tau(bdbd^{-1}) = S_{1,b}S_{b,d}S_{bd,b}S_{1,d}^{-1} = S_{bd,b} = S_{d,b}^{-1}$
$gh(S_{d,b}) = g(S_{b,d}^{-1}) = S_{d,b}$.
$g(S_{bd,b}) = \tau(bbdbd^{-1}b^{-1}) = \tau(dbd^{-1}b^{-1}) = S_{b,d} = S_{bd,b}^{-1}$
$h(S_{bd,b}) = \tau(dbdbd^{-1}d^{-1}) = \tau(dbdb) = S_{d,b} = S_{bd,b}^{-1}$
We have computed all of the images of the b-words under the group.

7.3. c-generators.

$S_{1,c} = cd^{-1}$
$S_{b,c} = bc(bd)^{-1} = bcd^{-1}b^{-1}$
$S_{d,c} = dc\overline{(cd)}^{-1} = dcd^{-2} = dc$
$S_{bd,c} = bdc\overline{bdc}^{-1} = bdc(bd^2)^{-1} = bdcb^{-1}$
Images of the c words under the group:

$g(S_{1,c}) = \tau(bS_{1,c}b^{-1}) = \tau(bcd^{-1}b^{-1} = S_{1,b}S_{b,c}S_{b,d}^{-1}S_{1,b}^{-1} = S_{b,c}$
$h(S_{1,c}) = \tau(dS_{1,c}d^{-1}) = \tau(dcd^{-1}d^{-1}) = S_{d,c}$
$gh(S_{1,c}) = \tau(bdcd^{-1}d^{-1}b^{-1}) = \tau(bdcb^{-1}) = S_{1,b}S_{b,d}S_{bd,c}S_{1,b}^{-1} = S_{bd,c}$
We note $S_{b,c} = S_{bd,c}^{-1}$ and $S_{1,c} = S_{d,c}^{-1}$. $g(S_{d,c}) = \tau(bdcb^{-1}) = S_{1,b}S_{b,d}S_{bd,c}S_{1,b}^{-1} = S_{bd,c}$
$h(S_{d,c}) = \tau(ddcd^{-1}) = S_{1,c}$
$gh(S_{d,c}) = g(S_{1,c}) = S_{b,c}$
$g(S_{b,c}) = \tau(bbcd^{-1}b^{-1}b^{-1}) = \tau(cd^{-1}) = S_{1,c}S_{1,d}^{-1} = S_{1,c}$
$h(S_{b,c}) = \tau(dbcd^{-1}b^{-1}d^{-1}) = \tau(dbcd^{-1}b^{-1}d^{-1}) = S_{1,d}S_{d,b}S_{bd,c}S_{bd,d}^{-1}S_{d,b}^{-1}S_{1,d}^{-1} = S_{d,b}S_{db,c}S_{d,b}^{-1}$
$gh(S_{b,c}) = g(S_{d,b})g(S_{db,c})g(S_{d,b})^{-1}) = S_{d,b}^{-1}g(S_{bd,c})S_{d,b}$
$g(S_{bd,c}) = \tau(bbdc(bd^2)-1b^{-1}) = \tau(bbdcd^{-1}d^{-1}b^{-1}b^{-1}) = S_{d,c}$

7.4. d-generators.
$S_{1,d} = d \cdot d^{-1} = 1$
$S_{b,d} = bd(bd)^{-1} = 1$
$S_{d,d} = dd^{-1} = 1$
$S_{bd,d} = bd^2(b)^{-1} = 1$

7.5. τ images.
$\tau(a^2) = S_{1,a}S_{b,a} = 1$
$\tau(ba^2b^{-1}) = S_{1,b}S_{b,a}S_{1,a}S_{1,b}^{-1} = S_{b,a}S_{1,a} = 1$
$\tau(da^2d^{-1})S_{1,d}S_{d,a}S_{bd,a}S_{1,d}^{-1} = S_{d,a}S_{bd,a} = 1.$
$\tau(bda^2d^{-1}b^{-1}) = S_{1,b}S_{b,d}S_{bd,a}S_{d,a}S_{b,d}{-}1S_{1,b}^{-1} = S_{bd,a}S_{d,a} = 1.$

7.6. $\tau(Kb^2K^{-1})$. $\tau(b^2) = S_{1,b}S_{b,b} = 1$
$\tau(bb^2b^{-1}) = 1$
$\tau(db^2d^{-1}) = S_{1,d}S_{d,b}S_{bd,b}S_{1,d}^{-1} = S_{d,b}S_{bd,b} = 1.$
$\tau(bdb^2d^{-1}b^{-1}) = S_{1,b}S_{b,d}S_{bd,b}S_{d,b}S_{*,d}{-}1S_{1,b}^{-1} = S_{bd,b}S_{d,b} = 1.$

7.7. $\tau(Kc^2K^{-1})$. $\tau(c^2) = S_{1,c}S_{b,c} = 1$
$\tau(bc^2b^{-1}) = S_{1,b}S_{b,c}S_{bd,c}S_{1,b}^{-1} = S_{b,c}S_{bd,c}$
$\tau(dc^2d^{-1}) = S_{1,d}S_{d,c}S_{1,c}S_{1,d}^{-1} = S_{d,c}S_{1,c}.$
$\tau(bdc^2d^{-1}b^{-1}) = S_{1,b}S_{b,d}S_{bd,c}S_{d,c}S_{*,d}^{-1}S_{1,b}^{-1} = S_{bd,c}S_{d,c} = 1.$

7.8. $\tau(Kd^2K^{-1}) = 1 \forall K.$

7.9. $\tau(KRK^{-1})$. where $R = abcd$

(1) $\tau(abcd) = S_{1,a}S_{\overline{a},b}S_{\overline{ab},c}S_{\overline{abc},d} = S_{1,a}S_{b,b}S_{1,c}S_{d,d} = S_{1,a}S_{1,c} = 1$

(2) $\tau(babcdb^{-1}) = S_{1,b}S_{b,a}S_{\overline{ba},b}S_{\overline{bab},c}S_{\overline{babc},d}S_{1,b}^{-1} = S_{b,a}S_{b,c} = 1$

(3) $\tau(dabcdd^{-1}) = S_{1,d}S_{d,a}S_{\overline{da},b}S_{\overline{dab},c}S_{\overline{dabc},d}S_{1,d}^{-1} = S_{d,a}S_{bd,b}S_{b,c} = 1$

(4) $\tau(bdabcdd^{-1}b^{-1}) = S_{1,b}S_{b,d}S_{bd,a}S_{\overline{bda},b}S_{\overline{bdab},c}S_{\overline{bdabc},b}^{-1} = S_{bd,a}S_{d,b}S_{bd,c} = 1$

7.10. Combining.

Let $M =: S_{d,a}$ so that $M^{-1} =: S_{bd,a}$ and let $T =: S_{b,c}$ so that $T^{-1} =: S_{bd,c}$. Let $Z = S_{b,d}$. Since $S_{d,b} = S^{-1}_{bd,b}$, we have by 3 $MZT = 1$ and by 4 $M^{-1}Z^{-1}T^{-1} = 1$ so that $Z^{-1} = MT$ or $Z = T^{-1}M^{-1}$, yielding

$$MTM^{-1}T^{-1} = 1.$$

We note:

$$g(M) = \tau(bS_{d,a}b^{-1}) = \tau(bdad^{-1}b^{-1}b^{-1}) = S_{1,b}S_{b,d}S_{bd,a}S^{-1}_{1,d} = S_{bd,a} = S^{-1}_{d,a} = M^{-1}.$$

$$h(M) = \tau(dS_{d,a}d^{-1}) = \tau(ddad^{-1}b^{-1}d^{-1}) = S_{1,a} = S^{-1}_{b,a} = M^{-1}$$

$$gh(M) = M$$

$$g(T) = \tau(bS_{b,c}b^{-1}) = \tau(bbcd^{-1}b^{-1}b^{-1}) = \tau(cd^{-1}) =$$

$$g(T) = S_{1,c}S^{-1}_{1,d} = S_{1,c} = S^{-1}_{b,c} = T^{-1}.$$

$$h(T) = \tau(dS_{b,c}d^{-1}) = \tau(dbcd^{-1}b^{-1}d^{-1}) = S_{1,d}S_{b,d}S_{bd,c}S^{-1}_{b,d}S^{-1}_{b,b}S^{-1}_{1,d} = S_{bd,c}$$

$$= S^{-1}_{d,c} = T^{-1} \text{ and } gh(T) = T.$$

We are left with two generators M and T each of type (4) definition 3.1. Finally we can double check that the elements in our group fix the correct number of points on the surface using the fact that the number of fixed points of an element is the same as 2 - the trace of its action on homology by the Lefshetz fixed point formula [**23**].

8. Example #3: different Γ_0, but the same $G = \mathbb{Z}_2 \times \mathbb{Z}_3$

The Fuchsian group Γ_0*:* $\Gamma_0 = \langle a, b, c, d, e \mid abcde; a^2 = b^2; c^3 = d^3 = e^3 \rangle$

The finite group G: $G = \mathbb{Z}_2 \times \mathbb{Z}_3 = \langle g, h | g^2 = h^3 = 1, gh = hg \rangle$

The surface kernel map ϕ*:* $\phi(a) = \phi(b) = g$, $\phi(c) = \phi(d) = \phi(e) = h$,

The generating vector: $(\phi(a), \phi(b), \phi(c), \phi(d), \phi(e)) = (g, g, h, h, h)$

Coset representatives: $1, b, d, d^2, bd, bd^2$

Use the Riemann-Hurwitz theorem to compute that since $g_0 = 0$, $g = 4$, so that a homology basis will have 8 elements.

8.1. a-generators.

$S_{1,a} = ab^{-1} =: A$

$S_{b,a} = ba$

$\tau(bab^{-1}a^{-1}) = S_{1,b}S_{b,a}S^{-1}_{b,b}S^{-1}_{1,b} = S_{b,a}$. Later we will write $S_{b,a} = g(S_{1,a})$.

$\tau(aa) = S_{1,a}S_{b,a}$

$\implies S_{b,a} = S^{-1}_{1,a}$

$S_{d,a} = da\overline{bd}^{-1}$

$\tau(dab^{-1}d^{-1}0 = S_{1,d}S_{d,a}S^{-1}_{d,d}S^{-1}_{1,d} = S_{d,a}$. Later we will write $h(S_{1,a}) = S_{d,a}$.

$S_{d^2,a} = d^2a(bd^2)^{-1} = d^2ad^{-2}b^{-1} = [d^2, a]ab^{-1}$.

Note $S_{d^2,a} =\approx_h S_{1,a}$

$S_{bd,a}$

$S_{bd^2,a}$

8.2. b-generators.

$S_{1,b} = bb^{-1} = 1$

$S_{b,b} = b^2$

Since this is a relator. We can remove $S_{b,b}$ from the generating set if we replace each of its occurrences by 1.

$S_{d,b} = db(bd)^{-1} = dbd^{-1}b^{-1} = [b,d]$. Thus $S_{d,b} \approx_h 1$.

$S_{d^2,b} = d^2b(bd^2)^{-1} = [b,d^2]$. Thus $S_{d^2,b} \approx_h 1$.

$S_{bd,b} = bdbd^{-1} = b[d,b]b$. Thus $S_{bd,b} \approx_h 1$.

$S_{bd^2,b} = bd^2bd^-2 = [b,d^2] = b[d^2,b]b$. Thus $S_{bd^2,b} \approx_h 1$.

8.3. c-generators.

$S_{1,c} = cd^{-1}$

$S_{b,c} = bc(bd)^{-1} = bcd^{-1}b^{-1}$. Since $S_{b,c} = [b,c]cb^{-1}$, $S_{b,c} \approx_h 1 \approx cb^{-1}$. and $\tau(S_{b,c}) = \tau(bbc\overline{bc}^{-1}b^{-1} = \tau(bbcd^{-1}b^{-1}b^{-1})\tau(cd^{-1} = S1,cS1,d^{-1} = S_{1,c}$.

Thus $S_{d,c} = dc(cd)^{-1} = dcd^{-2}$

Thus $S_{d,c} = dcd^{-1}c^{-1}cd^{-1} = [d,c^{-1}]cd^{-1} =\approx_h 1cd^{-1} \approx_h cd^{-1} = S_{1,c}^{-1}$

$S_{d^2,c} = d^2c\overline{d^2c}^{-1} = d^2c$

and

$h(S_{d^2,c})\tau(dd^2cd^{-1}) = \tau(cd^{-2} = S_{1,c}S_{1,d}^{-1} = S_{1,c}$.

$S_{bd,c} = bdc\overline{bdc}^{-1} = bdc(bd^2)^{-1} = bdcd^{-2}b^{-1} = bdcd^{-1}b^{-1} =$

$b[d,c]cd^{-1}b^{-1} = [b,[d,c]][d,c]bcd^{-1}b^{-1}$

$= [b,[d,c]][d,c]bcb^{-1}c^{-1}cbd^{-1}b^{-1}$

$= [b,[d,c]][d,c]cd^{-1}dbd^{-1}b^{-1}b^{-1} = [b,[d,c]][d,c]cd^{-1}[d,b]$.

Thus $S_{bd,c} \approx_h cd^{-1}$

$S_{bd^2,c} = bd^2cb^{-1} = bd^{-1}cb^{-1}$

$S_{bd^2,c} = bd^{-1}cdc^{-1}cd^{-1}b^{-1} = b^{-1}[d^{-1},c]cd^{-1}b$

$S_{bd^2,c} \approx_h b^{-1}cd^{-1}b = (cd^{-1})[cd^{-1},b^{-1}]$

$h(S_{bd^2,c}) = [b,d]S_{b,c}[b,d]^{-1}$

$h(S_{b,c}) \approx_h \tau(dcb^{-1}d^{-1} =\approx_h S_{1,d}S_{d_2,c}S_{bd^2,b}^{-1}S_{bd,d}^{-1} \approx_h S_{d^2,c}$

8.4. d-generators.

$S_{1,d} = dd^{-1} = 1$

$S_{b,d} = bd(bd)^{-1} = 1$

$S_{d,d} = dd^{-1} = 1$

$S_{d^2,d} = d^3 = 1$

$S_{bd,d} = bd^2(bd^2)^{-1} = 1$

$S_{bd^2,d} = bd^ddb^{-1} = bd^3b^{-1}$

8.5. e-generators.

$S_{1,e} = ed^{-1}$

$S_{b,e} = be(bd)^{-1} = bed^{-1}b^{-1}$.

Thus $g(S_{1,e}) = \tau(bS_{1,b}b^{-1}) = \tau(bed^{-1}b^{-1}) = S_{b,e}$

$S_{d,e} = de(ed)^{-1} = ded^{-2}$

Thus $h(S_{1,e}) = \tau(ded^{-2}) = S_{1,d}S_{d,e}S_{d,d}^{-1}S_{1,d}^{-1} = S_{d,e}$,

and we have also: $S_{d,e} = ded^{-1}e^{-1}ed^{-1}) = ([d,e]ed^{-1}) \approx_h S_{1,e}$

$S_{d^2,e} = d^2e = d^{-1}e$

Thus $h(S_{d,e}) = \tau(dS_{d,e}d^{-1}) = \tau(dded^{-2}d^{-1} = \tau(dde) = S_{1,d}S_{d,d}S_{d^2,e} = S_{d^2,e}$.

$h(S_{d^2,e}) = \tau(ddded\overline{d^2e}^{-1}d^{-1}) = \tau(ed^{-1}) = S_{1,d}S_{1,d}^{-1} = S_{1,e}$.

$S_{bd,e} = bde\overline{bde}^{-1} = bde(bd^2){-}1 = bded^{-2}b^{-1}$.

Thus $S_{bd,e} = bded^{-1}e^{-1}ed^{-1}b^{-1} = bd[e,d^{-1}]ed^{-1}b^{-1} \approx_h bded^{-1}b^{-1}$ $= bded^{-2}db^{-1} = bd\cdot ed^{-1}e^{-1}d\cdot d^{-1}e\cdot d^{-1}b^{-1} = bd[e,d^{-1}]d^{-1}ed^{-1}b^{-1} =$ product of some commutators $d^{-1}e \approx_h d^{-1}e = S_{d^2,e}$.

$S_{bd^2,e} = bd^2eb^{-1} = bd^{-1}eb^{-1}$

$S_{bd^2,e} = bd^{-1}ede^{-1}ed^{-1}b^{-1} = b[d^{-1},e]ed^{-1}b^{-1} = b[d^{-1},e]S_{1,e}b^{-1}$ $= [b,[d^{-1}e,e]S_{1,e}]S_{1,e} \approx_h S_{1,e}$

8.6. τ images.

$\tau(a^2) = S_{1,a}S_{b,a} = 1$

$\tau(ba^2b^{-1}) = S_{1,b}S_{b,a}S_{1,a}S_{1,b}^{-1} = S_{b,a}S_{1,a} = 1$

$\tau(da^2d^{-1})S_{1,d}S_{d,a}S_{bd,a}S_{1,d}{-}1 = S_{d,a}S_{bd,a} = 1$.

$\tau(d^2a^2d^{-2}) = S_{1,d}S_{d,d}S_{d^2,a}S_{bd^2,a}S_{d,d}^{-1}S_{1,d}^{-1} = S_{d^2,a}S_{bd^2,a} = 1$

$\tau(bda^2d^{-1}b^{-1}) = S_{1,b}S_{b,d}S_{bd,a}S_{d,a}S_{b,d}{-}1S_{1,b}^{-1} = S_{bd,a}S_{d,a} = 1$.

$\tau(bd^2a^2d^{-2}b^{-1}) = S_{1,b}^{-1}S_{b,d}S_{bd,d}S_{bd^2,a}S_{d^2,a}S_{bd,d}^{-1}S_{b,d}{-}1S_{1,b}^{-1} = S_{bd^2,a}S_{d^2,a} = 1$

Note we can thus eliminate $a-$ words at some point and combine the 6 relators into 3.

8.7. $\tau(Kb^2K^{-1})$.

$\tau(bb) = S_{1,b}S_{b,b} = 1$ but each these two words are already the identity, thus the relator is $S_{b,b} = 1$

$\tau(bbbb^{-1})$ yields nothing that is, $\tau(b^2) = t(bbbb^{-1}) = 1$.

$\tau(dbbd^{-1}) = S_{1,d}S_{d,b}S_{bd,b}S_{1,d}^{-1} \implies S_{d,b} = S_{bd,b}^{-1}$

$\tau(bdbbd^{-1}b^{-1}) = S_{1,b}S_{b,d}S_{bd,b}S_{d,b}S_{b,d}^{-1}S_{1,b} \implies S_{d,b} = S_{bd,b}^{-1}$

$\tau(ddbbd^{-1}) = S_{1,d}S_{d,d}S_{d^2,b}S_{bd^2,b}S_{d,d}^{-1} = 1 \implies S_{d^2,b} = S_{bd^2,b}^{-1}$.

8.8. $\tau(KcccK^{-1})$.

$\tau(ccc) = S_{1,c}S_{d,c}S_{d^2,c} = 1$

I.e. $Ch(C)h^2(C) = 1$.

$\tau(bcccb^{-1}) = 1 \implies g(C)gh(C)gh^2(C) = 1$.

8.9. $\tau(KeeeK^{-1})$. $\tau(eee) = S_{1,e}S_{d,e}S_{d^2,e} = 1$

Eventually we will have, $Eh(E)h^2(E) = 1$ if we set $E = S_{1,e}$ and

$\tau(beeeb^{-1}) = 1 \implies g(E)gh(E)gh^2(E) = 1$.

8.10. $\tau(ddd)$.

$$\tau(ddd) = S_{1,d}S_{d,d}S_{d^2,d}$$

$$\tau(bdddb^{-1}) = S_{1,b}S_{b,d}S_{bd,d}S_{bd^2,d}S_{1,b}^{-1} = S_{b,d}S_{bd,d}S_{bd^2,d}$$

8.11. $\tau(KRK^{-1})$. where $R = abcde$.

(1) $\tau(abcde) = S_{1,a}S_{\overline{a},b}S_{\overline{ab},c}S_{\overline{abc},d}S_{\overline{abcd},e}$.
Note that all of the $S_{*,d}$ *are* 1. *Let* $A = S_{1,a}, C = S_{1,c},$ *and* $E = S_{1,e}$.

$$\tau(abcde) = A \cdot S_{b,b} \cdot C \cdot S_{d,d} \cdot h^2(E) = 1.$$

Continue computing.

(2) $\tau(babcdeb^{-1}) = 1$ gives

$$S_{1,b}S_{b,a}S_{\overline{ba},b}S_{\overline{bab},c}S_{\overline{babc},d}S_{\overline{babcd},e}S_{1,b}^{-1}$$

(3) $\tau(dabcded^{-1}) = 1$ gives

$$S_{1,d}S_{d,a}S_{\overline{da},b}S_{\overline{dab},c}S_{\overline{dabc},d}S_{\overline{dabcd},e}S_{1,d}^{-1}$$

(4) $\tau(d^2abcded^{-2}) = 1$

$$S_{1,d}S_{d,d}S_{d^2,a}S_{\overline{d^2a},b}S_{\overline{d^2ab},c}S_{\overline{d^2abc},d}S_{\overline{d^2abcd},e}S_{d,d}^{-1}S_{1,d}^{-1}$$

(5) $\tau(bdabcded^{-1}b{-}1) = 1$ gives

$$S_{1,b}S_{b,d}S_{bd,a}S_{\overline{bda},b}S_{\overline{bdab},c}S_{\overline{bdabc},d}S_{\overline{bdabcd},e}S_{b,d}^{-1}S_{1,b}^{-1}$$

(6) $\tau(bd^2abcded^{-2}b^{-1}) = 1$ gives

$$S_{1,b}S_{b,d}S_{bd,d}S_{bd^2,a}S_{\overline{bd^2a},b}S_{\overline{bd^2ab},c}S_{\overline{bd^2abc},d}S_{\overline{bd^2abcd},e}S_{bd,d}^{-1}S_{b,d}^{-1}S_{1,b}^{-1}$$

We solve the above relations and observe

(1) $\tau(abcde) = 1$ gives

$$S_{1,a}^{-1} = S_{\overline{a},b}S_{\overline{ab},c}S_{\overline{abc},d}S_{\overline{abcd},e}$$

(2) $\tau(babcdeb^{-1}) = 1$ gives (since $S_{1,b} = 1$),

$$S_{b,a}^{-1} = S_{\overline{ba},b}S_{\overline{bab},c}S_{\overline{babc},d}S_{\overline{babcd},e}$$

(3) $\tau(dabcded^{-1}) = 1$ gives (since $S_{1,d} = 1$)

$$S_{d,a}^{-1} = S_{\overline{da},b}S_{\overline{dab},c}S_{\overline{dabc},d}S_{\overline{dabcd},e}$$

(4) $\tau(d^2abcded^{-2}) = 1$ gives (since $S_{1,d} = 1$ and $S_{d,d} = 1$,)

$$S_{d^2,a}^{-1} = S_{\overline{d^2a},b}S_{\overline{d^2ab},c}S_{\overline{d^2abc},d}S_{\overline{d^2abcd},e}$$

(5) $\tau(bdabcded^{-1}b{-}1) = 1$ gives (since $S_{1,b} = 1$ and $S_{b,d} = 1$),

$$S_{bd,a}^{-1} = S_{\overline{bda},b}S_{\overline{bdab},c}S_{\overline{bdabc},d}S_{\overline{bdabcd},e}$$

(6) $\tau(bd^2abcded^{-2}b^{-1}) = 1$ gives (since $S_{1,b} = 1$, $S_{b,d} = 1$ and $S_{bd,d} = 1$),

$$S_{bd^2,a}^{-1} = S_{\overline{bd^2a},b}S_{\overline{bd^2ab},c}S_{\overline{bd^2abc},d}S_{\overline{bd^2abcd},e}$$

8.12. Elimination to three relations:

Now we have relators

$$S_{1,a}S_{b,a} = 1$$

$$S_{bd,a}S_{d,a} = 1$$

$$S_{bd^2,a}S_{d^2,a} = 1$$

We can, therefore, eliminate the $\tau(Ka^2K^{-1}) = 1$ relations for all K. and obtain three relators from the six long relators and eliminate the six a^2-relators and the six $a-$generators.

(1)

$$S_{\overline{ba},b}S_{\overline{bab},c}S_{\overline{babc},d}S_{\overline{babcd},e}S_{\overline{a},b}S_{\overline{ab},c}S_{\overline{abc},d}S_{\overline{abcd},e}$$

(2)

$$S_{\overline{bda},b}S_{\overline{bdab},c}S_{\overline{bdabc},d}S_{\overline{bdabcd},e}S_{\overline{da},b}S_{\overline{dab},c}S_{\overline{dabc},d}S_{\overline{dabcd},e}$$

(3)

$$S_{\overline{bd^2a},b}S_{\overline{bd^2ab},c}S_{\overline{bd^2abc},d}S_{\overline{bd^2abcd},e}S_{\overline{d^2a},b}S_{\overline{d^2ab},c}S_{\overline{d^2abc},d}S_{\overline{d^2abcd},e}$$

Use the fact that $S_{1,d}S_{d,d}S_{d^2,d}$ is a relator to combine these three relators into one relator and eliminate the generators $S_{1,d}$, $S_{d,d}$ and $S_{d\cdot d}$, but First rewrite without the $\overline{x}$

(1) $S_{1,b}S_{b,c}S_{bd,d}S_{bd^2,e}S_{b,b}S_{1,c}S_{d,d}S_{d^2,e}$
(2) $S_{d,b}S_{bd,c}S_{bd^2,d}S_{b,e}S_{bd,b}S_{d,c}S_{d^2,d}S_{1,e}$
(3) $S_{d^2,b}S_{bd^2,c}S_{b,d}S_{bd,e}S_{bd^2,b}S_{d^2,c}S_{1,d}S_{d,e}$

and then

(1) $S_{d,d}^{-1} = S_{d^2,e}S_{1,b}S_{b,c}S_{bd,d}S_{bd^2,e}S_{b,b}S_{1,c}$
(2) $S_{d^2,d}^{-1} = S_{1,e}S_{d,b}S_{bd,c}S_{bd^2,d}S_{b,e}S_{bd,b}S_{d,c}$
(3) $S_{1,d}^{-1} = S_{d,e}S_{d^2,b}S_{bd^2,c}S_{b,d}S_{bd,e}S_{bd^2,b}S_{d^2,c}$

We obtain

$$S_{d^d,d}^{-1}S_{d,d}^{-1}S_{1,d}^{-1} = S_{d^2,e}S_{1,b}S_{b,c}S_{bd,d}S_{bd^2,e}S_{b,b}S_{1,c}S_{1,e}S_{d,b}S_{bd,c}S_{bd^2,d}$$
$$S_{b,e}S_{bd,b}S_{d,c}S_{d,e}S_{d^2,b}S_{bd^2,c}S_{b,d}S_{bd,e}S_{bd^2,b}S_{d^2,c}$$

Finally, replace $S_{d^2,e}$ by $(S_{1,e}S_{d,e})^{-1}$, $S_{bd^2,e}$ by $(S_{b,e}S_{bd,e})^{-1}$, $S_{d^2,c}$ by $(S_{1,c}S_{d,c})^{-1}$, $S_{bd^2,c}$ by $(S_{b,c}S_{bd,c})^{-1}$.

Eliminate the $b-$ generators and relations, to obtain one single relation (displayed on two lines, but a product):

$$(S_{1,e}S_{d,e})^{-1}S_{b,c}S_{bd,d}(S_{b,e}S_{bd,e})^{-1}S_{1,c}S_{1,e}S_{bd,c}S_{b,e}S_{d,c}S_{d,e}(S_{b,c}S_{bd,c})^{-1}S_{bd,e}(S_{1,c}S_{d,c})^{-1}$$
$$\times S_{d^2,e}S_{b,c}S_{bd,d}S_{bd^2,e}S_{1,c}S_{1,e}S_{bd,c}S_{b,e}S_{d,c}S_{d,e}S_{bd^2,c}S_{bd,e}S_{d^2,c}$$

And lastly eliminate the remaining $d-$generators and relations to obtain one long relation involving eight generators for the fundamental group of Γ and their inverses

$$(S_{1,e}S_{d,e})^{-1}S_{b,c}(S_{b,e}S_{bd,e})^{-1}S_{1,c}S_{1,e}S_{bd,c}S_{b,e}S_{d,c}S_{d,e}(S_{b,c}S_{bd,c})^{-1}S_{bd,e}(S_{1,c}S_{d,c})^{-1}$$

8.13. Final homology computation.

Recall that g and h denote the action of elements of Γ_0 on Γ by conjugation.

Now we may assume that E is the image of $S_{1,e}$ in the first homology group so that $h(E), h^2(E)$ denote the images under h with $g(E), hg(E) = gh(E)$, $h^2g(E) = gh^2(E)$ the other images of E.

Similarly let C be the image in the first homology group of $S_{1,c}$.

We have a set of generators for homology of the form:

$$E, h(E), g(E), gh^2(E), gh(E), C, h(C), g(C), gh^2(C), gh(C)$$

We turn this into a homology basis with

E and $g(E)$ are in the basis and C and $g(C)$ are in the basis. The same holds for E^{-1} and C^{-1} and their images under g.

E and $h(E)$ are in the basis with $h^2(E) \approx_h -(E + h(E))$ $g(E)$ and $gh(E)$ are in the basis with $gh^2(E) \approx_h -(g(E) + h^2g(E))$

C and $h(C)$ are in the basis with $h^2(C) \approx_h -(C + h(C))$

$g(C)$ and $gh(C)$ are in the basis with $gh^2(C) \approx_h -(g(C) + hg(C))$

with appropriate images relations for the curves under the other generators for Γ_0/Γ and $g(C)$

9. The fixed points and matrices

We compute the matrices of the action on homology with respect the *ordered* basis $E, g(E), h(E), gh(E), C, g(C), h(C), gh(C)$

$$\text{Let } P = \begin{pmatrix} 0 & 1 & 0 & 0 \\ 1 & 0 & 0 & 0 \\ 0 & 0 & 0 & 1 \\ 0 & 0 & 1 & 0 \end{pmatrix}, \quad Q = \begin{pmatrix} 0 & 0 & 1 & 0 \\ 0 & 0 & 0 & 1 \\ -1 & 0 & -1 & 0 \\ 0 & -1 & 0 & -1 \end{pmatrix},$$

and

$$T = \begin{pmatrix} 0 & 0 & 0 & 1 \\ 0 & 0 & 1 & 0 \\ 0 & -1 & 0 & -1 \\ -1 & 0 & -1 & 0 \end{pmatrix}$$

The matrices of g, h and gh with respect to the basis

$$E, g(E), h(E), gh(E), C, g(C), h(C), gh(C)$$

break up into blocks where 0_n represents the $n \times n$ the matrix of all zeros for an integer n and thus 0_4 the 4×2 such matrix.

$$M_g = \begin{pmatrix} P & 0_4 \\ 0_4 & P \end{pmatrix}, M_h = \begin{pmatrix} Q & 0_4 \\ 0_4 & Q \end{pmatrix}, \text{and } M_{gh} = \begin{pmatrix} T & 0_4 \\ 0_4 & T \end{pmatrix}$$

Using the Lefschetz fixed point formula **[23]** we can double check that these transformations have the correct number of fixed points (as defined by the original surface kernel map) by computing $2 - \mathrm{TrM_h} = 6$, $2 - \mathrm{TrM_g} = 2$, and $2 - \mathrm{TrM_{gh}} = 0$ where Tr denotes the trace of a matrix.

Observe that M_g has four non-zero blocks that are 2×2 permutation matrices. The matrix M_h is built upon an element of type (2) and if the ordered basis is rearranged to be $E, h(E), g(E), gh(E), C, h(C), g(C), gh(C)$ we have if

$$Q_0 = \begin{pmatrix} 0 & 1 & 0 & 0 \\ -1 & -1 & 0 & 0 \\ 01 & 0 & 0 & 1 \\ 0 & 01 & -1 & -1 \end{pmatrix}, \text{ then } h \text{ has the matrix } \begin{pmatrix} Q_0 & 0_4 \\ 0_4 & Q_0 \end{pmatrix}.$$

If the ordered basis is $E, h^2(E), gh(E), g(E), C, h^2(C), gh(C), g(C)$, let

$$T_0 = \begin{pmatrix} 0 & 0 & 1 & 0 \\ 0 & 0 & 0 & 1 \\ 0 & 1 & 0 & 0 \\ -1 & -1 & 0 & 0 \end{pmatrix} \text{ so that } gh \text{ then has the matrix } \begin{pmatrix} T_0 & 0_4 \\ 0_4 & T_0 \end{pmatrix}.$$

Acknowledgements

The author thanks Marston Conder, Gareth Jones and David Singerman for some useful conversations about connections between Schreier-Reidemeister Theory and conformal automorphism groups. The author also thanks the referee for some useful suggestions.

References

[1] S. Allen Broughton, *Classifying finite group actions on surfaces of low genus*, J. Pure Appl. Algebra **69** (1991), no. 3, 233–270, DOI 10.1016/0022-4049(91)90021-S. MR1090743 (92b:57021)

[2] Marston Conder, *Experimental algebra*, Math. Chronicle **20** (1991), 1–11. MR1137869 (92k:20007)

[3] M. Condor, *Group actions on surfaces*, lists, http://www.math.auckland.ac.nz/ conder/, (2013).

[4] Marston D. E. Conder and Ravi S. Kulkarni, *Infinite families of automorphism groups of Riemann surfaces*, Discrete groups and geometry (Birmingham, 1991), London Math. Soc. Lecture Note Ser., vol. 173, Cambridge Univ. Press, Cambridge, 1992, pp. 47–56, DOI 10.1017/CBO9780511565793.007. MR1196915 (93k:30073)

[5] Antonio F. Costa and Peter Turbek, *Lifting involutions to ramified covers of Riemann surfaces*, Arch. Math. (Basel) **81** (2003), no. 2, 161–168, DOI 10.1007/s00013-003-4709-x. MR2009558 (2004k:30097)

[6] J. Gilman, *Relative Modular Groups in Teichmüller Spaces*, Thesis, Colombia University (1971).

[7] Jane Gilman, *Compact Riemann surfaces with conformal involutions*, Proc. Amer. Math. Soc. **37** (1973), 105–107. MR0308390 (46 #7504)

[8] Jane Gilman, *On the moduli of compact Riemann surfaces with a finite number of punctures*, Discontinuous groups and Riemann surfaces (Proc. Conf., Univ. Maryland, College Park, Md., 1973), Princeton Univ. Press, Princeton, N.J., 1974, pp. 181–205. Ann. of Math. Studies, No. 79. MR0361052 (50 #13498)

[9] Jane Gilman, *A matrix representation for automorphisms of compact Riemann surfaces*, Linear Algebra and Appl. **17** (1977), no. 2, 139–147. MR0499135 (58 #17077)

[10] Jane Gilman, *Canonical symplectic representations for prime order conjugacy classes of the mapping-class group*, J. Algebra **318** (2007), no. 1, 430–455, DOI 10.1016/j.jalgebra.2007.06.009. MR2363143 (2008k:20100)

[11] Jane Gilman, *Prime order automorphisms of Riemann surfaces*, Teichmüller theory and moduli problem, Ramanujan Math. Soc. Lect. Notes Ser., vol. 10, Ramanujan Math. Soc., Mysore, 2010, pp. 229–246. MR2667558 (2011e:30106)

[12] J. Gilman, in preparation.

[13] Jane Gilman and David Patterson, *Intersection matrices for bases adapted to automorphisms of a compact Riemann surface*, Riemann surfaces and related topics: Proceedings of the 1978 Stony Brook Conference (State Univ. New York, Stony Brook, N.Y., 1978), Ann. of Math. Stud., vol. 97, Princeton Univ. Press, Princeton, N.J., 1981, pp. 149–166. MR624812 (83a:30046)

[14] William M. Goldman, *Representations of fundamental groups of surfaces*, Geometry and topology (College Park, Md., 1983/84), Lecture Notes in Math., vol. 1167, Springer, Berlin, 1985, pp. 95–117, DOI 10.1007/BFb0075218. MR827264 (87j:32068)

[15] William M. Goldman, *Mapping class group dynamics on surface group representations*, Problems on mapping class groups and related topics, Proc. Sympos. Pure Math., vol. 74, Amer. Math. Soc., Providence, RI, 2006, pp. 189–214, DOI 10.1090/pspum/074/2264541. MR2264541 (2007h:57020)

[16] W. J. Harvey, *Cyclic groups of automorphisms of a compact Riemann surface*, Quart. J. Math. Oxford Ser. (2) **17** (1966), 86–97. MR0201629 (34 #1511)

[17] W. J. Harvey, *On branch loci in Teichmüller space*, Trans. Amer. Math. Soc. **153** (1971), 387–399. MR0297994 (45 #7046)

[18] D. L. Johnson, *Presentations of groups*, 2nd ed., London Mathematical Society Student Texts, vol. 15, Cambridge University Press, Cambridge, 1997. MR1472735 (98e:20001)

[19] Gareth Jones and David Singerman, *Maps, hypermaps and triangle groups*, The Grothendieck theory of dessins d'enfants (Luminy, 1993), London Math. Soc. Lecture Note Ser., vol. 200, Cambridge Univ. Press, Cambridge, 1994, pp. 115–145. MR1305395 (95m:20055)

[20] Gareth A. Jones and David Singerman, *Complex functions*, Cambridge University Press, Cambridge, 1987. An algebraic and geometric viewpoint. MR890746 (89b:30001)

[21] Ravi S. Kulkarni, *Riemann surfaces admitting large automorphism groups*, Extremal Riemann surfaces (San Francisco, CA, 1995), Contemp. Math., vol. 201, Amer. Math. Soc., Providence, RI, 1997, pp. 63–79, DOI 10.1090/conm/201/02610. MR1429195 (98g:30070)

[22] Roger C. Lyndon and Paul E. Schupp, *Combinatorial group theory*, Classics in Mathematics, Springer-Verlag, Berlin, 2001. Reprint of the 1977 edition. MR1812024 (2001i:20064)

[23] A. M. Macbeath, *Action of automorphisms of a compact Riemann surface on the first homology group*, Bull. London Math. Soc. **5** (1973), 103–108. MR0320301 (47 #8840)

[24] W. Magnus, *Non-Euclidean Tesselations and their groups* Acdamic press (1974).

[25] Wilhelm Magnus, Abraham Karrass, and Donald Solitar, *Combinatorial group theory*, Second revised edition, Dover Publications, Inc., New York, 1976. Presentations of groups in terms of generators and relations. MR0422434 (54 #10423)

[26] David Singerman, *Symmetries of Riemann surfaces with large automorphism group*, Math. Ann. **210** (1974), 17–32. MR0361059 (50 #13505)

[27] Richard A. Wentworth, *The action of the mapping class group on representation varieties*, Transformation groups and moduli spaces of curves, Adv. Lect. Math. (ALM), vol. 16, Int. Press, Somerville, MA, 2011, pp. 277–299. MR2883691

Mathematics Department, Rutgers University, Newark, New Jersey 07102
E-mail address: gilman@rutgers.edu

Fields of definition of uniform dessins on quasiplatonic surfaces

Ernesto Girondo, David Torres-Teigell, and Jürgen Wolfart

ABSTRACT. In previous work the authors introduced certain Shimura curves that possess different uniform dessins d'enfants (equivalently, uniform Belyi functions). They are all quasiplatonic and therefore they can be defined over its field of moduli. In this paper the authors determine fields of moduli and some fields of definition of these curves and their related uniform dessins.

1. Introduction

Dessins can be considered as *hypermaps* on compact oriented two–manifolds, that is bipartite graphs cutting the underlying surface in finitely many simply connected cells. In the special case that all white (or all black) vertices have valency 2, we may omit these vertices and therefore we are in the classical group theoretic and geometric–topological theory of *maps* which can be traced back to the classification of platonic solids.

More recently, these objects attracted wider interest by two observations of Grothendieck [**10**] who invented the term *dessins d'enfants* for them.

(1) Dessins arise in a very natural way on all smooth complex projective algebraic curves defined over number fields: by a result of Belyi [**1**], there exist non–constant meromorphic functions
$$\beta : C \to \mathbb{P}^1(\mathbb{C})$$
on a curve C (equivalently, a compact Riemann surface) ramified at most over $0, 1, \infty$ precisely if C can be defined over a number field. Such functions are called Belyi functions and, consequently, algebraic curves defined over $\overline{\mathbb{Q}}$ are also called *Belyi curves*. In this case, the β–preimage of the real interval $[0, 1]$ defines a bipartite graph on C which is a dessin.

(2) Conversely, if we start with an arbitrary topological hypermap on an oriented compact two–manifold, there is a unique conformal structure on the manifold, corresponding to an algebraic curve defined over a number

2010 *Mathematics Subject Classification.* Primary 11G32, Secondary 14G35, 14H57, 30F10, 30F35.

Key words and phrases. Compact Riemann surfaces, dessins d'enfants, fields of moduli, fields of definition, Shimura curves.

The first and second author's were partially supported by a MICINN grant MTM2012-31973.
The second author was partially supported by the Alexander von Humboldt Foundation.
The third author was supported by DFG project Wo 199/4-1.

field, together with a Belyi function corresponding to this dessin. A first construction of this conformal structure – at least in the map case – was given by Singerman [**15**] even before the term dessin was coined.

(3) A third important aspect gives a link to the uniformisation of Riemann surfaces, see for example [**20**]: Belyi functions on a curve C correspond to Fuchsian group inclusions $\Gamma < \Delta(r,s,t)$, where Γ is a finite index subgroup of a Fuchsian triangle group $\Delta = \Delta(r,s,t)$ such that C can be written as the quotient $\Gamma\backslash\mathbb{H}$, where $\mathbb{H}$ denotes the upper half plane. With a suitable identification of the target quotient space with the Riemann sphere (the fixed points of Δ have to be identified with $0, 1, \infty$), the Belyi function is then the canonical quotient mapping

$$\begin{array}{cccc} \beta : & \Gamma\backslash\mathbb{H} & \to & \Delta\backslash\mathbb{H} \\ & \Gamma z & \mapsto & \Delta z \,. \end{array}$$

Dessins should therefore encode all relevant properties of the underlying Riemann surfaces, in particular their defining equations as algebraic curves and the explicit form of the Belyi function of the dessin. Finding these data from the combinatorial structure of the dessin is however a very hard problem, so one is already satisfied to determine the field of moduli and the minimal field of definition for C and the *Belyi pair* (C, β), if such minimal field exists; for precise definitions see the next section. The present article focuses on these objects in the case of certain Shimura curves containing different uniform dessins. Why are these dessins so interesting?

Although dessins define their underlying curves uniquely, on a Belyi curve one can always construct an infinity of Belyi functions, hence an infinity of dessins. How are these different dessins linked to each other? Can one get uniqueness of dessins by imposing additional conditions?

One can in fact show [**9**] that *regular dessins* – those for which the group of colour– and orientation–preserving automorphisms acts transitively on the edges – on surfaces of genus $g > 1$ are almost uniquely determined: their surfaces C are called *quasiplatonic*, and they can be characterized by the fact that their surface groups Γ are normal subgroups of triangle groups Δ. It is nontrivial but not too surprising that the existence of several non–isomorphic regular dessins on the same curve C is always induced by the finitely many and well–known inclusion relations between different triangle groups.

Regular dessins are in particular *uniform*, that is (due to the symmetry under the automorphism group) all white vertices have the same valency, and also all black vertices and all faces. Uniform dessins can be characterized by the fact that the surface groups Γ of their surfaces C are contained in triangle groups Δ, but they are no longer necessarily normal subgroups; the signature (r,s,t) of Δ is determined by the valencies of the dessin. In the terminology of point (3) of the above enumeration, the uniform dessins correspond to inclusions $\Gamma < \Delta$ with torsion–free subgroup Γ. In this case, inclusions between triangle groups always induce new uniform dessins, so one may ask the more interesting question of when and how many different uniform dessins of the *same signature* may exist on one curve. The answer is given in [**8**]: an unexpected large number of different uniform dessins of the same signature exists if and only if Γ is contained in a congruence subgroup of an arithmetic triangle group Δ, for details see Section 3.

The present paper is in some sense a continuation of [**8**]. We determine fields of moduli and some fields of definition of the curves involved and of these exotic uniform dessins – at least in the case of *Shimura curves* whose surface groups are principal congruence subgroups $\Gamma = \Delta(\mathfrak{p}^n)$ of prime power level in an arithmetic triangle group Δ.

2. Fields of moduli and fields of definition

Let S be a smooth algebraic curve and $k \subseteq \mathbb{C}$ a field. We say that k is a *field of definition* of S if there exist homogeneous polynomials

$$F_j \in k[x_0, x_1, \dots, x_n]\,, \quad j = 1, \dots, J$$

such that for $F := \{F_1, \dots, F_J\}$, S and

$$S_F = \{[x_0, x_1, \dots, x_n] \in \mathbb{P}^n(\mathbb{C}) : F_j(x_0, \dots, x_n) = 0\,, \text{ for } j = 1, \dots, J\}$$

are isomorphic. Throughout the paper we will deal with Belyi curves, and hence we may always assume that $k < \overline{\mathbb{Q}}$.

Consider now a compact Riemann surface S, an algebraic model $S_F \cong S$ with $F = \{F_1, \dots, F_J \in \overline{\mathbb{Q}}[x_0, \dots, x_n]\}$ and a Galois element $\sigma \in \mathrm{Gal}\,(\overline{\mathbb{Q}}) := \mathrm{Gal}\,(\overline{\mathbb{Q}}/\mathbb{Q})$. One can construct the Galois conjugate curve $S_F^\sigma = S_{F^\sigma}$, where F^σ is obtained from F by applying σ to all coefficients of all F_j, and the inertia group

$$I_S = \{\sigma \in \mathrm{Gal}(\overline{\mathbb{Q}}) : S_F^\sigma \cong S_F\}$$

which clearly does not depend on the choice of model. The fixed field

$$\overline{\mathbb{Q}}^{I_S} = \mathrm{Fix}(I_S) = \{\alpha \in \overline{\mathbb{Q}} : \sigma(\alpha) = \alpha, \text{ for all } \sigma \in I_S\}$$

is called the *field of moduli* of S, and we will denote it by $M(S)$. It is a well–known fact that the field of moduli is always contained in any field of definition, but in general $M(S)$ is not a field of definition of S, see for example [**6**] or [**14**]. However, a quasiplatonic curve can always be defined over its field of moduli ([**20**]).

Similarly, if $G < \mathrm{Aut}S$ is a group of automorphisms of S, we will say that k is a field of definition of (S, G) if there exists a model S_F defined over k and an isomorphism $\varphi : S_F \longrightarrow S$ such that the group of automorphisms $\varphi^{-1}G\varphi < \mathrm{Aut}S_F$ is also defined over k. The inertia group is now defined as

$$I_{(S,G)} = \left\{\sigma \in \mathrm{Gal}\,(\overline{\mathbb{Q}}) : \begin{array}{l} \text{there exists an isomorphism } f_\sigma : S \longrightarrow S^\sigma \\ \text{such that } \;\; \alpha^\sigma \circ f_\sigma = f_\sigma \circ \alpha, \; \forall \alpha \in G \end{array}\right\},$$

in other words, for every $\sigma \in I_{(S,G)}$ there is an isomorphism $f_\sigma : S \longrightarrow S^\sigma$ such that the diagram

$$\begin{array}{ccc} S & \xrightarrow{f_\sigma} & S^\sigma \\ {\scriptstyle\alpha}\downarrow & & \downarrow{\scriptstyle\alpha^\sigma} \\ S & \xrightarrow{f_\sigma} & S^\sigma \end{array}$$

is commutative for every $\alpha \in G$. Accordingly, the field of moduli of the pair (S, G) is $M(S, G) = \overline{\mathbb{Q}}^{I_{(S,G)}}$.

Consider now a Belyi function $\beta : S \longrightarrow \mathbb{P}^1(\mathbb{C})$ defining a dessin d'enfant $\mathcal{D}$ on S. The group of automorphisms of β is defined as the subgroup of $\mathrm{Aut}S$ given by $\mathrm{Aut}(S, \beta) = \{\, f \in \mathrm{Aut}S \,:\, \beta = \beta \circ f\,\}$. We will write $\mathrm{Aut}\mathcal{D} := \mathrm{Aut}(S, \beta)$. For every model S_F as above $\beta \circ \varphi$ is a rational function on S_F. We will say that k is a field of definition of (S,β) if there exists a model S_F of S such that both F and the covering $\beta\circ\varphi$ are defined over k. Accordingly, we will define the field of moduli $M(S,\beta)$ as the field fixed by

$$I_{(S,\beta)} = \left\{\sigma\in\mathrm{Gal}\,(\overline{\mathbb{Q}}) : \begin{array}{l}\text{there exists an isomorphism } f_\sigma : S \longrightarrow S^\sigma \\ \text{such that} \quad \beta^\sigma\circ f_\sigma = \beta\end{array}\right\}.$$

We will also call $M(\mathcal{D}) := M(S,\beta)$ the *field of moduli* of the dessin $\mathcal{D}$ and, correspondingly, a *field of definition* of $\mathcal{D}$ will be a field of definition for the Belyi pair (S,β). Clearly, $I_{(S,\beta)} < I_S$ and therefore $M(S) < M(S,\beta)$.

In many important cases these fields of moduli coincide. For instance, the fields of moduli of regular Belyi functions often agree with the field of definition of the surface in which they are defined. Recall that a Belyi function β is called *regular* if it defines a normal covering $S\to S/G\cong\mathbb{P}^1(\mathbb{C})$ for some group of automorphisms $G<\mathrm{Aut}S$. In this case the corresponding dessins are also called regular and the surface S is called quasiplatonic.

LEMMA 1. *If S is quasiplatonic with surface group $\Gamma\lhd\Delta$ where Δ is a maximal triangle group, then we have $M(S) = M(S,\beta)$ for the Belyi function*

$$\beta\ :\ \Gamma\backslash\mathbb{H}\ \to\ \Delta\backslash\mathbb{H}\quad .$$

PROOF. We may suppose that S is defined over $M(S)$, so β^σ is another regular Belyi function of the same degree on S for all $\sigma\in I_S$. But, since Δ is maximal, one has $\mathrm{Aut}(S,\beta) = \mathrm{Aut}S = \mathrm{Aut}(S,\beta^\sigma)$ and, being β and β^σ regular, one has $\beta=\beta^\sigma$. □

REMARK 1. *It is useful to notice what may happen if the triangle group considered is not maximal. Take for example the possibility $r=s\neq t$; then there may exist a $\sigma\in I_S$ fixing S but exchanging the zero set of its Belyi function B and the zero set of $1-B$ such that $M(S,B)$ can be a quadratic extension of $M(S)$. The same problem may occur for uniform non–regular Belyi functions, of course.*

A refined argument shows how to extend Lemma 1 in some cases to non–maximal triangle groups:

LEMMA 2. *Let S be a quasiplatonic surface with surface group $\Gamma\lhd\Phi$ where Φ is a non–maximal triangle group of signature (r,r,t), $r\neq t$, with regular Belyi function B. Suppose moreover that Γ is also normal in the maximal triangle group Δ of signature $(2,r,2t)$, containing Φ with index $|\Delta:\Phi|=2$. Then there is a model for (S,B) defined over $M(S)$.*

PROOF. We can assume S to be defined over $M(S)$, and that all zeros of B and of $1-B$ have order r. Any $\sigma\in\mathrm{Gal}(\overline{\mathbb{Q}}/M(S))$ fixes S and either fixes B or maps it to $B^\sigma = 1-B$, see Remark 1. Both B and $1-B$ take the value $\frac{1}{2}$ at the same points $x\in S$, images of the Φ–orbit (and also Δ–orbit) of the fixed point of the canonical order 2 generator of Δ under the projection $\mathbb{H}\to\Gamma\backslash\mathbb{H}$. We take such a point $x\in B^{-1}(\frac{1}{2})$; for any $\sigma\in\mathrm{Gal}(\overline{\mathbb{Q}}/M(S))$, also $\sigma(x)\in B^{-1}(\frac{1}{2})$. Since it is not a ramification point of B nor $1-B$, there is a unique g_σ in the covering group Φ/Γ of B (and of $1-B$) with the property $g_\sigma(x)=\sigma(x)$. Now we distinguish two cases:

First, if $B^\sigma = B$, we put $f_\sigma := g_\sigma\in\mathrm{Aut}\,S$.

Second, if $B^\sigma = 1-B$, there is an involution $i_\sigma\in\mathrm{Aut}\,S$ induced by an order 2 generator of Δ fixing x, interchanging the zero sets of B and $1-B$ and fixing their

pole sets. In this case, we put $f_\sigma := g_\sigma \circ i_\sigma$.
In both cases, the $f_\sigma \in \operatorname{Aut} S$ are uniquely determined by σ and make the following diagrams commutative:

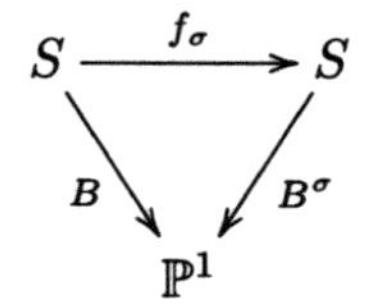

By uniqueness, these f_σ trivially satisfy Weil's cocycle condition [**19**], whence (S, B) can also be defined over $M(S)$. □

We learned from Rubén Hidalgo another application of Weil's criterion which turns out to be very useful for proving the following theorem – for more sophisticated versions see [**4**].

THEOREM 1. *Let S be a quasiplatonic curve of genus $g > 1$ with full automorphism group* $\operatorname{Aut} S = G$. *If G contains its centre $Z(G)$ as a direct factor, that is if $G \cong G' \times Z(G)$, the pair (S, G) can be defined over its field of moduli $M(S, G)$.*

PROOF. We can suppose that S is defined over its field of moduli $M(S)$. Let $\sigma \in I_{(S,G)}$. Every automorphism f_σ satisfying the condition

$$f_\sigma \circ \alpha \circ f_\sigma^{-1} = \alpha^\sigma \qquad \text{for all} \qquad \alpha \in G$$

must be of the form $(g_\sigma, h) \in G' \times Z(G)$ for some $g_\sigma \in G'$. Then the set $\{(g_\sigma, 1)\}$ satisfies Weil's cocycle condition and we can define (S, G) over $M(S, G)$. □

A remark in the preprint [**2**] suggests that the result may still be true without the hypothesis about the centre. We will not use such a stronger version since the automorphism groups considered here are of type PSL_2, PGL_2 or direct products of such groups with cyclic factors C_2, so Theorem 1 applies.

COROLLARY 1. *Under the same hypotheses, let U be a subgroup of G and $C := U \backslash S$ the quotient curve. Then, C can be defined over the field $M(S, G)$. Moreover, all such quotient curves C can be simultaneously defined over $M(S, G)$ in the sense that all their function fields $M(S, G)(C)$ are subfields of $M(S, G)(S)$.*

PROOF. Recall that k is a field of definition for C if the function field $\mathbb{C}(C)$ can be obtained by a constant field extension of a function field $k(C)$, that is tensoring some ring of functions $k[C]$ with $\mathbb{C}$ and taking the quotient field. Now, we know that the function field of S is a constant field extension of $M(S, G)(S)$ and that moreover all automorphisms of S are defined over $M(S, G)$. Therefore, $M(S, G)(S)$ is a Galois extension of a rational function field $M(S, G)(\beta)$, and $M(S, G)(C)$ is the fixed field of the subgroup U. This construction works simultaneously for all quotient curves of S. □

3. Congruence subgroups and uniform dessins

A finite group G is called a *Hurwitz group* if it acts as automorphism group on a compact Riemann surface S of genus $g > 1$ and has order $84(g - 1)$, which is Hurwitz's universal upper bound for all such automorphism group orders. In this case the surface is moreover quasiplatonic: it is well known that S is uniformised in this case by a normal subgroup K of the triangle group $\Delta(2, 3, 7)$ and, in particular, one has $G \cong \Delta(2,3,7)/K$. A classical theorem by Macbeath [12] shows that $\mathrm{PSL}(2,\mathbb{F}_q)$ is a Hurwitz group exactly in the following cases

(i) $q = 7$,
(ii) $q = p$ prime for $p \equiv \pm 1 \bmod 7$,
(iii) $q = p^3$ for p prime and $p \equiv \pm 2$ or $\pm 3 \bmod 7$.

The corresponding Riemann surfaces are usually known as *Macbeath–Hurwitz curves* and the first examples are found in genus three (Klein's Riemann surface, with automorphism group isomorphic to $\mathrm{PSL}(2,\mathbb{F}_7)$), genus seven (the Fricke–Macbeath curve with automorphism group $\mathrm{PSL}(2,\mathbb{F}_8)$) and genus fourteen (where there exist three non–isomorphic but Galois conjugate Macbeath–Hurwitz curves with automorphism group isomorphic to $\mathrm{PSL}(2,\mathbb{F}_{13})$).

It was proved in [5] by A. Džambić that all Macbeath–Hurwitz curves can be constructed arithmetically as follows. Given a number field k, denote by $\mathcal{O}_k$ its ring of integers. The triangle group $\Delta(2,3,7)$ is the norm 1 group of a maximal order in a quaternion algebra A over the field $k = \mathbb{Q}(\cos \pi/7)$ (more precisely its image under the canonical homomorphism $\mathrm{SL}(2,\mathbb{R}) \to \mathrm{PSL}(2,\mathbb{R})$), and it can be seen as a subgroup of $\mathrm{PSL}(2,\mathcal{O}_L)$, the projective group of determinant 1 matrices over the ring of integers $\mathcal{O}_L$ of an at most quadratic extension L of k. Any rational prime p defines an ideal $p\mathcal{O}_k$ in $\mathcal{O}_k$ such that

(i) if $p = 7$ then p is ramified and $p\mathcal{O}_k = \mathfrak{p}^3$ for a prime ideal $\mathfrak{p} \subset \mathcal{O}_k$ of norm $q = N(\mathfrak{p}) = 7$;
(ii) if $p \equiv \pm 1 \bmod 7$ then p splits, i.e. $p\mathcal{O}_k = \mathfrak{p}_1\mathfrak{p}_2\mathfrak{p}_3$ for prime ideals $\mathfrak{p}_1, \mathfrak{p}_2, \mathfrak{p}_3 \subset \mathcal{O}_k$ of norm $q = N(\mathfrak{p}_i) = p$;
(iii) if $p \equiv \pm 2$ or $\pm 3 \bmod 7$ then p is inert, i.e. $p\mathcal{O}_k$ is a prime ideal in $\mathcal{O}_k$ of norm $q = N(\mathfrak{p}) = p^3$.

For every prime $\mathfrak{p}$ in $\mathcal{O}_k$ we can define the subgroup of matrices of $\Delta(2,3,7)$ congruent to the identity modulo $\mathfrak{p}$. This is a normal torsion-free subgroup of $\Delta(2,3,7)$ with quotient group isomorphic to $\mathrm{PSL}(2,\mathbb{F}_q)$ where $q = N(\mathfrak{p})$, yielding therefore a Macbeath–Hurwitz curve.

3.1. Principal congruence subgroups. One can generalise the construction above in the following way. Consider any arithmetic triangle group $\Delta = \Delta(r,s,t)$ which is the norm 1 group $\mathcal{M}^1$ of a maximal order $\mathcal{M}$ of a quaternion algebra A over a field k. In the last section, we will extend our assumptions to triangle groups containing (always with index 2 or 4) such a norm 1 group. Up to an extension of the quotient groups $\mathrm{PSL}(2,\mathbb{F}_q)$ to $\mathrm{PGL}(2,\mathbb{F}_q)$ and/or adding one or two direct factors C_2, most results of the present section remain valid. It is well known [17] that in the case of arithmetic triangle groups, the invariant trace field k has always class number 1, and therefore any prime ideal in $\mathcal{O}_k$ is principal, hence of the form $\mathfrak{p} = \pi\mathcal{O}_k$ for some prime $\pi \in \mathcal{O}_k$. Given such a prime $\mathfrak{p}$ in $\mathcal{O}_k$ one can define the local quaternion algebra $A_\mathfrak{p}$ over the $\mathfrak{p}$–adic field $k_\mathfrak{p}$.

For each prime $\mathfrak{p}$ not dividing the discriminant of A, the $\mathfrak{p}$–adic completion $A_\mathfrak{p}$ is isomorphic to $M_2(k_\mathfrak{p})$. As in the previous case one can define the *principal congruence subgroup* of level $\mathfrak{p}$. It is defined as the (normal) subgroup $\Delta(\mathfrak{p})$ of Δ whose localisation with respect to a prime $\mathfrak{q} \in \mathcal{O}_k$ coincides with

$$\Phi(\mathfrak{p}) = \left\{ \begin{pmatrix} a & b \\ c & d \end{pmatrix} \in \mathcal{M}_\mathfrak{q} \cong M_2(\mathcal{O}_\mathfrak{q}) \ : \ \begin{pmatrix} a & b \\ c & d \end{pmatrix} \equiv \begin{pmatrix} 1 & 0 \\ 0 & 1 \end{pmatrix} \bmod \mathfrak{q} \right\}$$

(here the congruences have to be read coefficient–wise mod $\pi\mathcal{O}_{\mathfrak{q}}$) if $\mathfrak{q} = \mathfrak{p}$ and with $\mathcal{M}_{\mathfrak{q}}^1$ otherwise, where $\mathcal{M}_{\mathfrak{p}}$ stands for the localisation of $\mathcal{M}$ and $\mathcal{O}_{\mathfrak{q}}$ denotes the ring of integers of the local field $k_{\mathfrak{q}}$. The existence and uniqueness of such a subgroup is granted by the strong approximation theorem for arithmetic groups (see for example [**18**] or [**11**]). We have $\Delta/\Delta(\mathfrak{p}) \cong \mathrm{PSL}(2, \mathbb{F}_q)$ where $q = N(\mathfrak{p}) := |\mathcal{O}_k/\mathfrak{p}|$.

The first consequence is that all these surfaces $S = \Delta(\mathfrak{p})\backslash\mathbb{H}$ have a regular Belyi function $\beta : \Delta(\mathfrak{p})\backslash\mathbb{H} \longrightarrow \Delta\backslash\mathbb{H}$, given by the normal inclusion $\Delta(\mathfrak{p}) \lhd \Delta$, and an automorphism group of order $|\mathrm{Aut}S| = |N(\Delta(\mathfrak{p}))/\Delta(\mathfrak{p})| \geq |\mathrm{PSL}(2, \mathbb{F}_q)|$ where N denotes the normaliser in $\mathrm{PSL}(2, \mathbb{R})$.

Analogously, for each positive integer n we can consider the principal congruence subgroup $\Delta(\mathfrak{p}^n)$ of level $\mathfrak{p}^n$, whose localisation in $\mathfrak{p}$ corresponds to

$$\Phi(\mathfrak{p}^n) = \left\{ \begin{pmatrix} a & b \\ c & d \end{pmatrix} \in \mathcal{M}_{\mathfrak{p}} \cong M_2(\mathcal{O}_{\mathfrak{p}}) \ : \ \begin{pmatrix} a & b \\ c & d \end{pmatrix} \equiv \begin{pmatrix} 1 & 0 \\ 0 & 1 \end{pmatrix} \bmod \mathfrak{p}^n \right\}$$

This time the quotient $\Delta/\Delta(\mathfrak{p}^n)$ is isomorphic to $\mathrm{PSL}(2, \mathcal{O}_{\mathfrak{p}}/\mathfrak{p}^n\mathcal{O}_{\mathfrak{p}})$. Once again the normal inclusion $\Delta(\mathfrak{p}^n) \lhd \Delta$ yields a regular Belyi function on $S = \mathbb{H}/\Delta(\mathfrak{p}^n)$.

From the point of view of quaternion algebras, principal congruence subgroups of level $\mathfrak{p}^n$ with $n \geq 1$ correspond to the intersection of certain maximal orders in the local quaternion algebra $A_{\mathfrak{p}}$. More precisely, maximal orders in a split local quaternion algebra $A_{\mathfrak{p}}$ can be represented as vertices of a regular Bruhat–Tits–tree of valency $q+1$, where $q = N(\mathfrak{p})$ (see Figure 1). The principal congruence subgroup of level $\mathfrak{p}^n$ corresponds then to the intersection of all the vertices at distance $\leq n$ from some point, which represents the maximal order $\mathcal{M}$ whose norm 1 group is Δ.

FIGURE 1. Part of the tree of maximal orders of a local algebra $A_{\mathfrak{p}}$ corresponding to $\Delta(\mathfrak{p}^3)$ (in the case $N(\mathfrak{p}) + 1 = 8$).

3.2. Multiple uniform dessins. In [**8**] we studied under which conditions a surface S contains different uniform Belyi functions of a given type (r, s, t). This is equivalent to determine when the uniformising group K of S is contained in different triangle groups of that signature. In the case when $\Delta = \Delta(r, s, t)$ is the norm 1 group of a maximal order in a quaternion algebra as above, this happens if and only if K is contained in a group conjugate in Δ to the congruence subgroup $\Delta_0(\mathfrak{p})$, where $\mathfrak{p}$ is a prime in $\mathcal{O}_k$ not dividing the discriminant of A, and $\Delta_0(\mathfrak{p})$ is defined as the subgroup of Δ whose localisation with respect to the prime $\mathfrak{p}$ is

$$\Phi_0(\mathfrak{p}) = \left\{ \begin{pmatrix} a & b \\ c & d \end{pmatrix} \in \mathcal{M}_{\mathfrak{p}}^1 \cong \mathrm{SL}(2, \mathcal{O}_{\mathfrak{p}}) \ : \ c \equiv 0 \bmod \mathfrak{p} \right\}.$$

REMARK 2. *Even if Δ is strictly larger than the norm 1 group of a maximal order (by index 2 or 4, as already mentioned) we define the congruence subgroups $\Delta(\mathfrak{p})$ and $\Delta_0(\mathfrak{p})$ always as subgroups only consisting of norm 1 elements. The same convention will be used for higher power levels $\mathfrak{p}^j$.*

In order to explain this multiple inclusion of K in different conjugate triangle groups one has to introduce the so called *Fricke* (or *Atkin–Lehner) extension* $\Delta_{\mathrm{Fr}}(\mathfrak{p})$. In the local algebra $A_{\mathfrak{p}}$, the group $\Phi_0(\mathfrak{p})$ is the norm 1 group of an Eichler order $\mathcal{E}_{\mathfrak{p}}$, that is the intersection of two maximal orders $\mathcal{M}_{\mathfrak{p}}$ and $\gamma^{-1}\mathcal{M}_{\mathfrak{p}}\gamma$, where $\gamma = \left(\begin{smallmatrix} \pi & 0 \\ 0 & 1 \end{smallmatrix}\right)$. The norm 1 groups of these two (local) maximal orders correspond in the global case to two triangle groups $\Delta = \Delta_1$ and Δ_2 of the same type. The (Fricke) element $\left(\begin{smallmatrix} 0 & \pi^{-1} \\ -1 & 0 \end{smallmatrix}\right) \in A_{\mathfrak{p}}$ conjugates one maximal order into the other, and therefore induces an isomorphism of $\Phi_0(\mathfrak{p})$ of order two, called the Fricke involution. In the global case, this Fricke element corresponds to a matrix in $\mathrm{GL}(2,\mathbb{R})$ interchanging Δ_1 and Δ_2 by conjugation. By the rigidity of triangle groups, this conjugation can be realised inside $\mathrm{PSL}(2,\mathbb{R})$ by an element $\alpha_{\mathfrak{p}}$, which generates an index two extension $\Delta_{\mathrm{Fr}}(\mathfrak{p}) = \langle \Delta_0(\mathfrak{p}), \alpha_{\mathfrak{p}} \rangle$ called the Fricke extension.

Since $\alpha_{\mathfrak{p}}$ normalises $\Delta_0(\mathfrak{p})$, but not Δ, and $\alpha_{\mathfrak{p}}^2 \in \Delta_0(\mathfrak{p})$, conjugation by $\alpha_{\mathfrak{p}}$ induces an involution on the curve $\Delta_0(\mathfrak{p})\backslash\mathbb{H}$. We call it therefore the *Fricke involution* for $\Delta_0(\mathfrak{p})$ (even if it might not satisfy $\alpha_{\mathfrak{p}}^2 = 1$ in $\mathrm{PSL}(2,\mathbb{R})$). As a consequence every group $K < \Delta_0(\mathfrak{p})$ is included in both Δ and $\Delta_2 = \alpha_{\mathfrak{p}}\Delta\alpha_{\mathfrak{p}}^{-1}$, yielding two different uniform dessins in $K\backslash\mathbb{H}$.

In the same way, for every integer $j > 1$ one can introduce the congruence group $\Delta_0(\mathfrak{p}^j)$ which corresponds in the local algebra $A_{\mathfrak{p}}$ to

$$\Phi_0(\mathfrak{p}^j) = \left\{ \begin{pmatrix} a & b \\ c & d \end{pmatrix} \in \mathrm{SL}(2, \mathcal{O}_{\mathfrak{p}}) \ : \ c \equiv 0 \bmod \mathfrak{p}^j \right\}.$$

Note that $\Delta_0(\mathfrak{p}^j) < \Delta_0(\mathfrak{p})$. The Fricke involution is in this case an element $\alpha_{\mathfrak{p}^j}$ corresponding in the local case to the element

$$\begin{pmatrix} 0 & \pi^{-j} \\ -1 & 0 \end{pmatrix},$$

and $\Delta_0(\mathfrak{p}^j)$ is normal in the Fricke extension $\Delta_{\mathrm{Fr}}(\mathfrak{p}^j) = \langle \Delta_0(\mathfrak{p}^j), \alpha_{\mathfrak{p}^j} \rangle$. However, none of the groups $\Delta_0(\mathfrak{p}^l)$ is normal in $\Delta_{\mathrm{Fr}}(\mathfrak{p}^j)$, for $j < l$, and neither is any of its Δ–conjugates $\Delta_0^i(\mathfrak{p}^l)$. Existence and uniqueness of all these Fricke involutions is well known in the case of the elliptic modular group $\Delta(2,3,\infty) = \mathrm{PSL}(2,\mathbb{Z})$ and is probably known to the experts in the cases we need here. However, by lack of a good reference we include a statement and a proof:

LEMMA 3. *Let Δ be an arithmetic Fuchsian triangle group, a PSL–image of a norm 1 group of a maximal order M_0 in the quaternion algebra A with totally real centre field k of class number 1, and $\mathfrak{p}$ a prime ideal in k not dividing the discriminant of A. We consider A as embedded in the matrix algebra $M_2(\mathbb{R})$. For* *all $j \geq 1$, the congruence subgroup $\Delta_0(\mathfrak{p}^j)$ is the norm 1 subgroup of the intersection $M_0 \cap M_j$ of two maximal orders M_0 and M_j in A. These two maximal orders are conjugate under a (Fricke) involution $\alpha_{\mathfrak{p}^j}$; we can suppose that – as an element of A – $\alpha_{\mathfrak{p}^j}$ has norm π^{-j}. Then the class $\alpha_{\mathfrak{p}^j}\Delta_0(\mathfrak{p}^j)$ is uniquely determined.*

PROOF. Since we are considering only congruence subgroups of levels $\mathfrak{p}^j$, we can see the maximal orders as intersections of their $\mathfrak{p}$–adic completions with A and visualise the local maximal orders by the vertices of the Bruhat–Tits tree already introduced above, see Figure 1.

For the first claim about the role of $\Delta_0(\mathfrak{p}^j)$ as norm 1 subgroup of the intersection $M_0 \cap M_j$ of two maximal orders M_0 and M_j one may consult **[8]** to see that these orders correspond to two vertices which are at distance j from each other in the Bruhat–Tits tree.

Existence of $\alpha_{\mathfrak{p}^j}$: since k has class number 1, all maximal orders are conjugate in A **[18]**, so we have some $\gamma \in A^*$ such that $M_j = \gamma M_0 \gamma^{-1}$. It is easy to see that we may even suppose $\gamma \in \mathrm{SL}(2,\mathbb{R})$. If $M_0 = \gamma M_j \gamma^{-1}$, the element γ^2 fixes both M_0, M_j, hence belongs to $\Delta_0(\mathfrak{p}^j)$, and γ normalises this subgroup. If not, we have $\gamma M_j \gamma^{-1} = M' \neq M_0$, both having distance j from M_j. But in this case there is some δ in the norm 1 group of M_j (hence fixing M_j) and sending M' to M_0, see the arguments in the proof of the next lemma. Then, $\delta\gamma$ exchanges M_0 and M_j, so we can consider this element as $\alpha_{\mathfrak{p}^j}$.

Uniqueness: suppose we have two such elements $\gamma, \delta \in \mathrm{PSL}(2,\mathbb{R})$ exchanging M_0 and M_j under conjugation. They extend to automorphisms of A, so we can assume by the Skolem–Noether Theorem that they are PSL–images of elements of A. Then $\delta^{-1}\gamma$ fixes M_0 and M_j and – as an element of A – has norm 1, hence belongs to $\Delta_0(\mathfrak{p}^j)$ and induces the identity on the quotient curve $\Delta_0(\mathfrak{p}^j)\backslash\mathbb{H}$. □

Now, the principal congruence subgroups $\Delta(\mathfrak{p}^n)$ are obviously included in $\Delta_0(\mathfrak{p}^j)$ for each $j = 1, \ldots, n$, hence we deduce that, in particular, the corresponding surfaces $\Delta(\mathfrak{p}^n)\backslash\mathbb{H}$ contain several uniform dessins. In fact, one has the following

LEMMA 4. *Under the same hypothesis as in Lemma 3, for each $j = 1, \ldots, n$ there are $q^{j-1}(q+1)$ congruence subgroups $\Delta_0^i(\mathfrak{p}^j)$ conjugate to $\Delta_0(\mathfrak{p}^j)$ in Δ, for $i = 0, \ldots, q^{j-1}(q+1) - 1$. Each of them is contained in Δ and in j different triangle groups conjugate to Δ, in which $\Delta(\mathfrak{p}^n)$ is included non-normally. Every $\Delta_0^i(\mathfrak{p}^j)$ is the intersection of Δ with a conjugate triangle group $\Delta^{j,i}$, and for fixed j, the different $\Delta^{j,i}$ form an orbit under conjugation by Δ.*

PROOF. The proof proceeds by induction on j. The group $\Delta_0(\mathfrak{p})$ is an index $q+1$ subgroup of Δ, and for each class of elements $\rho_i \in \Delta$ modulo $\Delta_0(\mathfrak{p})$, $i = 0, \ldots, q$, we can construct the group $\Delta_0^i(\mathfrak{p}) \equiv \rho_i \cdot \Delta_0(\mathfrak{p}) \cdot \rho_i^{-1}$ for $\rho_0 = \mathrm{Id}, \ldots, \rho_q$, such that

$$\Delta(\mathfrak{p}^n) \lhd \Delta_0^i(\mathfrak{p}) < \Delta$$

For each of them we have the Fricke involution $\alpha_i := \rho_i \alpha_{\mathfrak{p}} \rho_i^{-1}$, the Fricke extension $\Delta_{\mathrm{Fr}}^i(\mathfrak{p}) = \langle \Delta_0^i(\mathfrak{p}), \alpha_i \rangle \not< \Delta$ and therefore a non-normal inclusion of $\Delta(\mathfrak{p}^n)$ in

$\alpha_i \Delta \alpha_i^{-1}$, yielding the following diagram of inclusions

$$\begin{array}{ccc} \Delta & & \alpha_i \Delta \alpha_i^{-1} \\ & \diagdown \quad \Delta^i_{\mathrm{Fr}}(\mathfrak{p}) \quad \diagup & \\ & | & \\ & \Delta^i_0(\mathfrak{p}) & \\ & | & \\ & \Delta(\mathfrak{p}^n) & \end{array}$$

Now suppose that there are $q^{j-1}(q+1)$ congruence subgroups conjugate to $\Delta_0(\mathfrak{p}^j)$ in Δ. Inside this subgroup lies $\Delta_0(\mathfrak{p}^{j+1})$ with index q. Its normaliser in $\Delta_0(\mathfrak{p}^j)$ is trivial, and therefore there are q subgroups of $\Delta_0(\mathfrak{p}^j)$ conjugate to $\Delta_0(\mathfrak{p}^{j+1})$. Now the Fricke element $\alpha_{\mathfrak{p}^{j+1}}$ conjugates Δ into a different triangle group $\Delta^{j,1}$ in which $\Delta(\mathfrak{p}^n)$ is included.

The claim about the conjugacy of the different $\Delta^i_0(\mathfrak{p}^j)$ for fixed j follows from the action of Δ on the *fake projective line* $\mathbb{P}^1(\mathcal{O}_k/\mathfrak{p}^j)$ consisting of pairs $(x,y) \in (\mathcal{O}_k/\mathfrak{p}^j)^2$, not both coordinates divisible by $\mathfrak{p}$, modulo the diagonal action of $(\mathcal{O}_k/\mathfrak{p}^j)^*$. The action of Δ on this fake projective line is transitive again because the $\mathfrak{p}$–adic completion of Δ is isomorphic to the $\mathrm{PSL}(2,\mathcal{O}_\mathfrak{p})$ of the $\mathfrak{p}$–adic integers in $k_\mathfrak{p}$, and the subgroups $\Delta^i_0(\mathfrak{p}^j)$ are the stabiliser subgroups of the different points on this fake projective line. They are obviously different and conjugate under the action of Δ by transitivity. By the same conjugations, Δ acts on the $\Delta^{j,i}$ giving the final claim. □

The representation in the tree of local maximal orders is the following. The $q+1$ groups $\Delta^i_0(\mathfrak{p})$ correspond to the $q+1$ paths joining the middle vertex with each of its neighbours at distance 1, which correspond to the $q+1$ groups $\alpha_i \Delta \alpha_i^{-1}$. Similarly, for each $j = 2, \ldots, n$ the groups $\Delta^i_0(\mathfrak{p}^j)$, $i = 1, \ldots, q^{j-1}(q+1)$ are represented by the paths joining the middle vertex with each of its distance j neighbours, which correspond precisely to the triangle groups mentioned in the previous lemma.

Example 1. The principal congruence subgroup $\Delta(\mathfrak{p}_7) < \Delta(2,3,7)$, for a prime $\mathfrak{p}_7$ in $\mathbb{Q}(\cos \pi/7)$ dividing the rational prime 7, uniformises Klein's quartic $\mathcal{K}$ and, accordingly, one has one regular Belyi function and $N(\mathfrak{p}_7) + 1 = 8$ uniform dessins of type $(2,3,7)$ on $\mathcal{K}$.

Example 2. Consider Bring's curve $\mathcal{B}$, given by the equations

$$\begin{cases} x_0 + x_1 + x_2 + x_3 + x_4 = 0 \\ x_0^2 + x_1^2 + x_2^2 + x_3^2 + x_4^2 = 0 \\ x_0^3 + x_1^3 + x_2^3 + x_3^3 + x_4^3 = 0 \end{cases}$$

in $\mathbb{P}^4(\mathbb{C})$. It is known that $\mathcal{B}$ is the only Riemann surface of genus 4 admitting the symmetric group on 5 elements $\mathfrak{S}_5$ as a group of automorphisms (see for example [**13**]), and that this action is simply given by permutation of the projective coordinates.

Now, the triangle group $\Delta := \Delta(2,5,5)$ is the norm 1 group of a maximal order in a quaternion algebra defined over $\mathbb{Q}(\sqrt{5})$, and the principal congruence subgroup $\Delta(\sqrt{5}) \triangleleft \Delta(2,5,5)$ uniformises a surface of genus 4. This group is also normal in the maximal triangle group $\Delta(2,4,5)$, and it can be proved that $\Delta(2,4,5)/\Delta(\sqrt{5}) = \mathrm{PGL}(2,\mathbb{F}_5) \cong \mathfrak{S}_5$. As a consequence the surface $\mathcal{B}$ has the principal congruence subgroup $\Delta(\sqrt{5})$ as uniformising group.

By the results in the previous sections, it has a regular Belyi function and $N(\sqrt{5})+1 = 6$ subgroups $\Delta_0^i(\sqrt{5})$. Note that in this case, these subgroups lead to 12 uniform Belyi functions of type $(2,5,5)$ on $\mathcal{B}$, if we take into account the renormalizations B_i and $1 - B_i$ as in Remark 1 and Lemma 2.

In the following, we will always suppose that Δ is an arithmetic Fuchsian triangle group containing a PSL–image of a norm 1 group of a maximal order M_0 in the quaternion algebra A with totally real centre field.

We will first focus our attention on principal congruence subgroups of level $\mathfrak{p}$. By the discussion above, there are $q+1$ surfaces $S_i \cong \mathbb{H}/\Delta_0^i(\mathfrak{p})$, $i = 0,\dots,q$, and for each of them there are uniform Belyi functions

$$\begin{array}{rlcl} r_i: & \Delta_0^i(\mathfrak{p})\backslash\mathbb{H} & \longrightarrow & \Delta\backslash\mathbb{H} \\ u_i: & \Delta_0^i(\mathfrak{p})\backslash\mathbb{H} & \longrightarrow & \alpha_i\Delta\alpha_i^{-1}\backslash\mathbb{H} \end{array}$$

such that β and all β_i decompose via the intermediate coverings $\tau_i : S \cong \Delta(\mathfrak{p})\backslash\mathbb{H} \longrightarrow S_i \cong \Delta_0^i(\mathfrak{p})\backslash\mathbb{H}$ as

$$\beta = r_i \circ \tau_i, \qquad \beta_i = u_i \circ \tau_i \ .$$

So, for each $i = 0,\dots,q$ one has the following diagram

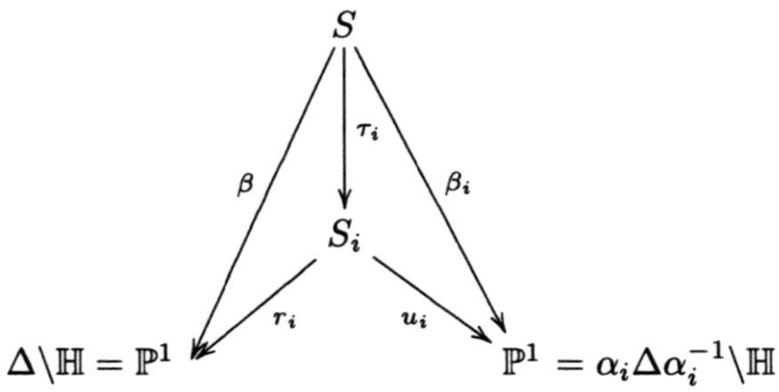

LEMMA 5. *Let a_i be the automorphism of the intermediate curve $S_i := \Delta_0^i(\mathfrak{p})\backslash\mathbb{H}$ induced by the Fricke involution α_i. Then, $u_i \circ a_i = r_i$, hence $\beta = r_i \circ \tau_i = u_i \circ a_i \circ \tau_i$ and $\beta_i = u_i \circ \tau_i = r_i \circ a_i \circ \tau_i$.*

PROOF. Remember that r_i and u_i are the Belyi functions on S_i induced by the Belyi functions β and β_i on S, and that their dessins $\mathcal{D}$ and $\mathcal{D}_i$ come from the tessellations of $\mathbb{H}$ by the fundamental domains F, F_i for Δ and $\alpha_i\Delta\alpha_i^{-1}$. Since for F_i we can take $\alpha_i(F)$, the Belyi functions are linked as indicated. □

Lemma 5 generalises to higher levels as follows.

LEMMA 6. *Let $S_{j,i} := \Delta_0^i(\mathfrak{p}^j)\backslash\mathbb{H}$, $\tau_{j,i}$ the quotient map $S_{\mathfrak{p}^n} \to S_{j,i}$ and*

$$r_{j,i} : S_{j,i} \to \Delta\backslash\mathbb{H}\ , \quad u_{j,i} : S_{j,i} \to \alpha_{j,i}\Delta\alpha_{j,i}^{-1}\backslash\mathbb{H} = \Delta^{j,i}\backslash\mathbb{H}$$

the two quotient maps giving the regular and the nonregular uniform Belyi functions

$$\beta = r_{j,i} \circ \tau_{j,i} \quad , \qquad \beta_{j,i} = u_{j,i} \circ \tau_{j,i} \ : \ S_{\mathfrak{p}^n} \to \mathbb{P}^1(\mathbb{C})\ .$$

Then the involution $a_{j,i}$ induced by $\alpha_{j,i}$ on $S_{j,i}$ satisfies

$$u_{j,i} = r_{j,i} \circ a_{j,i} \quad \textit{and} \quad r_{j,i} = u_{j,i} \circ a_{j,i} \quad .$$

4. Shimura congruence curves of prime power level

In this section we will get some information about the fields of moduli and the fields of the definition of the curves and dessins described in Section 3.

4.1. Fields of moduli. In order to determine explicitly the field of moduli $M(S_\mathfrak{p})$ of our Shimura curves we could rely on Clark and Voight's Theorem A and Proposition 5.1 in [**2**], at least in the case of prime levels. We will however prove parts of these results in another way, more similar to the methods already used here.

First we need the following useful lemma.

LEMMA 7. *Let Δ be an arithmetically defined cocompact triangle group, containing the norm 1 group of a maximal order of a quaternion algebra A with centre field k, and let $\mathfrak{p}$ be a prime ideal in k not dividing the discriminant of A. Then, for each positive integer n one has $M(S_{\mathfrak{p}^n}) = M(S_\mathfrak{p})$.*

PROOF. We restrict to the case $n = 2$, the higher powers behave similarly. Suppose $S_{\mathfrak{p}^2}$ to be defined over $M(S_{\mathfrak{p}^2})$. First,

$$M(S_{\mathfrak{p}^2}) > M(S_\mathfrak{p})$$

because otherwise we would have a $\sigma \in \mathrm{Gal}\,(\overline{\mathbb{Q}})$ leaving invariant $S_{\mathfrak{p}^2}$ but having two non–isomorphic quotient congruence curves

$$S_\mathfrak{p} \qquad \text{and} \qquad (S_\mathfrak{p})^\sigma$$

which is impossible.

Second, we get $M(S_{\mathfrak{p}^2}) < M(S_\mathfrak{p})$ by the following argument. For all $\sigma \in \mathrm{Gal}\,(\overline{\mathbb{Q}}/M(S_\mathfrak{p}))$, the Galois conjugate curve $(S_{\mathfrak{p}^2})^\sigma$ has as many uniform dessins as $S_{\mathfrak{p}^2}$, and they are of the same types and have the same automorphism groups as them (up to isomorphism, of course). By the main results of [**8**], this can occur only if this Galois conjugate curve has a surface group Γ contained in a principal congruence subgroup $\Delta(\mathfrak{q}^2)$ with a prime ideal $\mathfrak{q}$ in k of the same norm q as $\mathfrak{p}$. Moreover, since also the index $(\Delta : \Delta(\mathfrak{p}^2))$ is invariant under Galois conjugation, we have even $\Gamma = \Delta(\mathfrak{q}^2)$. Now, by the same argument as in the first part of the proof, their common quotient curve $S_\mathfrak{p}$ shows

$$\Delta(\mathfrak{q}^2) \triangleleft \Delta(\mathfrak{p}) \triangleright \Delta(\mathfrak{p}^2)$$

which is possible only for $\mathfrak{q} = \mathfrak{p}$. Therefore, σ sends $S_{\mathfrak{p}^2}$ to an isomorphic curve, hence $M(S_{\mathfrak{p}^2}) = M(S_\mathfrak{p})$. □

LEMMA 8. *Let Δ be an arithmetically defined cocompact triangle group, containing the norm 1 group of a maximal order of a quaternion algebra A with centre field k (abelian over $\mathbb{Q}$), and let p be a rational prime coprime to the discriminant of A, $\mathfrak{p}$ one of the prime ideals of the ring of integers $\mathcal{O}_k$ of k in the prime decomposition*

$$p\mathcal{O}_k = (\mathfrak{p}_1 \cdot \ldots \cdot \mathfrak{p}_g)^e \quad ,$$

g the splitting number *and e the* ramification index *of p in the field extension $k/\mathbb{Q}$. For every positive integer n let $S = S_{\mathfrak{p}^n}$ be the quasiplatonic surface with surface group $\Delta(\mathfrak{p}^n)$. Then $[M(S) : \mathbb{Q}] \leq g$.*

PROOF. By Lemma 7 we can restrict ourselves to the case $n = 1$. For the action of an arbitrary $\sigma \in \mathrm{Gal}(\overline{\mathbb{Q}})$ on S we know that S^σ is still quasiplatonic, that regularity, uniformity, and the type of all dessins are preserved. Therefore, if we denote by $\Gamma = \Delta(\mathfrak{p})$ the surface group of S and by Γ_σ the surface group of S^σ, then Γ_σ is again normal in Δ and S^σ has one regular dessin and $N(\mathfrak{p}) + 1$ uniform ones of type (r, s, t) if $\Delta = \Delta(r, s, t)$. By [**8**] we know that this is only possible if the surface group of S^σ is contained in one of the principal congruence subgroups $\Delta(\mathfrak{p}_j)$. Because also degrees of Belyi functions are Galois invariant, we have even $\Gamma_\sigma = \Delta(\mathfrak{p}_j)$ for some j, since

$$(\Delta : \Delta(\mathfrak{p}_j)) = (\Delta : \Delta(\mathfrak{p})) = (\Delta : \Gamma_\sigma) \quad .$$

Therefore the well–known group action of $\mathrm{Gal}(\overline{\mathbb{Q}})$ on the set of prime ideals $\mathfrak{p}_j$, $j = 1, \ldots, g$, induces an action on the principal congruence subgroups $\Delta(\mathfrak{p}_j)$ and their corresponding surfaces $S_{\mathfrak{p}_j}$ whose stabiliser subgroup I_S has an index at most g in $\mathrm{Gal}(\overline{\mathbb{Q}})$. So Galois theory implies the claim about the field degree. □

COROLLARY 2. *Under the same assumptions, we have* $[M(S) : \mathbb{Q}] \leq [k : \mathbb{Q}]$. *Moreover,* $M(S) = \mathbb{Q}$ *in all cases with splitting number* $g = 1$, *that is in particular for all rational primes* p *inert or totally ramified in* k.

This applies in particular to the examples we have already seen: for Bring's curve we have $k = \mathbb{Q}(\sqrt{5})$ with $p = 5$ totally ramified, for Klein's quartic $k = \mathbb{Q}(2\cos\frac{2\pi}{7})$ and $p = 7$ is totally ramified too. The next Hurwitz curve (with the same $\Delta = \Delta(2, 3, 7)$, hence with the same k) is the *Fricke–Macbeath curve* S_2 in genus 7 with surface group $\Delta(2)$, 2 inert in k, hence with $M(S_2) = \mathbb{Q}$ as well.

Not only in these cases the estimate of Lemma 8 is sharp. Streit's method used in [**16**] extends to many other arithmetic triangle groups; the technical difficulties in his approach are however not smaller than those in the preprint [**2**] by Clark and Voight.

4.2. Fields of definition. In this section we will study the fields of definition of some of the curves involved in the construction and we will find simultaneous fields of definition for the uniform dessins constructed in the previous sections.

We start by stating some facts about the fields of definition of the intermediate curves S_i.

LEMMA 9. *Let* $S = S_{\mathfrak{p}}$ *be the quasiplatonic surface with surface group* $\Delta(\mathfrak{p})$, *with* Δ *a maximal arithmetic triangle group containing the norm 1 group of a maximal order of* A *and* $\mathfrak{p}$ *a prime not dividing the discriminant of* A, *and let* $S_i \cong \Delta_0^i(\mathfrak{p})\backslash\mathbb{H}$ *be as above. Then* S_i *can be defined over* $M(S)$.

PROOF. Let $g_{ij} : S_i \to S_j = S_i^\sigma$ be the unique isomorphism between S_i and S_j induced by an element δ_{ij} in the norm 1 group Φ contained in Δ. This choice is unique because there are precisely $q + 1$ residue classes $\delta\Delta_0^i(\mathfrak{p}) \in \Phi/\Delta_0^i(\mathfrak{p})$ which give by conjugation the $q + 1$ different $\Delta_0^j(\mathfrak{p})$. Composition of such isomorphisms gives again an isomorphism induced by an element of Φ, and Galois conjugation by $\tau \in \mathrm{Gal}(\overline{\mathbb{Q}}/M(S))$ preserves also this collection of isomorphisms induced by the norm 1 group Φ because $\Phi/\Delta(\mathfrak{p})$ is the commutator subgroup of the full automorphism group $G := \Delta/\Delta(\mathfrak{p})$ of S. Therefore, the collection of these g_{ij} satisfies Weil's cocycle condition whence all S_i can be defined over $M(S)$. □

THEOREM 2. *Under the same hypotheses as in Lemma 9,* $M(S,\beta_i) = M(S)$ *for all* i, *in other words: the moduli field of* S *is also a moduli field of all uniform non–regular dessins on it.*

PROOF. We may suppose S to be defined over $M(S)$. For all i and all $\sigma \in \mathrm{Gal}(\overline{\mathbb{Q}}/M(S))$, β_i^σ is another uniform Belyi function β_j of the same type, and for them the same δ_{ij} as in Lemma 9 induce automorphisms $d_{ij} := \delta_{ij}\Delta(\mathfrak{p}) \in G = \mathrm{Aut}\, S$ with the property $\beta_j \circ d_{ij} = \beta_i$. □

Unfortunately, in contrast to the isomorphisms g_{ij} in the proof of Lemma 9, the automorphisms d_{ij} are uniquely determined only modulo the subgroup $\Delta_0^j(\mathfrak{p})/\Delta(\mathfrak{p})$ of G, so Weil's criterion does not apply immediately. The question of whether the uniform dessins can be defined over their field of moduli $M(S)$ does not have therefore an obvious answer. However, some partial answers can be given.

THEOREM 3. *Under the same hypotheses as in Lemma 9, let* S_i *be defined over* $M(S)$. *Suppose moreover that there exists a* $M(S)$*–rational point* $x \in S_i$, *not critical for the canonical covering* $\tau_i : S \to S_i$. *Then* (S,β_i) *can be defined over* $M(S)$.

PROOF. By Lemma 5 we know that $\beta_i = r_i \circ a_i \circ \tau_i$. First, let us prove that both τ_i and r_i can defined over $M(S)$. Since S_i and x are $\mathrm{Gal}(\overline{\mathbb{Q}}/M(S))$–invariant and τ_i is a normal covering, one can apply the method of Coombes and Harbater [**3**]: take a preimage y of x under τ_i. For all $\sigma \in \mathrm{Gal}(\overline{\mathbb{Q}}/M(S))$, we have $\sigma(y)$ in the preimage of x under τ_i^σ, hence there is a unique isomorphism

$$d_\sigma : S \to S^\sigma \qquad \text{with} \quad d_\sigma(y) = \sigma(y) \quad \text{and} \quad \tau_i = \tau_i^\sigma \circ d_\sigma\,,$$

so Weil's criterion applies to the pair (S,τ_i). The function r_i is also $\mathrm{Gal}(\overline{\mathbb{Q}}/M(S))$–invariant: otherwise we would have several different regular Belyi functions $\beta = r_i \circ \tau_i$ and $\beta^\sigma = r_i^\sigma \circ \tau_i$ on S.

Now, the Fricke involution a_i of S_i is not necessarily fixed by all Galois elements $\sigma \in \mathrm{Gal}(\overline{\mathbb{Q}}/M(S))$. If $a_i^\sigma = a_i$, then $\beta_i^\sigma = r_i^\sigma \circ a_i^\sigma \circ \tau_i^\sigma = r_i \circ a_i \circ \tau_i = \beta_i$.

Let us suppose that $a_i^\sigma \neq a_i$. Note that, if a_i is induced by conjugation by $\alpha_i \in N(\Delta_0^i(\mathfrak{p}))$, then $a_i^\sigma = a_i \circ d$ is induced by conjugation by $\alpha_i\delta$ for some $\delta \in N(\Delta_0^i(\mathfrak{p})) \cap \alpha_i\Delta\alpha_i^{-1}$, since $\alpha_i\delta$ must interchange the maximal triangle groups Δ and $\alpha_i\Delta\alpha_i^{-1}$. But the automorphism group $\mathrm{Aut}\,(S_i,u_i)$ of the Belyi function $u_i = r_i \circ a_i$ is induced by the normaliser of $\Delta_0^i(\mathfrak{p})$ in $\alpha_i\Delta\alpha_i^{-1}$. As a consequence one has $u_i \circ d = u_i$ and $\beta_i^\sigma = r_i^\sigma \circ a_i^\sigma \circ \tau_i^\sigma = r_i \circ a_i \circ d \circ \tau_i = \beta_i$. □

If we ask for fields over which all relevant functions are defined simultaneously, the results become slightly weaker, of course.

THEOREM 4. *Under the same hypotheses as in Lemma 9, the regular and all uniform Belyi functions on the Shimura curve* $S = S_\mathfrak{p}$ *can be defined simultaneously over the field of moduli* $M(S,G)$.

PROOF. By Corollary 1, the function fields of S, $\mathbb{P}^1$ and all S_i can be defined over $M(S,G)$, and all quotient maps τ_i and r_i in Lemma 5 correspond to the respective embeddings of these function fields into each other. As a consequence all these maps can be defined over $M(S,G)$ simultaneously. Now, by the same argument as in the proof of Theorem 3, for all $\sigma \in \mathrm{Gal}(\overline{\mathbb{Q}}/M(S,G))$ one has $u_i^\sigma = u_i$, hence also $\beta_i^\sigma = \beta_i$. □

Example 3. Let $\mathcal{B} \cong \mathbb{H}/\Delta(\sqrt{5})$ be Bring's curve already introduced in Example 2. Let us recall that its full automorphism group $\operatorname{Aut}\mathcal{B} \cong \mathfrak{S}_5$ acts on the point $[x_0, \ldots, x_4] \in \mathcal{B}$ by permutation of the projective coordinates. Since both $\mathcal{B}$ and its automorphism group are defined over $\mathbb{Q}$, by Theorem 4 its regular dessin and all its uniform dessins can be defined over $\mathbb{Q}$ (even simultaneously).

Example 4. In the affine model $\mathcal{K}_F := \{y^7 = x(x-1)^2\}$ of Klein's quartic $\mathcal{K}$ (see Example 2), the function

$$\beta_0(x,y) = -\frac{\left(x^6 - 235x^5 + 1430x^4 - 1695x^3 + 270x^2 + 229x + 1\right)^3 \left(x^2 - x + 1\right)^3}{1728x(x-1)\left(x^3 + 5x^2 - 8x + 1\right)^7}$$

is one of the $(2,3,7)$ uniform Belyi functions (see Example 4.44 in [**7**]) and it is defined over $M(\mathcal{K}) = \mathbb{Q}$. However, in this model there are automorphisms not defined over $\mathbb{Q}$, and therefore we cannot expect all uniform Belyi functions in $\mathcal{K}_F$ to be defined over $\mathbb{Q}$ simultaneously.

However, there is a model for Klein's quartic in which all uniform Belyi functions are simultaneously defined over the field $\mathbb{Q}(\sqrt{-7})$. This follows from a result in Clark and Voight's preprint [**2**] saying that in this case $M(\mathcal{K}, G) = \mathbb{Q}(\sqrt{-7})$. A relatively easy proof for this claim is the following.

Another model for $\mathcal{K}$ defined over the rationals is the projective equation $x^3y + y^3z + z^3x = 0$, and in this model all automorphisms are defined over the cyclotomic field $\mathbb{Q}(\zeta_7)$, therefore $M(\mathcal{K}, G)$ has to be a subfield of this cyclotomic field. The absolute Galois group acts as automorphism group on $G \cong \mathrm{PSL}(2, \mathbb{F}_7)$ whose automorphism group is isomorphic to $\mathrm{PGL}(2, \mathbb{F}_7)$, an index 2 extension of its inner automorphism group $\cong G$. For inner automorphisms, that is for those σ acting on G as

$$\alpha \mapsto \alpha^\sigma = \gamma^{-1}\,\alpha\,\gamma \quad \text{for some} \quad \gamma = f_\sigma \in G$$

we have obviously $\sigma \in I_{(\mathcal{K},G)}$, therefore $M(\mathcal{K}, G)$ can be an extension of $\mathbb{Q}$ of degree at most 2. Because $\mathbb{Q}(\sqrt{-7})$ is the only quadratic subfield of $\mathbb{Q}(\zeta_7)$, the claim follows.

In a similar way – using Lemma 6 instead of Lemma 5 – we can prove

THEOREM 5. *For all prime ideal powers coprime to the discriminant of A, the regular and all uniform Belyi functions on the Shimura curve $S = S_{\mathfrak{p}^n}$ can be defined simultaneously over the field of moduli $M(S, G)$.*

References

[1] G. V. Belyĭ, *Galois extensions of a maximal cyclotomic field* (Russian), Izv. Akad. Nauk SSSR Ser. Mat. **43** (1979), no. 2, 267–276, 479. MR534593 (80f:12008)

[2] Clark, P., Voight, J., *Algebraic curves uniformized by congruence subgroups of triangle groups*, preprint, available at the authors' web pages.

[3] Kevin Coombes and David Harbater, *Hurwitz families and arithmetic Galois groups*, Duke Math. J. **52** (1985), no. 4, 821–839, DOI 10.1215/S0012-7094-85-05243-3. MR816387 (87g:14012)

[4] Pierre Dèbes and Jean-Claude Douai, *Algebraic covers: field of moduli versus field of definition* (English, with English and French summaries), Ann. Sci. École Norm. Sup. (4) **30** (1997), no. 3, 303–338, DOI 10.1016/S0012-9593(97)89922-3. MR1443489 (98k:11081)

[5] Džambić, A., *Macbeath's infinite series of Hurwitz groups*. In: Arithmetic and geometry around hypergeometric functions, ed. Holzapfel, R.–P., Uludag, A.M., Yoshida, M. (Progr. Math., 260, Birkhäuser, Basel, 2007), pp. 101–108.

[6] Clifford J. Earle, *On the moduli of closed Riemann surfaces with symmetries*, Advances in the theory of riemann surfaces (Proc. Conf., Stony Brook, N.Y., 1969), Princeton Univ. Press, Princeton, N.J., 1971, pp. 119–130. Ann. of Math. Studies, No. 66. MR0296282 (45 #5343)

[7] Girondo, E., González-Diez, G., *Introduction to compact Riemann surfaces and dessins d'enfants,* London Mathematical Society Student Texts **79**, Cambridge University Press, Cambridge, 2011.

[8] Ernesto Girondo, David Torres-Teigell, and Jürgen Wolfart, *Shimura curves with many uniform dessins*, Math. Z. **271** (2012), no. 3-4, 757–779, DOI 10.1007/s00209-011-0889-4. MR2945583

[9] Ernesto Girondo and Jürgen Wolfart, *Conjugators of Fuchsian groups and quasiplatonic surfaces*, Q. J. Math. **56** (2005), no. 4, 525–540, DOI 10.1093/qmath/hah054. MR2182464 (2006g:30072)

[10] Alexandre Grothendieck, *Esquisse d'un programme* (French, with French summary), Geometric Galois actions, 1, London Math. Soc. Lecture Note Ser., vol. 242, Cambridge Univ. Press, Cambridge, 1997, pp. 5–48. With an English translation on pp. 243–283. MR1483107 (99c:14034)

[11] Colin Maclachlan and Alan W. Reid, *The arithmetic of hyperbolic 3-manifolds*, Graduate Texts in Mathematics, vol. 219, Springer-Verlag, New York, 2003. MR1937957 (2004i:57021)

[12] A. M. Macbeath, *Generators of the linear fractional groups*, Number Theory (Proc. Sympos. Pure Math., Vol. XII, Houston, Tex., 1967), Amer. Math. Soc., Providence, R.I., 1969, pp. 14–32. MR0262379 (41 #6987)

[13] Gonzalo Riera and Rubí E. Rodríguez, *The period matrix of Bring's curve*, Pacific J. Math. **154** (1992), no. 1, 179–200. MR1154738 (93e:14033)

[14] Goro Shimura, *On the field of rationality for an abelian variety*, Nagoya Math. J. **45** (1972), 167–178. MR0306215 (46 #5342)

[15] David Singerman, *Automorphisms of maps, permutation groups and Riemann surfaces*, Bull. London Math. Soc. **8** (1976), no. 1, 65–68. MR0393209 (52 #14019)

[16] Manfred Streit, *Field of definition and Galois orbits for the Macbeath-Hurwitz curves*, Arch. Math. (Basel) **74** (2000), no. 5, 342–349, DOI 10.1007/s000130050453. MR1753011 (2001e:14028)

[17] Kisao Takeuchi, *Commensurability classes of arithmetic triangle groups*, J. Fac. Sci. Univ. Tokyo Sect. IA Math. **24** (1977), no. 1, 201–212. MR0463116 (57 #3077)

[18] Marie-France Vignéras, *Arithmétique des algèbres de quaternions* (French), Lecture Notes in Mathematics, vol. 800, Springer, Berlin, 1980. MR580949 (82i:12016)

[19] André Weil, *The field of definition of a variety*, Amer. J. Math. **78** (1956), 509–524. MR0082726 (18,601a)

[20] J. Wolfart, *ABC for polynomials, dessins d'enfants and uniformization—a survey*, Elementare und analytische Zahlentheorie, Schr. Wiss. Ges. Johann Wolfgang Goethe Univ. Frankfurt am Main, 20, Franz Steiner Verlag Stuttgart, Stuttgart, 2006, pp. 313–345. MR2310190 (2009d:14031)

Departamento de Matemáticas, Universidad Autónoma de Madrid, Cantoblanco Universidad, Madrid 28049 — and — Instituto de Ciencias Matemáticas, ICMAT-CSIC-UAM-UCM-UC3M

E-mail address: `ernesto.girondo@uam.es`

Departamento de Matemáticas, Universidad Autónoma de Madrid, Cantoblanco Universidad, Madrid 28049 — and — Math. Seminar der Goethe Universität, Postfach 111932, D-60054 Frankfurt a. M., Germany

E-mail address: `david.torres@uam.es`

Math. Seminar der Goethe Universität, Postfach 111932, D-60054 Frankfurt a. M., Germany

E-mail address: `wolfart@math.uni-frankfurt.de`

Surfaces as complete intersections

Vinicio Gómez Gutiérrez and Santiago López de Medrano

The paper is dedicated to Emilio Bujalance, on his 60th birthday.

ABSTRACT. The topology of real algebraic varieties is an active field of research and there are open problems starting from the 1-dimensional case. In this article we present some open problems and results about certain infinite families of real surfaces that are intersections of affine quadrics and are related with families of complex curves that are projective complete intersections. The surfaces in these families admit actions of 2-groups with quotient a polygon.

Introduction

For many years we have been interested in the intersections of certain affine real quadrics:

We call *homogeneous quadric* an affine quadric in $\mathbb{R}^n$ given by a quadratic form

$$\Sigma a_{ij} x_i x_j = 0$$

where a_{ij} are the entries of a real symmetric $n \times n$ matrix. We then consider the intersection V of any number of such quadrics in $\mathbb{R}^n$, where we will always assume that 0 *is the only singular point of* V. The topology of V is completely determined by its intersection with the unit sphere:

$$Y = V \cap S^{n-1}$$

Most of the time we will say, briefly, that Y is an *intersection of homogeneous quadrics*, short for *generic (regular, complete) intersection of real homogeneous quadrics in the sphere in* $\mathbb{R}^n$.

There has been work also on the *deformations* of V (intersections of quadrics of the form $\Sigma a_{ij} x_i x_j = c$) as well as on variants of Y such as their complex or projective versions, some related semi-algebraic varieties and also on geometric structures on all of them. All of these objects appear in many fields of Mathematics. Some results about their topology were obtained almost thirty years ago and many more were obtained in the last five, but many questions are still open. See [**LdM1**] and [**LdM2**] for details of all this.

One of the open problems is that of describing the topology of Y in the case it has dimension 2. It seems way out of reach to give a complete answer of the type: If you give me any (generic) collection of $k = n - 3$ homogeneous quadrics in $\mathbb{R}^n$,

---

2010 *Mathematics Subject Classification.* Primary 14P25, 14M10.

Key words and phrases. Surfaces, actions of 2-groups, complete intersections, complex projective curves.

I can tell you the topology of Y from their equations. Starting with three quadrics the problem is difficult since there are no known general methods for handling three quadratic forms.

But a reasonable question is the following:

Which topological types of compact surfaces can appear as such intersections?

After all, it is not difficult to prove that every compact orientable surface is a transversal intersection of *arbitrary* (i.e., not necessarily homogeneous) affine quadrics in some euclidean space $\mathbb{R}^n$ (see [**Go-LdM1**]), even if we do not know how to tell the topology of a given complete intersection from its equations. A closely related problem is to characterize the compact surfaces that are complete intersections of real *projective* quadrics, where we have to include now also non-orientable ones. One can produce examples of these projective intersections from the affine ones given in this article, but for the moment we do not give them explicitly.

In the first section we describe the present state of our knowledge about the topology of intersections of homogeneous quadrics in general.

In the second section we describe the topology of surfaces that are intersections of diagonal homogeneous quadrics, and relate it to the topology of projective complex curves that are complete intersections of quadrics.

In the third section we present a construction of surfaces with 2-group actions whose quotient is a polygon.

In the last section we show how to realize a two parameter family of the previous group actions by intersections of homogeneous quadrics and relate them to the projective complex curves that are complete intersections of quadrics and quartics. We end the section by stating some open problems suggested by our results.

We have presented partial results on this question at conferences in Gdansk (2012) and Linksöping (2013) and in both cases we have received very useful information from many of the participants. Especially enlightening have been comments by Antonio Costa, Víctor González, Rubén Hidalgo and Rubí Rodríguez. We are also in debt with Grzegorz Gromadzki and Milagros Izquierdo for their kind invitations to participate in those meetings and their warm reception.

We hope that this written presentation of our results and open questions will bring us new useful comments and suggestions.

1. Topology of Intersections of Quadrics, in general

We describe now our present knowledge of the topology of the regular, generic variety Y (joint work with Yadira Barreto, Samuel Gitler and Alberto Verjovsky, see [**B-LdM-V**], [**Gi-LdM**], [**Go-LdM2**], [**LdM1**], [**LdM2**]:

Much of the work has been done in the special but interesting case where each component is *diagonal*, that is, when the quadrics are simultaneously diagonalizable. In this case Y admits a $\mathbb{Z}_2^n$ action with quotient a simple convex polytope P which essentially determines the manifold Y and the action.

In this case our present knowledge is as follows: when the dimension or the number of quadrics is not bigger than 2 we have a complete classification.

The cases where the dimension and the number of quadrics is bigger than 2 is related to the problem of classifying all the corresponding simple polytopes: a task that looks impossible for the moment. Nevertheless, in joint work with Samuel Gitler ([**Gi-LdM**], following roads opened in [**B-M**]), we were able to describe the topology of infinitely many infinite families of them. This seems to cover a huge number of cases: just the polytopes corresponding to the roots of these infinite families are considered by some experts to be *most of simple polytopes*. The upper figure 1 shows all this diagrammatically: the solid region indicates complete knowledge and crosses indicate partial but rich knowledge.

Together with this we have a good knowledge of the topology of associated manifolds with boundary, deformations, open book structures, ...

In the non-diagonal case we have complete knowledge when the number of quadrics is at most two and partial knowledge when the dimension is at most two. In the first case this is because, for two quadrics, the non-diagonal cases are iterated double covers of diagonal ones, a situation that can be handled. When the dimension and the number of quadrics is bigger than 2 we only know the diagonal cases, so proportionally there is a lot more of unknown territory and in any case the intersection is invariant under a smaller group of reflections whose role is less well-understood. This seems to be one of the obstacles to the complete description of the 2-dimensional ones.

The lower figure on page 174 shows the situation in the non-diagonal case: the solid region indicates complete knowledge and dark crosses indicate partial knowledge, while white crosses indicate that the there are some advances (reported in this article) that may hopefully lead to the knowledge of all the possible topological types.

In many of the cases the intersections are connected sums of manifolds, each of which is a product of two spheres. This means that in those cases their topology is really no more complicated than that of surfaces. The same is true about some related manifolds with boundary. One obtains also arbitrary products of any number of spheres and any number of those connected sums, as well as other simple constructions with them. Nevertheless, there are cases where the topology can be extremely complicated.

2. Surfaces which are Intersections of Homogeneous Diagonal quadrics

These surfaces are completely characterized. At the Gdansk conference we learned that the same surfaces appear as the topological types of *generalized Humbert curves*, giving the following remarkable coincidence:

PROPOSITION 2.1. *For a compact connected orientable surface S the following are equivalent:*

(i) *S can be realized as the intersection of homogeneous diagonal quadrics in $\mathbb{R}^n$ (and has a $\mathbb{Z}_2^n$ action with quotient an n-gon).*
(ii) *S can be realized as the complete intersection of projective quadrics in $\mathbb{C}P^n$ (and has a $\mathbb{Z}_2^{n-1}$ action with quotient a sphere and a $\mathbb{Z}_2^n$ action with quotient a polygon).*
(iii) *The genus of S is $2^{n-3}(n-4)+1$.*

The surfaces (i) were constructed by Hirzebruch as a by-product of his work on arrangements of lines [**H2**][1]. The same surfaces were obtained in 1938 by Coxeter in [**C**] by a construction that now is recognized as a predecessor of the *universal*

[1]Private conversation, 1986. They did not appear in the published version.

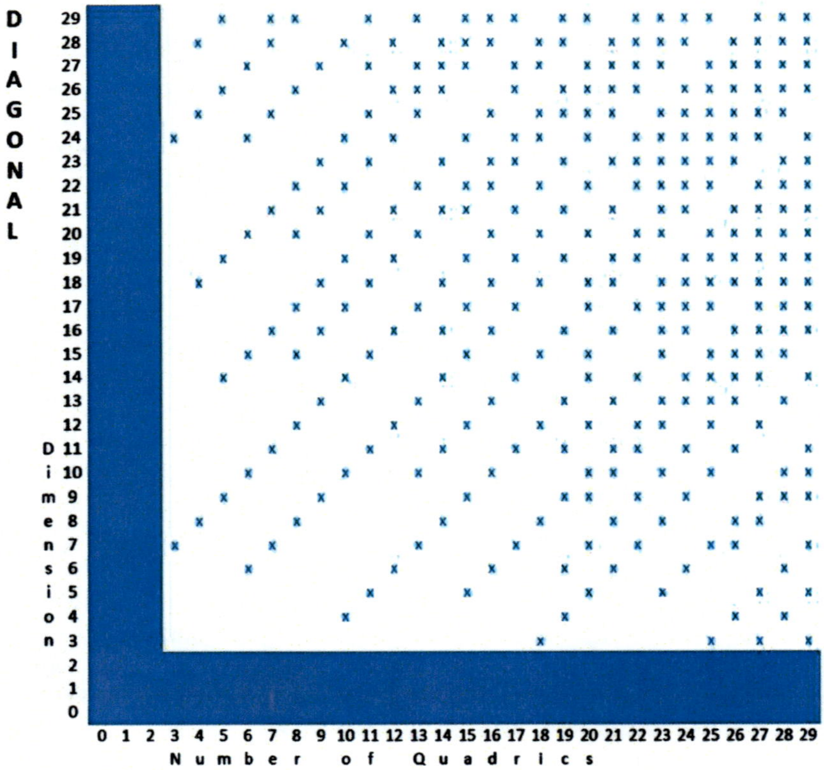
DIAGONAL
Dimension
Number of Quadrics

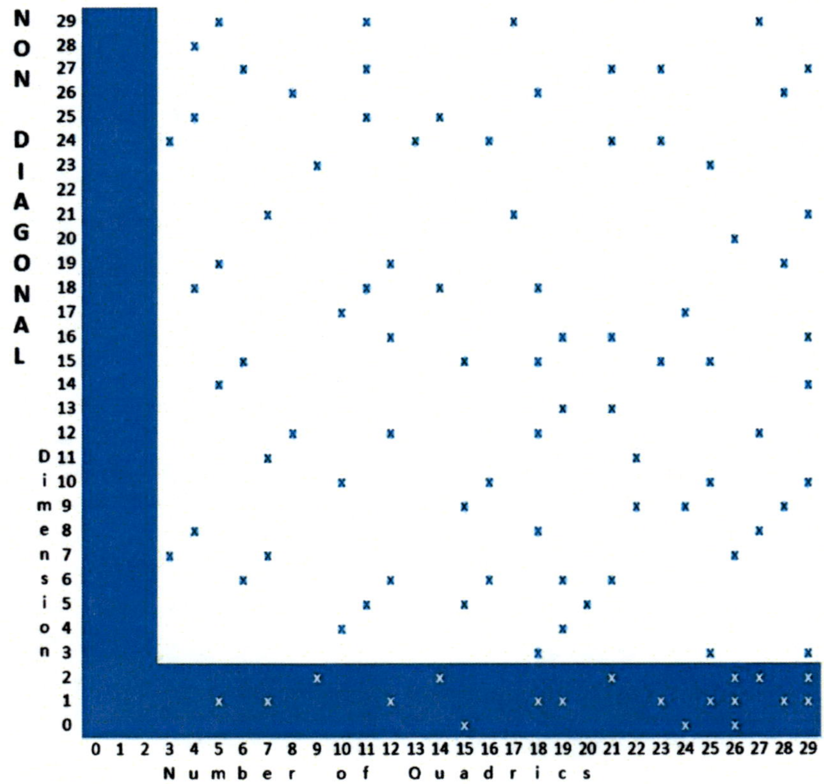
NON DIAGONAL
Dimension
Number of Quadrics

abelian covers ([**D-J**]), moment-angle manifolds ([**B-P**]) and polyhedral products ([**B-B-C-G**]) and can be shown to be equivalent to an intersection of homogeneous diagonal quadrics. The same family of surfaces appears in R^3 as the boundaries of regular neighborhoods of embeddings of the *graph* (i.e., 1-skeleton) of the $(n-2)$-cube[2] and also in [**M-S-W**] as certain polyhedral manifolds.

The surfaces (ii) were constructed in [**C-G-H-R**] as *generalized Humbert curves* and many remarkable properties of them were proved. In section 4.1 of [**C-G-H-R**] these curves in $\mathbb{C}P^n$ are given by $n-1$ equations of the form

$$\lambda_i z_1^2 + z_2^2 + z_{i+3}^2 = 0, \quad i = 0, \ldots, n-2$$

where all the λ_i are different and non-zero. To the natural $\mathbb{Z}_2^{n-1}$ action one can add, when the coefficients are real, another $\mathbb{Z}_2$ summand acting by simultaneous conjugation of all the variables.

By weakening the condition in (i) that the quadrics be diagonal (which is related to the fact that there is a smaller 2-group of symmetries) we can obtain many surfaces with genera different from those in (iii), starting from that of genus 3 (see section 4).

A natural question would be to find ways to weaken condition (ii) to obtain new families of algebraic curves corresponding to the intersections of non-diagonal quadrics. For example, there is a concept of *almost complete intersection* ([**K**]), so one can imagine various degrees of *incompleteness* of intersections of quadrics. We have not developed this idea since the concept of *almost complete intersection* has been considered, as far as we know, only in a very abstract setting and we have not found in the literature any algebraic curves which are given concretely as almost complete intersections.

Or we may consider surfaces which are complete intersections of quadrics and other hypersurfaces of higher degree such as quartics: for example the surface of genus 3 is the regular quartic. We will see in section 4 that a certain family of intersections of homogeneous non-diagonal quadrics corresponds exactly with the family of curves which are complete intersections of quadrics and quartics, in the spirit of the Proposition.

3. Actions of 2-groups on Surfaces with Quotient a Polygon

Consider the unit disk in the plane and divide its boundary into m arcs. We think of it as an m-gon P. Now paint the arcs with c colors in such a way that

(a) no contiguous arcs have the same color
(b) every color appears at least once.
(c) a point in the intersection of two arcs is considered to have both colors.

Now make the natural identifications on the product $P \times \mathbb{Z}_2^c$: if p is a point of an arc, collapse $\{p\} \times \mathbb{Z}_2^c$ by taking the quotient by the $\mathbb{Z}_2$ factor corresponding to the color of the arc. It is understood that if p is a vertex belonging to two arcs we take the quotient by both subgroups.

It is easy to see that:

[2] It is another curious fact that the same numbers in (iii) appear as the *genus* of the graph of the n-cube, which means the minimal genus of a surface where the graph embeds (see [**B-H**]), and in many combinatorial problems (see Sloane, http://oeis.org/, A000337).

(i) The result is a compact, connected and orientable surface S with an action of Z_2^c with quotient P and isotropy subgroups corresponding to the coloring (so there are no ramification points in the interior of P).
(ii) The Euler characteristic of S is $\chi = 2^{c-2}(4-m)$ and its genus $g = 2^{c-3}(m-4)+1$. ($c = m$ corresponds to the diagonal case).
(iii) Every action on a surface with the properties listed in (i) can be realized by this construction.
(iv) These actions are classified by combinatorial equivalence classes of colorings of the arcs. For m big, there is clearly a huge number of non-equivalent actions on the same surface, since the genus is determined by the number of colors only.

One can consider the subgroup of orientation preserving elements of the group. The quotient in this case is a sphere with certain marked points corresponding with the original points that are in two arcs.

The above construction can be generalized by taking, instead of the disk, any compact connected manifold with boundary and a colored map on its boundary. We are still working on this general construction and its implication for the realization of surfaces as complete intersections of homogeneous quadrics. All these actions are related to the theory of *non-Euclidean crystallographic (N.E.C.) groups* developed by Emilio Bujalance [**Buj**]. This was remarked to us by Antonio Costa, but we still have not studied this relation in detail.

4. A family of intersections of non-diagonal homogeneous quadrics

We will describe a way to realize as intersections of homogeneous quadrics some of the surfaces constructed in the previous section:

Take the sphere $x_1^2 + x_2^2 + x_3^2 = 1$ in $\mathbb{R}^3$ and project it into the (x_1, x_2) plane obtaining the unit disc. The sphere can be recovered from the disc by the construction of the previous section (only one *arc* and one color which we label by the number 3). Now remove from the disc the points satisfying $x_1^2 - x_2^2 > d$ where d is slightly smaller than 1.

We obtain a disc with four arcs with alternating colors where we paint the arcs which are part of the hyperbola $x_1^2 - x_2^2 = d$ with the color 4. The surface S obtained from this coloring of arcs can be realized in $\mathbb{R}^4$ by the intersection of quadrics:

$$x_1^2 - x_2^2 + x_4^2 = d \tag{4.1}$$

$$x_1^2 + x_2^2 + x_3^2 = 1 \tag{4.2}$$

Adding to the first equation the second one multiplied by $d+2$ we obtain the equation of an ellipsoid:

$$(d+3)x_1^2 + (d+1)x_2^2 + (d+2)x_3^2 + x_4^2 = 2d+2 \tag{4.3}$$

And multiplying the first equation by $(2d+2)/(d+2)$ and substracting from the result the third one multiplied by $d/(d+2)$ we obtain that of an non-degenerate homogeneous quadric:

$$(1-d)x_1^2 - (d+1)x_2^2 - dx_3^2 + x_4^2 = 0 \tag{4.4}$$

The intersection of this homogeneous quadric with the ellipsoid being diffeomorphic (by radial projection) to its intersection with the unit sphere, we conclude that S is diffeomorphic to a homogeneous intersection of quadrics.

We can proceed by cutting-off interiors of generic homogeneous hyperbolas in the x_1, x_2 plane and adding, for each of them, two new arcs with a new color labelled with a number. In terms of the equations we have, for each of them, a new equation with a new variable. Adding to the equation of the ellipsoid in the old variables a multiple of the new equation we get that of an ellipsoid which includes the old and new variables. Subtracting a multiple of the latter from the new equation we can make it homogeneous. When we use a hyperbola with oblique axes, the new equation will have crossed terms in the variables x_1, x_2 but will be diagonal *in the rest of them.*

There are many possibilities for the sequence of hyperbolas, but we choose the following pattern[3]: first take $a > 0$ disjoint hyperbolas that bite the circle, each producing two new hyperbolic arcs with a new color and increasing the number of circular arcs by 2. Then add a chain of b of them, the first of which cuts-off two opposite vertices with colors 3 and 4, introducing two new hyperbolic arcs with a new color and each of the following $b-1$ ones does the same with respect of the previous one and the color 3. The figure below illustrates the case $a = 4$, $b = 3$: starting from a disc with color 3 on its boundary, 4 hyperbolas bite off the disc disjointly introducing new arcs with colors $4, 5, 6, 7$. Then a hyperbola bites off two vertices colored 3 and 4, introducing two new arcs coloured 8, the next one does the same with respect to the colors 8 and 3 and introduces two arcs with color 9 and the same does the last one with respect to the colors 9 and 3 and introduces two arcs with color 10.

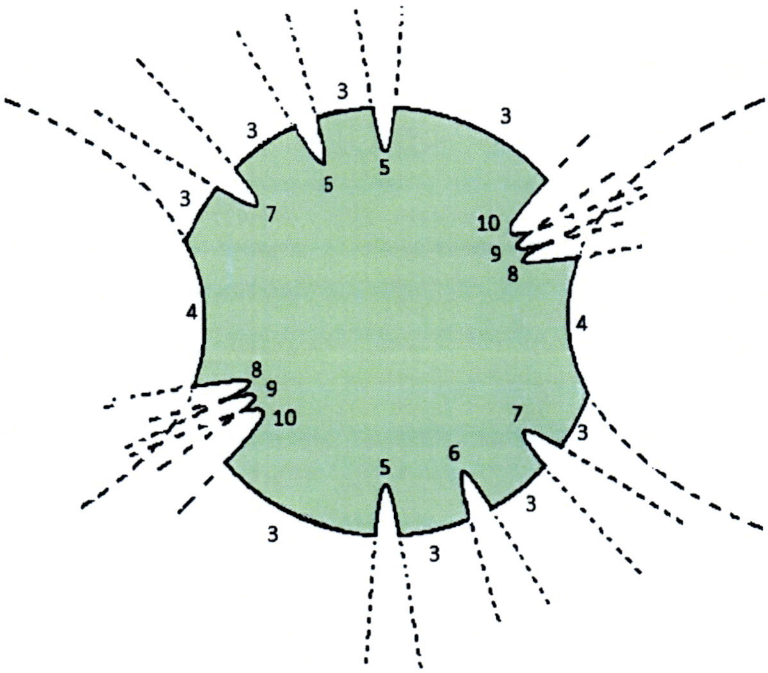

[3]It seems that other patterns do not give other genera than the ones so obtained.

In this case we get (unless $a = b = 0$) a disk with $m = 4a + 2b$ arcs in its boundary, painted with $c = 1 + a + b$ colors. The corresponding surface lies in $\mathbb{R}^{a+b+3}$ and admits an action of $\mathbb{Z}_2^{a+b+2}$, since to the involutions $x_i \mapsto -x_i$ corresponding to the construction in the previous section, we can add the involution $(x_1, x_2) \mapsto (-x_1, -x_2)$ corresponding to the fact that our polygon is symmetric. The quotient is another polygon but it has a ramification point in its interior, so it not of the type considered in section 3.

From the formula in the previous section we get that the surface has genus $g = 2^{a+b-1}(2a + b - 2) + 1$.

This gives many more genera than in the diagonal case (but they still form a sparse set of non-negative integers). For $b = 0$ we get the same genera as in the diagonal case but it seems that these never reappear for $b > 0$ (this is easy to prove for b odd), but already for $b = 1$ we get an infinite sequence of genera that do not appear in the diagonal case:

$$3, 13, 41, 113, 289, 705, 1665, 3841, 8705, 19457, 43009, 94209, 204801, \ldots$$

It would be interesting to see if the genera that appear for a given $b > 0$ correspond to some special type of complex curves, as in the case $b = 0$. For the whole collection we have an interesting partial correspondence with complete intersections of complex quadrics and *quartics*:

A curve which is the complete intersection of $n - 1$ hypersurfaces in $\mathbb{C}P^n$ of degrees $d_1, \ldots, d_{n-1}$ has Euler characteristic[4]

$$d_1 \ldots d_{n-1}(n + 1 - (d_1 + \cdots + d_{n-1}))$$

So the complete intersection of q quadrics and r quartics in $\mathbb{C}P^{q+r+1}$ has genus $g = 2^{q+2r-1}(q + 3r - 2) + 1$.

This genus can be realized by the previous construction by taking $a = r$ and $b = q + r$, and we get all of these satisfying $b \geq a$, so we have:

THEOREM 4.1. *Let S be the surface of genus $g = 2^{a+b-1}(2a + b - 2) + 1$, $a > 0, b \geq 0$. Then,*

(i) *S can be realized as an intersection of $a+b$ homogeneous quadrics and the unit sphere of $\mathbb{R}^{a+b+3}$. These quadrics can be assumed to have all its variables but 2 in diagonal form and so admit an action of $\mathbb{Z}_2^{a+b+2}$ with quotient a polygon with a ramification point in its interior.*

(ii) *Those with $b \geq a$ coincide with the topological types of complex projective curves which are complete intersection of quadrics and quartics in $\mathbb{C}P^{b+1}$. It can be assumed that these are all in diagonal form (i.e., all monomials in their equations are powers of one variable) and so the curve admits an action of $\mathbb{Z}_2^{b+1-a} \oplus \mathbb{Z}_4^a$ with quotient a sphere and an action of a group of order 2^{a+b+2} with quotient a polygon.*

To complete the proof we only need to show that in (ii) we can assume diagonal form: If we substitute a of the equations of the generalized Humbert curves (see

[4]This formula appears in the first paragraph of [**L-W**] without proof or reference. It also appears in [**G**], claiming no originality, with a proof based on Hirzebruch [**H1**]. So it must be well-known, yet we have not found it elsewhere. It should probably be attributed to Hirzebruch. It implies, in particular that the curves of genus 2 and 7 are not complete intersections.

section 2) by their quartic versions

$$\lambda_i z_1^4 + z_2^4 + z_{i+3}^4 = 0$$

we obtain a regular curve if the λ_i are sufficiently independent. For each of the a variables appearing only in degree 4 the corresponding $\mathbb{Z}_2$ summand in the group acting can be replaced by $\mathbb{Z}_4$, and the quotient is a again a sphere. When the coefficients are real we can add again the conjugation to get a group as stated.

By a variation of the construction, involving an annulus instead of a disc, one obtains besides the genera in the Theorem, also those of the form $= 2^p(2p+1)+1$, $p > 0$, which correspond formally with the case $a = p+2, b = -1$ of the Theorem. This gives a new sequence of genera $7, 21, 57, 145, 353, 833, 1921, \ldots$ of intersections of homogeneous quadrics, not covered by the Theorem. Observe that the curve of genus 7 is definitely not a complete intersection of complex hypersurfaces. We plan to develop this idea in a future article.

We end by stating some of the many unsolved problems:

(1) Our original goal: Characterize the topological type of surfaces that are intersections of homogeneous quadrics in some $\mathbb{R}^n$. (A sharp answer to this question will characterize the topological type of surfaces that are complete intersections of projective quadrics in $\mathbb{R}P^n$). Are all of them of the type constructed in section 2?

(2) For an intersection of homogeneous quadrics, it is natural how to define its degree of *diagonalizability* which should be related to its group of symmetries of order 2. For a complex projective curve one should be able to give a notion of its *degree of incompleteness*, which should be 0 when the curve is a complete intersection. Relate these concepts in the spirit of the Proposition and the Theorem. When the surface can be realized by a curve which is a complete intersection, relate the degrees of the curve with other invariants of the surface as an intersection of homogeneous quadrics.

(3) Can the surfaces in Theorem 4.1 with $b < a$ be realized as complete intersection curves? One can check that this is true for the two simpler ones.

References

[B-B-C-G] A. Bahri, M. Bendersky, F. R. Cohen, and S. Gitler, *The polyhedral product functor: a method of decomposition for moment-angle complexes, arrangements and related spaces*, Adv. Math. **225** (2010), no. 3, 1634–1668, DOI 10.1016/j.aim.2010.03.026. MR2673742 (2012b:13053)

[B-H] Lowell W. Beineke and Frank Harary, *The genus of the n-cube*, Canad. J. Math. **17** (1965), 494–496. MR0175805 (31 #81)

[B-LdM-V] Barreto, Y., López de Medrano, S. and Verjovsky, A., *Open Book Structures on Moment-Angle Manifolds $Z^{\mathbb{C}}(\Lambda)$ and Higher Dimensional Contact Manifolds*, arXiv:1303.2671.

[B-M] Frédéric Bosio and Laurent Meersseman, *Real quadrics in* $\mathbf{C}^n$, *complex manifolds and convex polytopes*, Acta Math. **197** (2006), no. 1, 53–127, DOI 10.1007/s11511-006-0008-2. MR2285318 (2007j:32037)

[B-P] V.M. Buchstaber and T.E. Panov, *Torus actions and their applications in Topology and Combinatorics*, University Lecture Series, AMS (2002).

[Buj] E. Bujalance, *Proper periods of normal N.E.C. subgroups with even index*, Proceedings of the Eighth Portuguese-Spanish Conference on Mathematics, Vol. I (Coimbra, 1981), Univ. Coimbra, Coimbra, 1981, pp. 81–83. MR672966 (84a:20054)

[C-G-H-R] Angel Carocca, Víctor González-Aguilera, Rubén A. Hidalgo, and Rubí E. Rodríguez, *Generalized Humbert curves*, Israel J. Math. **164** (2008), 165–192, DOI 10.1007/s11856-008-0025-2. MR2391145 (2009b:30086)

[C] H. S. M. Coxeter, *Regular Skew Polyhedra in Three and Four Dimension, and their Topological Analogues*, Proc. London Math. Soc. **S2-43**, no. 1, 33, DOI 10.1112/plms/s2-43.1.33. MR1575418

[D-J] Michael W. Davis and Tadeusz Januszkiewicz, *Convex polytopes, Coxeter orbifolds and torus actions*, Duke Math. J. **62** (1991), no. 2, 417–451, DOI 10.1215/S0012-7094-91-06217-4. MR1104531 (92i:52012)

[Gi-LdM] Samuel Gitler and Santiago López de Medrano, *Intersections of quadrics, moment-angle manifolds and connected sums*, Geom. Topol. **17** (2013), no. 3, 1497–1534, DOI 10.2140/gt.2013.17.1497. MR3073929

[Go-LdM1] V. Gómez Gutiérrez and S. López de Medrano, *Stably parallelizable compact manifolds are complete intersections of quadrics.*, to appear.

[Go-LdM2] V. Gómez Gutiérrez and S. López de Medrano, *Topology of the Intersection of Quadrics II.*, to appear in Boletín de la Sociedad Matemática Mexicana, volume dedicated to Francisco González Acuña.

[G] Vladimir N. Grujić, *χ_y-characteristics of projective complete intersections*, Publ. Inst. Math. (Beograd) (N.S.) **74(88)** (2003), 19–23, DOI 10.2298/PIM0374019G. MR2066988 (2005a:14066)

[H1] F. Hirzebruch, *Topological methods in algebraic geometry*, Third enlarged edition. New appendix and translation from the second German edition by R. L. E. Schwarzenberger, with an additional section by A. Borel. Die Grundlehren der Mathematischen Wissenschaften, Band 131, Springer-Verlag New York, Inc., New York, 1966. MR0202713 (34 #2573)

[H2] F. Hirzebruch, *Arrangements of lines and algebraic surfaces*, Arithmetic and geometry, Vol. II, Progr. Math., vol. 36, Birkhäuser, Boston, Mass., 1983, pp. 113–140. MR717609 (84m:14037)

[K] Ernst Kunz, *Almost complete intersections are not Gorenstein rings*, J. Algebra **28** (1974), 111–115. MR0330158 (48 #8496)

[L-W] Anatoly S. Libgober and John W. Wood, *Differentiable structures on complete intersections. I*, Topology **21** (1982), no. 4, 469–482, DOI 10.1016/0040-9383(82)90024-6. MR670748 (84a:57037)

[LdM1] Santiago López de Medrano, *Topology of the intersection of quadrics in $\mathbf{R}^n$*, Algebraic topology (Arcata, CA, 1986), Lecture Notes in Math., vol. 1370, Springer, Berlin, 1989, pp. 280–292, DOI 10.1007/BFb0085235. MR1000384 (90f:58015)

[LdM2] López de Medrano, S. *Singularities of real homogeneous quadratic mappings*, Revista de la Real Academia de Ciencias Exactas, Físicas y Naturales. Electronic Publication: doi:10.1007/s13398-012-0102-6.

[M-S-W] P. McMullen, Ch. Schulz, and J. M. Wills, *Polyhedral 2-manifolds in E^3 with unusually large genus*, Israel J. Math. **46** (1983), no. 1-2, 127–144, DOI 10.1007/BF02760627. MR727027 (84m:52011)

FACULTAD DE CIENCIAS
E-mail address: vgomez@ciencias.unam.mx

INSTITUTO DE MATEMÁTICAS
E-mail address: santiago@matem.unam.mx

Conjugacy classes of symmetries of compact Kleinian 3-manifolds

G. Gromadzki and R. A. Hidalgo

Dedicated to Emilio Bujalance for his sixtieth anniversary

ABSTRACT. Let M be a compact Kleinian 3-manifold with non-empty and connected boundary S. Then S carries the structure of a closed Riemann surface and by symmetries of S or M we understand their antiholomorphic involutions. In this paper we provide upper bounds for the number of conjugacy classes of symmetries of M in terms of the genus of S. Furthermore, we show that our bounds are sharp for hyperbolic handlebodies for infinitely many genera, by explicit constructions of finite normal extensions of certain Schottky groups, using at decisive stage of construction, quasi-conformal deformation theory of Riemann surfaces and Teichmüller theory of Fuchsian groups. In particular, we obtain that a Kleinian 3-manifold of even genus g has at most four non-conjugate symmetries and that this bound is achieved for arbitrary even g. Motivated by the behavior of Riemann surfaces, we propose the problem of the validity of our bounds, obtained for hyperbolic handlebodies, in a purely topological setting. Finally, since handlebodies can be seen as fattening-up of graphs, the results of our paper may be interpreted in terms of certain automorphisms of graphs and we propose another problem concerning it at the end of the paper.

1. Introduction and Preliminaries

A *symmetry* of a closed Riemann surface is an anti-conformal involution; in this case we say that the surface is *symmetric*. The theory of symmetric Riemann surfaces is related to the theory of real forms of complex algebraic curves and its study is by nowadays essentially well understood. Precise values of the number of pairwise non-conjugated symmetries of closed Riemann surfaces are known from [**5**] and [**3**]. Most of achievements in this area (see for example [**2**] viewed as a survey) were possible, mainly due to a good understanding of algebraic structure of the classical Fuchsian groups and non-Euclidean crystallographic groups introduced in

---

2010 *Mathematics Subject Classification.* Primary 30F10, 30F40.

Key words and phrases. Riemann surfaces, hyperbolic handlebodies, Kleinian 3-manifolds, symmetries of Riemann surfaces and Kleinian 3-manifolds, Fuchsian groups, Kleinian groups, Schottky groups.

The first author's research was supported by the Research Grant of Polish National Center of Sciences NCN 2012/05/B/ST1/02171. Part of this work was done during his staying in Max Planck Mathematical Institute in Bonn.

The second author was supported by Fondecyt 1110001 and UTFSM 12.13.01.

[**12**] by Macbeath and well developed Teichmülller theory for these groups [**4, 13, 16**].

Here we shall deal with a class of 3-manifolds M, called Kleinian manifolds, which are defined as, possibly bordered, 3-manifolds with a Möbius structure given by the maximal atlas $\mathcal{M}$, modeled on the closed 3-ball, for which the transition functions of charts are Möbius transformations. Given such a manifold, there is a torsion free Kleinian group Γ being a discrete subgroup of $\mathrm{PSL}(2,\mathbb{C})$ acting on the Riemann sphere $\widehat{\mathbb{C}}$ with the region of discontinuity $\Omega \subset \widehat{\mathbb{C}}$ so that $M = (\mathbb{H}^3 \cup \Omega)/\Gamma$ and its conformal boundary S is given by Ω/Γ being a closed Riemann surface if $\Omega \neq \emptyset$. The interior $M^0 = \mathbb{H}^3/\Gamma$ is a hyperbolic 3-manifold which means that it carries a complete Riemannian metric of constant negative curvature.

An *automorphism* of M is a self-homeomorphism whose restriction to the interior M^0 is a hyperbolic isometry and is said to be *conformal* or *anti-conformal* if it preserves orientation of M^0 or not. A *symmetry* of M is an anti-conformal involution. We denote by $\mathrm{Aut}(M)$ the group of automorphisms of M and by $\mathrm{Aut}^+(M)$ the subgroup of its conformal automorphisms. There is a natural injective group homomorphism $\mathrm{Aut}(M) \to \mathrm{Aut}(S)$ given by restriction to the conformal boundary and clearly, two symmetries of M which are conjugated in $\mathrm{Aut}(M)$ induce symmetries of S which are conjugated in $\mathrm{Aut}(S)$. But Example 3.1 shows that there may be symmetries of M which are non-conjugate in $\mathrm{Aut}(M)$ but whose restrictions to S are conjugate in $\mathrm{Aut}(S)$. Therefore, hypothetically, it may happen that the number of conjugacy classes of symmetries of M is larger than the number of conjugacy classes of symmetries of its conformal boundary S and so the results in [**3, 5**] do not provide immediately the ones for Kleinian 3-manifolds for which the former stand as the conformal boundaries. Nevertheless we prove, in Theorem 2.1, that actually this is not the case. Unfortunately, this is not the end of the story. As seen in Example 3.2, there may be a symmetry of S which cannot be extended to a symmetry of M. So again, hypothetically, it may occur that the known sharp upper bounds for surfaces are not any longer sharp for Kleinian 3-manifolds. However again we prove in Theorem 2.2, that actually this is not the case, even for handlebodies, which means that our bounds are attained even for this particular class of Kleinian 3-manifolds.

As a particular consequence of the Theorem 2.1, we obtain that a Kleinian manifold homeomorphic to a handlebody of even genus g has at most four conjugacy classes of symmetries and by Theorem 2.3, that for arbitrary even g there is a handlebody of genus g which has 4 conjugacy classes of symmetries.

Motivating by the behavior of Riemann surfaces, for which the bounds hold in more general topological settings, we ask the question on the validity of our bounds, obtained for hyperbolic handlebodies, in a purely topological setting. Finally, since handlebodies can be seen as fattening-up of graphs, the results of our paper may be interpreted in terms of certain automorphisms of graphs and we propose a problem concerning it at the end of the paper.

The case of Kleinian 3-manifolds whose conformal boundaries fail to be a closed Riemann surfaces is more involved. Nevertheless, we have some reasons to hope, that it can be successfully dealt, by subtly using the Heegaard decomposition and the Klein-Maskit decomposition theorems, which will be pursued elsewhere in the future.

2. Conjugacy classes of symmetries

In [**3**] the maximal number of conjugacy classes of symmetries of closed Riemann surfaces in terms of their fixed genus has been found. Here we deal with the similar problem for Kleinian 3-manifolds whose conformal boundaries are closed Riemann surfaces of genus $g \geq 2$. Finding precise number of conjugacy classes of symmetries seems to be a rather difficult problem but we find some upper bounds for them which we shall show to be sharp for infinitely many genera.

THEOREM 2.1. *Let M be a compact Kleinian 3-manifold whose conformal boundary is a closed Riemann surface of genus $g = 2^{r-1}u + 1$, where $u \geq 1$ is odd. Then the number of its pairwise non-conjugated symmetries is at most 2^{r+1}. In particular, if $g \geq 2$ is even, then M admits at most 4 conjugacy classes of symmetries.*

Proof. Since we are looking for the number k of conjugacy classes of symmetries of M we can assume, by Sylow theory, that their representatives $\tau_1, \dots, \tau_k$ generate a 2-subgroup G of $\mathrm{Aut}(M)$ say of order 2^s. If $s \leq r$, then obviously, the number of conjugacy classes of orientation reversing elements of order 2 in G is at most $2^{s-1} < 2^{r+1}$ for arbitrary abstract orientation of G, by which we understand any epimorphism $\alpha : G \to \mathbb{Z}_2$ with the obvious meaning of orientation-reversing element. On the other hand, if $s \geq r+1$, then G^+ acts with fixed points on S. Indeed, if we denote by g' the genus of S/G^+, then the Riemann-Hurwitz formula asserts that

$$\frac{g-1}{2^{r-1}} = 2^{s-r}(g'-1) + \frac{B}{2^r}.$$

But, since $u = (g-1)/2^{r-1}$ is odd and since $s - r > 0$, we obtain that $B > 0$ since for $B = 0$ the right hand side would be even and we are done. Now, Lemma 4.1 in [**3**] asserts that G contains either a cyclic or a dihedral subgroup of index 2^r. But then, by Lemma 3.1 and Corollary 3.2 in [**3**], the number of conjugacy classes of orientation reversing involutions is at most 2^{r+1} which completes the proof. □

Now we shall study the sharpness of the upper bound given in the above theorem.

THEOREM 2.2. *Given $r \geq 1$, there is infinitely many odd numbers u for which there exist compact Kleinian 3-manifolds homeomorphic to handlebodies of genus $g = 2^{r-1}u + 1$ having 2^{r+1} pairwise non-conjugated symmetries; the maximum possible, by the above theorem.*

Proof. We shall construct explicit examples to see that this upper bound is sharp for infinitely many values of g. For, let $u \geq 2^{r+1} - 3$ be an odd integer and let N be a power of 2 so that $s = 2(u+1)/N + 2 \geq 2^{r+1}$ is even. Set $g = 2^{r-1}u + 1$ and consider the finite group

$$G = \mathrm{D}_N \oplus \mathbb{Z}_2 \oplus \overset{r}{\ldots} \oplus \mathbb{Z}_2 = \langle x_0, y_0 \rangle \oplus \langle x_1 \rangle \oplus \ldots \oplus \langle x_r \rangle,$$

where D_N is a dihedral group of order $2N$ generated by two involutions x_0, y_0. Let $\{a_1, \dots, a_{2^{r+1}}\}$ be representatives of all conjugacy classes of elements of order 2 in G which are of odd length in $y_0, x_0, x_1, \dots, x_r$ and assume that $a_1 = y_0, a_2 = x_0$.

Let $\mathcal{S}$ be the unit circle, $\mathcal{R}$ the real line and $\mathcal{L}$ the line passing through 0 and $e^{i\pi/N}$. Let also, $\mathcal{C}_1, \ldots, \mathcal{C}_{s/2}$ be pairwise disjoint circles orthogonal to $\mathcal{S}$, contained in the upper half plane in the sector determined by $\mathcal{R}$ and $\mathcal{L}$ and bounding a domain of connectivity $s/2$. Let σ_i for $i = 0, 1, 2$ be the reflection with respect to $\mathcal{S}, \mathcal{R}$ an $\mathcal{L}$ respectively and let η_i be the reflections with respect to $\mathcal{C}_i$. Then the group K generated by all of the above reflections is an extended Kleinian group with connected region of discontinuity Ω so that Ω/K is a closed disc with exactly $s+3$ conical points on the boundary; one of order N and the remaining ones of order 2. Consider a surjective homomorphism $\varphi : K \to G$ so that $\varphi(\sigma_1) = x_0$, $\varphi(\sigma_2) = y_0$, $\varphi(\sigma_0) = x_1$ and $\{a_3, \ldots, a_{s/2}\} \subseteq \{\varphi(\eta_1), \ldots, \varphi(\eta_{s/2})\}$, where $\varphi(\eta_i)$ are required to be nontrivial. This is possible by our assumptions on s and N. Now, $\Gamma = \ker \varphi$ is torsion free and purely loxodromic and so it is a Schottky group of rank g. The handlebody M uniformized by Γ admits 2^{r+1} symmetries, represented by $a_1, \ldots, a_{2^{r+1}}$, which are pairwise non-conjugated in the group $G < \mathrm{Aut}(M)$.

Now, using quasi-conformal deformation theory and Teichmüller theory for Fuchsian groups [**4**, **16**], we shall modify the Kleinian structure on the topological support of M so that for a new hyperbolic handlebody M', $G' = \mathrm{Aut}(M')$ is topologically conjugate to G. Let us first restrict the action of G to the conformal boundary of M, being the Riemann surface $S = \Omega/\Gamma$ and let G^+ be the canonical subgroup of index 2 of G so restricted. G^+ consists of orientation preserving automorphisms. The signature of S/G^+ is $(0; 2, \overset{s+2}{\ldots}, 2, N)$ and so it is finitely maximal, by [**4**] and [**16**]. Therefore, there exists a closed Riemann surface S', and a quasi-conformal homeomorphism $q : S \to S'$ so that $qGq^{-1} = \mathrm{Aut}(S') = G'$. Let us denote by $\pi : \Omega \to S$, the Galois covering with Γ as the group of covering transformations and lift the Beltrami differential of q to obtain a Beltrami differential μ for the group Γ. In fact, μ is a Beltrami differential for the group K since qGq^{-1} is a group of conformal and anti-conformal automorphisms of S'. Next, we consider a quasi-conformal homeomorphism $\tilde{q} : \widehat{\mathbb{C}} \to \widehat{\mathbb{C}}$ with the dilation given by μ. Then $K_\mu = \tilde{q} K \tilde{q}^{-1}$ is an extended Kleinian group, topologically equivalent to K, so that the Schottky group $\Gamma_\mu = \tilde{q} \Gamma \tilde{q}^{-1}$ is a normal subgroup of it. Therefore, the handlebody $M' = M_\mu$ uniformized by Γ_μ has the desired property that $\mathrm{Aut}(M') = qGq^{-1}$, which completes the proof since $qa_1q^{-1}, \ldots, qa_{2^{r+1}}q^{-1}$ are orientation reversing involutions of M', pairwise non-conjugated in $\mathrm{Aut}(M')$. □

For even g our bound turns out to precise as shows the next Theorem.

THEOREM 2.3. *For every even value of $g \geq 2$ there is a handlebody of genus g with a Kleinian structure, admitting* 4 *pairwise non-conjugated symmetries.*

Proof. Let $\mathcal{S}_0$ be the unit circle and $\mathcal{S}_1, \mathcal{C}_1, \ldots, \mathcal{C}_{g/2}$ pairwise disjoint circles, each of which is orthogonal to $\mathcal{S}_0$ and all of them bounding a common domain of connectivity $g/2$. Assume also that $\mathcal{S}_1$ is orthogonal to the real line $\mathcal{R}$ and the circles $\mathcal{C}_1, \ldots, \mathcal{C}_{g/2}$ are contained in the upper half-plane $\Im(z) > 0$. Let σ_i be the reflection with respect to $\mathcal{S}_i$, for $i = 0, 1$, let σ_2 be the reflection with respect to $\mathcal{R}$, and let η_i be the reflection with respect to $\mathcal{C}_i$. Let finally K be the group generated by all of these reflections. Then, K is an extended Kleinian group with a fundamental domain given by the common domain bounded by $\mathcal{S}_0, \mathcal{S}_1, \mathcal{C}_1, \ldots, \mathcal{C}_{g/2}$ and $\mathcal{R}$ contained in the closed unit disc. If Ω is the region of discontinuity of K, then Ω/K is a closed disc with exactly $g+3$ conical points on its boundary; all of them of order 2. Let $\Gamma = \langle \gamma_1, \ldots, \gamma_g \rangle$, where $\gamma_i = \sigma_1 \eta_i$ and $\gamma_{i+g/2} = \sigma_2 \gamma_i \sigma_2$, for $i = 1, \ldots, g/2$. It is not difficult to see that Γ is a Schottky group of rank g with $\mathcal{C}_i$ and $\sigma_1(\mathcal{C}_i)$ for $i = 1, \ldots, g/2$ as a fundamental set of loops. Let M be the handlebody of genus g with a Schottky structure provided by Γ. Then the reflections σ_0, σ_1 and σ_2 induce three symmetries τ_0, τ_1 and τ_2 generating the group $\mathbb{Z}_2 \oplus \mathbb{Z}_2 \oplus \mathbb{Z}_2$. The symmetry τ_0 has a sphere with $g+1$ holes as the set of fixed points, the symmetry τ_1 has exactly $g+1$ connected components of fixed points each of which is a disc and τ_2 has a disc as the set of fixed points. The product $\tau_0\tau_1\tau_2$ is a symmetry having exactly one fixed point in the interior of M. Observe that for the handlebody M constructed in this way, its conformal boundary S is a hyperelliptic Riemann surface with $\tau_0\tau_1$ as the hyperelliptic involution.

Obviously the four symmetries of M, constructed above, are pairwise non-conjugated in the group $\mathbb{Z}_2 \oplus \mathbb{Z}_2 \oplus \mathbb{Z}_2 \cong G < \mathrm{Aut}(M)$. Furthermore, using quasi-conformal deformation theory and Teichmüller theory for Fuchsian groups, as in the previous theorem we can assume that, up to topological conjugacy, $G = \mathrm{Aut}(M)$ which completes our construction of four pairwise non-conjugated symmetries of a handlebody of even genus g. □

3. Examples, Remarks and Problems

The following examples allow one to see that the information about symmetries at the level of closed Riemann surfaces do not provide in general a complete information about symmetries at the level of handlebodies with a Kleinian structure, for which the former stand as conformal boundaries.

In the first example, we shall construct a handlebody M, with a Kleinian structure, admitting two different symmetries which are non-conjugate in $\mathrm{Aut}(M)$, but whose restrictions to the conformal boundary S of M are conjugated in $\mathrm{Aut}(S)$.

EXAMPLE 3.1. Let $\mathcal{S}_0$ be the unit circle, $\mathcal{S}_1$ a circle orthogonal to $\mathcal{S}_0$ and to the real line $\mathcal{R}$, and $\mathcal{L}$ the line passing through 0 and $e^{i\pi/3}$ and not intersecting $\mathcal{S}_1$. Let $\sigma_0, \sigma_1, \sigma_2$ and σ_3 be the reflections with respect to $\mathcal{S}_0, \mathcal{S}_1, \mathcal{R}$ and $\mathcal{L}$ and let K be the group generated by them. Then K is an extended Kleinian group with a fundamental domain contained in the closed unit disc and bounded by the four lines mentioned above. If Ω denotes the region of discontinuity of K, then Ω/K is a closed disc with exactly 4 conical points on its boundary having orders $2, 2, 2, 3$. Let Γ be generated by two loxodromic transformations $\gamma_1 = (\sigma_1\sigma_3)^2$ and $\gamma_2 = \sigma_2\gamma_1\sigma_2$. A fundamental domain for Γ is the region $\mathcal{D}$ bounded by the circles $\mathcal{C}_1 = \sigma_3(\mathcal{S}_1), \mathcal{C}_1' = \sigma_1(\sigma_3(\mathcal{S}_1)), \mathcal{C}_2 = \sigma_2(\sigma_3(\mathcal{S}_1)), \mathcal{C}_2' = \sigma_1(\sigma_2(\sigma_3(\mathcal{S}_1)))$. Furthermore, for the above circles $\gamma_1(\mathcal{C}_1) = \mathcal{C}_1', \gamma_2(\mathcal{C}_2) = \mathcal{C}_2', \gamma_1(\mathcal{D}) \cap \mathcal{D} = \emptyset$ and $\gamma_2(\mathcal{D}) \cap \mathcal{D} = \emptyset$. So Γ is a Schottky group of rank $g = 2$. Let M be the handlebody with a Schottky structure given by Γ and, as always, let S be its conformal boundary Riemann surface. The reflections σ_0 and σ_1 induce symmetries τ_0 and τ_1 of M. The symmetry τ_0 has a three-holed sphere as the set of fixed points and τ_1 has exactly 3 connected components of fixed points, each of which is a disc. So in particular, these two symmetries are not conjugate in $\mathrm{Aut}(M)$. Observe that for a suitable choice of the circle $\mathcal{S}_1$, the surface S is given by $y^2 = x^6 - 1$ and the two symmetries are conjugated by a conformal automorphism of S of order 6 which cannot be continuously extended to a conformal automorphism of M in virtue of [7]. □

In the second example we recall a construction, from [8], of a closed Riemann surface S admitting a couple of symmetries τ_1, τ_2 each of which can be extended to a symmetry $\widetilde{\tau}_i$ of some handlebody $\widetilde{S}_i$ with a Keinian structure for which S stands as the conformal boundary but there is no such handlebody $\widetilde{S}$ on which both symmetries can be extended, that is the mentioned handlebodies $\widetilde{S}_i$ can not be chosen to be equal.

EXAMPLE 3.2. Consider an orbifold $\widetilde{S}$ with one handle, one crosscap, and three orbifold points of order 3. It has a fundamental group $\pi_1^{\rm orb}(\widetilde{S})$, which we regard as a group of isometries of the hyperbolic plane $\mathbb{H}^2$, with the following presentation

$$\langle A, B, C, E_1, E_2, E_3 : ABA^{-1}B^{-1}E_1E_2E_3C^2 = E_1^3 = E_2^3 = E_3^3 = 1\rangle.$$

Let α, β be two involutions generating D_3 and define a homomorphism $\varphi : \pi_1^{\rm orb}(\widetilde{S}) \to \mathrm{D}_3$ by

$$\varphi(A) = \varphi(B) = 1, \varphi(C) = \alpha, \varphi(E_1) = \varphi(E_2) = \varphi(E_3) = \alpha\beta.$$

Then $\Gamma = \ker\varphi$ is purely hyperbolic and, by the Riemann-Hurwitz formula, $S = \mathbb{H}^2/\Gamma$ is a closed Riemann surface of genus 10. We see that D_3 acts on S, where α and β act as imaginary reflections which clearly are conjugate because their product has odd order. However, since $\widetilde{S} = S/\mathrm{D}_3$ has three orbifold points of order 3, it follows from results in [8] that there is no Schottky covering of S to which they both lift which is equivalent to desired statement. □

REMARK 3.3. It is easy to see that the bounds mentioned in the Introduction for the number of conjugacy classes of symmetries of Riemann surfaces hold in more general setting. Namely, given a compact topological surface X of genus g and its topologically non-conjugated symmetries (orientation reversing involutions) $\tau_1, \ldots, \tau_k$ generating a finite group G of self-homeomorphisms of X, the bounds in question hold for k. This is due to the Kerekjarto's [11] generalization of classical result of Hurwitz [9] by which each finite group of self-homeomorphisms of a compact topological surface X can be seen as a group of conformal automorphisms of X for some structure of Riemann surface imposed on it. □

The above Remark suggests the following question

PROBLEM 3.4. Do the bounds obtained here for symmetries of hyperbolic handlebodies hold true in the topological setting described in the above remark?

REMARK 3.5. A handlebody can be retracted to a finite connected graph and, conversely, any finite connected graph can be fattening-up to obtain a handlebody. Furthermore a finite group of self-homeomorphisms of a handlebody produces, under this retraction, a finite group of automorphisms of the graph. Unfortunately, the converse is not true in general which means that there are examples of groups of self-homeomorphisms of finite graphs which do not proceed from groups of automorphisms of the corresponding fattening handlebodies. All the above examples of groups of automorphisms of handlebodies with a Kleinian structure may be produced from graphs and certain subgroups of its automorphisms and the reader is invited to figure-out the graph description of them. □

We finish the paper with the following problem, motivated by the last Remark.

PROBLEM 3.6. Find an intrinsic characterization of groups of automorphisms of connected graphs which come from groups of self-homeomorphisms of the corresponding fattened-up handlebodies. □

References

[1] Lars V. Ahlfors and Leo Sario, *Riemann surfaces*, Princeton Mathematical Series, No. 26, Princeton University Press, Princeton, N.J., 1960. MR0114911 (22 #5729)

[2] Emilio Bujalance, Francisco Javier Cirre, José Manuel Gamboa, and Grzegorz Gromadzki, *Symmetries of compact Riemann surfaces*, Lecture Notes in Mathematics, vol. 2007, Springer-Verlag, Berlin, 2010. MR2683160 (2011h:30062)

[3] E. Bujalance, G. Gromadzki, and M. Izquierdo, *On real forms of a complex algebraic curve*, J. Aust. Math. Soc. **70** (2001), no. 1, 134–142, DOI 10.1017/S1446788700002329. MR1808396 (2001k:14108)

[4] Leon Greenberg, *Maximal Fuchsian groups*, Bull. Amer. Math. Soc. **69** (1963), 569–573. MR0148620 (26 #6127)

[5] G. Gromadzki and M. Izquierdo, *Real forms of a Riemann surface of even genus*, Proc. Amer. Math. Soc. **126** (1998), no. 12, 3475–3479, DOI 10.1090/S0002-9939-98-04735-2. MR1485478 (99b:30062)

[6] Axel Harnack, *Ueber die Vieltheiligkeit der ebenen algebraischen Curven* (German), Math. Ann. **10** (1876), no. 2, 189–198, DOI 10.1007/BF01442458. MR1509883

[7] Rubén A. Hidalgo, *Automorphism groups of Schottky type*, Ann. Acad. Sci. Fenn. Math. **30** (2005), no. 1, 183–204. MR2140306 (2006b:30074)

[8] Ruben A. Hidalgo and Bernard Maskit, *Fixed points of imaginary reflections on hyperbolic handlebodies*, Math. Proc. Cambridge Philos. Soc. **148** (2010), no. 1, 135–158, DOI 10.1017/S0305004109990272. MR2575379 (2011a:57040)

[9] A. Hurwitz, *Ueber Riemann'sche Flächen mit gegebenen Verzweigungspunkten* (German), Math. Ann. **39** (1891), no. 1, 1–60, DOI 10.1007/BF01199469. MR1510692

[10] John Kalliongis and Darryl McCullough, *Orientation-reversing involutions on handlebodies*, Trans. Amer. Math. Soc. **348** (1996), no. 5, 1739–1755, DOI 10.1090/S0002-9947-96-01515-2. MR1329535 (96h:57012)

[11] B. Kerejarto. *Vorlesungen über Topologie. I Flächentopologie*, Springer- Verlag 1923.

[12] A. M. Macbeath, *The classification of non-euclidean plane crystallographic groups*, Canad. J. Math. **19** (1967), 1192–1205. MR0220838 (36 #3890)

[13] A. M. Macbeath and D. Singerman, *Spaces of subgroups and Teichmüller space*, Proc. London Math. Soc. (3) **31** (1975), no. 2, 211–256. MR0397022 (53 #882)

[14] Bernard Maskit, *Kleinian groups*, Grundlehren der Mathematischen Wissenschaften [Fundamental Principles of Mathematical Sciences], vol. 287, Springer-Verlag, Berlin, 1988. MR959135 (90a:30132)

[15] Bernard Maskit, *On Klein's combination theorem. IV*, Trans. Amer. Math. Soc. **336** (1993), no. 1, 265–294, DOI 10.2307/2154347. MR1137258 (93e:30088)

[16] David Singerman, *Finitely maximal Fuchsian groups*, J. London Math. Soc. (2) **6** (1972), 29–38. MR0322165 (48 #529)

INSTITUTE OF MATHEMATICS, GDAŃSK UNIVERSITY, POLAND
E-mail address: grom@mat.ug.edu.pl

DEPARTAMENTO DE MATEMÁTICA, UTFSM, VALPARAÍSO, CHILE
E-mail address: ruben.hidalgo@usm.cl

Connectivity and dimension of the p-locus in moduli space

G. Gromadzki, A. Weaver, and A. Wootton

Dedicated to Emilio Bujalance on the occasion of his 60th birthday.

ABSTRACT. The singular locus $\mathcal{S}_g$ of the moduli space $\mathcal{M}_g$ of compact Riemann surfaces of genus $g > 1$ consists of points representing Riemann surfaces with nontrivial groups of holomorphic automorphisms. There are various coverings of this locus by certain natural, canonical subsets. Each such covering gives rise to a certain simplicial complex, called its nerve, in a similar way as in the construction of Čech homology for a topological space at the stage of its particular covering by open subsets. Here, given an odd prime p, we consider the nerve $\mathcal{N}_g^p$ for the sublocus $\mathcal{S}_g^p$ of $\mathcal{M}_g$ covered by subsets $\mathcal{S}_g^{p,n}$ consisting of points representing Riemann surfaces admitting a (p,n)-gonal automorphism and we study its connectivity and geometric dimension.

1. Introduction

The singular (or branch) locus $\mathcal{S}_g$ in the moduli space $\mathcal{M}_g$, $g > 1$, consists of conformal equivalence classes of compact Riemann surfaces of genus g with a non-trivial automorphism group. It is covered by its p-loci, consisting of surfaces admitting an automorphism of prime order p. Here a *covering* simply means a decomposition into a non-disjoint union of subsets. The covering of $\mathcal{S}_g$ by its p-loci is finite, since, in any given genus, there are only finitely many primes that could divide the order of a group of automorphisms **[11]**. A finer, but still finite, covering is given by (p,n)*-gonal* subloci $\mathcal{S}_g^{p,n}$, which we shall shortly denote by $(p,n)_g$ or just by (p,n), consisting of points of $\mathcal{M}_g$ corresponding to surfaces for which the quotient modulo a p-action has given genus $n \geq 0$.

This covering gives rise to a certain simplicial complex $\mathcal{N}_g^p$, called its nerve, in a similar way as in the construction of Čech homology for a topological space at the stage of its particular covering by open subsets. More precisely, (p,n) subloci are its vertices and a collection of non-empty loci (p,n_1), (p,n_2), ..., (p,n_{d+1}) constitute a d-simplex if and only if there is a surface of genus g which is simultaneously (p,n_1)-, (p,n_2)-, ..., (p,n_{d+1})-gonal.

2010 *Mathematics Subject Classification.* Primary 14J10, 14J50, 30F10, 30F20.

Key words and phrases. Automorphism groups of compact Riemann surfaces, singular locus of moduli space of compact Riemann surfaces, p-gonal Riemann surfaces, Fuchsian groups, Uniformization theorem.

The first author was supported by the grant NCN 2012/05/B/ST1/02171 and part of his contribution to this paper was made within the project granted by the Max Planck Mathematical Institute in Bonn.

The second author was supported by PSC-CUNY Research Award No. 65432-00 43.

Here, we determine an upper bound on the *geometric dimension* (highest dimension of a simplex) of $\mathcal{N}_g^p$, and show that it is attained for sufficiently large g. We also show that $\mathcal{N}_g^p$ (when non-empty) has at most one non-trivial path-connected component, and possibly some isolated vertices. A path between (p,n) and (p,m) (consisting of 1-simplices) exists whenever there is a sequence $n = k_1, k_2, \ldots, k_s = m$, such that the intersection of the (p,k_i)- and (p,k_{i+1})-loci, $i = 1,\ldots,m-1$, is nonempty. We show that the non-trivial path-connected component, when it exists, is spanned by a *star-like* tree - any two vertices are joined by a path of length at most two passing through a unique central vertex as in Figure 1.

There are other coverings of the singular locus in $\mathcal{M}_g$. One of the most important and useful is the *equisymmetricic stratification* [**3**, **4**, **11**]: each strata is an irreducible complex algebraic variety consisting of conformal equivalence classes of surfaces with isomorphic full automorphism groups whose actions are topologically equivalent. In this set-up, notions of dimension and connectivity come from Teichmüller theory; strata correspond to conjugacy classes of finite subgroups of the mapping class group. The strata we consider here are coarser since we do not distinguish topologically inequivalent actions of the same group. For example, in genus 3, there are two inequivalent $\mathbb{Z}_7$ actions with quotient genus 0, one extending to a $\mathbb{Z}_{14}$ action on the Wiman curve of type I, and the other to a $\mathrm{PSL}(2,7)$ action on the Klein quartic. In our complex, these curves are not distinguished – the $(7,0)$-gonal locus is a single vertex – while in the equisymmetric stratification this locus is neither connected nor irreducible in the Zariski topology.

We believe however, that our coarser stratification will serve as a useful stepping-stone on the path toward attaining complete equisymmetric stratifications in arbitrary genus. At present, this seems a very distant goal. Knowledge of the connectivity and geometric dimension of the p-loci should shed some light on corresponding properties of the full singular locus, questions of long-standing and still-current interest (see, e.g., [**2**, **16**]). Our approach could be globalized to $\mathcal{N}_g$ in various ways, for example, by considering d-simplices defined by intersections of (p,n)-loci in $\mathcal{M}_g$, where both p and n are allowed to vary. Furthermore, it opens up some potential new avenues for future study. What significance, for example, might one attach to homological properties of the p-local nerve?

For some related work on the p- and (p,n)-gonal loci, see, e.g., [**6–10**, **12**, **18**].

2. Preliminaries

In discussing group actions on compact Riemann surfaces of genus $g > 1$, we use the uniformization approach originating with Klein, Poincaré and Koebe, and reinvigorated by Macbeath in 1961 [**17**], which describes the actions in terms of covering actions by Fuchsian groups on the upper half-plane $\mathcal{H}$ endowed with the hyperbolic metric $ds^2 = (dx^2 + dy^2)/y^2$.

A *Fuchsian group* is a discrete group of orientation-preserving isometries of $\mathcal{H}$, the full group being isomorphic to $\mathrm{PSL}(2,\mathbb{R})$. We consider only co-compact Fuchsian groups, having compact orbit (quotient) space, and henceforth this is what we mean when we use the term Fuchsian group. The orbit space inherits the complex structure from $\mathcal{H}$ and hence becomes a compact Riemann surface. A Fuchsian group Λ has a *signature* of the form $(h; m_1, \ldots, m_r)$ indicating that the orbit genus (genus of $\mathcal{H}/\Lambda$) is $h \geq 0$, and the quotient map $\pi\colon \mathcal{H} \to \mathcal{H}/\Lambda$ branches over r points with ramification indices $m_1, \ldots, m_r > 1$. When $r = 0$, the signature is written $(h; -)$ and Λ is called a *surface group*. The signature determines a presentation for Λ as follows:

$$\left\langle \alpha_1, \beta_1, \ldots, \alpha_h, \beta_h, \gamma_1, \ldots, \gamma_r \middle| \gamma_1^{m_1}, \ldots, \gamma_r^{m_r}, \prod_{i=1}^{h}[\alpha_j, \beta_j] \prod_{j=1}^{r} \gamma_i \right\rangle.$$

The generators $\alpha_1, \beta_1, \ldots, \alpha_g, \beta_g$ are *hyperbolic* isometries, of infinite order and having no fixed points in $\mathcal{H}$ (but two on the ideal boundary $y = 0$), while $\gamma_1, \ldots, \gamma_r$ are *elliptic* isometries of maximal finite order having unique fixed points in $\mathcal{H}$. Any element of finite order in Λ is conjugate to a power of one of the elliptic generators (compactness of the quotient space rules out parabolic isometries.) Together, $\alpha_1, \beta_1, \ldots, \alpha_h, \beta_h, \gamma_1, \ldots, \gamma_r$ comprise a set of *canonical generators* for Λ.

Any compact Riemann surface X of genus $g > 1$ is conformally equivalent to the orbit space $\mathcal{H}/\Gamma$ where Γ is a surface group of genus g. X admits a group G of conformal automorphisms if and only if there is a Fuchsian group Λ, containing Γ as a normal subgroup, such that $G \cong \Lambda/\Gamma$. Equivalently, there exists an epimorphism $\theta : \Lambda \to G$ with Γ as the kernel. Such epimorphisms are called *smooth* or *surface-kernel* to indicate that no element of finite order is mapped to an element of smaller order, or (equivalently), that the kernels are torsion-free. Λ is called the *covering* or *uniformizing* group of the G action. By a famous theorem of Hurwitz [**13**], G is in fact a finite group of order $\leq 84(g-1)$.

The *Riemann-Hurwitz relation* ties together all the topological data associated with a G action on a surface of genus g, as follows. If the covering Fuchsian group has signature $(h; m_1, \ldots, m_r)$, and $|G|$ denotes the order of G, then

$$2g - 2 = |G|\Big[2h - 2 + \sum_{i=1}^{r}\Big(1 - \frac{1}{m_i}\Big)\Big].$$

Since $g > 1$, the rational number in brackets must be positive. In fact this number is proportional to the hyperbolic area of a fundamental region for the action of the covering Fuchsian group on $\mathcal{H}$. Signatures for which this number is 0 or negative define non-Fuchsian groups which, if not trivial, are isomorphic to groups of euclidean or spherical isometries, respectively.

We shall use a theorem due to Accola (Theorem 2.2 below) which yields a remarkably simple characterization of the possible sets of nonnegative integers which can be simultaneous orbit genera for p-actions on a single surface (Lemma 4.4). Accola's theorem treats *groups with a partition.* Such groups possess two or more proper non-trivial subgroups whose union is the whole group and whose pairwise intersections are the trivial group. For example, the elementary abelian p-group $\mathbb{Z}_p^2 = \mathbb{Z}_p \oplus \mathbb{Z}_p = \langle x, y\rangle$ has the partition

$$\langle xy\rangle, \langle xy^2\rangle, \ldots, \langle xy^{p-1}\rangle, \langle x\rangle, \langle y\rangle \tag{1}$$

into $p+1$ proper non-trivial subgroups. This is a special case of a more general result. Let

$$\sigma_N = 1 + p + p^2 + \cdots + p^{N-1}. \tag{2}$$

LEMMA 2.1. *$\mathbb{Z}_p^{e+1}$, $e \geq 1$, admits a partition into σ_{e+1} distinct cyclic subgroups of order p.*

PROOF. By (1), the statement is true for $e = 1$. Proceeding by induction on e, suppose it is true for $e = k \geq 1$. Let g_i, $i = 1, \ldots, \sigma_{k+1}$ be a generator for the ith group in the partition of $\mathbb{Z}_p^{k+1}$. Adjoin a new generator h of order p which commutes with all the g_i, to produce a group isomorphic to $\mathbb{Z}_p^{k+2}$. There are now $p-1$ new subgroups $\langle hg_i \rangle$, $\langle h^2 g_i \rangle$, ..., $\langle h^{p-1} g_i \rangle$, for each $i = 1, \ldots, \sigma_{k+1}$, in addition to the new subgroup $\langle h \rangle$. It is straightforward to verify that the new and old subgroups taken together form a partition of $\mathbb{Z}_p^{k+2}$. The partition has the original σ_{k+1} subgroups plus $1 + (p-1)\sigma_{k+1}$ new ones. The fact that

$$\sigma_{k+1} + 1 + (p-1)\sigma_{k+1} = \sigma_{k+2}$$

follows from the identity $p\sigma_N + 1 = \sigma_{N+1}$. □

When a group with partition acts on a surface, the total ramification is the sum of the separate ramifications of the induced actions of the subgroups. This is the crucial point in Accola's proof.

THEOREM 2.2 (Accola [**1**]). *Let G be a finite group with partition $\{G_1, \ldots, G_t\}$. If G acts on a compact Riemann surface of genus g with orbit genus m, and the restricted G_i actions have orbit genera m_i, $i = 1, \ldots, t$, then*

$$(t-1)g + |G| \cdot m = \sum_{i=1}^{t} |G_i| \cdot m_i.$$

PROOF. The Riemann-Hurwitz relation for the G_i action has the form

$$2g - 2 = |G_i|(2m_i - 2) + r_i, \tag{3}$$

where $r_i \geq 0$ is the ramification term. If a non-trivial element $g \in G$ fixes a point, it generates a cyclic subgroup which is contained in exactly one of the subgroups G_i (since the G_i form a partition). Thus, in the Riemann-Hurwitz relation for the full G action,

$$2g - 2 = |G|(2m - 2) + r, \tag{4}$$

the ramification term r is equal to $\sum_{i=1}^{t} r_i$. After summing (3) for $i = 1, 2, \ldots, t$ we obtain two expressions for r, the second coming from (4). Equating them yields $t(2g-2) - \sum_{i=1}^{t} |G_i|(2m_i - 2) = 2g - 2 - |G|(2m - 2)$. The relation in the theorem follows from this and the element count $\sum_{i=1}^{t} |G_i| = |G| + t - 1$. □

3. An upper bound on the geometric dimension of $\mathcal{N}_g^p$

In this section we prove that the geometric dimension of $\mathcal{N}_g^p$ is at most $\sigma_{e+1} - 1$, where p^e is the highest power of p dividing $g-1$. This requires a brief excursion into the theory of finite p-groups. We are grateful to Andrei Jaikin for the statements and proofs of Lemmas 3.2 and 3.3.

Let G be a group acting on a surface of genus $g > 1$. Conjugate subgroups of G act with identical signatures since their uniformizing Fuchsian groups are conjugate in $\mathrm{PSL}(2, \mathbb{R})$. If the subgroups in question are cyclic of prime order p, this is equivalent to the quotient genera being equal. Therefore, the largest integer d such that a surface of genus g is simultaneously (p, n_1)-, (p, n_2)-, ..., (p, n_{d+1})-gonal, for pairwise distinct $n_1, n_2, \ldots n_{d+1}$ (the geometric dimension of $\mathcal{N}_g^p$) is no larger than the largest number of distinct conjugacy classes of subgroups of order p that can occur within G, where G varies over all possible automorphism groups of surfaces of genus g. It is correct to say 'no larger' rather than 'equal to' since non-conjugate cyclic groups of order p may nonetheless have the same orbit genus.

Non-conjugate elements in a p-Sylow subgroup may be conjugate in the full group G, but, for the purpose of determining an upper bound on the number of distinct conjugacy classes in G, it suffices to assume that G is a p-group.

3.1. Kulkarni's result. A group has *exponent* E if it contains elements of order E but no elements of larger order. In the seminal paper [**15**], Kulkarni showed that a p-group P, where p is odd, of order p^k and exponent p^n, $n \leq k$, acts on a surface of genus g only if $g \equiv 1 \pmod{p^{k-n}}$. (A modified result is given in the case $p = 2$.) If the integer $k - n$ (the so-called *cyclic p-deficiency* of P) is small, then P contains a large cyclic subgroup. In the extreme case $k - n = 0$, P is itself cyclic (possibly trivial), containing (if not trivial) a unique cyclic subgroup of order p. So if $g \not\equiv 1 \pmod{p}$, no surface is both (p, n)- and (p, m)-gonal for $n \neq m$, that is, $\mathcal{N}_g^p$ is either empty or has geometric dimension 0. On the other hand, if P is far from cyclic, i.e., if $k - n$ is rather large, there is the potential for a large number of non-conjugate cyclic groups of order p. Kulkarni's result suggests that the geometric dimension of $\mathcal{N}_g^p$ should increase in accordance with the highest power of p dividing $g - 1$. To make this precise we need an upper bound on the number of conjugacy classes of cyclic subgroups of order p in a finite p-group.

3.2. Conjugacy classes of subgroups of order p in a p-group. Let P be a finite p-group. Let $K_p(P)$ be the number of conjugacy classes of elements of order p, and $K_{\mathbb{Z}_p}(P)$ the number of conjugacy classes of subgroups of order p in P.

LEMMA 3.1. $K_{\mathbb{Z}_p}(P) = K_p(P)/(p-1)$.

PROOF. First, for any element x of order p the conjugacy classes of the powers of x, $C_P(x)$, $C_P(x^2), \ldots C_P(x^{p-1})$ are all distinct. For if $\alpha x \alpha^{-1} = x^s$ for some $\alpha \in P$, then the order of α is a divisor of the order of $\mathrm{Aut}(\langle x \rangle)$, which is $p - 1$; thus α is the identity and $s = 1$. Next, an element y is conjugate to a power of x if and only if $\langle y \rangle$ and $\langle x \rangle$ are conjugate subgroups. Therefore there is a one-to-one correspondence between distinct sets of conjugacy classes of the form $\{C_P(x), C_P(x^2), \ldots C_P(x^{p-1})\}$, of which there are $K_p(P)/(p-1)$, and conjugacy classes of subgroups of order p. □

For the next two lemmas, whose proofs are quite technical (and might be skipped on first reading), the following additional notation is convenient. If G is a finite p-group, H a subgroup, and S an H-invariant subset of G, then $K^H(S)$ denotes the number of H-conjugacy classes of elements in S; and $K_p^H(S)$ the number of H-conjugacy classes of elements of order p in S. For simplicity, we write $K(G)$ for $K^G(G)$.

LEMMA 3.2 (Jaikin [**14**]). *If P contains a central element x of order p^2, then $K_p(P) \leq K_p(P/\langle x^p \rangle)$.*

PROOF. Let $\mathcal{T} = \{T_i/\langle x \rangle \mid T_i \leq P\}$ be a set of representatives of the conjugacy classes of subgroups of order p in the factor group $P/\langle x \rangle$. For each $T_i/\langle x \rangle \in \mathcal{T}$, let $C_i = N_P(T_i)$, the normalizer of T_i in P. Each T_i is an abelian group of order p^3. $T_i \backslash \langle x \rangle$ (note: backslash) denotes the elements of T_i not belonging to $\langle x \rangle$. If T_i is cyclic, then there are no elements of order p in this set. If T_i is not cyclic, fix an element $y_i \in T_i \backslash \langle x \rangle$ of order p. Then

$$K_p^{C_i}(T_i \backslash \langle x \rangle) = K_p^{C_i}(\langle y_i, x^p \rangle \backslash \langle x^p \rangle) = \begin{cases} p^2 - p & \text{if } C_i \text{ centralizes } y_i \\ p - 1 & \text{if } C_i \text{ does not centralize } y_i. \end{cases}$$

Let $\tilde{T}_i = T_i/\langle x^p\rangle$, $\tilde{C}_i = C_i/\langle x^p\rangle$, and let $\bar{x}$ and $\bar{y}_i$ be the images of x and y_i in $\tilde{G} = G/\langle x^p\rangle$. If T_i is cyclic there are no elements of order p in $\tilde{T}_i/\langle \bar{x}\rangle$, and if T_i is not cyclic, then

$$K_p^{\tilde{C}_i}(\tilde{T}_i\backslash\langle \bar{x}\rangle) = p^2 - p,$$

because $\tilde{C}_i$ centralizes $\bar{y}_i$. Thus we obtain

$$K_p(P) = p-1+\sum_{T_i/\langle x\rangle\in\mathcal{T}} K_p^{C_i}(T_i\backslash\langle x\rangle) \leq p-1+\sum_{T_i/\langle x\rangle\in\mathcal{T}} K_p^{\tilde{C}_i}(\tilde{T}_i\backslash\langle \bar{x}\rangle) = K_p(P/\langle x^p\rangle).$$

□

LEMMA 3.3 (Jaikin [**14**]). *For a p-group P of order p^k and exponent p^n, $K_p(P) \leq p^{k-n+1} - 1$.*

PROOF. The result is clearly true when $n = 1$, so assume $n > 1$. The lemma is proved by induction on n. Let $a \in P$ be an element of order p^n. Put $b = a^{p^{n-2}}$, an element of order p^2. Let $H = C_P(b)$, the centralizer of b in P. Note that if $g \notin H$, then the $\langle a\rangle$-conjugacy class of g has at least p^{n-1} elements. Thus,

$$K^{\langle a\rangle}(P\backslash H) \leq \frac{|P\backslash H|}{p^{n-1}}.$$

Also, by Lemma 3.1, $K_p(H) \leq K_p(H/\langle b^p\rangle)$. Since $a \in H$, $H/\langle b^p\rangle$ has exponent at least p^{n-1}. By induction,

$$K_p(H) \leq K_p(H/\langle b^p\rangle) \leq \frac{|H/\langle b^p\rangle|}{p^{n-2}} - 1 = \frac{|H|}{p^{n-1}} - 1.$$

Thus we obtain

$$K_p(P) \leq K^{\langle a\rangle}(P\backslash H) + K_p(H) \leq \frac{|P\backslash H|}{p^{n-1}} + \frac{|H|}{p^{n-1}} - 1 = p^{k-n+1} - 1.$$

□

Now we need only a short argument to obtain the upper bound on the geometric dimension of $\mathcal{N}_g^p$. Let p^e be the highest power of p dividing $g-1$. For odd p, if $e = 0$, as already already noted, $\mathcal{N}_g^p$ is either empty or has geometric dimension 0. So suppose $e > 0$. Then $g \equiv 1 \pmod{p^e}$. By Kulkarni's result, a p-group P of order p^k acting in genus g has exponent at least p^{k-e}. It follows from Lemma 3.1 combined with Lemma 3.3 that

$$K_{\mathbb{Z}_p}(P) \leq \frac{p^{k-(k-e)+1} - 1}{p-1} = \frac{p^{e+1} - 1}{p-1} = \sigma_{e+1}.$$

Thus the geometrical dimension is at most $\sigma_{e+1} - 1$. Note that this formula holds also in the case $e = 0$.

See [**5**] for related results in the case $p = 2$.

4. Sharpness of the upper bound

Henceforth, we assume p is an odd prime.

4.1. Existence of $\mathbb{Z}_p^{e+1}$ actions. The elementary abelian p group $\mathbb{Z}_p^{e+1}$ contains σ_{e+1} distinct (conjugacy classes of) subgroups of order p, hence when it acts on a surface of genus g, the upper bound on the geometric dimension of $\mathcal{N}_g^p$ is attained. In this case, by Kulkarni's result, $g \equiv 1 \pmod{p^e}$. The following lemma provides a stronger necessary condition for the existence of such an action, and a *canonical signature* for the covering Fuchsian group, having minimal orbit genus and maximal number of periods.

LEMMA 4.1. *Let $g \equiv 1 \pmod{p^e}$, and let $p' = (p-1)/2$. Let $0 \le \kappa < p'$ be the residue of $g \pmod{p'}$. There exists a $\mathbb{Z}_p^{e+1}$ action on a surface of genus g only if*

$$\frac{g-1-p^{e+1}(\kappa-1)}{p'p^e} = R \tag{5}$$

is a nonnegative integer. The covering Fuchsian group of a $\mathbb{Z}_p^{e+1}$ action has signature $(\kappa+tp'; p, \overset{R-tp}{\ldots}, p)$ for some $t = 0, 1, 2, \ldots$, such that $R-tp \ge 0$ and $R-tp \ne 1$.

PROOF. R is an integer because of the congruences $1 \equiv p \equiv g-1 \equiv \kappa - 1 \pmod{p'}$. If there exists an $H = \mathbb{Z}_p^{e+1}$ action in genus g, the signature of the covering Fuchsian group Λ_H is of the form $(k; p, \overset{r}{\ldots}, p)$ for some k, $r \ge 0$. The Riemann-Hurwitz relation is

$$(g-1)/p^e = p(k-1) + p'r. \tag{6}$$

This together with the fact that $p \equiv 1 \pmod{p'}$ implies $k \equiv g \pmod{p'}$. Hence if $k < p'$, then $k = \kappa$, and otherwise, there exists $t \ge 1$ such that $k = \kappa + tp'$. When $k = \kappa$, (6) coincides with (5) if we put $r = R$. For general $k = \kappa + tp'$, (6) yields $r = R - tp$. Clearly $t \le R/p$, and, of course, if $R < 0$, or if none of the signatures is Fuchsian, there can be no H action. Assuming $R \ge 0$ and the signature is Fuchsian, let $\rho : \Lambda_H \to H$ be a surface-kernel epimorphism corresponding to an H-action. Each of the k commutators $[\alpha_i, \beta_i] \in \Lambda_H$ maps to the trivial element (since H is abelian), hence the product $\gamma_1\gamma_2 \ldots \gamma_r$ of the elliptic generators must be the trivial element. None of the elliptic generators is itself mapped to the trivial element (ρ being surface-kernel), hence $r \ne 1$. □

REMARK 4.2. Necessary and sufficient conditions for the existence of a $\mathbb{Z}_p^{e+1}$ action in terms of R and κ can be given (see [**19**], §7). These conditions are satisfied if g is sufficiently large.

4.2. Induced (p, n)-gonal actions. An $H = \mathbb{Z}_p^{e+1}$ action on a surface X induces (p, n)-gonal actions by its proper non-trivial subgroups. The next lemma gives the possible values of n. Let $\mathbb{Z}_{p,n}$ denote a proper non-trivial subgroup of H whose induced action on X is (p, n)-gonal. The quotient surface $X/\mathbb{Z}_{p,n}$ is a branched covering of the quotient surface X/H, so $n \ge k$, where k is the genus of X/H. To allow for all possible n, we assume k is minimal, that is, we assume the H action has the canonical signature $(\kappa; p, \overset{R}{\ldots}, p)$ with minimal orbit genus κ.

LEMMA 4.3. *Let X be surface of genus $g \equiv 1 \pmod{p^e}$ on which $H = \mathbb{Z}_p^{e+1}$ acts. Let Λ_H be the covering Fuchsian group with signature $(\kappa; p, \overset{R}{\ldots}, p)$ and $\rho\colon \Lambda_H \to H$ the corresponding surface-kernel epimorphism. Let $\mathbb{Z}_{p,n}$ denote a proper, non-trivial subgroup of H whose induced action on X is (p, n)-gonal. Then the possible values of n are*

$$n = n(s) = 1 + p^e(\kappa - 1) + p^{e-1}p'(R - s), \tag{7}$$

where

$$s = \begin{cases} 0, 1, 2, \dots, R-2 & \text{if } \kappa = 0; \text{ or} \\ 0, 1, 2, \dots, R-2, R & \text{if } \kappa > 0. \end{cases}$$

The parameter s is the number of elliptic generators of Λ_H in the kernel of the map $\chi_s \circ \rho$, where $\chi_s : H \to H/\mathbb{Z}_{p,n(s)} \simeq \mathbb{Z}_p^e$ is the canonical quotient map.

PROOF. Let $\Lambda_{p,n}$, with signature $(n; p, \overset{q}{\dots}, p)$, be the covering Fuchsian group of the induced $\mathbb{Z}_{p,n}$ action. We first show that q is a multiple of p^e. Since the $\mathbb{Z}_{p,n}$ action has q fixed points on X, it follows that the factor group $H/\mathbb{Z}_{p,n} = \mathbb{Z}_p^e$ freely permutes q corresponding points on the quotient surface $X/\mathbb{Z}_{p,n}$, in cycles of length p^e. (Otherwise there would be a point in X fixed by non-cyclic subgroup of $\mathbb{Z}_p^{e+1}$.) Hence $q = sp^e$ for some $s \geq 0$. Now the Riemann-Hurwitz relation for the induced $\mathbb{Z}_{p,n}$ action reduces to

$$n = 1 + (g-1)/p - sp'p^{e-1}. \tag{8}$$

Substituting for $(g-1)/p$ using (6), with $k = \kappa$ and $r = R$, we obtain equation (7). $s \leq R$, since R is the total number of H-orbits on X with a non-trivial isotropy subgroup and s is the number with a particular one, $\mathbb{Z}_{p,n(s)}$. The composition $\chi_s \circ \rho : \Lambda_H \to H/\mathbb{Z}_{p,n(s)}$ maps the elliptic generators of Λ_H fixing those s orbits to the trivial element in $H/\mathbb{Z}_{p,n}$. The relation $\gamma_1 \dots \gamma_R[\alpha_1, \beta_1] \dots [\alpha_\kappa, \beta_\kappa] = \mathrm{id}$ in Λ_H induces the relation

$$\chi_s \circ \rho(\gamma_1 \dots \gamma_r[\alpha_1, \beta_1] \dots [\alpha_\kappa, \beta_\kappa]) = \chi_s \circ \rho(\gamma_1 \dots \gamma_R) = \mathrm{id}$$

in $H/\mathbb{Z}_{p,n}$. If $s = R$, the induced relation holds trivially. If $s < R$, we may suppose that the first s canonical elliptic generators are in the kernel of $\chi_s \circ \rho$. Then the induced relation is

$$\chi_s \circ \rho(\gamma_{s+1}\gamma_{s+2} \cdot \dots \cdot \gamma_R) = \mathrm{id}.$$

This is not possible if $s = R-1$, for then only γ_R has non-trivial image under $\chi_s \circ \rho$, and $\chi_s \circ \rho(\gamma_R) = \mathrm{id}$ is a contradiction. If $\kappa = 0$, $s = R$ is also not possible, since, in this case, $n(s) = n(R) < 0$. □

4.3. Solutions of Accola's relation. Specializing Accola's Theorem 2.2 to $H = \mathbb{Z}_p^{e+1}$, $e \geq 1$, acting with canonical signature $(\kappa; p, \overset{R}{\dots}, p)$ on a surface of genus $g > 1$, yields the relation

$$g\sigma_e + \kappa p^e = \sum_{i=1}^{\sigma_{e+1}} n(i), \tag{9}$$

where as before $\sigma_N = 1 + p + p^2 + \dots + p^{N-1}$. The summands $n(i)$ are the orbit-genera of the σ_{e+1} subgroups in the partition of H (Lemma 2.1). By Lemma 4.3, the indices i, with possible repeats, come from the set $\{0, 1, 2, \dots, R\}$.

LEMMA 4.4. *$\{n(i) \mid i \in I\}$, where the index set I has cardinality $|I| = \sigma_{e+1}$, is a solution of Accola's relation (9) if and only if $\sum_{i \in I} i = R$.*

PROOF. We claim that $\{n(i) \mid i \in I_0\}$, with index set $I_0 = \{0^{(-1+\sigma_{e+1})}, R\}$, is a solution, where the superscript in parentheses denotes the multiplicity of the corresponding index. (This convention is used henceforth.) Note that I_0 has cardinality σ_{e+1}. To see that

$$n(0) \cdot (-1 + \sigma_{e+1}) + n(R)$$

is equal to the left-hand side of (9), use $n(0) = 1 + (g-1)/p$ (from (8)) and the formula for $n(R)$ given in Lemma 4.3. We leave the details to the reader.

If two distinct index sets I and J determine solutions of (9), equality of the sums

$$\sum_{i\in I} n(i) = \left(1 + \frac{g-1}{p}\right)|I| - p'p^{e-1}\left(\sum_{i\in I} i\right)$$

$$\sum_{j\in J} n(j) = \left(1 + \frac{g-1}{p}\right)|J| - p'p^{e-1}\left(\sum_{j\in J} j\right),$$

implies, since $|I| = |J| = \sigma_{e+1}$, that $\sum_{i\in I} i = \sum_{j\in J} j$. The common sum is R, since that is the case for I_0. □

Lemma 4.4 shows that solutions of (9) are in one-to-one correspondence with additive partitions of R into σ_{e+1} parts chosen from $\{0, 1, \ldots, R\}$. For example, the index sets

$$I_s = \{0^{(-2+\sigma_{e+1})}, s, R-s\}, \quad 0 \le s \le R \tag{10}$$

provide $\lfloor R/2 \rfloor$ distinct solutions in which any possible orbit-genus $n(s)$ appears together with the maximal one $n(0)$. We shall make use of this solution in Section 5. If $j < \sigma_{e+1}$, the index set

$$I_j = \{0^{(-j-1+\sigma_{e+1})}, 1, 2, \ldots, j, R - T_j\}, \tag{11}$$

where T_j is the triangular number $\sum_{i=1}^{j} i = j(j+1)/2$, determines a solution to (9), provided $T_j \le R$. If we define $T_0 = 0$, then for all $R \ge 0$, there is a unique maximal j such that $T_j \le R$ and $0 \le j < -1 + \sigma_{e+1}$; with this j, (11) has the maximum possible number of distinct parts, since the smallest indices are used a minimal number of times. Indeed, the $j+2$ indices (including 0) are distinct if $R - T_j > j$; otherwise there is an index j', $0 \le j' \le j$, such that $R - T_j = j'$ and there are only $j+1$ distinct indices.

4.4. Surface-kernel epimorphisms. This section is devoted to the proof of the following theorem.

THEOREM 4.5. *Let $g \equiv 1 \pmod p$, and let $e \ge 1$ be the largest positive integer such that $g \equiv 1 \pmod{p^e}$. Let R be the integer defined at (5), and let d be the geometric dimension of $\mathcal{N}_g^p$. If g is sufficiently large, then $d \ge j$ when $T_j \le R < T_{j+1}$ and $d = -1 + \sigma_{e+1}$ when $R \ge T_{-1+\sigma_e}$.*

We produce an action of $H = \mathbb{Z}_p^{e+1} = \langle x_1, x_2, \ldots, x_{e+1}\rangle$ on a surface of genus g which induces $(p, n(i))$-gonal actions for each i in the index set I_j at (11), where $j \ge 0$ is the largest integer such that $T_j \le R$ and $j < -1 + \sigma_{e+1}$. If such an action exists, it has covering Fuchsian group Λ_H with signature $(\kappa; p, \overset{R}{\ldots}, p)$, hyperbolic generators $\alpha_1, \beta_1, \ldots, \alpha_\kappa, \beta_\kappa$, and elliptic generators $\gamma_1, \gamma_2, \ldots, \gamma_R$. To produce the desired action, we define a surface-kernel epimorphism $\rho\colon \Lambda_H \to H$ so that exactly i elliptic generators of Λ_H map into the subgroup $H_i < H$, where $H_i \simeq \mathbb{Z}_p$ is one of the σ_{e+1} subgroups of the partition of H, and $H_i \ne H_k$ for $i \ne k$. By the last sentence of Lemma 4.3, the induced action of H_i is $(p, n(i))$-gonal, as desired.

We first treat the case $e = 1$. Since $\sigma_2 = p + 1$, we show that $d = p$ when $R \ge T_p$, and $d = j$ when $T_j \le R < T_{j+1}$, $j < p$. For $R > 3$, Table 1 gives a schema defining a surface-kernel epimorphism $\rho\colon \Lambda_H \to H = \mathbb{Z}_p^2 = \langle x_1, x_2\rangle$. (We put $x_1 = x$ and $x_2 = y$ to improve legibility.) Because $\mathrm{id} = \prod_{i=1}^{\kappa}[\alpha_i, \beta_i]\prod_{k=1}^{R}\gamma_k$ in Λ_H, the nontrivial powers of x and y must be chosen so that the product of all R elements of H is the identity. We assert that this is always possible under the stated restrictions (see Remark 4.6 below). We do not specify the images of the 2κ hyperbolic generators of Λ_H (if any); they can be assigned arbitrarily, but it is convenient to take them to be κ x's and κ y's. If $R < T_p$, then $j \leq p-1$, and in I_j, the index 0 appears together with the indices $1, 2, \ldots, j$. $R - T_j \leq j$ by the definition of j, hence there are $j+1$ distinct indices and therefore $d \geq j$. If $R \geq T_p$, we use the index set $I_{p-1} = \{0, 1, 2, \ldots, p-1, R-T_{p-1}\}$, which has $p+1$ distinct indices, since $R - T_{p-1} \geq p$. Hence, again, $d \geq p$.

Index in I_j	Elliptic Generators of Λ_H	Images in H if $T_j < R$:	Images in H if $T_j = R$:
1	γ_1	xy	xy
2	γ_2, γ_3	xy^2, xy^2	xy^2, xy^2
3	γ_4, γ_5, γ_6	xy^3, xy^3, xy^3	xy^3, xy^3, xy^3
$\vdots$	$\vdots$	$\vdots$	$\vdots$
$j-1$	$\gamma_{1+T_{j-2}}, \ldots, \gamma_{T_{j-1}}$	$xy^{j-1}, \ldots, xy^{j-1}$	powers of x
j	$\gamma_{1+T_{j-1}}, \ldots, \gamma_{T_j}$	powers of x	powers of y
$R - T_j$	$\gamma_{1+T_j}, \ldots, \gamma_R$	powers of y	-

TABLE 1. Schema for defining $\rho : \Lambda_H \to H$, $H = \mathbb{Z}_p^2$, $R > 3$

We next show how the schema in Table 1 generalizes to define actions of $\mathbb{Z}_p^{e+1}$, $e \geq 1$ of the desired type. We make an inductive construction: Let $g_1, g_2, \ldots, g_{\sigma_e}$ be a set of generators for the distinct subgroups of a partition of $\mathbb{Z}_p^e = \langle x_1, \ldots, x_e \rangle$, and assume that

$$g_{\sigma_e} = x_e, \quad g_{\sigma_e - 1} = x_{e-1}, \quad \ldots \quad g_{\sigma_e - e+1} = x_1.$$

Adjoin a new generator x_{e+1} to form $H = \mathbb{Z}_p^{e+1} = \langle x_1, \ldots, x_{e+1} \rangle$. Form a set of σ_{e+1} generators for the distinct subgroups of a partition of the larger group as follows. Let $h_{ij} = g_i x_{e+1}^j$, $i = 1, \ldots, \sigma_e$, $j = 0, \ldots, p-1$. This amounts to $p\sigma_e$ generators, one less than the necessary $\sigma_{e+1} = p\sigma_e + 1$. Relabel the generators using a single index, as h_t, $t = 1, \ldots, p\sigma_e$. Replace the last e generators by $x_1, \ldots, x_e$, and adjoin one more, defining $h_{\sigma_{e+1}} = x_{e+1}$. We now have a complete set of generators, and a schema for constructing $\rho : \Lambda_H \to H = \mathbb{Z}_p^{e+1}$ is given in Table 2. (Table 1 is the special case $e = 1$.)

REMARK 4.6. In assuming that g is 'sufficiently large' we exclude cases where $R = R(e) < 0$. Since R grows with g, we may further assume $R > T_{e+1}$, which implies $j - e \geq 1$. The latter is the minimal assumption needed for the schema in Table 2 to succeed even in the most restrictive case $\kappa = 0$. For example, taking $e = 2$ and $R = 7 = 1 + T_3$, the schema allows the following surface-kernel epimorphism

Index in I_j	Elliptic Generators of Λ_H	Images in H if $T_j < R$:	Images in H if $T_j = R$:
1	γ_1	h_1	h_1
2	γ_2, γ_3	h_2, h_2	h_2, h_2
3	$\gamma_4, \gamma_5, \gamma_6$	h_3, h_3, h_3	h_3, h_3, h_3
$\vdots$	$\vdots$	$\vdots$	$\vdots$
$j-e$	$\gamma_{1+T_{j-e-1}}, \ldots, \gamma_{T_{j-e}}$	$h_{j-e}, \ldots, h_{j-e}$	powers of x_1
$j-e+1$	$\gamma_{1+T_{j-e}}, \ldots, \gamma_{T_{j-e+1}}$	powers of x_1	powers of x_2
$\vdots$	$\vdots$	$\vdots$	$\vdots$
$j-1$	$\gamma_{1+T_{j-2}}, \ldots, \gamma_{T_{j-1}}$	powers of x_{e-1}	powers of x_e
j	$\gamma_{1+T_{j-1}}, \ldots, \gamma_{T_j}$	powers of x_e	powers of x_{e+1}
$R-T_j$	$\gamma_{1+T_j}, \ldots, \gamma_R$	powers of x_{e+1}	-

TABLE 2. Schema for defining $\rho : \Lambda_H \to H$, $H = \mathbb{Z}_p^{e+1}$, $e \geq 1$, $R > T_{e+1}$

onto $\mathbb{Z}_p^3$: $\gamma_1 \mapsto x_1x_2x_3$, $\gamma_2, \gamma_3 \mapsto x_1, x_1^{-2}$, $\gamma_4, \gamma_5, \gamma_6 \mapsto x_2, x_2, x_2^{-3}$, $\gamma_7 \to x_3^{-1}$. But with $R = 6 = T_3$, no construction is possible.

5. A spanning tree for $\mathcal{N}_g^p$

The results of the previous section show there is a simplex in $\mathcal{N}_g^p$ for almost any additive partition of R into σ_{e+1} parts. Instead of attempting to draw every simplex, we conclude by constructing a minimal spanning tree. This determines the connected components, and, along with the geometric dimension, provides a rather complete picture of the general structure of $\mathcal{N}_g^p$. The proposition below shows that the connected components of $\mathcal{N}_g^p$ can be determined by considering only $\mathbb{Z}_p^2$ actions.

PROPOSITION 5.1. *Let X be a surface of genus $g \equiv 1 \pmod p$ which is both (p,n)-gonal and (p,m)-gonal, $n \neq m$. Then either there exists a (p,n)-gonal automorphism ϕ and a (p,m)-gonal automorphism τ such that $\langle \phi, \tau \rangle \simeq \mathbb{Z}_p^2$, or there exists a (p,l)-gonal automorphism η, where $l \notin \{n,m\}$, and a (p,n)-gonal automorphism ϕ and (p,m)-gonal automorphism τ, such that $\mathbb{Z}_p^2 \simeq \langle \phi, \eta \rangle \neq \langle \tau, \eta \rangle \simeq \mathbb{Z}_p^2$.*

PROOF. If there exists a (p,n)-gonal automorphism ϕ and a (p,m)-gonal automorphism τ which commute, then trivially $\langle \phi, \tau \rangle \simeq \mathbb{Z}_p^2$. Hence assume no (p,m)-gonal automorphism commutes with any (p,n)-gonal automorphism. Let G be the full automorphism group of the surface, and let P be a p-Sylow subgroup of G. P has a non-trivial center which contains a (p,l)-gonal group for some $l \notin \{n,m\}$. This central (p,l)-gonal group commutes with any element in any (p,n)- and (p,m)-gonal group in P, generating, with each in turn, a distinct subgroup isomorphic to $\mathbb{Z}_p^2$. □

Thus a surface which is both $(p, n(i))$- and $(p, n(j))$-gonal, $i \neq j$, admits a $\mathbb{Z}_p^2$ action which induces the $(p, n(i))$-gonal action, and a (possibly distinct) $\mathbb{Z}_p^2$ action inducing the $(p, n(j))$-gonal action. Solutions (10) to Accola's relation allow for a $\mathbb{Z}_p^2$ action which induces a $(p, n(0))$-gonal action and a $(p, n(s))$-gonal automorphism, for every non-excluded $s \in \{1, \ldots, R\}$. We use the (maximum) value of R determined by putting $e = 1$ in (5), even if $g \equiv 1 \pmod{p^k}$ for some $k > 1$. The simple schema in Table 3 shows these actions can be realized. (Nontrivial powers of x and y must be chosen so that the total exponent on each is a multiple of p.) By symmetry of I_s in s and $R - s$, we may assume $s \leq \lfloor R/2 \rfloor$. $s = 1$ requires an adjustment, since $\{0^{(p-1)}, 1, R - 1\}$ contains the excluded index $R - 1$. One can simply replace the index set by $\{0^{(p-2)}, 1^2, R - 2\}$. It is easy to see that an appropriate surface-kernel epimorphism can be constructed in this case.

Index in I_s	Elliptic Generators of Λ_H	Images in H
s	$\gamma_1, \ldots, \gamma_s$	powers of x
$R - s$	$\gamma_{s+1}, \ldots, \gamma_R$	powers of y

TABLE 3. Schema for $\rho : \Lambda_H \to H = \mathbb{Z}_p^2, \quad 1 < s \leq \lfloor R/2 \rfloor$

It follows that any path from $(p, n(i))$ to $(p, n(j))$ in $\mathcal{N}_g^p$, $i \neq j$, $i, j \neq 0$, can be replaced, if necessary, by a path of length 2 passing through $(p, n(0))$. So, if $g \equiv 1 \pmod p$, there is a path-connected component of $\mathcal{N}_g^p$ with a star-like spanning tree centered at $(p, n(0))$ having an edge from $(p, n(0))$ to each of the vertices $(p, n(s))$. If $\kappa > 0$, there are other, isolated vertices in $\mathcal{N}_g^p$. This is because (7), with $e = 1$, yields positive values of n for some excluded values of s, corresponding to (p, n)-gonal groups not induced by any $H = \mathbb{Z}_p^2$ action. In particular, $n = 1 + p'$ is obtained by taking $s = R - 1$ (provided $R > 0$); in addition, if $\kappa > 1$, there are $2\kappa - 2 > 0$ smaller positive values of n obtained by taking $s > R$:

$$n(R + t) = \kappa + p'(2\kappa - 2 - t), \quad t = 1, 2, \ldots, 2\kappa - 2.$$

This information is summarized in Figure 1, for $R \geq 3$.

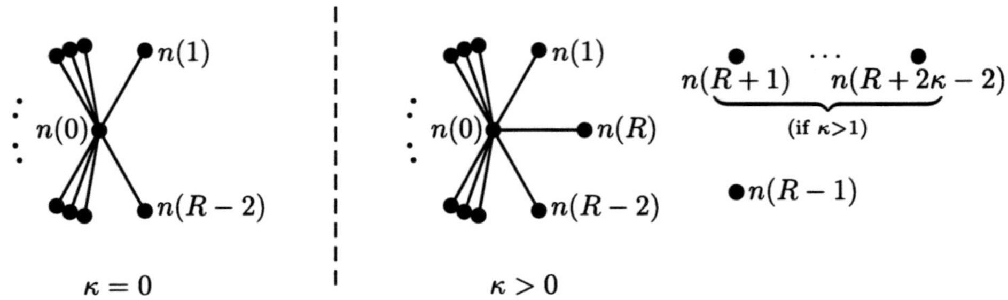

FIGURE 1. Spanning trees for $\mathcal{N}_g^p$, $g \equiv 1 \pmod p$, $R \geq 3$

If $R = 0$, the star-like component reduces to the singleton $n(0)$; if $\kappa > 1$, the additional isolated vertices are $n(2), \ldots, n(2\kappa - 2)$. If $R = 2$, the starlike component is just the edge from $n(0)$ and $n(2)$; if $\kappa > 1$, the additional isolated vertices are $n(1), n(3), \ldots, n(2\kappa - 1)$. If $R = 1$, $\mathcal{N}_g^p$ is empty. Finally, if $g \not\equiv 1 \pmod p$, $\mathcal{N}_g^p$ is either empty or consists of isolated vertices.

In Table 4 we show some spanning trees for $p = 7$.

g	κ	R	Spanning Tree	g	κ	R	Spanning Tree
50	2	0		92	2	2	
57	0	5		99	0	7	
64	1	3		106	1	5	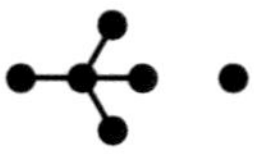
71	2	1	–	113	2	3	
78	0	6		120	0	8	
85	1	4	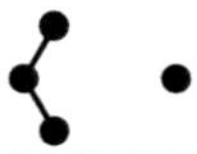	127	1	6	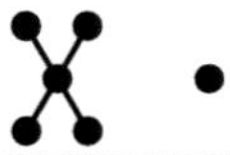

TABLE 4. Spanning trees for some $\mathcal{N}_g^7$'s

Acknowledgements

The authors are grateful to the referee for comments and questions which led us to improve the exposition of our results.

References

[1] Robert D. M. Accola, *Riemann surfaces with automorphism groups admitting partitions*, Proc. Amer. Math. Soc. **21** (1969), 477–482. MR0237764 (38 #6045)

[2] Gabriel Bartolini and Milagros Izquierdo, *On the connectedness of the branch locus of the moduli space of Riemann surfaces of low genus*, Proc. Amer. Math. Soc. **140** (2012), no. 1, 35–45, DOI 10.1090/S0002-9939-2011-10881-5. MR2833515 (2012f:14059)

[3] S. Allen Broughton, *Classifying finite group actions on surfaces of low genus*, J. Pure Appl. Algebra **69** (1991), no. 3, 233–270, DOI 10.1016/0022-4049(91)90021-S. MR1090743 (92b:57021)

[4] S. Allen Broughton, *The equisymmetric stratification of the moduli space and the Krull dimension of mapping class groups*, Topology Appl. **37** (1990), no. 2, 101–113, DOI 10.1016/0166-8641(90)90055-7. MR1080344 (92d:57013)

[5] E. Bujalance, G. Gromadzki, and M. Izquierdo, *On real forms of a complex algebraic curve*, J. Aust. Math. Soc. **70** (2001), no. 1, 134–142, DOI 10.1017/S1446788700002329. MR1808396 (2001k:14108)

[6] Maurizio Cornalba, *On the locus of curves with automorphisms* (English, with Italian summary), Ann. Mat. Pura Appl. (4) **149** (1987), 135–151, DOI 10.1007/BF01773930. MR932781 (89b:14038)

[7] Maurizio Cornalba, *Erratum: "On the locus of curves with automorphisms" [Ann. Mat. Pura Appl. (4)* **149** *(1987), 135–151; MR0932781]*, Ann. Mat. Pura Appl. (4) **187** (2008), no. 1, 185–186, DOI 10.1007/s10231-006-0031-0. MR2346015 (2008h:14026)
[8] Gabino González-Diez, *On prime Galois coverings of the Riemann sphere*, Ann. Mat. Pura Appl. (4) **168** (1995), 1–15, DOI 10.1007/BF01759251. MR1378235 (97c:14033)
[9] Grzegorz Gromadzki, Anthony Weaver, and Aaron Wootton, *On gonality of Riemann surfaces*, Geom. Dedicata **149** (2010), 1–14, DOI 10.1007/s10711-010-9459-x. MR2737675 (2011m:14046)
[10] Grzegorz Gromadzki, *On conjugacy of p-gonal automorphisms of Riemann surfaces*, Rev. Mat. Complut. **21** (2008), no. 1, 83–87. MR2408038 (2009a:30088)
[11] W. J. Harvey, *On branch loci in Teichmüller space*, Trans. Amer. Math. Soc. **153** (1971), 387–399. MR0297994 (45 #7046)
[12] Ruben A. Hidalgo, *A remark on (p, n)-gonal quasiplatonic Riemann surfaces*, Geom. Dedicata **160** (2012), 309–312, DOI 10.1007/s10711-011-9683-z. MR2970056
[13] A. Hurwitz, Über algebraische Gebilde mit eindeutigen Transformationen in sich, Matt. Annalen **41** (1893) 403-42.
[14] A. Jaikin, personal communication.
[15] Ravi S. Kulkarni, *Symmetries of surfaces*, Topology **26** (1987), no. 2, 195–203, DOI 10.1016/0040-9383(87)90059-0. MR895571 (88m:57051)
[16] Ravi S. Kulkarni, *Isolated points in the branch locus of the moduli space of compact Riemann surfaces*, Ann. Acad. Sci. Fenn. Ser. A I Math. **16** (1991), no. 1, 71–81. MR1127697 (93a:30048)
[17] A.M. Macbeath, Fuchsian groups, in: *Proceedings of the Summer School in Geometry and Topology*, Queens College, Dundee, 1962.
[18] Andreas Schweizer, *On the uniqueness of (p, h)-gonal automorphisms of Riemann surfaces*, Arch. Math. (Basel) **98** (2012), no. 6, 591–598, DOI 10.1007/s00013-012-0397-8. MR2935665
[19] C. Maclachlan and Y. Talu, *p-groups of symmetries of surfaces*, Michigan Math. J. **45** (1998), no. 2, 315–332, DOI 10.1307/mmj/1030132185. MR1637662 (99f:57018)

Gdańsk University, Poland
E-mail address: grom@mat.ug.edu.pl

Bronx Community College of CUNY
E-mail address: anthony.weaver@bcc.cuny.edu

University of Portland
E-mail address: wootton@up.edu

Free degree of periodic self-homeomorphisms of compact bordered orientable surfaces

G. Gromadzki and X. Zhao

Dedicated to Emilio Bujalance for his sixtieth birthday

ABSTRACT. The free degree $\mathfrak{fr}(X)$ of a topological space X is the minimum positive integer n with the property that for any self-homeomorphism φ of X, at least one of the iterates $\varphi^i, i \leq n$ has a fixed point, or ∞ if such minimum does not exists. In a recent paper Wu and Zhao have showed that for compact orientable surface $F_{g,b}$ of topological genus $g \geq 2$ and having b boundary components $\mathfrak{fr}^+(F_{g,b}) \leq 24g - 24$ while for non-orientable surface $N_{g,b}$ of topological genus $g \geq 3$ with b boundary components $\mathfrak{fr}(N_{g,b}) = 12g - 24$. Here we show that, for periodic self-homeomorphisms, these bounds can be essentially improved. We prove that periodic free degree $\mathfrak{fr}_{\mathrm{p}}^+(F_{g,b})$ is a periodic function of b, which allow us to find its explicit values for $g \leq 5$ and for all b. We observe that $\mathfrak{fr}_{\mathrm{p}}^+(F_{g,b}) \leq -\chi(F_{g,b})$ and, given g we characterize the set $\mathcal{B}_g$ of those b for which equality holds. Finally we prove that if $\mathfrak{fr}_{\mathrm{p}}^+(F_{g,b}) > 2g-2$ then $b' = \mathfrak{fr}_{\mathrm{p}}^+(F_{g,b}) - (2g-2)$ belongs to $\mathcal{B}_g$ and $\mathfrak{fr}_{\mathrm{p}}^+(F_{g,b}) = -\chi(F_{g,b'})$, except for $g = 8$ and certain b congruent to 1 mod 15, which we explicitly describe.

1. Introduction

The free degree $\mathfrak{fr}(\varphi)$ of a self-homeomorphism φ of a topological space X is the maximal finite period of φ or infinity if such maximum does not exist, which can be caused either by the fact that the orders of finite orbits with respect to the action of φ form an unlimited set or also by the fact that no point of X generates a finite orbit under the action of φ. The free degree $\mathfrak{fr}(X)$ of X is the maximum of $\mathfrak{fr}(\varphi)$, for φ running over all self-homeomorphisms of X. Moreover, one may consider $\mathfrak{fr}^+(X)$, $\mathfrak{fr}^-(X)$, $\mathfrak{fr}_{\mathrm{p}}(X)$, and $\mathfrak{fr}_{\mathrm{p}}^+(X)$ etc, where the super index $+$ (or $-$) means that only orientation-preserving (or orientation-reversing) self-homeomorphisms are considered, and the subscript p means that only periodic self-homeomorphisms are considered.

2010 *Mathematics Subject Classification.* Primary 55M20, 37E30, 30F10.

Key words and phrases. Fixed point, Periodic point, Periodic self-homeomorphism of closed surface, Automorphism of a Riemann surface.

The first author was supported by Research Grant of Polish National Center of Sciences NCN 2012/05/B/ST1/02171 and by Max-Planck Mathematical Institute in Bonn.

The second author was supported by the NSF of China (10931005) and a project of Beijing Municipal Education Commission (KZ201310028030).

The study of free degree of closed surfaces has a long and rich history and go back to Nielsen who showed in [**13**] that $\mathfrak{fr}^+(F_2) \leq 3$ and $\mathfrak{fr}^+(F_g) = 2g-2$ for $g > 2$ and later Dicks and Llibre [**7**] proved $\mathfrak{fr}^+(F_2) = 2$. In 1992 Wang [**15**] proved $\mathfrak{fr}(F_2) = 4$ and $\mathfrak{fr}(F_g) = 2g-2$ for $g > 2$.

In [**19**], Wu and Zhao have shown that for compact orientable surfaces $F_{g,b}$ of topological genus $g \geq 2$ having b boundary components $\mathfrak{fr}^+(F_{g,b}) \leq 24g-24$ while for compact non-orientable surfaces $N_{g,b}$ of topological genus $g \geq 3$ with b boundary components $\mathfrak{fr}(N_{g,b}) = 12g-24$. They used Thurston's classification of self-homeomorphisms of compact surfaces and showed that these bounds hold both for reducible and pseudo-Anosov self-homeomorphisms φ. There was no necessity of considering the periodic case $\mathfrak{fr}^+_{\mathrm{p}}(F_{g,b})$ in details since Wiman's bound $4g+2$ [**17**] for orders of periodic self-homeomorphisms of $F_{g,b}$ is smaller than $12g-24$.

We prove that $\mathfrak{fr}^+_{\mathrm{p}}(F_{g,b})$ is periodic with respect to b, which allows us to find explicit values of $\mathfrak{fr}^+_{\mathrm{p}}(F_{g,b})$ for $g = 2,3,4,5$ and for all b. The crucial point for our results is the Euler characteristic $\chi(F_{g,b})$ of $F_{g,b}$. It is easy to see that $\mathfrak{fr}^+_{\mathrm{p}}(F_{g,b}) \leq -\chi(F_{g,b})$ and given g, we characterize the set $\mathcal{B}_g$ of those b for which $\mathfrak{fr}^+_{\mathrm{p}}(F_{g,b}) = -\chi(F_{g,b})$. We show that for $\mathfrak{fr}^+_{\mathrm{p}}(F_{g,b}) > 2g-2$ and $b' = \mathfrak{fr}^+_{\mathrm{p}}(F_{g,b}) - 2g + 2$, we have $\mathfrak{fr}^+_{\mathrm{p}}(F_{g,b}) = -\chi(F_{g,b'})$, and for $b' > 0$, $b' \in \mathcal{B}_g$, except for $g = 8$ and certain b with $16 \leq b \equiv 1\ (15)$. This, roughly speaking, means that either $\mathfrak{fr}^+_{\mathrm{p}}(F_{g,b}) \leq 2g-2$ or $\mathfrak{fr}^+_{\mathrm{p}}(F_{g,b})$ is determined by certain element of $\mathcal{B}_g$. We give also some explicit calculations of $\mathcal{B}_g$ for low values of g, obtained with the aid of computer. A sequel concerning similar results for $\mathfrak{fr}^-_{\mathrm{p}}(F_{g,b})$ and for $\mathfrak{fr}_{\mathrm{p}}(N_{g,b})$ for non-orientable surfaces $N_{g,b}$ is planned by the authors.

After publishing [**19**] and having most of results presented here already obtained, we have learned (c.f. [**18**] and Acknowledgements at the end of this paper) that, a sharper bound $4g+2$ for $\mathfrak{fr}^+(F_{g,b})$ was obtained by Chas in her unpublished thesis [**6**]. However, in both mentioned works, the periodic case was not taken with any special, particular care while from our paper follows that $\mathfrak{fr}^+_{\mathrm{p}}(F_{g,b})$ and $\mathfrak{fr}^+(F_{g,b})$ behave in essentially different manner. The difference can be seen, already for $g = 2$ in Table 1.1 of [**6**] and in our Table 6 which also show the contrast with the case of closed surfaces, where free degree is attained by periodic self-homeomorphisms [**7, 13, 15**]. Finally, the periodicity of $\mathfrak{fr}^+_{\mathrm{p}}(F_{g,b})$ with respect to b discovered in our paper show much finer character of $\mathfrak{fr}^+_{\mathrm{p}}(F_{g,b})$ than $\mathfrak{fr}^+(F_{g,b})$ which was proven to be constant for $b \geq 6g+6$ (see synthetic data in Table 5 and details in Tables 6-10).

2. Preliminaries

We shall use a combinatorial approach based on Fuchsian and non-euclidean crystallographic groups (NEC-groups in short); we refer the reader to the monographs [**1**] and [**3**] for detailed expositions of the whole theory.

2.1. Fuchsian and non-euclidean crystallographic groups. An NEC-group is a discrete and co-compact subgroup of the group $\mathcal{G}$ of isometries of the hyperbolic plane $\mathcal{H}$ including those which reverse orientations. If such a subgroup contains only orientation-preserving isometries, it is called a Fuchsian group. Macbeath [**12**] and Wilkie [**16**] associated to every NEC-group Λ a signature $s = s(\Lambda)$, which determines its algebraic structure. It has the form

$$(1)\qquad (g; \pm; [m_1, \ldots, m_r]; \{C_1, \ldots, C_k\}).$$

The numbers $m_i \geq 2$ are called the *proper periods*, the brackets $C_i = (n_{i1}, \ldots, n_{is_i})$, the *period cycles*, the numbers $n_{ij} \geq 2$, the *link periods* and $g \geq 0$ is said to be the *orbit genus* of Λ. The orbit space $\mathcal{H}/\Lambda$ is a surface having k boundary components, orientable or not according to the sign being plus or minus and having topological genus g. A Fuchsian group can be regarded as an NEC-group with the signature

$$(2)\qquad (g; +; [m_1, \ldots, m_r]; \{-\}),$$

which shortly shall be denoted by $(g; m_1, \ldots, m_r)$; a Fuchsian group without periods is called a surface Fuchsian group and has signature of the form $(g; -)$. The group with the signature (1) has a presentation given by generators:

(a) $x_i,\ i = 1, \ldots, r,$ (rotations)
(b) $c_{ij},\ i = 1, \ldots, k; j = 0, \ldots, s_i,$ (reflections)
(c) $e_i,\ i = 1, \ldots, k,$ (connecting generators)
(d) $a_i, b_i,\ i = 1, \ldots, g$ if the sign is $+$, (hyperbolic translations)
$d_i,\ i = 1, \ldots, g$ if the sign is $-$, (glide reflections)

and relations:

(A) $x_i^{m_i} = 1,\ i = 1, \ldots, r,$
(B) $c_{is_i} = e_i^{-1} c_{i0} e_i,\ i = 1, \ldots, k,$
(C) $c_{ij-1}^2 = c_{ij}^2 = (c_{ij-1}c_{ij})^{n_{ij}} = 1,\ i = 1, \ldots, k; j = 1, \ldots, s_i,$
(D) $x_1 \ldots x_r e_1 \ldots e_k a_1 b_1 a_1^{-1} b_1^{-1} \ldots a_g b_g a_g^{-1} b_g^{-1} = 1$, if the sign is $+$
$x_1 \ldots x_r e_1 \ldots e_k d_1^2 \ldots d_g^2 = 1$ if the sign is $-$.

Any set of generators of an NEC-group satisfying the above relations will be called *a canonical set of generators.* Reflections $c_{i,j-1}, c_{i,j}$ are said to be consecutive or neighbouring.

Consider an NEC-group Λ with signature (1), any fundamental region of Λ has normalized area (that is the hyperbolic area divided by 2π) $\mu(\Lambda)$ given by

$$(3)\qquad \varepsilon g + k - 2 + \sum_{i=1}^{r}\left(1 - \frac{1}{m_i}\right) + \frac{1}{2}\sum_{i=1}^{k}\sum_{j=1}^{s_i}\left(1 - \frac{1}{n_{ij}}\right),$$

where $\varepsilon = 2$ if the sign is plus and $\varepsilon = 1$ otherwise.

It is known that an abstract group with the presentation given by the generators (a)-(d) and the relations (A)-(D) can be realized as an NEC-group with signature (1) if and only if the above expression (3) is positive. If Γ is a subgroup of finite index in an NEC-group Λ then it is an NEC-group itself and the Riemann-Hurwitz formula says that

$$(4)\qquad [\Lambda : \Gamma] = \frac{\mu(\Gamma)}{\mu(\Lambda)}.$$

2.2. Riemann and Klein surfaces and their groups of automorphisms. A Klein surface is a compact topological surface with dianalytic structure which, in contrast to the classical analytic structure, allows the conjugation $z \to \bar{z}$ for transition maps between charts and for the local forms of automorphisms between such surfaces. Furthermore, the maps $X \to Y$ between Klein surfaces may allow, over boundary points of Y, the foldings $a + bi \to a + |b|i$ in their local forms (see [1] for precise definitions) though these phenomena obviously do not appear in the case of isomorphisms. Therefore a classical Riemann surface can be considered as a Klein surface simply by adding the conjugate charts to the existing ones. Now, by the Riemann uniformization theorem, a classical compact Riemann surface of genus $g \geq 2$ can be represented as the orbit space $\mathcal{H}/\Gamma$ for some Fuchsian surface group Γ with the signature $(g; -)$. Furthermore, a group of conformal automorphisms G of a surface given in such a way can be represented as the factor group Λ/Γ, where Λ is a proper NEC or a Fuchsian group, according to whether G contains orientation-reversing automorphisms or not. Combinatorial study of groups of automorphisms of bordered Klein surfaces X is possible essentially due to the same facts. This time, however, a surface group Γ uniformizing X has a signature $(g; \pm; [-]; \{(-), \overset{k}{\ldots}, (-)\})$, where $k \geq 1$ and is called bordered surface NEC-group. These groups are characterized between all NEC groups as the ones having no elliptic elements. An epimorphism $\theta : \Lambda \to G$, whose kernel Γ is a bordered surface group will be called *smooth*.

2.3. Periodic self-homeomorphisms. Any orientation-preserving periodic self-homeomorphism φ of order N on an orientable closed surface F_g gives a Fuchsian group Λ with signature $(h; m_1, \ldots, m_r)$ and an epimorphism $\Lambda \to \mathbb{Z}_N$ with a kernel being a Fuchsian surface group of genus g. Conversely, any epimorphism $\Lambda \to \mathbb{Z}_N$ from a Fuchsian group, with this signature, to a cyclic group having a kernel being Fuchsian surface group determines a periodic self-homeomorphism with r singular orbits, which have lengths N/m_i, $i = 1, 2, \ldots, r$. Moreover, the kernel of an epimorphism $\theta : \Lambda \to \mathbb{Z}_N$ is a surface group if and only if $\theta(x_i) = a^{\alpha_i(N/m_i)}$, where α_i is coprime to m_i, i.e. x_i and $\theta(x_i)$ have the same order.

3. Compact surfaces with larger periodic free degrees

The most important tool in a combinatorial study of finite cyclic groups of self-homeomorphisms of closed orientable surfaces is the following result of Harvey [8]

THEOREM 3.1. *Let Λ be a Fuchsian group with signature $(h; m_1, \ldots, m_r)$ and let $M = \operatorname{lcm}(m_1, \ldots, m_r)$. Then there exists a smooth epimorphism from Λ onto $\mathbb{Z}_N$ if and only if*

(1) *$M = \operatorname{lcm}(m_1, \ldots, m_{i-1}, m_{i+1}, \ldots, m_r)$ for all i;*
(2) *M divides N and if $h = 0$ then $M = N$;*
(3) *$r \neq 1$ and if $h = 0$ then $r \geq 3$;*
(4) *if N is even then the number of periods m_i such that N/m_i is odd is also even.* □

It gives at once (c.f. [9])

LEMMA 3.2. *There is an orientation-preserving periodic self-homeomorphism φ of order N of a closed surface F_g if and only if there are $\beta_i \in \mathbb{Z}_N$ for $i = 1, 2, \ldots, r$ for some r such that*

(1) $\beta_1 + \cdots + \beta_r \equiv 0\ (N)$,
(2) $\gcd(\beta_1, \ldots, \beta_r) \equiv 1\ (N)$ *if $F_g/\langle\varphi\rangle$ is the sphere,*

(3) $2g-2=N\Big(2h-2+\sum_{i=1}^{r}\Big(1-\dfrac{1}{m_i}\Big)\Big)$, *where h is the genus of $F_g/\langle\varphi\rangle$, and $m_i=N/\gcd(N,\beta_i)$ for $i=1,2,\dots,r$.* □

In this lemma, r is the number of singular orbits of φ, and for each $i=1,2,\dots,r$, the integers $N/m_i=\gcd(N,\beta_i)$ are their lengths. The next lemma is a consequence of Hirose's classification [**9**] of periodic self-homeomorphisms of order $\geq 3g$ of closed orientable surfaces of genus $g\geq 2$; in [**9**], Hirose actually proved more - namely that these orders determine cyclic actions up to the topological type.

LEMMA 3.3. *The singular orbit lengths and the ramification types of all orientation-preserving periodic self-homeomorphisms of order $N\geq 3g$ of a closed orientable surface of genus $g\geq 2$ are given in* Table 1. □

Order	Singular orbit lengths	Ramification type	Condition
$4g+2$	$[1,2,2g+1]$	$[4g+2,2g+1,2]$	
$4g$	$[1,1,2g]$	$[4g,4g,2]$	
$3g+3$	$[1,3,g+1]$	$[3g+3,g+1,3]$	$g\not\equiv 2\ (3)$
$3g$	$[1,1,g]$	$[3g,3g,3]$	
$3g=6$	$[2,2,3,3]$	$[3,3,2,2]$	$g=2$
$3g=12$	$[1,2,3]$	$[12,6,3]$	$g=4$
$3g+2=20$	$[1,4,5]$	$[20,5,4]$	$g=6$
$3g+1=28$	$[1,4,7]$	$[28,7,4]$	$g=9$
$3g=30$	$[1,5,6]$	$[30,6,5]$	$g=10$
$3g=36$	$[1,4,9]$	$[36,9,4]$	$g=12$

TABLE 1. Ramification of self-homeomorphisms of F_g of order $\geq 3g$

PROPOSITION 3.4. *Let $g\geq 2$. Then $\mathfrak{fr}_{\mathrm{p}}^{+}(F_{g,b})\geq 3g$ if and only if it is one of cases listed in the* Table 2.

PROOF. Let us consider the first row of the tabulate. By Lemma 3.3 (cf. [**14**, Lemma 5 (1)]), there is an orientation-preserving periodic self-homeomorphism φ on F_g of order $4g+2$. Moreover, we know that there are exactly three singular orbits whose lengths are 1, 2 and $2g+1$, respectively. Remove a small disc centered at each point in these three orbits, we shall obtain a compact surface $F_{g,2g+4}$. If these discs are chosen so that their union is invariant under φ, the restriction $\varphi|$ on $F_{g,2g+4}$ is a periodic self-homeomorphism of order $4g+2$. Note that each regular orbit have length $4g+2$. By definition of free degree, we obtain $\mathfrak{fr}_{\mathrm{p}}^{+}(F_{g,2g+4})\geq 4g+2$. Since $4g+2$ is the maximum of the orders, we have that $\mathfrak{fr}_{\mathrm{p}}^{+}(F_{g,2g+4})=4g+2$. Moreover, if $b\equiv 2-2g\ (4g+2)$, one may remove $b-(2g+4)$ more discs, which are neighbors of $\big(b-(2g+4)\big)/\big(4g+2\big)$ regular φ orbits. The restriction of φ on $F_{g,b}$ gives a free action. Hence, we still have that $\mathfrak{fr}_{\mathrm{p}}^{+}(F_{g,b})=4g+2$.

Conversely, since $4g+2$ is the maximum of the order, the equality $\mathfrak{fr}_{\mathrm{p}}^{+}(F_{g,b})=4g+2$ implies that there is a $\mathbb{Z}_{4g+2}$ free action on $F_{g,b}$. By Riemann-Hurwitz

$\mathfrak{fr}_{\mathrm{p}}^{+}(F_{g,b})$	g	b
$4g+2$	arbitrary	$b \equiv 2-2g\ (4g+2)$
$4g$	arbitrary	$b \equiv 2-2g\ (4g)$ $b \not\equiv 2-2g\ (4g+2)$
$3g+3$	$g \not\equiv 2\ (3)$	$b \equiv 2-2g\ (3g+3)$ $b \not\equiv 2-2g\ (4g, 4g+2)$
$3g$	$g \equiv 2\ (3)$	$b \equiv 2-2g\ (3g)$ $b \not\equiv 2-2g\ (4g, 4g+2)$
$3g$	$g \not\equiv 2\ (3),\ g \neq 6, 9$	$b \equiv 2-2g\ (3g)$ $b \not\equiv 2-2g\ (3g+3, 4g, 4g+2)$
$3g+2=20$	$g=6$	$b \equiv -10\ (20)$ $b \not\equiv -10\ (21, 24, 26)$
$3g=18$	$g=6$	$b \equiv -10\ (18)$ $b \not\equiv -10\ (20, 21, 24, 26)$
$3g+1=28$	$g=9$	$b \equiv -16\ (28)$ $b \not\equiv -16\ (30, 36, 38)$
$3g=27$	$g=9$	$b \equiv -16\ (27)$ $b \not\equiv -16\ (28, 30, 36, 38)$

TABLE 2. (g, b) for which $\mathfrak{fr}_{\mathrm{p}}^{+}(F_{g,b}) \geq 3g$

formula, we have that $4g+2$ divides $\chi(F_{g,b})$. Hence, $2-2g-b \equiv 0\ (4g+2)$. It follows that $b \equiv 2-2g\ (4g+2)$.

Now consider that case $\mathfrak{fr}_{\mathrm{p}}^{+}(F_{g,b}) = 4g$. By Lemma 3.3, there is an orientation-preserving periodic self-homeomorphism φ on F_g of order $4g$ having three singular orbits of lengths 1, 1 and $2g$ respectively. If $b \equiv 2-2g\ (4g)$, we have that $b = 2g+2+4gk$ for some non-negative integer k. Remove a small disc centered at each point in these three orbits and k regular orbits, we shall obtain a compact surface $F_{g,b}$. If these discs are chosen so that their union is invariant under φ, the restriction $\varphi|$ on $F_{g,2g+4}$ is a periodic homeomorphism of order $4g$ without singular orbits. It follows that $\mathfrak{fr}_{\mathrm{p}}^{+}(F_{g,b}) \geq 4g$. Note that $F_{g,b}$ has no orientation-preserving periodic self-homeomorphism of order $4g+1$. Since $b \not\equiv 2-2g\ (4g+2)$, by (1) we have that $\mathfrak{fr}_{\mathrm{p}}^{+}(F_{g,b}) = 4g$.

Conversely, since $4g+2$ is the maximum of the order and $4g$ is greater than any proper divisor of $4g+2$, the equality $\mathfrak{fr}_{\mathrm{p}}^{+}(F_{g,b}) = 4g$ implies that there is a $\mathbb{Z}_{4g}$ free action on $F_{g,b}$. By Riemann-Hurwitz formula, we have that $4g$ divides $\chi(F_{g,b})$. Hence, $2-2g-b \equiv 0\ (4g)$. It follows that $b \equiv 2-2g\ (4g)$. The proof of the remaining cases are similar. □

COROLLARY 3.5. *If* $\mathfrak{fr}_{\mathrm{p}}^{+}(F_{g,b}) \geq 3g$ *with* $g \geq 2$, *then* $\mathfrak{fr}_{\mathrm{p}}^{+}(F_{g,b})$ *divides* $\chi(F_{g,b})$, *and* $\mathfrak{fr}_{\mathrm{p}}^{+}(F_{g,b}) = -\chi(F_{g,b_0})$ *for the smallest positive integer* b_0 *with the property that* $b_0 \equiv b\ (2g + b_0 - 2)$. □

4. Periodic free degrees and Euler characteristics

The Proposition 3.4 can be seen as the calculation of precise values of $\mathfrak{fr}_{\mathrm{p}}^{+}(F_{g,b})$ for all (g, b) for which $\mathfrak{fr}_{\mathrm{p}}^{+}(F_{g,b}) \geq 3g$. The principal point there was the knowledge of singular orbit lengths and the ramification types of all orientation-preserving periodic self-homeomorphisms of order $N \geq 3g$ of a closed orientable surface of genus $g \geq 2$ from Lemma 3.3. Here we made a step forward, finding similar results provided $\mathfrak{fr}_{\mathrm{p}}^{+}(F_{g,b}) > 2g - 2$. The point is however, that, using the result of Harvey mentioned in Theorem 3.1, we see that between $3g$ and $2g - 2$ there are infinitely many values

(5) $4g + 2 > 4g > 3g + 3 > 3g + 2 > 3g + 1 > 3g > 8g/3 + 4 > 8g/3 + 4/3 > \dots$

each of which can be the the order of self-homeomorphisms of a closed surfaces of some genera g. Observe that the first six elements are mentioned orders of Hirose which can be found in Table 1 in Lemma 3.3. It seems to be difficult to decide, in general, for given rationales α, β, whether the expression $\alpha g + \beta$ appears in (5) or not. But we shall characterize elements of (5) in terms of periodic free degrees or the Euler characteristics of, associated to them, compact bordered surfaces (see Remark 4.6).

LEMMA 4.1. *Let* $g \geq 2, b \geq 1$. *Then* $\mathfrak{fr}_{\mathrm{p}}^{+}(F_{g,b}) \leq 2g - 2 + b$.

PROOF. Let φ be an orientation-preserving periodic self-homeomorphism of $F_{g,b}$, of order N. It is sufficient to show that $\mathfrak{fr}(\varphi) \leq 2g - 2 + b$. Note that $\langle\varphi\rangle \cong \Lambda/\Gamma$, where Γ and Λ are NEC-groups with the signatures

(6) $(g; +; [-]; \{(-), \overset{b}{\dots}, (-)\})$ and $(h; +; [m_1, \dots, m_r]; \{(-), \overset{k}{\dots}, (-)\})$

respectively and the Riemann-Hurwitz formula gives

$$N = \frac{2g - 2 + b}{\mu(\Lambda)}.$$

where

$$\mu(\Lambda) = 2h + k - 2 + \sum_{i=1}^{r}(1 - 1/m_i).$$

Consider the case $r \geq 1$. We may assume that and that $m_1 \geq \dots \geq m_r$. Thus, $\mathfrak{fr}(\varphi) = N/m_1$ by definition. Observe that $k \geq 1$ because $r \geq 1$. So if $h \geq 1$, then $N \leq 2g - 2 + b$ and similar inequality holds for $h = 0$ and $k \geq 3$. For $h = 0, k = 2$, $r \geq 1$ and so $\mu(\Lambda) \geq 1/2$. This in turn gives $N \leq 2(2g - 2 + b)$ and consequently $\mathfrak{fr}_{\mathrm{p}}^{+}(F_{g,b}) = N/m_1 \leq 2g - 2 + b$.

Then let $h = 0, k = 1$. Observe that here $r \geq 2$ since otherwise $\mu(\Lambda) < 0$. But for $r \geq 3$, $\mu(\Lambda) \geq 1/2$ which as before gives our assertion. So finally let $h = 0, k = 1, r = 2$. Then for $m_2 \geq 3$, $\mu(\Lambda) \geq 1/3$. So $N \leq 3(2g - 2 + b)$ and therefore $\mathfrak{fr}(\varphi) = N/m_1 \leq 2g - 2 + b$. Then for $m_2 = 2$, observe that $m_1 > 2$ and $\mu(\Lambda) = (m_1 - 2)/2m_1$ which give $N \leq 2m_1(2g - 2 + b)/(m_1 - 1)$ and in consequence

$$\mathfrak{fr}(\varphi) \leq \frac{N}{m_1} \leq \frac{2}{m_1 - 1}\Big(2g - 2 + b\Big) \leq 2g - 2 + b.$$

Finally, we consider the case $r = 0$. Since φ gives a free action on $F_{g,b}$, we have that

$$\mathfrak{fr}(\varphi) = N = \frac{\chi(F_{g,b})}{\chi(F_{g,b}/\langle\varphi\rangle)} = \frac{2g-2+b}{-\chi(F_{g,b}/\langle\varphi\rangle)} \leq 2g-2+b.$$

□

The above Lemma serves as a good motivation for studying the set $\mathcal{B}_g$ of those b for which $\mathfrak{fr}_{\mathrm{p}}^{+}(F_{g,b}) = 2g-2+b$. What makes such a study more interesting is that the right hand side is the absolute value $-\chi(F_{g,b})$ of the Euler characteristic. Now we shall characterize elements of $\mathcal{B}_g$.

THEOREM 4.2. *Given $g \geq 2, b \geq 1$, $\mathfrak{fr}_{\mathrm{p}}^{+}(F_{g,b}) = -\chi(F_{g,b})$ if and only if $b = 2$ or else there are divisors n_1, n_2, n_3 of $N = 2g-2+b$ and integers $\alpha_1, \alpha_2, \alpha_3$, relatively prime to n_i so that*

(i) $N = \mathrm{lcm}(n_1, n_2, n_3)$,
(ii) $b = N/n_1 + N/n_2 + N/n_3$,
(iii) $\alpha_1(N/n_1) + \alpha_2(N/n_2) + \alpha_3(N/n_3) \equiv 0\ (N)$.

PROOF. To prove the sufficiency, consider first the cyclic action of order $4g$ on a closed surface F_g of genus g, described in the second row of Table 1. It has exactly three singular orbits whose lengths are 1, 1 and $2g$. Removing a small disc centered at points forming orbits of length 1 (being fixed points of the action) in such a way that they are invariant, we obtain a cyclic action of order $4g$ on $F_{g,2}$ with unique singular orbit in its interior which has the length $2g$. So

$$\mathfrak{fr}_{\mathrm{p}}^{+}(F_{g,2}) = 2g = -\chi(F_{g,2}) \tag{7}$$

for arbitrary $g \geq 2$.

Now assume, that conditions (i)-(iii) are satisfied and let Λ be an NEC group with signature

$$(0; +; [-]; \{(-), (-), (-)\}).$$

Let $\theta : \Lambda \to \mathbb{Z}_N = \langle x \rangle$ be an epimorphism which sends all canonical reflection to the identity and $\theta(e_i) = x^{\alpha_i(N/n_i)}$. Observe first that $b \geq 3$, so we have here $N \geq 2g+1$. Then θ is surjective by (i), θ preserve the relation $e_1 \ldots e_k = 1$ by (iii), $\theta(e_i)$ has order n_i since $(\alpha_i, n_i) = 1$ and so each e_i produce N/n_i empty period cycles in $\Gamma = \ker\theta$ by the results in [**3**] and there are no other period cycles in Γ. So by (ii) Γ has b empty period cycles and by the Riemann-Hurwitz formula the orbit genus g as required. Summing up we have constructed $F_{g,b}$ and its freely acting self-homeomorphism φ of order $2g-2+b$ which complete the proof of sufficiency.

The converse is more involved. Let φ be an arbitrary periodic self-homeomorphism of $F_{g,b}$ of order N, let us represent $\langle\varphi\rangle$ as Λ/Γ like in the proof of Lemma 4.1, and let $\theta : \Lambda \to \mathbb{Z}_N$ be the canonical projection. Let n_j be the orders of $\theta(e_j)$ for $j = 1, \ldots, k$ and let the number s be chosen so that

$$n_1 \geq n_2 \geq \cdots \geq n_s > 1 = n_{s+1} = \cdots = n_k.$$

Notice that to prove this implication, it is enough to show that for $M = \mathfrak{fr}_{\mathrm{p}}^{+}(\varphi)$ and $b \neq 2$ we have either:

(a) $M \leq 2g-2$ or
(b) $2g-2 < M < 2g-2+b$ (and additionally, for later needs, we list all such cases explicitly by marked equations) or else

(c) $r = h = 0, k \geq s = 3$.

We find it convenient to prove it for $g = 2$ and $g = 3$, by inspecting Tables 3, 6 and 7 which were obtained independently of this theorem. So let $g \geq 4$ and consider, first the case $r \geq 1$. Then, by definition of free degree, $M = N/m_1$. If $m_1 \geq 4$, then $M \leq (4g+2)/m_1 \leq 2g-2$. If $m_1 = 3$ then $M \leq (4g+2)/3$ which is smaller than $2g-2$ for $g \geq 4$. Finally let $m_1 = 2$. Then from Lemma 3.3 we see that either $N < 3g$ and so $M \leq 2g-2$ or $N = 4g$ or else $N = 4g+2$. If $N = 4g$, then $M = 2g$ and $b \equiv 2\ (4g)$ while if $N = 4g+2$, then $M = 2g+1$ and $b \equiv 3\ (4g+2)$. So for $r \geq 1$ we have to mark only the case

$$M = 2g < 2g - 2 + b \tag{8}$$

for $b > 2, b \equiv 2\ (4g)$ since φ realizing $\mathfrak{fr}_{\mathrm{p}}^{+}(F_{g,b}) = 2g+1$ for $b = 3 + (2g+1)t$ with arbitrary t, can be obtained for $k = t+3, s = 3, N = n_1 = n_2 = n_3 = 2g+1$, by taking $\alpha_1 = \alpha_2 = 1, \alpha_3 = 2g-1$ in the proof of sufficiency.

So, consider the case $r = 0$. Then $N = M$ and φ induces an unbranched N-sheeted covering $F_{g,b} \to F_{g,b}/\langle\varphi\rangle \cong F_{h,k}$. We also have

$$b = N(k-s) + \sum_{i=1}^{s} \frac{N}{n_i}. \tag{9}$$

Let $\bar{F}_{g,b}$ be the closed marked surface obtained by shrinking each boundary component to one point, and $\bar{\varphi}$ for the map on $\bar{F}_{g,b}$ induced by φ. Applying the Riemann-Hurwitz formula to the branched covering $\bar{F}_{g,b} \to \bar{F}_{g,b}/\langle\bar{\varphi}\rangle \cong F_h$, we obtain that

$$\frac{2g-2}{N} = 2h - 2 + \sum_{j=1}^{k}\left(1 - \frac{1}{n_j}\right) = 2h - 2 + \sum_{j=1}^{s}\left(1 - \frac{1}{n_j}\right). \tag{10}$$

Thus if $h \geq 2$, then $\mathfrak{fr}(\varphi) \leq g-1$ while for $h = 1$, $\mathfrak{fr}(\varphi) \leq 2g-2$ except for the case

$$0 < \sum_{j=1}^{s}\left(1 - \frac{1}{n_j}\right) < 1.$$

But since any n_j is a positive integer, the above happens if and only if $s = 1$ and $n_1 \geq 2$. This means that $\bar{\varphi}$ has unique singular orbit on $\bar{F}_{g,b}$, which is an impossible case by (3) of Theorem 3.1.

So we assume that the $h = 0$. The equality (10) implies that

$$N = \frac{2g-2}{\sum_{j=1}^{s}\left(1 - \frac{1}{n_j}\right) - 2}. \tag{11}$$

Since $N > 2g-2$, we have that

$$0 < \sum_{j=1}^{s}\left(1 - \frac{1}{n_j}\right) - 2 < 1.$$

Note that for each j, $1/2 < (1 - 1/n_j) < 1$ because n_j is an integer greater than 1. It follows that s must be $3, 4$ or 5. This also implies that $k \geq 3$.

For $s = 5$ assume first that $n_5 \geq 3$. Then by (11), $N \leq (2g-2)/(5\times(2/3)-2) = 3(g-1)/2$. So let $n_5 = 2$. If $n_4 \geq 3$, then by (11), $N \leq (2g-2)/(4\times(2/3)+1/2-2) = 12(g-1)/7 < 2g-2$.

If $n_5 = n_4 = 2$ and $n_3 \geq 3$, then by (11), $N \leq (2g-2)/(3\times(2/3)+2\times(1/2)-2) = 2g-2$. So let $n_5 = n_4 = n_3 = 2$. If $n_2 = 3$ then $n_1 = 6$ and $N = 2g-2$ while for $n_2 \geq 4$, $N \leq 2g-2$ again by (11). This finish the case $s = 5$ since for $n_2 = 2$, we would have $n_1 = 1$ as it is the order of the product of four elements of order 2 in Z_N.

Now let $s = 4$. If $n_4 \geq 4$ then by (11), $N \leq (2g-2)/(4\times(3/4)-2) = 2g-2$. So let $n_4 = 3$. If $n_3 \geq 5$ then by (11), $N \leq (2g-2)/(3\times(4/5)+2/3-2) = 15(g-1)/8 < 2g-2$. So let $n_4 = 3, n_3 = 4$. If $n_2 \geq 5$ then $N \leq 120(g-1)/61 < 2g-2$ while for $n_2 = 4$ we have $n_1 \geq 6$, so $N \leq 2g-2$.

Now let $n_4 = n_3 = 3$. If $n_2 \geq 6$ or $n_2 = 5$ and $n_1 \geq 6$ then $N \leq 2g-2$. Observe however that for $n_2 = n_1 = 5$, we obtain $15 = N = 15(g-1)/7$ and hence $g = 8$. Furthermore, by (9), $b \equiv 16\ (15)$ and $b \geq 16$. So

$$N = 15 < 30 \leq 2g-2+b. \tag{12}$$

Next for $n_2 = 4$, we have either $n_1 = 12$ and so $N = 2g-2$ or $n_1 = 4$ which means that $g = 6$, $b \equiv 14\ (12)$ and $b \geq 14$, by (9). So

$$N = 12 < 24 \leq 2g-2+b. \tag{13}$$

Now for $n_4 = n_3 = n_2 = 3$, we have $n_1 = 3 = N$ also, which give $N = 3(g-1)$ and so $g = 2$.

So finally let $n_4 = 2$. Then for $n_3 \geq 6$, $N \leq 2g-2$ by (11). For $n_3 = 5$, if $n_2 \geq 7$, then (11) gives that $N \leq 140(g-1)/71 < 2g-2$. Let $n_4 = 2$, $n_3 = 5$ and $n_2 = 6$, then $n_1 = 30$, and hence $N = 20(g-1)/11 < 2g-2$. We have similar situation for $n_2 = 7, 6$. Let $n_4 = 2$ and $n_3 = n_2 = 5$, then $n_1 = 10$. It follows that $N = 2g-2$. So let $n_4 = 2$, $n_3 = 4$. If now, $n_2 \geq 8$, then $N \leq 2g-2$. While for $n_2 = 5$ we have $n_1 = 20$ and therefore $N \leq 2g-2$ by (11). Finally for $n_2 = 4$, $N = 4$ which is impossible since in such case the product of three elements of Z_4 of order 4 would have order 2.

Now let $n_4 = 2, n_3 = 3$. We consider n_2 according to the value $(n_2, 6)$ which can be $1, 2, 3$ or 6, and gives respectively $n_1 = N = 6n_2, 3n_2, 2n_2$ or n_2. Assume that $(n_2, 6) = 1$, the equality (11) gives $N = 12n_2(g-1)/7(n_2-1)$ which in our case is bounded above by $2g-2$ except for $n_2 = 5$ which give $g = 15$, $b \equiv 32\ (30)$ and $b \geq 32$, by (9). So

$$N = 30 < 60 \leq 2g-2+b. \tag{14}$$

For $(n_2, 6) = 2$, $b = N(k-2) - N(7n_2-8)/6n_2$ which give $N = 12n_2(g-1)/(7n_2-8)$ which in our case is bounded above by $2g-2$ except for $n_2 = 4$ which give $g = 6$ and like in (13), $b \equiv 14\ (12)$ and $b \geq 14$. So

$$N = 12 < 24 \leq 2g-2+b. \tag{15}$$

For $(n_2, 6) = 3$, $b = N(k-2) - N(7n_2-9)/6n_2$ which give $N = 12n_2(g-1)/(7n_2-9)$ which in our case is bounded above by $2g-2$ except for $n_2 = 3$ which give $g = 3$, $b \equiv 8\ (6)$ and $b \geq 8$ and so

$$N = 6 < 12 \leq 2g-2+b. \tag{16}$$

Finally for $(n_2, 6) = 6$, $b = N(k-2) - N(7n_2 - 12)/6n_2$ which give $N = 12n_2(g-1)/(7n_2 - 12)$ which in our case is bounded above by $2g-2$ except for $n_2 = 6$. Observe however that this is impossible since the product of two elements of order 6 in Z_6 is an element of order 3 while the product of two elements of order 3 can not be an element of order 2.

Now, let $n_4 = n_3 = 2$. Then $n_2 = n_1 = n$. So $N = n$ or $N = 2n$ if n is even or odd. Here $b = N(k-2) - N(n-2)/n$ and so $N = 2n(g-1)/(n-2)$. Thus $N = 2g$ if n is even and so for arbitrary g or $N = 2g+2$ if n is odd and so for even g and respectively $2 < b \equiv 2\ (2g)$ and $4 < b \equiv 4\ (2g+2)$. So

$$N = 2g < 2g + 1 \leq 2g - 2 + b \tag{17}$$

and

$$N = 2g + 2 < 2g + 3 \leq 2g - 2 + b. \tag{18}$$

Finally let $s = 3$. Here $\theta(c_i) = 1$ and we have $\theta(e_i) = \varphi^{(N/n_i)\alpha_i}$ for $i = 1, 2, 3$ and some α_i and clearly (i)-(iii) are satisfied. □

In (7) we have seen that $\mathfrak{fr}_{\mathrm{p}}^{+}(F_{g,2}) = 2g$ for arbitrary g. We shall need also

COROLLARY 4.3. *For arbitrary $g \geq 2$, we have*

$$\mathfrak{fr}_{\mathrm{p}}^{+}(F_{g,3}) = 2g + 1 \tag{19}$$

$$\mathfrak{fr}_{\mathrm{p}}^{+}(F_{g,4}) = 2g + 2 \tag{20}$$

PROOF. For the first equality take for arbitrary $g \geq 2$, $b = 3, N = n_1 = n_2 = n_3 = 2g + 1$ and $\alpha_1 = \alpha_2 = 1$ and $\alpha_3 = 2g - 1$ while for the second take $b = 4, N = n_1 = n_2 = 2g + 2, n_3 = g + 1$ and $\alpha_1 = \alpha_2 = 1$ and $\alpha_3 = 2g + 1$ and observe that they satisfy (i)-(iii). □

As a corollary from the proof of Theorem 4.2 we obtain our principal result.

THEOREM 4.4. *Let $g \geq 2$ and $\mathfrak{fr}_{\mathrm{p}}^{+}(F_{g,b}) > 2g-2$. Then $b' = \mathfrak{fr}_{\mathrm{p}}^{+}(F_{g,b}) - 2g + 2 \in \mathcal{B}_g$ which means $\mathfrak{fr}_{\mathrm{p}}^{+}(F_{g,b'}) = -\chi(F_{g,b'})$ and in addition $\mathfrak{fr}_{\mathrm{p}}^{+}(F_{g,b}) = \mathfrak{fr}_{\mathrm{p}}^{+}(F_{g,b'})$, except for $g = 8$ and certain $b \equiv 1\ (15)$ which are given in the* Remark 4.5 *for which $\mathfrak{fr}_{\mathrm{p}}^{+}(F_{8,b}) = 15$*

PROOF. Let $\mathfrak{fr}_{\mathrm{p}}^{+}(F_{g,b}) > 2g - 2$. Then, in terms of the proof of Theorem 4.2, (g, b) is described in (8),(12)-(18) or it is obtained for $h = 0, r = 0, k \geq s = 3$ considered at the very end of the proof. Observe that the case (17) "covers" cases (8), (13)-(16) in the sense that for any of them, a self-homeomorphism of a surface $F_{g,b}$ obtained there $\mathfrak{fr}_{\mathrm{p}}^{+}(\varphi)$ can be obtained from (17). And so we have to deal with the cases (12), (17) and (18). We have two possibilities for the order N of a self-homeomorphism φ of $F_{g,b}$ obtained in (12) and (18):

$$\text{(a)}\ N < \mathfrak{fr}_{\mathrm{p}}^{+}(F_{g,b}) \quad \text{or} \quad \text{(b)}\ N = \mathfrak{fr}_{\mathrm{p}}^{+}(F_{g,b})$$

The (a) means that a configuration of φ realizing $\mathfrak{fr}_{\mathrm{p}}^{+}(F_{g,b})$ can be obtained for $h = 0, r = 0, k \geq s = 3$ since, as we just have mentioned, (12), (17) and (18) exhaust all cases for which $2g - 2 < N < 2g - 2 + b$ and in addition they are disjoint. So consider (b) and assume first (12) in which $N = 2g = \mathfrak{fr}_{\mathrm{p}}^{+}(F_{g,b})$. Then $b' = 2$. But (12) itself does not provide a self-homeomorphism of a surface $F_{g,2}$ of order $2g$. Fortunately we have that $\mathfrak{fr}_{\mathrm{p}}^{+}(F_{g,2}) = 2g$ for arbitrary g by (7). Next observe that for (12), $b' = 1$ and so for $(8, b)$ for which $N = 15 = \mathfrak{fr}_{\mathrm{p}}^{+}(F_{8,b})$ we obtain the mentioned exceptions. Finally for (18) we have $b' = 4$. But for φ constructed in (18), $b > 4$. Fortunately we have $\mathfrak{fr}_{\mathrm{p}}^{+}(F_{g,4}) = -\chi(F_{g,4})$ for arbitrary g by (20).

So let $k \geq s = 3$ and let $\mathfrak{fr}_{\mathrm{p}}^{+}(F_{g,b})$ be defined by integers n_1, n_2, n_3 and denote $\mathfrak{fr}_{\mathrm{p}}^{+}(F_{g,b})$ by N. Then

$$b = N(k-3) + \frac{N}{n_1} + \frac{N}{n_2} + \frac{N}{n_3}$$

and we have also $N(k-2) = 2g - 2 + b$ by the Riemann-Hurwitz formula (4). So $N(k-3) = b - b'$ which gives therefore

$$b' = \frac{N}{n_1} + \frac{N}{n_2} + \frac{N}{n_3}$$

and clearly (i)-(iii) are satisfied which means that $\mathfrak{fr}_{\mathrm{p}}^{+}(F_{g,b'}) = -\chi(F_{g,b'})$. □

REMARK 4.5. To find a formula for all j for which $\mathfrak{fr}_{\mathrm{p}}^{+}(F_{8,15j+16}) = 15$ seems to be difficult task. The first few values obtained by means of computer are: $0, 1, 3, 5, 7, 8, 9, 11, 12, 13, 17, 19, 20, 21, 23, 24, 25, 27, 29$. Observe that it seems to be practically impossible to list them all even if we know that $\mathfrak{fr}_{\mathrm{p}}^{+}(F_{8,b})$ is a periodic function of b since $M(8) = 171360$ by Table 5.

REMARK 4.6. We see that, the triples (n_1, n_2, n_3) from Theorem 4.2 are in bijective correspondence with the elements of the set $\mathcal{B}_g$. On the other side they are in such correspondence with the numbers

$$\frac{2n_1n_2n_3}{n_1n_2n_3 - n_1n_2 - n_1n_3 - n_2n_3}(g-1) \tag{21}$$

being, the bigger than $2g-2$, orders of self-homeomorphisms of a closed topological surface of some genus g which are precisely elements of the sequence (5). This gives a promised characterization of its elements in terms of the Euler characteristic of certain bordered surfaces F; namely those whose free degree coincide with its absolute value. The sets $\mathcal{B}_g$ for $g = 2, \ldots, 10$ are given in Table 11.

5. Periodic property of periodic free degree

Fix a genus $g \geq 2$. If we are given ramification type of all orientation-preserving periodic self-homeomorphisms of F_g, we can compute out $\mathfrak{fr}_{\mathrm{p}}^{+}(F_{g,b})$ for any boundary component number b. Next lemma gives that the free degree $\mathfrak{fr}_{\mathrm{p}}^{+}(F_{g,b})$ is eventually periodic according to the number b of boundary components.

THEOREM 5.1. *Let $M(g)$ be the least common multiple of the periods of all orientation-preserving periodic self-homeomorphisms of F_g, $S(g)$ be the maximum of the sum of lengths of singular orbits for all orientation-preserving periodic self-homeomorphisms of F_g. If $b' \equiv b \mod M(g)$, and $b', b \geq S(g)$, then $\mathfrak{fr}_{\mathrm{p}}^{+}(F_{g,b'}) = \mathfrak{fr}_{\mathrm{p}}^{+}(F_{g,b})$.*

PROOF. Suppose that $b' = b + kM(g)$ for some non-negative integer k. By definition of $\mathfrak{fr}_{\mathrm{p}}^{+}(F_{g,b})$, there is a periodic self-homeomorphism $\varphi\colon F_{g,b} \to F_{g,b}$ such that φ has also free degree $\mathfrak{fr}_{\mathrm{p}}^{+}(F_{g,b})$, i.e. φ^j is fixed point free for $j = 1, 2, \ldots, \mathfrak{fr}_{\mathrm{p}}^{+}(F_{g,b}) - 1$. By definition of $M(g)$, the period order n of φ is a divisor of $M(g)$. Pick $kM(g)/n$ regular orbits of φ, and choose a small disc centered at each of these $kM(g)$ points so that the union of these $kM(g)$ discs is an invariant set under action φ. Removing all of $kM(g)$ discs, we obtain the surface $F_{g,b+kM(g)} = F_{g,b'}$. The restriction of φ has also free degree $\mathfrak{fr}_{\mathrm{p}}^{+}(F_{g,b})$. It follows that $\mathfrak{fr}_{\mathrm{p}}^{+}(F_{g,b'}) \geq \mathfrak{fr}_{\mathrm{p}}^{+}(F_{g,b})$.

On the other hand, by definition of $\mathfrak{fr}_{\mathrm{p}}^{+}(F_{g,b'})$, there is a periodic self-homeomorphism $\varphi\colon F_{g,b'} \to F_{g,b'}$ such that φ has free degree $\mathfrak{fr}_{\mathrm{p}}^{+}(F_{g,b'})$. Shrinking each boundary component of $F_{g,b'}$ into one point, we shall obtain a closed surface F_g. The self-homeomorphism φ induces a periodic self-homeomorphism φ' on F_g. The shrunk points may lie in a singular orbit of φ'. Since $b \geq S(g)$, the resulted points in singular orbits are no more than b. In other word, there are at least $b' - b = kM(g)$ shrunk points which are in regular orbits. Go back to the original self-homeomorphism φ and shrink $kM(g)$ boundary components of $F_{g,b'}$, we obtain a compact surface $F_{g,b}$. The self-homeomorphism φ' reduced by φ has also free degree $\mathfrak{fr}_{\mathrm{p}}^{+}(F_{g,b})$. It follows that $\mathfrak{fr}_{\mathrm{p}}^{+}(F_{g,b}) \geq \mathfrak{fr}_{\mathrm{p}}^{+}(F_{g,b'})$. □

LEMMA 5.2. *The sum of singular orbit lengths of any orientation-preserving periodic self-homeomorphism of a closed surface F_g of genus $g \geq 2$ have an upper bound.*

PROOF. Let φ be a periodic self-homeomorphism of F_g of order N and having ramification type $(m_1, \ldots, m_r)$. By Riemann-Hurwitz formula, we have

$$\frac{2-2g}{N} = 2 - 2h - \sum_{i=1}^{r}\left(1 - \frac{1}{m_i}\right),$$

where h is the genus of $F_g/\langle\varphi\rangle$ Hence,

$$\sum_{i=1}^{r}\left(1 - \frac{1}{m_i}\right) = \frac{2g-2}{N} + 2 - 2h.$$

Since $1 < m_i \leq N$ for each i, we have that $1 - 1/m_i \geq 1/2$. Note that $N \geq 2$ and $h \geq 0$. So $r/2 \leq (2g-2)/2 + 2$ and therefore $r \leq 2g+2$. Since $N \leq 4g+2$, we have a bound

$$\sum_{i=1}^{r} \frac{N}{m_i} \leq (4g+2)(2g+2)$$

for the sum of singular orbit lengths. □

6. Computation results for small genera

The following lemma can be get, from classification of conformal actions on surfaces of genera 2, 3 given by Broughton [**5**] and genus 4 obtained independently by Bogopolski [4] and by Kimura [**10**] and for $g = 5$ by Bartolini in [**2**].

PROPOSITION 6.1. *The orientation-preserving periodic self-homeomorphisms of F_g of order N with $3 \geq N < 3g$ for $g = 2, 3, 4, 5$ with different singular orbit lengths are listed in* Tables 3 *and* 4*, where h is the genus of quotient surface, and L is the sum of singular orbit lengths.* □

As in Theorem 5.1, we write $M(g)$ for the least common multiplication of the orders of all orientation-preserving periodic self-homeomorphisms of F_g, $S(g)$ for the maximum of the sum of length of singular orbits for all periodic self-homeomorphisms of F_g. From a direct computation, we obtain Table 5 with $S(g)$ and $M(g)$ for small values of g.

By Theorem 5.1, we know that $\mathfrak{fr}_{\mathrm{p}}^{+}(F_{g,b'}) = \mathfrak{fr}_{\mathrm{p}}^{+}(F_{g,b})$ if $b' \equiv b\ (M(g))$ and $b, b' > S(g)$. Thus, for any $g \geq 2$, the sequence $\{\mathfrak{fr}_{\mathrm{p}}^{+}(F_{g,b})\}_{b=0}^{M(g)+S(g)}$ tell us all of information of $\mathfrak{fr}_{\mathrm{p}}^{+}(F_{g,b})$.

Genus	Order	h	L	Singular orbit lengths	Ramification type
2	10	0	8	[1, 2, 5]	[10, 5, 2]
	8	0	6	[1, 1, 4]	[8, 8, 2]
	6	0	10	[2, 2, 3, 3]	[3, 3, 2, 2]
	6	0	4	[1, 1, 2]	[6, 6, 3]
	5	0	3	[1, 1, 1]	[5, 5, 5]
	4	0	6	[1, 1, 2, 2]	[4, 4, 2, 2]
	3	0	4	[1, 1, 1, 1]	[3, 3, 3, 3]
3	14	0	10	[1, 2, 7]	[14, 7, 2]
	12	0	8	[1, 3, 4]	[12, 4, 3]
	12	0	8	[1, 1, 6]	[12, 12, 2]
	9	0	5	[1, 1, 3]	[9, 9, 3]
	8	0	4	[1, 1, 2]	[8, 8, 4]
	7	0	3	[1, 1, 1]	[7, 7, 7]
	6	0	8	[1, 2, 2, 3]	[6, 3, 3, 2]
	6	0	8	[1, 1, 3, 3]	[6, 6, 2, 2]
	4	0	8	[1, 1, 2, 2, 2]	[4, 4, 2, 2, 2]
	4	0	4	[1, 1, 1, 1]	[4, 4, 4, 4]
	4	1	4	[2, 2]	[2, 2]
	3	0	5	[1, 1, 1, 1, 1]	[3, 3, 3, 3, 3]
	3	1	2	[1, 1]	[3, 3]
4	18	0	12	[1, 2, 9]	[18, 9, 2]
	16	0	10	[1, 1, 8]	[16, 16, 2]
	15	0	9	[1, 3, 5]	[15, 5, 3]
	12	0	6	[1, 1, 4]	[12, 12, 3]
	12	0	6	[1, 2, 3]	[12, 6, 4]
	10	0	4	[1, 1, 2]	[10, 10, 5]
	10	0	14	[2, 2, 5, 5]	[5, 5, 2, 2]
	9	0	3	[1, 1, 1]	[9, 9, 9]
	8	0	10	[1, 1, 4, 4]	[8, 8, 2, 2]
	6	0	12	[1, 2, 3, 3, 3]	[6, 3, 2, 2, 2]
	6	0	6	[1, 1, 2, 2]	[6, 6, 3, 3]
	6	0	6	[1, 1, 1, 3]	[6, 6, 6, 2]
	6	0	12	[2, 2, 2, 3, 3]	[3, 3, 3, 2, 2]
	6	1	4	[2, 2]	[3, 3]
	5	0	4	[1, 1, 1, 1]	[5, 5, 5, 5]
	4	1	2	[1, 1]	[4, 4]
	4	0	10	[1, 1, 2, 2, 2, 2]	[4, 4, 2, 2, 2, 2]
	4	0	6	[1, 1, 1, 1, 2]	[4, 4, 4, 4, 2]
	3	0	6	[1, 1, 1, 1, 1, 1]	[3, 3, 3, 3, 3, 3]
	3	1	3	[1, 1, 1]	[3, 3, 3]
	3	2	0	[–]	[–]

TABLE 3. Cyclic actions of order ≥ 3 on surfaces of genera $2, 3, 4$

Genus	Order	h	L	Singular orbit lengths	Ramification type
5	22	0	14	$[1,2,11]$	$[22,11,2]$
	20	0	12	$[1,1,10]$	$[20,20,2]$
	15	0	7	$[1,1,5]$	$[15,15,3]$
	12	0	4	$[1,1,2]$	$[12,12,6]$
	11	0	3	$[1,1,1]$	$[11,11,11]$
	10	0	12	$[1,1,5,5]$	$[10,10,2,2]$
	8	0	8	$[1,1,2,4]$	$[8,8,4,2]$
	8	1	8	$[4,4]$	$[2,2]$
	6	0	16	$[2,2,3,3,3,3]$	$[3,3,2,2,2,2]$
	6	0	10	$[1,2,2,2,3]$	$[6,3,3,3,2]$
	6	0	10	$[1,1,2,3,3]$	$[6,6,3,2,2]$
	6	0	4	$[1,1,1,1]$	$[6,6,6,6]$
	6	1	4	$[2,2]$	$[3,3]$
	5	1	2	$[1,1]$	$[5,5]$
	4	0	8	$[1,1,1,1,2,2]$	$[4,4,4,4,2,2]$
	4	0	14	$[1,1,2,2,2,2,2,2]$	$[4,4,2,2,2,2,2,2]$
	4	1	4	$[1,1,2]$	$[4,4,2]$
	4	1	8	$[2,2,2,2]$	$[2,2,2,2]$
	4	2	0	$[-]$	$[-]$
	3	0	7	$[1,1,1,1,1,1,1]$	$[3,3,3,3,3,3,3]$
	3	1	4	$[1,1,1,1]$	$[3,3,3,3]$

TABLE 4. Cyclic actions of order ≥ 3 on surfaces of genus 5

g	2	3	4	5	6	7	8	9	10
$S(g)$	10	10	14	16	18	18	22	24	26
$M(g)$	120	504	720	1320	65520	5040	171360	287280	277200

TABLE 5. $M(g)$ and $S(g)$ for small g

By checking directly the sequence $\{\mathfrak{fr}_{\mathrm{p}}^{+}(F_{g,b})\}_{b=0}^{M(g)+S(g)}$, we obtain that

$$
\begin{aligned}
&\mathfrak{fr}_{\mathrm{p}}^{+}(F_{2,b'}) = \mathfrak{fr}_{\mathrm{p}}^{+}(F_{2,b}), && \text{if } b' \equiv b\ (120),\\
&\mathfrak{fr}_{\mathrm{p}}^{+}(F_{3,b'}) = \mathfrak{fr}_{\mathrm{p}}^{+}(F_{3,b}), && \text{if } b' \equiv b\ (504) \text{ and } b',b>0,\\
&\mathfrak{fr}_{\mathrm{p}}^{+}(F_{4,b'}) = \mathfrak{fr}_{\mathrm{p}}^{+}(F_{4,b}), && \text{if } b' \equiv b\ (720) \text{ and } b',b>0,\\
&\mathfrak{fr}_{\mathrm{p}}^{+}(F_{5,b'}) = \mathfrak{fr}_{\mathrm{p}}^{+}(F_{5,b}), && \text{if } b' \equiv b\ (1320) \text{ and } b',b>1.
\end{aligned}
\tag{22}
$$

The excluded cases are: $\mathfrak{fr}_{\mathrm{p}}^{+}(F_{3,0}) = 2$ while $\mathfrak{fr}_{\mathrm{p}}^{+}(F_{3,504}) = 4$; $\mathfrak{fr}_{\mathrm{p}}^{+}(F_{4,0}) = 3$ while $\mathfrak{fr}_{\mathrm{p}}^{+}(F_{4,720}) = 6$; $\mathfrak{fr}_{\mathrm{p}}^{+}(F_{5,0}) = 4$ while $\mathfrak{fr}_{\mathrm{p}}^{+}(F_{5,1320}) = 8$; $\mathfrak{fr}_{\mathrm{p}}^{+}(F_{5,1}) = 2$ while $\mathfrak{fr}_{\mathrm{p}}^{+}(F_{5,1321}) = 3$. The sequences $\{\mathfrak{fr}_{\mathrm{p}}^{+}(F_{2,b})\}_{b=1}^{120}$, $\{\mathfrak{fr}_{\mathrm{p}}^{+}(F_{3,b})\}_{b=1}^{504}$, $\{\mathfrak{fr}_{\mathrm{p}}^{+}(F_{4,b})\}_{b=1}^{720}$ and $\{\mathfrak{fr}_{\mathrm{p}}^{+}(F_{5,b})\}_{b=1}^{1340}$ are given in Tables 6–10 (to be read left to right and then up to down).

	1	2	3	4	5	6	7	8	9	10	11	12	13	14	15	16	17	18	19	20
1	2	4	5	6	2	8	3	10	2	6	2	2	5	8	2	6	2	10	3	2
2	2	8	5	2	3	4	2	10	2	8	3	2	5	6	2	2	3	10	2	6
3	2	4	5	2	2	8	2	10	3	4	2	6	5	8	3	2	2	10	2	2
4	3	8	5	6	2	4	3	10	2	8	2	2	5	4	2	6	2	10	3	2
5	2	6	5	2	3	8	2	10	2	4	3	2	5	8	2	2	3	10	2	6
6	2	8	5	2	2	6	2	10	3	8	2	6	5	4	3	2	2	10	2	2

TABLE 6. $\{\mathfrak{fr}_{\mathrm{p}}^{+}(F_{2,b})\}_{b=1}^{120}$

	1	2	3	4	5	6	7	8	9	10	11	12	13	14	15	16	17	18	19	20
1	3	6	7	8	9	2	2	12	2	14	3	8	3	9	2	4	7	2	2	12
2	2	2	9	14	3	6	2	8	3	2	7	12	2	2	3	8	3	14	2	4
3	9	2	2	12	7	2	3	4	3	9	2	14	3	2	2	12	2	2	9	8
4	3	6	2	4	3	14	2	12	2	2	3	4	7	6	2	8	9	2	2	14
5	2	2	3	8	3	9	7	4	3	2	2	12	2	14	9	4	3	6	2	8
6	7	2	2	12	2	2	3	14	3	6	2	4	9	2	7	12	2	2	3	4
7	3	14	2	8	3	2	2	12	7	2	9	8	3	6	2	14	3	2	2	12
8	2	2	7	4	3	6	2	8	9	14	2	12	2	2	3	8	7	9	2	4
9	3	2	2	14	2	2	9	4	3	6	7	8	3	2	2	12	2	14	3	8
10	3	6	2	4	9	2	2	12	2	2	3	14	3	9	2	8	3	2	7	12
11	2	2	9	8	3	14	2	4	3	2	2	12	7	2	3	4	3	6	2	14
12	9	2	2	12	2	2	7	8	3	9	2	4	3	14	2	12	2	2	9	4
13	7	6	2	8	3	2	2	14	2	2	3	8	3	6	7	4	9	2	2	12
14	2	14	3	4	3	9	2	8	7	2	2	12	2	2	9	14	3	6	2	4
15	3	2	7	12	2	2	3	4	3	14	2	8	9	2	2	12	7	2	3	8
16	3	9	2	14	3	2	2	12	2	2	9	4	3	6	2	8	3	14	2	12
17	2	2	3	8	7	6	2	4	9	2	2	14	2	2	3	4	3	9	7	8
18	3	2	2	12	2	14	9	8	3	6	2	4	7	2	2	12	2	2	3	14
19	3	6	2	8	9	2	7	12	2	2	3	8	3	14	2	4	3	2	2	12
20	7	2	9	4	3	6	2	14	3	2	2	12	2	2	7	8	3	6	2	4
21	9	14	2	12	2	2	3	4	7	9	2	8	3	2	2	14	2	2	9	8
22	3	6	7	4	3	2	2	12	2	14	3	4	3	6	2	8	9	2	2	12
23	2	2	3	14	3	9	2	4	3	2	7	12	2	2	9	4	3	14	2	8
24	3	2	2	12	7	2	3	8	3	6	2	14	9	2	2	12	2	2	7	4
25	3	9	2	8	3	14	2	12	2	2,	9	8	7	6	2	4	3	2	2	14
26	2	2	3	4.																

TABLE 7. $\{\mathfrak{fr}_{\mathrm{p}}^{+}(F_{3,b})\}_{b=1}^{504}$

	1	2	3	4	5	6	7	8	9	10	11	12	13	14	15	16	17	18	19	20
1	3	8	9	10	2	12	2	2	15	16	2	18	2	10	3	3	2	12	5	2
2	9	4	2	15	2	16	3	2	5	18	3	2	3	10	2	6	2	4	15	2
3	2	16	2	10	3	4	2	18	5	8	3	2	2	15	2	2	9	16	5	6
4	3	4	3	10	2	18	2	2	15	4	2	6	2	16	9	3	2	12	5	2
5	3	8	2	18	2	4	3	2	5	16	3	2	9	10	2	6	2	8	15	2
6	2	18	2	10	3	16	2	6	5	4	9	2	2	15	2	2	3	4	5	18
7	3	16	3	10	2	12	2	2	15	8	2	6	2	10	3	3	2	18	5	2
8	3	4	2	15	2	8	9	2	5	12	3	2	3	16	2	18	2	4	15	2
9	2	12	2	10	9	4	2	6	5	16	3	2	2	18	2	2	3	8	5	6
10	3	4	9	10	2	16	2	2	15	4	2	18	2	10	3	3	2	12	5	2
11	9	16	2	15	2	4	3	2	5	18	3	2	3	10	2	6	2	16	15	2
12	2	12	2	10	3	8	2	18	5	4	3	2	2	16	2	2	9	4	5	6
13	3	8	3	10	2	18	2	2	15	16	2	6	2	10	9	3	2	12	5	2
14	3	4	2	18	2	16	3	2	5	12	3	2	9	10	2	6	2	4	15	2
15	2	18	2	10	3	4	2	6	5	8	9	2	2	15	2	2	3	16	5	18
16	3	4	3	10	2	12	2	2	15	4	2	6	2	16	3	3	2	18	5	2
17	3	8	2	15	2	4	9	2	5	16	3	2	3	10	2	18	2	8	15	2
18	2	12	2	10	9	16	2	6	5	4	3	2	2	18	2	2	3	4	5	6
19	3	16	9	10	2	12	2	2	15	8	2	18	2	10	3	3	2	16	5	2
20	9	4	2	15	2	8	3	2	5	18	3	2	3	16	2	6	2	4	15	2
21	2	12	2	10	3	4	2	18	5	16	3	2	2	15	2	2	9	8	5	6
22	3	4	3	10	2	18	2	2	15	4	2	6	2	10	9	3	2	12	5	2
23	3	16	2	18	2	4	3	2	5	12	3	2	9	10	2	6	2	16	15	2
24	2	18	2	10	3	8	2	6	5	4	9	2	2	16	2	2	3	4	5	18
25	3	8	3	10	2	12	2	2	15	16	2	6	2	10	3	3	2	18	5	2
26	3	4	2	15	2	16	9	2	5	12	3	2	3	10	2	18	2	4	15	2
27	2	16	2	10	9	4	2	6	5	8	3	2	2	18	2	2	3	16	5	6
28	3	4	9	10	2	12	2	2	15	4	2	18	2	16	3	3	2	12	5	2
29	9	8	2	15	2	4	3	2	5	18	3	2	3	10	2	6	2	8	15	2
30	2	12	2	10	3	16	2	18	5	4	3	2	2	15	2	2	9	4	5	6
31	3	16	3	10	2	18	2	2	15	8	2	6	2	10	9	3	2	16	5	2
32	3	4	2	18	2	8	3	2	5	12	3	2	9	16	2	6	2	4	15	2
33	2	18	2	10	3	4	2	6	5	16	9	2	2	15	2	2	3	8	5	18
34	3	4	3	10	2	16	2	2	15	4	2	6	2	10	3	3	2	18	5	2
35	3	16	2	15	2	4	9	2	5	12	3	2	3	10	2	18	2	16	15	2
36	2	12	2	10	9	8	2	6	5	4	3	2	2	18	2	2	3	4	5	6

TABLE 8. $\{\mathfrak{ft}_{\mathrm{p}}^{+}(F_{4,b})\}_{b=1}^{720}$

	1	2	3	4	5	6	7	8	9	10	11	12	13	14	15	16	17	18	19	20
1	2	10	11	12	2	2	15	8	2	6	2	20	3	22	2	12	5	2	3	4
2	2	15	2	8	11	2	5	12	2	2	3	20	2	6	2	22	15	2	2	12
3	2	10	3	4	2	6	11	8	3	2	2	20	2	2	3	8	5	22	2	4
4	3	10	2	12	2	2	15	4	11	6	2	20	3	2	2	12	5	2	3	22
5	2	15	2	4	3	2	5	12	2	2	11	20	2	6	2	8	15	2	2	12
6	2	22	3	8	2	6	5	4	3	2	2	20	11	2	3	4	5	6	2	8
7	3	10	2	22	2	2	15	8	2	6	2	20	3	2	11	12	5	2	3	4
8	2	15	2	8	3	22	5	12	2	2	3	20	2	6	2	4	15	2	2	12
9	2	10	3	4	2	6	5	22	3	2	2	20	2	2	3	8	5	6	11	4
10	3	10	2	12	2	2	15	4	2	22	2	20	3	2	2	12	5	2	3	8
11	11	15	2	4	3	2	5	12	2	2	3	22	2	6	2	8	15	2	2	12
12	2	10	11	8	2	6	5	4	3	2	2	20	2	22	3	4	5	6	2	8
13	3	10	2	12	11	2	15	8	2	6	2	20	3	2	2	22	5	2	3	4
14	2	15	2	8	3	2	11	12	2	2	3	20	2	6	2	4	15	22	2	12
15	2	10	3	4	2	6	5	8	11	2	2	20	2	2	3	8	5	6	2	22
16	3	10	2	12	2	2	15	4	2	6	11	20	3	2	2	12	5	2	3	8
17	2	22	2	4	3	2	5	12	2	2	3	20	11	6	2	8	15	2	2	12
18	2	10	3	22	2	6	5	4	3	2	2	20	2	2	11	4	5	6	2	8
19	3	10	2	12	2	22	15	8	2	6	2	20	3	2	2	12	11	2	3	4
20	2	15	2	8	3	2	5	22	2	2	3	20	2	6	2	4	15	2	11	12
21	2	10	3	4	2	6	5	8	3	22	2	20	2	2	3	8	5	6	2	4
22	11	10	2	12	2	2	15	4	2	6	2	22	3	2	2	12	5	2	3	8
23	2	15	11	4	3	2	5	12	2	2	3	20	2	22	2	8	15	2	2	12
24	2	10	3	8	11	6	5	4	3	2	2	20	2	2	3	22	5	6	2	8
25	3	10	2	12	2	2	15	8	2	6	2	20	3	2	2	12	5	22	3	4
26	2	15	2	8	3	2	5	12	11	2	3	20	2	6	2	4	15	2	2	22
27	2	10	3	4	2	6	5	8	3	2	11	20	2	2	3	8	5	6	2	4
28	3	22	2	12	2	2	15	4	2	6	2	20	11	2	2	12	5	2	3	8
29	2	15	2	22	3	2	5	12	2	2	3	20	2	6	11	8	15	2	2	12
30	2	10	3	8	2	22	5	4	3	2	2	20	2	2	3	4	11	6	2	8
31	3	10	2	12	2	2	15	22	2	6	2	20	3	2	2	12	5	2	11	4
32	2	15	2	8	3	2	5	12	2	22	3	20	2	6	2	4	15	2	2	12
33	11	10	3	4	2	6	5	8	3	2	2	22	2	2	3	8	5	6	2	4
34	3	10	11	12	2	2	15	4	2	6	2	20	3	22	2	12	5	2	3	8
35	2	15	2	4	11	2	5	12	2	2	3	20	2	6	2	22	15	2	2	12
36	2	10	3	8	2	6	11	4	3	2	2	20	2	2	3	4	5	22	2	8
37	3	10	2	12	2	2	15	8	11	6	2	20	3	2	2	12	5	2	3	22
38	2	15	2	8	3	2	5	12	2	2	11	20	2	6	2	4	15	2	2	12
39	2	22	3	4	2	6	5	8	3	2	2	20	11	2	3	8	5	6	2	4
40	3	10	2	22	2	2	15	4	2	6	2	20	3	2	11	12	5	2	3	8
41	2	15	2	4	3	22	5	12	2	2	3	20	2	6	2	8	15	2	2	12
42	2	10	3	8	2	6	5	22	3	2	2	20	2	2	3	4	5	6	11	8
43	3	10	2	12	2	2	15	8	2	22	2	20	3	2	2	12	5	2	3	4
44	11	15	2	8	3	2	5	12	2	2	3	22	2	6	2	4	15	2	2	12
45	2	10	11	4	2	6	5	8	3	2	2	20	2	22	3	8	5	6	2	4

TABLE 9. $\{\mathfrak{fr}_{\mathrm{p}}^{+}(F_{5,b})\}_{b=1}^{900}$

	1	2	3	4	5	6	7	8	9	10	11	12	13	14	15	16	17	18	19	20
46	3	10	2	12	11	2	15	4	2	6	2	20	3	2	2	22	5	2	3	8
47	2	15	2	4	3	2	11	12	2	2	3	20	2	6	2	8	15	22	2	12
48	2	10	3	8	2	6	5	4	11	2	2	20	2	2	3	4	5	6	2	22
49	3	10	2	12	2	2	15	8	2	6	11	20	3	2	2	12	5	2	3	4
50	2	22	2	8	3	2	5	12	2	2	3	20	11	6	2	4	15	2	2	12
51	2	10	3	22	2	6	5	8	3	2	2	20	2	2	11	8	5	6	2	4
52	3	10	2	12	2	22	15	4	2	6	2	20	3	2	2	12	11	2	3	8
53	2	15	2	4	3	2	5	22	2	2	3	20	2	6	2	8	15	2	11	12
54	2	10	3	8	2	6	5	4	3	22	2	20	2	2	3	4	5	6	2	8
55	11	10	2	12	2	2	15	8	2	6	2	22	3	2	2	12	5	2	3	4
56	2	15	11	8	3	2	5	12	2	2	3	20	2	22	2	4	15	2	2	12
57	2	10	3	4	11	6	5	8	3	2	2	20	2	2	3	22	5	6	2	4
58	3	10	2	12	2	2	15	4	2	6	2	20	3	2	2	12	5	22	3	8
59	2	15	2	4	3	2	5	12	11	2	3	20	2	6	2	8	15	2	2	22
60	2	10	3	8	2	6	5	4	3	2	11	20	2	2	3	4	5	6	2	8
61	3	22	2	12	2	2	15	8	2	6	2	20	11	2	2	12	5	2	3	4
62	2	15	2	22	3	2	5	12	2	2	3	20	2	6	11	4	15	2	2	12
63	2	10	3	4	2	22	5	8	3	2	2	20	2	2	3	8	11	6	2	4
64	3	10	2	12	2	2	15	22	2	6	2	20	3	2	2	12	5	2	11	8
65	2	15	2	4	3	2	5	12	2	22	3	20	2	6	2	8	15	2	2	12
66	11	10	3	8	2	6	5	4	3	2	2	22	2	2	3	4	5	6	2	8
67	3	10	11	12	2	2	15	8	2	6	2	20	3	22	2	12	5	2	3	4

TABLE 10. $\{\mathfrak{ft}_{\mathrm{p}}^{+}(F_{5,B})\}_{b=901}^{1340}$

g	$\mathcal{B}_g$	g	$\mathcal{B}_g$
2	$\{0,2,3,4,6,8\}$	7	$\{2,3,4,6,8,9,12,16,18\}$
3	$\{2,3,4,5,8,10\}$	8	$\{2,3,4,6,10,18,20\}$
4	$\{2,3,4,6,9,10,12\}$	9	$\{2,3,4,5,8,11,12,14,20,22\}$
5	$\{2,3,4,7,12,14\}$	10	$\{2,3,4,6,7,10,12,15,22,24\}$
6	$\{2,3,4,5,6,8,10,11,14,16\}$		

TABLE 11. $\mathcal{B}_g$ for small g

Acknowledgements

The authors grateful to W. Marzantowicz for turning attention of the first author to the paper [**19**], to W. Harvey and J. Gilman, who turned out the attention to an unpublished thesis of M. Chas [**6**] after talk on the results of this paper during the Conference to which these Proceedings are devoted and to J. Wu for commenting out results of [**6**].

References

[1] Norman L. Alling and Newcomb Greenleaf, *Foundations of the theory of Klein surfaces*, Lecture Notes in Mathematics, Vol. 219, Springer-Verlag, Berlin-New York, 1971. MR0333163 (48 #11488)

[2] G. Bartolini, *On the Branch Loci of Moduli Spaces of Riemann Surfaces of Low Genera*, Ph.D. Thesis University of Linköping (2009).

[3] Emilio Bujalance, José J. Etayo, José M. Gamboa, and Grzegorz Gromadzki, *Automorphism groups of compact bordered Klein surfaces*, Lecture Notes in Mathematics, vol. 1439, Springer-Verlag, Berlin, 1990. A combinatorial approach. MR1075411 (92a:14018)

[4] O. V. Bogopol′skiĭ, *Classifying the actions of finite groups on orientable surfaces of genus 4 [translation of Proceedings of the Institute of Mathematics, 30 (Russian), 48–69, Izdat. Ross. Akad. Nauk, Sibirsk. Otdel., Inst. Mat., Novosibirsk, 1996]*, Siberian Adv. Math. **7** (1997), no. 4, 9–38. Siberian Advances in Mathematics. MR1604157 (2000e:30083)

[5] S. Allen Broughton, *Classifying finite group actions on surfaces of low genus*, J. Pure Appl. Algebra **69** (1991), no. 3, 233–270, DOI 10.1016/0022-4049(91)90021-S. MR1090743 (92b:57021)

[6] M. Chas, Minimum periods of homeomorphisms of orientable surfaces, PhD thesis, Universitat Autonoma de Barcelona, 1998, 141+iii pp., see http://arxiv.org/abs/1204.0023.

[7] Warren Dicks and Jaume Llibre, *Orientation-preserving self-homeomorphisms of the surface of genus two have points of period at most two*, Proc. Amer. Math. Soc. **124** (1996), no. 5, 1583–1591, DOI 10.1090/S0002-9939-96-03131-0. MR1301020 (96g:55004)

[8] W. J. Harvey, *Cyclic groups of automorphisms of a compact Riemann surface*, Quart. J. Math. Oxford Ser. (2) **17** (1966), 86–97. MR0201629 (34 #1511)

[9] Susumu Hirose, *On periodic maps over surfaces with large periods*, Tohoku Math. J. (2) **62** (2010), no. 1, 45–53, DOI 10.2748/tmj/1270041026. MR2654302 (2011c:57052)

[10] Hideyuki Kimura, *Classification of automorphism groups, up to topological equivalence, of compact Riemann surfaces of genus 4*, J. Algebra **264** (2003), no. 1, 26–54, DOI 10.1016/S0021-8693(03)00138-8. MR1980684 (2004c:14055)

[11] Ravi S. Kulkarni, *Riemann surfaces admitting large automorphism groups*, Extremal Riemann surfaces (San Francisco, CA, 1995), Contemp. Math., vol. 201, Amer. Math. Soc., Providence, RI, 1997, pp. 63–79, DOI 10.1090/conm/201/02610. MR1429195 (98g:30070)

[12] A. M. Macbeath, *The classification of non-euclidean plane crystallographic groups*, Canad. J. Math. **19** (1967), 1192–1205. MR0220838 (36 #3890)

[13] Jakob Nielsen, *Fixed point free mappings* (Danish), Mat. Tidsskr. B. **1942** (1942), 25–41. MR0013308 (7,137c)

[14] Shi Cheng Wang, *Maximum orders of periodic maps on closed surfaces*, Topology Appl. **41** (1991), no. 3, 255–262, DOI 10.1016/0166-8641(91)90008-A. MR1135102 (92k:57023)

[15] Shi Cheng Wang, *Free degrees of homeomorphisms and periodic maps on closed surfaces*, Topology Appl. **46** (1992), no. 1, 81–87, DOI 10.1016/0166-8641(92)90041-W. MR1177165 (93g:55003)

[16] H. C. Wilkie, *On non-Euclidean crystallographic groups*, Math. Z. **91** (1966), 87–102. MR0185013 (32 #2483)

[17] A. Wiman, Über die hyperelliptischen Kurven und diejenigen vom Geschlecht $p = 3$, welche eindeutige Transformationen in sich zulassen, *Bihang Till. Kongl. Svenska Vetenskaps-Akademiens Handlingar* **21** (1) (1895), 23 pp.

[18] Jianchun Wu and Qiang Zhang, *A note on the free degrees of homeomorphisms on genus 2 orientable compact surfaces*, Topology Appl. **159** (2012), no. 12, 2841–2844, DOI 10.1016/j.topol.2012.04.017. MR2942656

[19] Jianchun Wu and Xuezhi Zhao, *Free degrees of homeomorphisms on compact surfaces*, Algebr. Geom. Topol. **11** (2011), no. 4, 2437–2452, DOI 10.2140/agt.2011.11.2437. MR2835235 (2012h:55002)

INSTITUTE OF MATHEMATICS, GDAŃSK UNIVERSITY, POLAND
E-mail address: grom@mat.ug.edu.pl

DEPARTMENT OF MATHEMATICS, CAPITAL NORMAL UNIVERSITY, PEOPLE'S REPUBLIC OF CHINA
E-mail address: zhaoxve@mail.cnu.edu.cn

p-adic origamis

Frank Herrlich

ABSTRACT. An origami is a finite covering of a torus which is ramified over only one point. Origamis can be defined algebraically over an arbitrary field. In these notes, after a short reminder of complex origamis, we focus on origamis over p-adic fields with special emphasis on those that can be represented by Mumford curves. These *p-adic origamis*, at least those which are normal coverings of the torus, have been classified by K. Kremer. The main goal of this paper is to give a little background on p-adic uniformization and thus to introduce the reader to Kremer's work, the main results of which are summarized in the last section.

1. Introduction

These notes aim at giving an introduction to the interplay between complex and p-adic origamis for people who are familiar with Riemann surfaces, but not so much with their p-adic counterpart, the Mumford curves.

"Classical" complex origamis are a special class of translation surfaces, often also called "square-tiled surfaces". They have been studied a lot during the last 15 years from quite different points of view, e.g. by Gutkin-Judge [**6**], Lochak [**14**] (who coined the name "origami"), Schmithüsen [**18**], Hubert-Lelièvre [**11**], Eskin-Kontsevich-Zorich [**3**], Matheus-Möller-Yoccoz [**15**] and many others. They have a very simple combinatorial description by gluing squares, see Section 2, but nevertheless encode a very rich and deep structure. In particular, every origami determines an algebraic curve in moduli space, a so called *Teichmüller curve.* Each of these curves contains point representing Riemann surfaces which are (as complex algebraic curves) defined over a number field, and thus also determine a projective curve over the field $\mathbb{C}_p$, the completion of the algebraic closure of the field of p-adic numbers.

Over p-adic fields like $\mathbb{C}_p$ and $\mathbb{Q}_p$, there is a meaningful notion of analytic functions and of analytic spaces. But in contrast to the complex case there is no uniformization theorem for one-dimensional p-adic analytic manifolds. At least the easy part of the uniformization theorem holds: any quotient of an open subset of the projective line $\mathbb{P}^1$ by a discontinuously acting group of Möbius transformations is an analytic manifold. These are considered as the analogues of Riemann surfaces.

If such a quotient is compact it carries, as in the complex case, the structure of a

2010 *Mathematics Subject Classification.* Primary 30F10, 30G06, 14G22, 14H30, 32G15, 20E08.

nonsingular projective algebraic curve. But unlike the classical complex situation, not every nonsingular projective algebraic curve over $\mathbb{C}_p$ can be obtained this way. An algebraic curve over $\mathbb{C}_p$ that can be represented as quotient of an open subset of $\mathbb{P}^1(\mathbb{C}_p)$ by a discontinuous subgroup of $\mathrm{PGL}_2(\mathbb{C}_p)$, is called a *Mumford curve.* We shall explain this concept from different points of view in Sections 4, 5 and 6.

We call an origami over $\mathbb{C}_p$ a *p-adic origami* if it is a covering of Mumford curves. The task of describing all p-adic origamis has been done to a large extent by Karsten Kremer in his PhD thesis [**12**] and the subsequent paper [**13**]: he succeeded in classifying all normal p-adic origamis. In Section 7, we shall discuss one of his two basic examples in some detail, and then sketch Kremer's general results.

2. Complex origamis

The most elementary definition of a complex origami is the following, which also explains the name "square-tiled surface":

DEFINITION 2.1. An *origami* is a closed surface X that can be obtained from finitely many squares in the euclidean plane by gluing each left side of a square to a right side and each upper side to a lower side in such a way that all gluings are performed by translations in the plane.

On such a surface there is an obvious notion of horizontal (and vertical) cylinders. Numbering the squares by the integers $1, \ldots, d$, the decomposition of X into horizontal (resp. vertical) cylinders corresponds to the cycle decomposition of a permutation σ_h (resp. σ_v) in the symmetric group S_d. That X is connected implies that the subgroup of S_d generated by σ_h and σ_v acts transitively on $\{1, \ldots, d\}$. Conversely every such pair of permutations determines an origami.

A surface X as in Definition 2.1 comes along with a surjective map $p : X \to E$ to the torus $E = \mathbb{R}^2/\mathbb{Z}^2$: on each square of X, the map p identifies the left and the right edge, and the top and the bottom edge. Clearly this is compatible with the gluings that define X. Note that outside the vertices of the squares, p is an unramified covering.

Identifying $\mathbb{R}^2$ with $\mathbb{C}$ in the usual way, the torus $E = \mathbb{C}/(\mathbb{Z} \oplus i\mathbb{Z})$ is not only a topological surface, but also a complex torus and a complex projective nonsingular curve of genus 1. Since translations are in particular holomorphic functions, the gluing rules in Definition 2.1 turn X into a compact Riemann surface and thus also a complex nonsingular projective curve. The map p described above is then a holomorphic map between Riemann surfaces, and a finite morphism between algebraic curves. This observation suggests a definition of origami over more general fields than the complex numbers:

DEFINITION 2.2. Let k be an algebraically closed field. An *origami* over k is a finite morphism $p : X \to E$ between nonsingular projective curves X and E over k such that E has genus 1 and p is ramified over at most one point of E.

REMARK 2.3. 1) To make sure that the morphism p is finite it suffices to require that it be nonconstant.
2) The genus of X is at least 1, and if it is 1 then p is unramified by the Riemann-Hurwitz formula (in this case p is an *isogeny*).
3) The assumption that k is algebraically closed can be dropped if an appropriate notion of "curve over k" is used. In fact we shall be interested later in Riemann

surfaces that "can be defined over a number field K", by which we mean that there is a a scheme X_0 over K from which X is obtained by base change from K to $\mathbb{C}$, i. e. $X = X_0 \otimes_K \mathbb{C}$.

3. Teichmüller curves

As indicated above, origamis are special translation surfaces. We now make this notion precise:

DEFINITION 3.1. A 2-dimensional manifold X with a given atlas of local charts is called a *strict translation surface* if all transition functions between the charts are translations of the euclidean plane.

Clearly, if X is an origami as in Definition 2.1 and $\Sigma \subset X$ is the finite set of vertices of the squares, then $X^* := X \setminus \Sigma$ is a strict translation surface. This observation motivates the following definition:

DEFINITION 3.2. A 2-dimensional manifold X is called a *translation surface* if there is a discrete subset $\Sigma \subset X$ such that $X^* := X \setminus \Sigma$ is a strict translation surface.

REMARK 3.3. Every translation surface is also a Riemann surface. This is obvious for strict translation surfaces since translations are holomorphic maps. The general case follows from the Riemann extension theorem.

REMARK 3.4. The differential dz in the complex plane pulls back via the translation atlas to a holomorphic differential on every translation surface.
Conversely, every translation surface X can be obtained from a Riemann surface X and the choice of a nonzero holomorphic differential ω on X: Let Σ be the set of zeroes of ω, and define charts on simply connected open subsets $U \subset X \setminus \Sigma$ by $P \mapsto \int_{P_0}^{P} \omega$ (for some $P_0 \in U$).

This remark shows that compact translation surfaces of genus $g \geq 1$ are classified by the space ΩM_g of pairs (S, ω), where S is a Riemann surface of genus g and ω is a nonzero holomorphic differential on S. Note that ΩM_g is a vector bundle of rank g on M_g from which the zero section is removed.
There is a natural action of $\mathrm{SL}_2(\mathbb{R})$ on the space ΩT_g of marked translation surfaces: given a matrix $A \in \mathrm{SL}_2(\mathbb{R})$ and a translation surface X, we obtain a new translation surface X_A by postcomposing all chart maps with the $\mathbb{R}$-linear map on $\mathbb{R}^2$ induced by A. If a point X in ΩT_g is considered as a pair (S, ω) with a marked Riemann surface S and a nonzero holomorphic 1-form ω on S, the point X_A corresponds to (S_A, ω_A), where $\omega_A = A \cdot \begin{pmatrix} \mathrm{Re}\,\omega \\ \mathrm{Im}\,\omega \end{pmatrix}$ and S_A is the unique complex structure on the surface S such that ω_A is holomorphic.
For given $X = (S, \omega) \in \Omega T_g$ this action induces a map from $\mathrm{SL}_2(\mathbb{R})$ to T_g by $A \mapsto S_A$. Note that S_A is the same marked Riemann surface as S if and only if $A \in \mathrm{SO}_2(\mathbb{R})$. Thus we obtain an embedding ι_X of $\mathbb{H} = \mathrm{SO}_2(\mathbb{R})\backslash\mathrm{SL}_2(\mathbb{R})$ into T_g. Such embeddings are called *Teichmüller embeddings*, and the image $\Delta_X = \iota_X(\mathbb{H})$ in T_g is called a *Teichmüller disk.*

PROPOSITION 3.5. *Teichmüller embeddings are holomorphic and isometric (for the hyperbolic metric on $\mathbb{H}$ and the Teichmüller metric on T_g).*

PROOF. A proof of this proposition can be found in [9], Thm. 3.4. □

Recall that the mapping class group Mod_g acts properly discontinuously and isometrically on T_g, and the quotient space $M_g = T_g/\mathrm{Mod}_g$ is the moduli space of Riemann surfaces of genus g, which is known to be a quasiprojective variety. We are interested in the images of Teichmüller disks in moduli space. In general, this is a weird (and dense) subset of M_g, but in certain cases it has a much nicer structure:

PROPOSITION 3.6. *Let X be a translation surface of genus $g \geq 1$ and Δ_X the corresponding Teichmüller disk. Denote by $\widetilde{\Gamma}_X$ the stabilizer of Δ_X in* Mod_g *and by Γ_X the factor group of $\widetilde{\Gamma}_X$ by the (finite) pointwise stabilizer of Δ_X.*
Then Γ_X acts as a Fuchsian group on $\mathbb{H} = \Delta_X$, and if Γ_X is a lattice in $\mathrm{SL}_2(\mathbb{R})$ *the following holds:*

(i) *Δ_X/Γ_X is a Riemann surface of finite type.*
(ii) *The image C_X of Δ_X in M_g is birational to Δ_X/Γ_X.*

Moreover $\widetilde{\Gamma}_X$ is isomorphic to the group $\mathrm{Aff}^+(X)$ *of orientation preserving affine diffeomorphisms of X, and Γ_X is isomorphic to the image of* $\mathrm{Aff}^+(X)$ *in* $\mathrm{SL}_2(\mathbb{R})$, *i. e. the group of linear parts of the affine diffeomorphisms.*

PROOF. For a proof of most of the claims in this proposition see [9]. □

The group Γ_X in the proposition is called the *Veech group* of the translation surface X. If Γ_X is a lattice in $\mathrm{SL}_2(\mathbb{R})$, the proposition implies that Δ_X/Γ_X and C_X are complex algebraic curves. In this case C_X, considered as a subvariety of the moduli space M_g, is called a *Teichmüller curve.* As already observed by Veech [20], Teichmüller curves are never projective, hence always have "cusps" at the boundary of M_g.
For an arbitrary translation surface X, C_X is rarely a Teichmüller curve. But for origamis, the situation is different:

PROPOSITION 3.7. *For any origami $p : X \to E$ of genus g, C_X is a Teichmüller curve in M_g. More precisely, the Veech group Γ_X is a finite index subgroup of* $\mathrm{SL}_2(\mathbb{Z})$.

This was already observed by Thurston and Veech in the late 80's. Gutkin and Judge [6] showed that the converse also holds: any translation surface whose Veech group is commensurable to $\mathrm{SL}_2(\mathbb{Z})$, is an origami. A nice proof of the proposition and moreover a very useful characterization of the Veech group of an origami in terms of automorphisms of the free group F_2 of rank 2 can be found in G. Schmithüsen's thesis [18].

PROPOSITION 3.8. *For any origami X, the Teichmüller curve C_X is defined over a number field.*

This follows from the previous proposition and Belyi's theorem: Since the only elliptic fixed points of $\mathrm{SL}_2(\mathbb{Z})$ in $\mathbb{H}$ are the orbits of i and of $\rho = e^{2\pi i/3}$, the modular map $j : \mathbb{H} \to \mathbb{H}/\mathrm{SL}_2(\mathbb{Z}) = \mathbb{C}$ is ramified only above two points (namely 0 and 1728). Since Γ_X is a finite index subgroup of $\mathrm{SL}_2(\mathbb{Z})$, it induces a finite covering $\mathbb{H}/\Gamma_X \to \mathbb{H}/\mathrm{SL}_2(\mathbb{Z})$ which can be extended to a finite covering $q : \overline{\mathbb{H}/\Gamma_X} \to \mathbb{P}^1(\mathbb{C})$ of compact Riemann surfaces. q is ramified over at most three points (0, 1728 and ∞). By Belyi's theorem this implies that $\overline{\mathbb{H}/\Gamma_X}$ and hence also C_X is defined over a number field.

Möller [16], Thm. 5.1 has shown that every Teichmüller curve can be defined over a number field. The proof of this result requires quite different techniques.

4. p-adic uniformization and Mumford curves

A key point in the classical theory of Riemann surfaces is the uniformization theorem which implies that every compact Riemann surface of genus $g \geq 2$ can be conformally represented as the quotient of the upper half plane $\mathbb{H}$ by a discontinuous and free action of a Fuchsian group. More generally, any compact Riemann surface can be obtained in many ways as the quotient of an open subdomain of the Riemann sphere $\mathbb{C} \cup \{\infty\} = \hat{\mathbb{C}}$, which we identify with the projective line $\mathbb{P}^1(\mathbb{C})$, by a discontinuous group of Möbius transformations.

In this section we discuss to what extent an analog of this uniformization is possible over a p-adic field. We begin quite elementary by recalling the construction of the fields $\mathbb{Q}_p$ and $\mathbb{C}_p$:

4.1. p-adic valuation and p-adic fields. The *p-adic valuation* of an integer $a \neq 0$ is defined as $v_p(a) = n$ if a can be written as $a = p^n \cdot a'$, where p does not divide a'. In other words $v_p(a) = n$ if p^n divides a, but p^{n+1} does not. For a nonzero rational number $z = \frac{a}{b}$ we define $v_p(z) = v_p(a) - v_p(b)$, and the *p-adic absolute value* as $|z|_p = p^{-v_p(z)}$. Extending the map $|.|_p$ to all of $\mathbb{Q}$ by $|0|_p = 0$ we obtain a *norm* $|.|_p : \mathbb{Q} \to \mathbb{R}$. Besides the usual properties of a norm ($|x|_p = 0$ if and only if $x = 0$, $|xy|_p = |x|_p \cdot |y|_p$), it satisfies the "ultrametric" triangle inequality $|x+y|_p \leq \max(|x|_p, |y|_p)$. As do all norms, $|.|_p$ induces a metric on $\mathbb{Q}$ by $d_p(x,y) = |x-y|_p$. This metric is non-archimedean in the sense that

$$d_p(x,z) \leq \max(d_p(x,y), d_p(y,z)) \text{ for all } x, y, z \text{ in } \mathbb{Q}.$$

Recall that any non-archimedean metric implies the following "simplified" geometry:
(a) Any triangle is isosceles, i. e. has two sides of equal length; the third side then also has the same length or is shorter.
(b) Any two disks $B_{r_i}(x_i) = \{y |\, d_p(x_i, y) < r_i\}$, $i = 1, 2$, are either disjoint, or one of them is contained in the other.

One very important property of a non-archimedean norm $|\,.\,|$ on a field k is that the closed unit disk $\mathbb{D} = \overline{B}_1(0) = \{x \in k |\, |x| \leq 1\}$ is a ring $\mathcal{O}_k$, called the *valuation ring* of $|\,.\,|$. This holds because for x, y with $|x| \leq 1$ and $|y| \leq 1$ we also have $|x+y| \leq 1$ and $|-x| = |x| \leq 1$. Note that $\mathcal{O}_k$ is a local ring with the unique maximal ideal $\mathfrak{m} = \{x \in k |\, |x| < 1\}$. The quotient field $\kappa = \mathcal{O}_k/\mathfrak{m}$ is called the *residue field*; for $k = \mathbb{Q}$ and $|\,.\,|$ the p-adic absolute value, the residue field is the prime field $\mathbb{F}_p$.

Like for the euclidean norm, also for the p-adic norm $|\,.\,|_p$ the metric space $(\mathbb{Q}, |\,.\,|_p)$ is not complete. Denote by $\mathbb{Q}_p$ the completion of $\mathbb{Q}$ with respect to $|\,.\,|_p$. It has the following properties:

REMARK 4.1. (i) $\mathbb{Q}_p$ is a field which contains $\mathbb{Q}$ as a subfield.
(ii) The p-adic norm on $\mathbb{Q}$ extends to an absolute value $|\,.\,|_p$ on $\mathbb{Q}_p$ which takes on the same values as on $\mathbb{Q}$, namely the integer powers of p.
(iii) $\mathbb{Q}_p$ is complete w. r. t. $|\,.\,|_p$.

The second property holds because, due to the non-archimedean triangle inequality, every Cauchy sequence either converges to 0 or else has constant absolute value for sufficiently large n.

Since the p-adic norm on $\mathbb{Q}_p$ has no other values than on $\mathbb{Q}$, $\mathbb{Q}_p$ cannot be as close to an algebraically closed field as $\mathbb{R}$, the completion of $\mathbb{Q}$ w. r. t. the euclidean norm: e. g. not all the roots $p^{1/k}$ of p can be contained in one single finite field extension of $\mathbb{Q}_p$.
On the other hand, the residue field of $\mathbb{Q}_p$ is still $\mathbb{F}_p$. The valuation ring is $\mathbb{Z}_p$, which can also be obtained as the completion of $\mathbb{Z}$ w. r. t. $|\,.\,|_p$.

Let $\overline{\mathbb{Q}}_p$ be the algebraic closure of $\mathbb{Q}_p$. By standard arguments the absolute value on $\mathbb{Q}_p$ extends in a unique way to an absolute value on $\overline{\mathbb{Q}}_p$, which is still denoted $|\,.\,|_p$. The residue field of $\overline{\mathbb{Q}}_p$ is $\overline{\mathbb{F}}_p$, the algebraic closure of $\mathbb{F}_p$. Unfortunately $\overline{\mathbb{Q}}_p$ is no longer complete for the absolute value $|\,.\,|_p$, so we pass to the completion $\mathbb{C}_p$ of $\overline{\mathbb{Q}}_p$ w. r. t. $|\,.\,|_p$. It turns out that $\mathbb{C}_p$ is the analogue of the complex field $\mathbb{C}$ in the following sense:

PROPOSITION 4.2. (i) *$\mathbb{C}_p$ is a field extension of $\mathbb{Q}$, and the p-adic norm $|\,.\,|_p$ on $\mathbb{Q}$ extends in a unique way to $\mathbb{C}_p$.*
(ii) *$\mathbb{C}_p$ is complete w. r. t. $|\,.\,|_p$ and algebraically closed. The residue field of $\mathbb{C}_p$ is $\overline{\mathbb{F}}_p$.*

The drawback of the big extension from $\mathbb{Q}_p$ to $\mathbb{C}_p$ is that $\mathbb{C}_p$ is no longer locally compact, and the valuation ring is no longer a noetherian ring. Nevertheless, $\mathbb{C}_p$ allows for a theory of analytic functions in the sense of H. Cartan.

4.2. p-adic analytic functions. As in the complex world, convergent power series are the building blocks of analytic functions also over p-adic fields. A power series $\sum_{n\geq 0} a_n z^n$ with coefficients $a_n \in \mathbb{C}_p$ has the usual radius of convergence $r = (\limsup \sqrt[n]{|a_n|})^{-1}$ and thus converges (if at all) on some open disk $B_r(0) = \{z \in \mathbb{C}_p|\, |z| < r\}$. But since p-adic fields are totally disconnected and thus a disk $B_r(0)$ is a disjoint union of disks of radius $r' < r$, the classical definition of a holomorphic function as a function that can locally be expressed as a convergent power series, would not lead to a satisfactory theory: e. g. a locally constant function, which is 0 on some of the smaller disks and 1 on the others, would satisfy this definition.

To obtain a notion which shares the basic properties of complex holomorphic functions (like the identity theorem), one defines the holomorphic functions on a closed disk $\overline{B}_r(a)$ as the power series in $z - a$ that converge on that disk. Due to the non-archimedean valuation this is equivalent to the condition on the coefficients a_n of the series, that $|a_n|r^n$ converges to 0. Thus in particular, the holomorphic functions on $\overline{B}_1(0)$ form the "Tate algebra"

$$\mathbb{C}_p\!<\!z\!>= \{\sum_{n\geq 0} a_n z^n |\, a_n \to 0\}.$$

Similarly the holomorphic functions on $\overline{B}_r(a)$ are the elements of $\mathbb{C}_p\!<\!\frac{z-a}{c}\!>$, where $c \in \mathbb{C}_p$ is such that $|c| = r$.

Next we define the holomorphic functions on a domain $D \subset \mathbb{C}_p$ which is a "disk with holes", i. e. of the form

$$D = \overline{B}_r(a) - \bigcup_{i=1}^{m} B_{r_i}(a_i)$$

for some $a_i \in \overline{B}_r(a)$ and $r_i \leq r$ as

$$A(D) = \mathbb{C}_p < \frac{z-a}{c}, \frac{c_1}{z-a_1}, \ldots, \frac{c_m}{z-a_m} >$$

with $|c_i| = r_i$.

Disks with holes are the simplest examples of (one-dimensional) *affinoid domains.* In general, affinoid domains correspond to "affinoid algebras", i. e. quotients of a Tate algebra $\mathbb{C}_p < z_1, \ldots, z_n >$ by certain ideals. The correspondence is such that the affinoid algebra is the ring of holomorphic functions on the affinoid domain. General p-adic analytic varieties are obtained by gluing affinoid domains in an "admissible" way. The admissible coverings in this sense include in particlar the finite ones. Good introductions to the one-dimensional theory can be found in [**5**] and [**4**], the general case is treated in [**1**].

Fortunately, in these notes we do not need more complicated affinoid domains than disks with holes. It should be noted however, that this is due to our limitation to Mumford curves: Exactly as in the complex case, every algebraic variety over $\mathbb{C}_p$ carries a natural structure as analytic variety. In dimension one, i. e. for projective algebraic curves over $\mathbb{C}_p$, it turns out, however, that almost no curve can "admissibly" be covered by disks, as is, in contrast, the case for all complex nonsingular projective curves ("admissibly" here would mean by finitely many of them). The basic insight, which goes back to Mumford [**17**], is that many, but by far not all curves over $\mathbb{C}_p$ can be covered by finitely many disks with holes. We call a projective nonsingular curve over $\mathbb{C}_p$ which, as an analytic variety, can be covered by finitely many disks with holes, a *Mumford curve.* In the following sections, we shall characterize these p-adic analogs of Riemann surfaces in different ways.

4.3. p-adic Schottky groups. The projective line $\mathbb{P}^1(\mathbb{C}_p) = \mathbb{C}_p \cup \{\infty\}$ is an analytic variety which can be covered by two disks. As usual, the group $\mathrm{PGL}_2(\mathbb{C}_p)$ of Möbius transformations acts on $\mathbb{P}^1(\mathbb{C}_p)$ by analytic automorphisms. If $\Gamma \subset \mathrm{PGL}_2(\mathbb{C}_p)$ is a finitely generated subgroup which acts discontinuously on some open subdomain $\Omega \subset \mathbb{P}^1(\mathbb{C}_p)$, then the quotient Ω/Γ inherits a structure of one-dimensional analytic variety exactly in the same way as it is a Riemann surface in the corresponding complex situation.

There is one kind of classical Kleinian groups that can be defined in the same way over p-adic fields, namely Schottky groups. Instead of recalling Schottky's classical construction we give the p-adic definition right away since it is literally the same:

DEFINITION 4.3. Let $g \geq 1$ and $D_i = \overline{B}_{r_i}(a_i)$, $i = 1, \ldots, 2g$ mutually disjoint closed disks in $\mathbb{C}_p$. Denote by C_i the "boundary" of D_i, i. e. $C_i = \overline{B}_{r_i}(a_i) - B_{r_i}(a_i) = \{z \in \mathbb{C}_p | \, |z - a_i| = r_i\}$ (note that, unlike the complex case, the "center" a_i of D_i is not unique and therefore C_i depends on the choice of a_i).
Choose $\gamma_i \in \mathrm{PGL}_2(\mathbb{C}_p)$, $i = 1, \ldots, g$, such that $\gamma_i(C_i) = C_{i+g}$ and $\gamma_i(D_i) = (\mathbb{P}^1(\mathbb{C}_p) - D_{i+g}) \cup C_{i+g} = \mathbb{P}^1(\mathbb{C}_p) - B_{r_{i+g}}(a_{i+g})$.
Then the subgroup Γ of $\mathrm{PGL}_2(\mathbb{C}_p)$ generated by $\gamma_1, \ldots, \gamma_g$ is called a *p-adic Schottky group.*

Clearly the definition can be extended to allow disks in $\mathbb{P}^1(\mathbb{C}_p)$ that contain ∞. A p-adic Schottky group has the same properties as a complex one:

THEOREM 4.4. *Let $\Gamma \subset \mathrm{PGL}_2(\mathbb{C}_p)$ be a Schottky group as in Definition 4.3. Then the following hold:*

(i) *Γ is a free group of rank g.*

(ii) *Let $F = \mathbb{P}^1(\mathbb{C}_p) - \bigcup_{i=1}^{2g} D_i$ and $\Omega = \bigcup_{\gamma\in\Gamma} \gamma(F)$. Then Γ acts properly discontinuously (and freely) on Ω, and F is a fundamental domain for this action.*

(iii) *The quotient space $X^{\mathrm{an}} = \Omega/\Gamma$ is the analytic space associated with a projective nonsingular algebraic curve X_Γ over $\mathbb{C}_p$ of genus g.*

(iv) *X_Γ is a Mumford curve.*

PROOF. (i) and (ii) are proved exactly in the same way as in the complex situation.

(iii) can be proved by constructing nonconstant Γ-invariant meromorphic functions on Ω, see [**5**]. The original result, in a more algebraic-geometric setting, is due to Mumford [**17**].

To prove (iv), note that X_Γ can be covered by one disk with $2g$ holes F' and $2g$ annuli: For $i = 1, \ldots, 2g$ choose positive real numbers r_i^-, r_i' and r_i^+ satisfying $r_i^- < r_i < r_i' < r_i^+$ such that the closed disks $\overline{B}_{r_i^+}(a_i)$ are still mutually disjoint and the projection map from Ω to X^{an} is injective on each annulus $\Delta_i = \{z \in \mathbb{C}_p \mid r_i^- \leq |z - a_i| \leq r_i^+\}$. Finally let $F' = \mathbb{P}^1(\mathbb{C}_p) - \bigcup_{i=1}^{2g} B_{r_i'}(a_i)$. □

The following result shows that p-adic Schottky groups play a very important role among the discontinuous subgroups of $\mathrm{PGL}_2(\mathbb{C}_p)$:

THEOREM 4.5. *Let $G \subset \mathrm{PGL}_2(\mathbb{C}_p)$ be a finitely generated subgroup which acts discontinuously on some nonempty open subset of $\mathbb{P}^1(\mathbb{C}_p)$. Then G contains a p-adic Schottky group Γ as a subgroup of finite index.*

This theorem can be obtained by looking at the action of G on the Bruhat-Tits tree and using the structure theorem for the fundamental group of a graph of groups from Bass-Serre theory. A different proof is contained in [**5**].

There is a second result that emphasizes the importance of Schottky groups; its proof relies on the results of the following section:

THEOREM 4.6. *Every Mumford curve can be obtained as the quotient of an open dense subset of $\mathbb{P}^1(\mathbb{C}_p)$ by a p-adic Schottky group.*

5. The Bruhat-Tits tree

There is a very useful tree associated with the field of p-adic numbers which at the same time helps to understand the action of $\mathrm{PGL}_2(\mathbb{Q}_p)$ and its subgroups on $\mathbb{P}^1(\mathbb{Q}_p)$ and also the "reduction mod p" of certain subsets of $\mathbb{P}^1(\mathbb{Q}_p)$ and analytic varieties. Algebraically it is a special case of the general concept of the Bruhat-Tits building for a reductive algebraic group over a local field. The group here is $\mathrm{PGL}_2(\mathbb{Q}_p)$, and the building is one-dimensional and turns out to be a tree.

The tree $T = T(\mathbb{Q}_p)$ is defined as follows: the vertex set $V(T)$ is the set of closed disks in $\mathbb{P}^1(\mathbb{Q}_p)$, i. e. $V(T) = \{\overline{B}_r(a) \mid a \in \mathbb{Q}_p), r \in p^{\mathbb{Z}}\}$. There is a (directed) edge between disks D and D' if $D \subset D'$ and D is maximal in D' w.r.t. inclusion, i. e. there is no disk D'' with $D \neq D'' \neq D'$ and $D \subset D'' \subset D'$.

Clearly this graph is connected since any two disks $D = \overline{B}_r(a)$ and $D' = \overline{B}_{r'}(a')$ are contained in a common larger disk, e. g. $\overline{B}_{|a-a'|}(a)$. Moreover, from every vertex

$\overline{B}_r(a)$ there is exactly one directed edge going out, namely the one to $\overline{B}_{r\cdot p}(a)$. Thus a path in T without backtracking can change its direction at most once. From this observation one deduces immediately that T is a tree.

REMARK 5.1. The ends of T correspond bijectively to the elements of $\mathbb{P}^1(\mathbb{Q}_p)$.

PROOF. An end is an equivalence class of rays in T, where "equivalence" in a tree means finite difference. From a given vertex v, there is a unique ray in T starting at v and going up all the time. All these rays (for different v) are equivalent and thus form a single end in T which we let correspond to $\infty \in \mathbb{P}^1(\mathbb{Q}_p)$. All other rays ultimately go down infinitely many times and thus correspond to a sequence of disks $D_1 \supset D_2 \supset \dots$ whose radii tend to 0. Therefore $\bigcap_{i=1}^{\infty} D_i$ is a single point in $\mathbb{Q}_p$, which corresponds to the end represented by the ray $D_1 \supset D_2 \supset \dots$. □

Note that T is a regular tree of valency $p+1$; for the particular vertex $\mathbb{Z}_p = \overline{B}_1(0)$, the p smaller neighbors correspond to the residue classes $p\mathbb{Z}_p, 1+p\mathbb{Z}_p, \dots, (p-1)+p\mathbb{Z}_p$, while the unique larger neighbor is $\overline{B}_p(0) = \overline{B}_{|\frac{1}{p}|}(0)$. Similar descriptions can be given for all other vertices.

Any two elements a, b of $\mathbb{P}^1(\mathbb{Q}_p)$ define an *axis* $A(a,b)$ in T, namely the straight line in T whose two ends correspond to the points a and b. Any three points a, b, c define a unique vertex called the *median* $m(a,b,c)$; it is the intersection of $A(a,b)$, $A(a,c)$ and $A(b,c)$.

PROPOSITION 5.2. $\mathrm{PGL}_2(\mathbb{Q}_p)$ *acts in a way on* T *which is compatible with its action on* $\mathbb{P}^1(\mathbb{Q}_p)$ *(considered as the set of ends of* T*).*

PROOF. For a closed disk D in $\mathbb{Q}_p$ and an element $\gamma \in \mathrm{PGL}_2(\mathbb{Q}_p)$, $\gamma(D)$ is again a closed disk in $\mathbb{Q}_p$ if (and only if) $\gamma^{-1}(\infty) \notin D$. In this case γ maps the vertex D of T to the vertex $\gamma(D)$. Clearly inclusion is preserved by γ, hence edges are mapped to edges.

If $\gamma^{-1}(\infty) \in D$, then $\gamma(D) = \mathbb{P}^1(\mathbb{Q}_p) \setminus B_r(a)$ for some open disk $B_r(a)$; we then map the vertex D to the vertex $\overline{B}_r(a)$. It is an easy exercise to verify that this definition is compatible with inclusions and defines an action of $\mathrm{PGL}_2(\mathbb{Q}_p)$ on T. It is clear from the definition that an element $a \in \mathbb{P}^1(\mathbb{Q}_p)$ corresponding to a ray R in T is mapped by $\gamma \in \mathrm{PGL}_2(\mathbb{Q}_p)$ to $\gamma(a)$, which corresponds to the ray $\gamma(R)$. □

REMARK 5.3. Let $\gamma \in \mathrm{PGL}_2(\mathbb{Q}_p)$ be an element with two fixed points a, b in $\mathbb{P}^1(\mathbb{Q}_p)$.

a) If γ is hyperbolic, then γ acts on the axis $A(\gamma) = A(a,b)$ by nontrivial translation.

b) If γ is elliptic, it fixes the axis $A(a,b)$ pointwise.

PROOF. Conjugating γ with a suitable $\delta \in \mathrm{PGL}_2(\mathbb{Q}_p)$ we may assume $a = 0$ and $b = \infty$. Then $\gamma(z) = \lambda \cdot z$ for some $\lambda \in \mathbb{Q}_p \setminus \{0\}$. γ is hyperbolic iff $|\lambda| \neq 1$, in which case γ acts on $A(0,\infty)$ by translation by $\log_p|\lambda|$. If $|\lambda| = 1$, γ is elliptic, and $\gamma(B_r(0)) = B_r(0)$ for every r. □

In the same way as for $\mathbb{Q}_p$, a tree $T(K)$ can be constructed for every finite field extension $K/\mathbb{Q}_p$. There is a unique way to extend the p-adic valuation to K. The value group of K is a finite extension, say of degree e, of the value group $p^{\mathbb{Z}}$ of $\mathbb{Q}_p$. Thus K is again a field with a discrete valuation, its field of integers $\mathcal{O}_K = \{z \in K |\, |z| \leq 1\}$ is a discrete valuation ring with the unique maximal ideal $\mathfrak{m}_K = \{z \in K |\, |z| < 1\}$. The residue field $\mathcal{O}_K/\mathfrak{m}_K$ is a finite field extension of

$\mathbb{Z}_p/p\mathbb{Z}_p \cong \mathbb{F}_p$, say of degree f. The degree $n = [K : \mathbb{Q}_p]$ of K over $\mathbb{Q}_p$ satisfies the famous formula $n = e \cdot f$.

The tree $T(K)$ arises from $T = T(\mathbb{Q}_p)$ by subdividing each edge into e edges (of equal length), subsequently adding $p^f - p$ edges to each of the original vertices, and finally completing this to a regular tree of valency $p^f + 1$.

We also have to consider infinite field extensions of $\mathbb{Q}_p$ which arise as limits of finite ones, e. g. $\mathbb{C}_p$. In this case one of the numbers e and f, or both of them, tends to ∞. As long as e stays bounded, we still have a simplicial tree $T(K)$, but it is not locally finite if f is not finite. Note that for any finitely generated subfield K of $\mathbb{C}_p$, the degree e of the extension of value groups is finite. If e tends to ∞ the corresponding limit of trees is no longer a simplicial tree, but it exists as an $\mathbb{R}$-tree.

It should be mentioned that there is an alternative way of constructing the tree $T(K)$ (which is in fact the original one): the vertices are homothety classes of $\mathcal{O}_K$-lattices in K^2, and the edges correspond to classes with representatives that are neighbors for inclusion. This approach is used e. g. in Serre's classical book [**19**].

PROPOSITION AND DEFINITION 5.4. *Let $\Gamma \subset \mathrm{PGL}_2(\mathbb{C}_p)$ be a finitely generated discontinuous subgroup and $K \subset \mathbb{C}_p$ the smallest subfield containing the fixed points of all elements of $\Gamma \setminus \{\mathrm{id}\}$. Then*
a) *K is a finitely generated field extension of $\mathbb{Q}_p$.*
b) *There is a minimal nonempty subtree $T(\Gamma)$ of $T(K)$ on which Γ acts; this action is without inversions of edges.*
c) *If Γ is finite, $T(\Gamma)$ is reduced to a single vertex.*
d) *If Γ is infinite, $T(\Gamma)$ is the union of the axes of the hyperbolic elements in Γ.*
e) *$T(\Gamma)/\Gamma$ is a finite graph.*

PROOF. **a)** Let $\gamma_1, \dots, \gamma_n \in \mathrm{GL}_2(\mathbb{C}_p)$ be representatives of generators of Γ and K_0 the subfield of $\mathbb{C}_p$ generated by the matrix entries of the γ_i. K_0 is finitely generated over $\mathbb{Q}_p$ and contains all matrix entries of all representatives of elements of Γ. The fixed points of an element of $\mathrm{PGL}_2(K_0)$ are solutions of quadratic equations over K_0. There is an extension of K_0 of degree 4 which contains the solutions to all quadratic equations over K_0: a quadratic extension of the value group and a quadratic extension of the residue field suffice.
b) follows from c) and d). These in turn follow from the following two lemmas valid in general for group actions on trees:

LEMMA 5.5. *Let γ_1, γ_2 be elliptic elements without common fixed point in T. Then the product $\gamma_1\gamma_2$ is hyperbolic.*

LEMMA 5.6. *Let $\gamma \in \mathrm{Aut}(T)$ and $v \in V(T)$ a vertex of T. Then the unique path in T connecting v and $\gamma(v)$ intersects the axis $A(\gamma)$.*

Lemma 5.5 shows that all elements of a finite subgroup of $\mathrm{PGL}_2(\mathbb{C}_p)$ must have a common fixed point; this proves **c)**.
Lemma 5.6 shows that any nonempty subtree of $T(K)$ on which Γ acts must contain the axes of all hyperbolic elements of Γ. Since their union is already a Γ-invariant subtree, **d)** follows.
e) A finite fundamental domain for the action of Γ on $T(\Gamma)$ is obtained as follows: let $\gamma_1, \dots, \gamma_n$ be generators of Γ; for each i, let F_i be a segment of length l_i of the axis $A(\gamma_i)$, where l_i is the translation length of γ_i. Then the smallest subtree F of $T(\Gamma)$ containing $F_1, \dots, F_n$ is finite and surjects onto $T(\Gamma)/\Gamma$. $\square$

The quotient graph $\bar{T}(\Gamma) = T(\Gamma)/\Gamma$ acquires the structure of a *graph of groups* as follows: For every vertex v of $\bar{T}(\Gamma)$ take a vertex $\tilde{v}$ in $T(\Gamma)$ representing v and let G_v be the stabilizer in Γ of the disk corresponding to $\tilde{v}$; since Γ is discontinuous, G_v is a finite group. For every edge e of $\bar{T}(\Gamma)$ let $\tilde{e}$ be an edge in T representing e and G_e the stabilizer of $\tilde{e}$ in Γ. Let v_1 and v_2 be the endpoints of e and $\tilde{v}_1'$, $\tilde{v}_2'$ the endpoints of $\tilde{e}$; then since Γ acts without inversions on $T(\Gamma)$, G_e is the intersection of the stabilizers of $\tilde{v}_1'$ and $\tilde{v}_2'$ in Γ. These stabilizers are conjugate to G_{v_1} resp. G_{v_2}, and thus G_e corresponds to well defined subgroups of G_{v_1} and G_{v_2}.
These data, the graph $\bar{T}(\Gamma)$ together with the vertex groups G_v, the edge groups G_e and the inclusions $G_e \hookrightarrow G_{v_i}$, constitute a graph of groups. The main theorem of Bass-Serre theory says that the group Γ can be recovered (as an abstract group) from these data as the "fundamental group" of the graph of groups, see **[19]**, Thm. 13. In our situation, Γ can even be recovered as a subgroup of $\mathrm{PGL}_2(K)$ up to conjugation.

Note that if Γ is a Schottky group, all vertex and edge groups are trivial. For an arbitrary discontinuous group, the vertex groups are finite subgroups of $\mathrm{PGL}_2(K)$. Since K is a field of characteristic 0, $\mathrm{PGL}_2(K)$ has the same finite subgroups as $\mathrm{PGL}_2(\mathbb{C})$, namely the cyclic groups $\mathbb{Z}/n\mathbb{Z}$ and the dihedral groups D_n for all $n \geq 1$, and the symmetry groups of the platonic solids, A_4, S_4 and A_5. It turns out that there are not too many possibilities for an edge group to be a proper subgroup of the vertex groups of both its end points. A complete list of all finite graphs of groups that come from finitely generated discontinuous subgroups of $\mathrm{PGL}_2(K)$ can be found in **[8]**.

6. Reduction

Another way of characterizing Mumford curves is via reduction. This can be done either in an algebraic-geometric way (which I shall mention briefly at the end of this section) or in an analytic way which I shall explain now.
The basic idea of analytic reduction is a geometric interpretation of reduction mod p of K-algebras, where K is a p-adic field. In the simplest situation, the geometric object to be reduced is the p-adic unit disk, which is the same as the ring $\mathbb{Z}_p$ of p-adic integers. Its algebraic reduction $\mathbb{Z}_p/p\mathbb{Z}_p \cong \mathbb{F}_p$ is geometrically interpreted as the points of the affine line over $\mathbb{F}_p$, thus in particular an affine variety.

Conceptually this interpretation runs as follows: recall from Section 4.2 that the holomorphic functions on the unit disk are the elements of the Tate algebra $\mathbb{Q}_p < z >$ of power series in the variable z with coefficients that tend to 0. For a power series $f = \sum a_n z^n$ we define its norm as $||f|| = \max\{|a_n| \mid n \geq 0\}$. Let $\mathbb{Q}_p^0 < z >$ be the subring of $\mathbb{Q}_p < z >$ of elements of norm ≤ 1, and $\mathbb{Q}_p^{00} < z >$ the ideal of elements of norm < 1. Then the quotient ring $\mathbb{Q}_p^0 < z > /\mathbb{Q}_p^{00} < z >$ is obviously isomorphic to the polynomial ring $\mathbb{F}_p[z]$, which should be seen as the ring of regular functions on the affine line over $\mathbb{F}_p$. If we replace, as we did in Section 4.2, $\mathbb{Q}_p$ by $\mathbb{C}_p$ we have to replace $\mathbb{F}_p$ by its algebraic closure $\bar{\mathbb{F}}_p$.

In Section 4.2 we have defined Mumford curves as those curves that can be covered by finitely many disks with holes. Therefore we need to understand the reduction of disks with holes.

DEFINITION 6.1. Let $D \subset \mathbb{C}_p$ be a disk with holes and $A(D)$ the $\mathbb{C}_p$-algebra of holomorphic functions on D. Let $||.||$ be the norm on $A(D)$ defined as the maximal

absolute value of the coefficients (where each of the generators is given norm 1). Finally let $A^0(D)$ be the subring of $A(D)$ of elements of norm ≤ 1 and $A^{00}(D)$ the ideal in $A^0(D)$ of elements of norm < 1.
Then $\bar{A}(D) = A^0(D)/A^{00}(D)$ is a finitely generated $\bar{\mathbb{F}}_p$-algebra and hence the ring of regular functions of a unique (up to isomorphism) affine variety $\bar{D}$ over $\bar{\mathbb{F}}_p$. $\bar{D}$ is called the *reduction* of D.

EXAMPLE 6.2. **1)** Let $D = \overline{B}_1(0) \setminus B_1(0) = \{z \in \mathbb{C}_p | \, |z| = 1\}$. Then $A(D) = \mathbb{C}_p < z, \frac{1}{z} >$ and $\bar{A}(D) = \bar{\mathbb{F}}_p[z, \frac{1}{z}]$. Hence $\bar{D} = \bar{\mathbb{F}}_p \setminus \{0\}$.
2) $D = \overline{B}_1(0) \setminus B_{1/p}(0) = \{z \in \mathbb{C}_p | \, \frac{1}{p} \leq |z| \leq 1\}$. Here $A(D) = \mathbb{C}_p < z, \frac{p}{z} >= \mathbb{C}_p < Z_1, Z_2 > /(Z_1 Z_2 - p)$. It follows that $\bar{A}(D) = \bar{\mathbb{F}}_p[Z_1, Z_2]/(Z_1 Z_2)$ is the affine coordinate ring of the union of two lines that intersect in one point (the Z_1-axis and the Z_2-axis in the plane).

In the first example, we removed a disk which is a whole residue class (in the ring $\bar{\mathbb{Z}}_p$ of integers in $\mathbb{C}_p$). The "missing" point in the reduction $\bar{D}$ corresponds to this residue class.
In the second example, the residue class of 0 is not completely removed, but only part of it (namely a residue class mod p^2). This is reflected in the affine variety $\bar{D}$ by "blowing up" the point corresponding to the residue class of 0 to a whole projective line. The points of this projective line correspond to the residue classes mod p^2 which lie within the class of 0 mod p, and one additional point ("∞") corresponding to the elements in $\bar{\mathbb{Z}}_p$ that are not 0 mod p. Out of this projective line one point is missing, namely the one corresponding to the removed disk $B_{1/p}(0)$.

PROPOSITION 6.3. *Let $D \subset \mathbb{C}_p$ be a disk with holes. Then the reduction $\bar{D}$ is a "tree of affine lines" over $\bar{\mathbb{F}}_p$, i. e. an affine variety such that all irreducible components of $\bar{D}$ are affine lines (over $\bar{\mathbb{F}}_p$) and the dual graph is a tree (the vertices of the dual graph are the irreducible components of $\bar{D}$ and an edge is drawn for every point of intersection of two components).*

PROOF. The proof of this proposition is a nice exercise, mainly in classical algebraic geometry. □

The next step is to define the reduction of an analytic variety that can be covered by disks with holes. This will be done by gluing the reductions of the covering sets. It turns out that the reduction depends to some extent on the chosen covering. Since Definition 6.1 carries over literally to more general affinoid domains it is possible to define reductions for arbitrary p-adic analytic varieties, but we shall explain the construction in detail only for varieties that can be covered by disks with holes since this is sufficient in the context of Mumford curves.

REMARK 6.4. Let D_1, D_2 be two disks with holes in $\mathbb{C}_p$. If $D_1 \cap D_2 \neq \emptyset$, the intersection Y is another disk with holes. The inclusions $Y \subset D_i$ induce restriction homomorphisms $A^0(D_i) \to A^0(Y)$ and thus morphisms $\bar{Y} \to \bar{D}_i$ $(i = 1, 2)$ which are in fact inclusions.

We define the reduction of $D_1 \cup D_2$ with respect to the covering $\{D_1, D_2\}$ as the algebraic variety obtained by gluing $\bar{D}_1$ and $\bar{D}_2$ along $\bar{Y}$. Observe that $Z = D_1 \cup D_2$ is itself a disk with holes if $Y \neq \emptyset$ and that the reduction of $\bar{Z}$ of this disk with holes may be different from the reduction with respect to $\{D_1, D_2\}$, as the following example shows:

EXAMPLE 6.5. Let $D_1 = \overline{B}_1(0) \setminus B_{1/p}(0)$ and $D_2 = \overline{B}_{1/p}(0) \setminus B_{1/p^2}(0)$. We have seen above that each of $\bar{D}_1$ and $\bar{D}_2$ is the union of two lines intersecting in one point, and the same holds for the union $D_1 \cup D_2 = \overline{B}_1(0) \setminus B_{1/p^2}(0)$. On the other hand, the intersection is $Y = D_1 \cap D_2 = \{z \in \mathbb{C}_p \mid |z| = \frac{1}{p}\}$; by the first example, $\bar{Y}$ is a line with one point removed. In the glued variety, $\bar{Y}$ is an open part of a line which has one additional point in $\bar{D}_1$ and one in $\bar{D}_2$, hence is a projective line. The reduction thus has 3 irreducible components, two affine lines and one projective line; each of the affine lines intersects the projective line in one point (not both in the same), and there are no other intersections.

DEFINITION 6.6. Let X be a p-adic analytic variety und $\mathfrak{U} = (U_i)_{i \in I}$ an open covering of X by disks with holes. Then we define the *reduction* $\bar{X}$ *of* X *with respect to* $\mathfrak{U}$ to be the gluing of the affine varieties $\bar{U}_i$ along their intersections as above.

REMARK 6.7. If X is an open analytic subset of $\mathbb{C}_p$, $\bar{X}$ is a tree of (affine and) projective lines.

This follows from Proposition 6.3 and a generalization of Example 6.5.

EXAMPLE 6.8. Let $X = \mathbb{C}_p \setminus \{0\}$ and $q \in \mathbb{C}_p$ with $|q| > 1$. Then the annuli $U_i = \{z \in \mathbb{C}_p \mid |q|^i \leq |z| \leq |q|^{i+1}\}$ $(i \in \mathbb{Z})$ cover X. The reduction is an infinite "cyclic" (i.e. 2-regular) tree of projective lines.

This example generalizes to

PROPOSITION 6.9. *Let* $\Gamma \subset \mathrm{PGL}_2(\mathbb{C}_p)$ *be a Schottky group,* F *a fundamental region as in Theorem 4.4 and* Ω *its region of discontinuity; the* $\gamma(F)$, $\gamma \in \Gamma$, *form an open covering of* Ω *by disks with holes. The reduction of* Ω *with respect to this covering is a tree of projective lines whose dual graph is isomorphic to the tree* $T(\Gamma)$ *defined in Proposition 5.4 (up to removing or inserting vertices of order 2).*

COROLLARY 6.10. *Let* X *be a Mumford curve of genus* g *over* $\mathbb{C}_p$. *Then* X *admits a finite open covering* $\mathfrak{U}$ *by disks with holes such that the reduction of* X *with respect to* $\mathfrak{U}$ *is a (singular) projective curve* $\bar{X}$ *over* $\bar{\mathbb{F}}_p$ *of arithmetic genus* g. *All irreducible components of* $\bar{X}$ *are rational curves, and all intersection points are ordinary double points.*

PROOF. Let $\Gamma \subset \mathrm{PGL}_2(\mathbb{C}_p)$ be a Schottky group with domain of discontinuity Ω such that $\Omega/\Gamma = X$. The reduction $\bar{\Omega}$ of Ω has a natural Γ-action, and $\bar{\Omega}/\Gamma$ is the reduction of X with respect to the covering used in the proof of Theorem 4.4 (iv).
Since Γ acts by translation on the irreducible components of $\bar{\Omega}$, identification of points on the same component can occur at most for intersection points with other components. Thus the irreducible components of $\bar{X}$ are projective lines on which possibly finitely many pairs of points are identified (at most g). This shows that all irreducible components of $\bar{X}$ have geometric genus 0. It follows that the arithmetic genus of $\bar{X}$ is just the genus (or first Betti number) of the intersection graph. By the proposition, this graph is (up to subdivision of edges) isomorphic to $\bar{T}(\Gamma)$. As the quotient graph of a tree by the free action of a free group of rank g, $\bar{T}(\Gamma)$ has genus g. □

Projective curves as in the corollary are called "totally degenerate". As mentioned before, reduction can be defined for all p-adic analytic varieties and thus in particular for all projective curves. Then Mumford curves can be characterized as those nonsingular projective curves over $\mathbb{C}_p$ whose reduction is totally degenerate.

The reduction of a variety is not unique. It is e. g. always possible to insert an additional annulus into the intersection of two disks with holes. In the reduction such an annulus corresponds to replacing the intersection point of two components L_1, L_2 by an additional component L which intersects L_1 and L_2 in the respective points where they originally intersected:

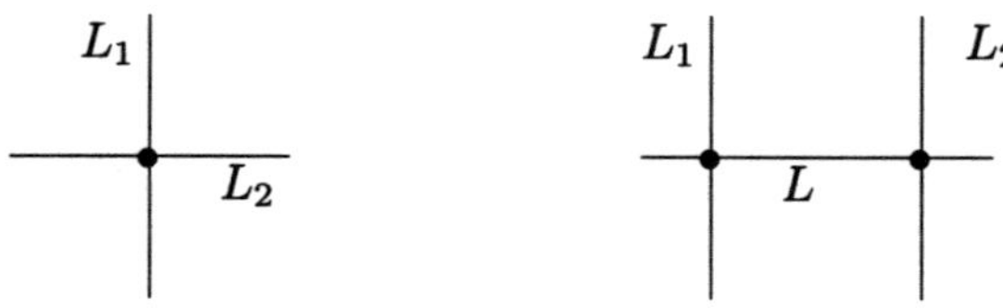

The inverse process is called "contracting the component L". It turns out that reduction is unique up to inserting or contracting components in this way. In particular, there is a unique "stable reduction" where no component can be contracted, i. e. all components are either singular or intersect other components in at least three points.

There is a purely algebraic way of defining the reduction of a projective curve X over $\mathbb{C}_p$: one has to find a "model" $\mathcal{X}$ of X over $\bar{\mathbb{Z}}_p$, the valuation ring of $\mathbb{C}_p$. $\mathcal{X}$ is a scheme over $\bar{\mathbb{Z}}_p$ such that X is obtained from $\mathcal{X}$ by base change: $X = \mathcal{X} \times_{\bar{\mathbb{Z}}_p} \mathbb{C}_p$. The reduction $\bar{X}$ is then defined as the base change with respect to the residue map $\bar{\mathbb{Z}}_p \to \bar{\mathbb{F}}_p$:

$$\bar{X} = \mathcal{X} \times_{\bar{\mathbb{Z}}_p} \bar{\mathbb{F}}_p.$$

Again, the model $\mathcal{X}$ and the reduction $\bar{X}$ are not unique, but there is a unique stable reduction. This follows from the famous work of Deligne-Mumford, see [**7**], Sect. 3C for an introduction. It is well known that the analytic and the algebraic stable reduction of a nonsingular curve agree, see e. g. [**2**].

If X is a plane curve given as the zero set of a homogeneous polynomial $F \in \mathbb{Q}_p[X, Y, Z]$, we may assume that the coefficients of F are in $\mathbb{Z}_p$, but not all in $p\mathbb{Z}_p$. Then the reduction $\bar{F} \in \mathbb{F}_p[X, Y, Z]$ of F mod p determines a projective curve over $\mathbb{F}_p$ which is a reduction of X (in most cases not the stable one).

7. p-adic origamis

In this section we combine the notion of an origami from Section 2 with the concept of Mumford curves. Recall that an origami over the field $\mathbb{C}_p$ is a finite morphism $p : X \to E$ of projective nonsingular curves over $\mathbb{C}_p$ such that E is of genus 1 and p is ramified over (at most) one point.

DEFINITION 7.1. An origami $p : X \to E$ over $\mathbb{C}_p$ is called a *p-adic origami* if X and E are Mumford curves.

As usual, a covering $p : Y \to X$ is called *normal* (or *Galois*) if the group $\mathrm{Deck}(Y/X)$ of deck transformations acts transitively on the fibre $p^{-1}(x)$ for each $x \in X$. In this case, X is the quotient of Y by the subgroup $\mathrm{Deck}(Y/X)$ of the automorphism group $\mathrm{Aut}(Y)$ of Y. In his thesis [**12**], the main results of which are published in [**13**], K. Kremer gave a complete classification and description of all normal p-adic origamis. This section gives a brief account of his work.

7.1. Normal p-adic origamis. The starting point is the following observation:

PROPOSITION 7.2. *Let $p : X \to E$ be a normal p-adic origami, where X is a Mumford curve of genus $g > 1$. Then there is a discontinuous subgroup G of $\mathrm{PGL}_2(\mathbb{C}_p)$ and a normal subgroup Γ of G of finite index which is a Schottky group of rank g, such that $\Omega/G \cong E$ and $\Omega/\Gamma \cong X$, where $\Omega \subset \mathbb{P}^1(\mathbb{C}_p)$ is the region of discontinuity of G (and hence also of Γ).*

PROOF. Since X is a Mumford curve of genus g, by Thm. 4.6 there is a Schottky group $\Gamma \subset \mathrm{PGL}_2(\mathbb{C}_p)$ of rank g with $\Omega(\Gamma)/\Gamma \cong X$. By assumption, E is the quotient of X by a subgroup $\bar{G}$ of its automorphism group $\mathrm{Aut}(X)$. It is well known that $\mathrm{Aut}(X) \cong N(\Gamma)/\Gamma$ where $N(\Gamma)$ is the normalizer of Γ in $\mathrm{PGL}_2(\mathbb{C}_p)$ (cf. [**5**], VII.2). Then the inverse image G of $\bar{G}$ in $N(\Gamma)$ satisfies all the required properties. □

The group G in the proposition cannot be a Schottky group: since E is of genus 1, G would have to be a free group of rank 1, i. e. isomorphic to $\mathbb{Z}$, and thus could not have a free subgroup of rank > 1. This observation corresponds to the fact that p has to be ramified since any unramified covering of an elliptic curve is again an elliptic curve (by the Riemann-Hurwitz formula).
Ramification of the covering $\Omega/\Gamma \to \Omega/G$ occurs precisely in the fixed points of elements of $G \setminus \Gamma$ that lie in Ω. Since fixed points of hyperbolic elements are limit points, G must contain elliptic elements of finite order. The condition that p is ramified over only one point requires that all fixed points in Ω of elliptic elements of G lie in the same G-orbit. This is a serious restriction on G and Γ, as we shall see. But first we give an example that satisfies all conditions:

7.2. An Example. In this subsection we construct a normal p-adic origami whose Galois group is a dihedral group D_n.
Let $n \geq 3$ be an odd integer and $\zeta \in \mathbb{C}_p$ a primitive n-th root of unity. Consider the Möbius transformation $z \mapsto \zeta \cdot z$ as an element of $\mathrm{PGL}_2(\mathbb{C}_p)$, represented by the matrix $\begin{pmatrix} \zeta & 0 \\ 0 & 1 \end{pmatrix}$. Furthermore let $\alpha \in \mathrm{PGL}_2(\mathbb{C}_p)$ be represented by $\begin{pmatrix} 0 & 1 \\ 1 & 0 \end{pmatrix}$, i. e. $\alpha(z) = \frac{1}{z}$. Then δ and α generate a dihedral group of order $2n$, and α exchanges the two fixed points 0 and ∞ of δ.
Next let $\gamma \in \mathrm{PGL}_2(\mathbb{C}_p)$ be a hyperbolic element with fixed points 1 and -1, i. e. with the same fixed points as α. Let G be the subgroup of $\mathrm{PGL}_2(\mathbb{C}_p)$ generated by γ, δ and α. There is an obvious homomorphism $\tau : G \to D_n$, where D_n is the dihedral group $< t, s \mid t^n = s^2 = (ts)^2 = 1 >$: τ is given by $\tau(\delta) = t$, $\tau(\alpha) = s$ and $\tau(\gamma) = 1$. The kernel Γ of τ is the normal subgroup of G generated by γ. Since γ commutes with α, Γ is generated as a group by the n elements $\gamma_i = \delta^{i-1}\gamma\delta^{1-i}$, $i = 1, \dots, n$.
Let $d = \min\{|\zeta^i - 1|, |\zeta^i + 1| \mid i = 1, \dots, n-1\}$. As n is odd, $\zeta^i \neq -1$ for all i and hence $d > 0$; on the other hand $d \leq 1$ since $|\zeta| = 1$.
Being a hyperbolic element, γ is conjugate to $z \mapsto \lambda z$ for some $\lambda \in \mathbb{C}_p$ with $|\lambda| < 1$. Choose γ in such a way that $|\lambda| < d$, and assume for simplicity $p \neq 2$. The generators γ_i of Γ are all conjugate to $z \mapsto \lambda z$, therefore on the tree $T(\Gamma)$, γ_i acts on its axis $A(\gamma_i)$ by translation by $\log_p|\lambda^{-1}|$. Recall that the endpoints of

$A(\gamma_i)$ correspond to the fixed points of γ_i in $\mathbb{P}^1(\mathbb{C}_p)$, which are $\delta^{i-1}(1) = \zeta^{i-1}$ and $\delta^{i-1}(-1) = -\zeta^{i-1}$. By our choice of λ (and p), the intersection of $A(\gamma_i)$ and $A(\gamma_j)$ for $i \neq j$, which is of length at most $-\log_p(d)$, is shorter than the translation length of γ_i (and γ_j).
Let $T_0 \subset T(\Gamma)$ be the union of all intersections $A(\gamma_i) \cap A(\gamma_j)$ for $i \neq j$. Since $|\zeta^{i-1} - (-\zeta^{i-1})| = |2\zeta^{i-1}| = 1$ for all i (here we use $p \neq 2$), the vertex v_0 of $T(K)$ corresponding to the disk $\overline{B}_1(\zeta^{i-1}) = \overline{B}_1(0)$ lies on each $A(\gamma_i)$ and hence T_0 is connected.
Since $|\lambda| < d$ we can find a vertex v_i on $A(\gamma_i) \setminus T_0$ such that the segment from v_i to $\gamma(v_i)$ has nonempty intersection with T_0 (thus in particular contains v_0) and $\gamma(v_i) \notin T_0$. Let D_i and D'_i be the closed disks in $\mathbb{C}_p$ corresponding to the vertices v_i and $\gamma(v_i)$. By construction the $2n$ disks D_i, D'_i, $i = 1, \dots, n$, are mutually disjoint. They thus satisfy the properties of Definition 4.3, which shows that Γ is a Schottky group. As a consequence, G is also discontinuous.
Denote, as usual, by Ω the set of discontinuity of G and Γ. So far we have seen that $X = \Omega/\Gamma$ is a Mumford curve of genus g. Moreover we know (because G is discontinuous) that $E = \Omega/G$ is a Mumford curve, too, and that $p : X \to E$ is a normal covering of degree $2n$ with deck transformation group D_n.
Next we determine the ramification of the covering p. Clearly p is ramified precisely in the fixed points in X of the elements of the deck transformation group. These in turn are the images in X of the fixed points in Ω of the elements of the group generated by δ and α. Since the fixed points of α, and also those of the other involutions in this dihedral group, are limit points of Γ (they are fixed points of the hyperbolic elements $\delta^{i-1}\gamma\delta^{1-i}$), only the orbits of the two fixed points of δ give rise to ramification points of p. They are both of order $n = \operatorname{ord}(\delta)$. Thus we can read off the genus g_E of E from the Riemann-Hurwitz formula:

$$2n - 2 = 2n(2g_E - 2) + 2 \cdot (n - 1).$$

This shows $g_E = 1$. Moreover we know that the two fixed points of δ are in the same G-orbit since they are exchanged by α. This means that p maps them both to the same point in E. We have proved:

PROPOSITION 7.3. *With notations as above, $p : X = \Omega/\Gamma \to E = \Omega/G$ is a normal p-adic origami with Galois group D_n.*

We can also easily find the quotient graphs $T(\Gamma)/\Gamma$ and $T(\Gamma)/G = T(G)/G$: up to contraction of some edges in the case $|1 - \zeta| < 1$, $T(\Gamma)/\Gamma$ consists of a single vertex $\bar{v}_0$ (the image of v_0), and one loop for each of the n free generators of Γ. D_n acts in the following way on this graph: δ fixes the vertex and cyclically rotates the n loops, whereas α fixes one of the loops (the one corresponding to the axis of γ) and exchanges the remaining $n - 1$ loops in pairs (remember that n is odd!). Therefore the quotient graph $T(\Gamma)/G$, which is also the quotient graph of $T(\Gamma)/\Gamma$ by the action of D_n, consists of a single vertex v and a single edge e. As a graph of groups, the vertex group is D_n and the edge group is cyclic of order 2 (generated by α, say); it is embedded into D_n both ways (i.e. via e and via $\bar{e}$) as the same element, which reflects the fact that $\gamma\alpha\gamma^{-1} = \alpha$.

7.3. Kremer's results. By similar reasoning as in Section 7.2, Kremer found the following example: Let δ and α be elements of $\mathrm{PGL}_2(\mathbb{C}_p)$ of order 3 and 2, resp., that generate a tetrahedral group A_4, and let γ be a hyperbolic element commuting with δ. The group G generated by δ, α and γ can be mapped homomorphically onto A_4 in such a way that the kernel Γ is generated by γ and its conjugates. Γ is a Schottky group of rank 4, and $p : X = \Omega/\Gamma \to E = \Omega/G$ is a normal p-adic origami of degree 12. That p is ramified over only one point is due to the fact that A_4 acts transitively on the 6 fixed points of the elements of order 2 in A_4 (in the same way as A_4 acts transitively on the six edges of a tetrahedron). The quotient graph $T(G)/G$ in this case has, as above, one vertex and one edge, the vertex group being A_4 and the edge group being cyclic of order 3 generated by δ.

The main result of Kremer's paper is that essentially all normal p-adic origamis can be obtained from these two examples:

THEOREM 7.4 (Thm. 5.1 in [**13**]). *Let $p : X = \Omega/\Gamma \to E = \Omega/G$ be a normal p-adic origami with genus $(X) > 1$. Then $T(G)/G$ can be contracted to a graph with one vertex and one edge, in other words, G is the fundamental group of a graph of groups with a graph of this type.*
If $p > 5$, the vertex group is isomorphic to either D_n for some $n \geq 3$ or to A_4. The edge group is cyclic of order 2 in the first case and of order 3 if the vertex group is A_4.
For $p = 2, 3$ and 5 there exist additional possibilities for the vertex group.
The group Γ can be any normal subgroup of G that has trivial intersection with the vertex group.

The proof of this theorem relies on precise knowledge of the possible discontinuous subgroups of $\mathrm{PGL}_2(\mathbb{C}_p)$ and on a careful analysis of the properties of the quotient graphs.

The theorem tells us, which finite groups occur as Galois groups of normal p-adic origamis: such a group has to be a homomorphic image of one of the groups G in the theorem, and it must contain a copy of the corresponding vertex group (i. e. of D_n or A_4 if $p > 5$). This still allows for a wide range of interesting groups, see [**12**], Ex. 4.17 for a few examples. On the other hand the characterization excludes many groups that occur as Galois groups of complex origamis since this is the case for every finite group that can be generated by two elements.

A very interesting question, which is also addressed in Kremer's paper, concerns the relation between complex and p-adic origamis. The question can be formulated in at least two ways:

1. Given a complex origami O, does there exist a p-adic origami on the Teichmüller curve determined by O?
2. Given a p-adic origami, is it possible to recover the combinatorial description by squares of the corresponding complex origami?

More precisely the first question should be formulated as follows: A complex origami $O = (p : X \to E)$ of genus g defines a Teichmüller curve $C(O)$ in the moduli space M_g of compact Riemann surfaces of genus g. Since M_g is an algebraic variety which can be defined over $\bar{\mathbb{Q}}$ (and even over $\mathbb{Q}$), and since $C(O)$ is defined over a number field, $C(O)$ also determines an algebraic curve in $M_g(\bar{\mathbb{Q}})$ and hence, by extension of coefficients, also in $M_g(\mathbb{C}_p)$. The question now is, whether this algebraic curve in $M_g(\mathbb{C}_p)$ contains a point that represents a p-adic origami.

There is no general method to decide this question for a given normal complex origami. Negative results can be obtained by showing that the Galois group of the origami is not a homomorphic image of one of the groups G in Thm. 7.4. Positive results are possible in the rare cases where explicit equations are known for the points on the Teichmüller curve $C(O)$: In **[13]**, Example 7.1, Kremer uses a result of A. Kappes (see **[10]**) that describes the Teichmüller curve to a specific origami with 6 squares by hyperelliptic equations; for a hyperelliptic curve over a p-adic field, it depends only on the relative position of the branch points whether it is a Mumford curve or not. With this criterion, Kremer shows that there are p-adic origamis on this special Teichmüller curve.

For the second question, there are also only very partial results available: Given a p-adic origami in the form of Thm. 6.4, i. e. in terms of the uniformizing groups, we may assume that the corresponding Mumford curves are defined over a number field and thus determine a complex origami (because also p-adic origamis come in 1-parameter families which necessarily contain $\bar{\mathbb{Q}}$-points). One would like to describe this complex origami e. g. as a translation surface by gluing squares. Since we consider only normal origamis, there is sometimes an indirect way to fix the complex origami, namely if there is only one normal origami with the given Galois group. In **[12]** Kremer shows that this is the case for all but 30 out of the 2386 groups of order up to 250 which can be generated by two elements. But of course, for larger order there are more groups which occur as Galois groups of different origamis. Kremer also shows that each group in the infinite series $D_n \times \mathbb{Z}/m\mathbb{Z}$ and $A_4 \times \mathbb{Z}/m\mathbb{Z}$ uniquely determies a normal origami and thus provides a way to associate a complex origami to the corresponding p-adic origami.

References

[1] Siegfried Bosch, Ulrich Güntzer, and Reinhold Remmert, *Non-Archimedean analysis*, Grundlehren der Mathematischen Wissenschaften [Fundamental Principles of Mathematical Sciences], vol. 261, Springer-Verlag, Berlin, 1984. A systematic approach to rigid analytic geometry. MR746961 (86b:32031)

[2] Siegfried Bosch and Werner Lütkebohmert, *Stable reduction and uniformization of abelian varieties. I*, Math. Ann. **270** (1985), no. 3, 349–379, DOI 10.1007/BF01473432. MR774362 (86j:14040a)

[3] Alex Eskin, Maxim Kontsevich, and Anton Zorich, *Lyapunov spectrum of square-tiled cyclic covers*, J. Mod. Dyn. **5** (2011), no. 2, 319–353, DOI 10.3934/jmd.2011.5.319. MR2820564 (2012h:37073)

[4] Jean Fresnel and Marius van der Put, *Rigid analytic geometry and its applications*, Progress in Mathematics, vol. 218, Birkhäuser Boston, Inc., Boston, MA, 2004. MR2014891 (2004i:14023)

[5] Lothar Gerritzen and Marius van der Put, *Schottky groups and Mumford curves*, Lecture Notes in Mathematics, vol. 817, Springer, Berlin, 1980. MR590243 (82j:10053)

[6] Eugene Gutkin and Chris Judge, *Affine mappings of translation surfaces: geometry and arithmetic*, Duke Math. J. **103** (2000), no. 2, 191–213, DOI 10.1215/S0012-7094-00-10321-3. MR1760625 (2001h:37071)

[7] Joe Harris and Ian Morrison, *Moduli of curves*, Graduate Texts in Mathematics, vol. 187, Springer-Verlag, New York, 1998. MR1631825 (99g:14031)

[8] Frank Herrlich, *p-adisch diskontinuierlich einbettbare Graphen von Gruppen* (German), Arch. Math. (Basel) **39** (1982), no. 3, 204–216, DOI 10.1007/BF01899527. MR682448 (84d:14013)

[9] Frank Herrlich, *Introduction to origamis in Teichmüller space*, Strasbourg master class on geometry, IRMA Lect. Math. Theor. Phys., vol. 18, Eur. Math. Soc., Zürich, 2012, pp. 233–253, DOI 10.4171/105-1/3. MR2931888

[10] Frank Herrlich and Gabriela Schmithüsen, *An extraordinary origami curve*, Math. Nachr. **281** (2008), no. 2, 219–237, DOI 10.1002/mana.200510597. MR2387362 (2008k:14059)

[11] Pascal Hubert and Samuel Lelièvre, *Prime arithmetic Teichmüller discs in* $\mathcal{H}(2)$, Israel J. Math. **151** (2006), 281–321, DOI 10.1007/BF02777365. MR2214127 (2008f:37073)

[12] Karsten Kremer, Invariants of Complex and p-adic Origami-Curves, *PhD thesis, Universität Karlsruhe*, 2010.

[13] Karsten Kremer, *Normal origamis of Mumford curves*, Manuscripta Math. **133** (2010), no. 1-2, 83–103, DOI 10.1007/s00229-010-0379-8. MR2672540 (2011m:14037)

[14] Pierre Lochak, *On arithmetic curves in the moduli spaces of curves*, J. Inst. Math. Jussieu **4** (2005), no. 3, 443–508, DOI 10.1017/S1474748005000101. MR2197065 (2006m:32010)

[15] Carlos Matheus, Martin Möller and Jean-Christophe Yoccoz, A criterion for the simplicity of the Lyapunov spectrum of square-tiled surfaces, *Preprint* 2013, arXiv:1305.2033.

[16] Martin Möller, *Variations of Hodge structures of a Teichmüller curve*, J. Amer. Math. Soc. **19** (2006), no. 2, 327–344, DOI 10.1090/S0894-0347-05-00512-6. MR2188128 (2007b:32026)

[17] David Mumford, *An analytic construction of degenerating curves over complete local rings*, Compositio Math. **24** (1972), 129–174. MR0352105 (50 #4592)

[18] Gabriela Schmithüsen, Veech Groups of Origamis. *PhD thesis, Universität Karlsruhe*, 2005.

[19] Jean-Pierre Serre, *Trees*, Springer-Verlag, Berlin-New York, 1980. Translated from the French by John Stillwell. MR607504 (82c:20083)

[20] William A. Veech, *Teichmüller curves in moduli space, Eisenstein series and an application to triangular billiards*, Invent. Math. **97** (1989), no. 3, 553–583, DOI 10.1007/BF01388890. MR1005006 (91h:58083a)

DEPARTMENT OF MATHEMATICS, KARLSRUHE INSTITUTE OF TECHNOLOGY (KIT), 76128 KARLSRUHE, GERMANY

E-mail address: herrlich@kit.edu

Regular dessins with a given automorphism group

Gareth A. Jones

Dedicated to Emilio Bujalance on the occasion of his 60th birthday

ABSTRACT. Dessins d'enfants are combinatorial structures on compact Riemann surfaces defined over algebraic number fields, and regular dessins are the most symmetric of them. If G is a finite group, there are only finitely many regular dessins with automorphism group G. It is shown how to enumerate them, how to represent them all as quotients of a single regular dessin $U(G)$, and how certain hypermap operations act on them. For example, if G is a cyclic group of order n then $U(G)$ is a map on the Fermat curve of degree n and genus $(n-1)(n-2)/2$. On the other hand, if $G = A_5$ then $U(G)$ has genus 274218830047232000000000000000001. For other non-abelian finite simple groups, the genus is much larger.

1. Introduction

Belyĭ's Theorem [**2**] states that a compact Riemann surface C, regarded as a complex projective algebraic curve, can be defined over the field $\overline{\mathbb{Q}}$ of algebraic numbers if and only if there is a non-constant meromorphic function $\beta : C \to \mathbb{P}^1(\mathbb{C})$ ramified over at most three points. This is equivalent to the complex structure on C being obtained, in a canonical way, from a combinatorial structure D on C called a dessin; this can be regarded as an oriented hypermap, a bipartite map, or a tripartite triangulation of C. The most symmetric dessins are the regular dessins, those for which β is a regular covering, in which case the automorphism group $\mathrm{Aut}\, D$ of D, the group of covering transformations of β, and the monodromy group of β are all isomorphic to a single 2-generator finite group G. In principle, understanding regular dessins is sufficient for an understanding of all dessins, since each dessin arises as the quotient of some regular dessin by a group of automorphisms.

The set $\mathcal{R}(G)$ of regular dessins associated in this way with a given 2-generator finite group G corresponds bijectively to the set $\mathcal{N}(G)$ of normal subgroups N of a free group F of rank 2 with $F/N \cong G$, or equivalently to the set of orbits of $\mathrm{Aut}\, G$ on generating pairs of G. Our aim here is to show how a method due to Hall [**20**] can be used to determine the number $r(G) = |\mathcal{R}(G)|$ of such dessins, to show how they can be represented as quotients of a single regular dessin $U(G)$, and to describe the automorphism group $\overline{G} = \mathrm{Aut}\, U(G)$ of $U(G)$. A number of illustrative examples are given, including cases where G is a cyclic group, a dihedral group or

2010 *Mathematics Subject Classification.* Primary 14H57; Secondary 14H37, 20B25, 30F10.
Key words and phrases. Dessin d'enfant, regular dessin, automorphism group.

a non-abelian finite simple group; in the latter case, $U(G)$ and $\overline{G}$ turn out to be surprisingly large.

The absolute Galois group $\Gamma = \mathrm{Gal}\,\overline{\mathbb{Q}}/\mathbb{Q}$ acts on dessins by acting on the coefficients of the polynomials and rational functions defining them over $\overline{\mathbb{Q}}$. By a recent result of González-Diez and Jaikin-Zapirain [**16**], it acts faithfully on regular dessins. It preserves automorphism groups, so it acts on $\mathcal{R}(G)$ for each G. A group $\Omega \cong \mathrm{Out}\,F \cong GL_2(\mathbb{Z})$ of hypermap operations, introduced by James [**23**], also preserves regularity and automorphism groups, and its orbits on $\mathcal{R}(G)$ correspond to the T-systems of generating pairs for G. We shall consider some examples of such actions of Γ and Ω, interpreting group-theoretic results of Neumann and Neumann [**38**], Dunwoody [**13**], and Garion and Shalev [**14**] on T-systems in terms of dessins.

Acknowledgements The author is very grateful to Martin Dunwoody, Pierre Guillot and Frank Herrlich for their helpful comments about T-systems, the absolute Galois group and origamis, and to the organisers of the conference on Riemann and Klein Surfaces, Symmetries and Moduli Spaces, in honour of Emilio Bujalance (Linköping, June, 2013), for the invitation to give a talk on which this paper is based.

2. Dessins

If C is a compact Riemann surface, or equivalently a nonsingular complex projective algebraic curve, then by Belyĭ's Theorem [**2**], as reinterpreted and extended by Grothendieck [**17**], Wolfart [**47**] and others, the following are equivalent:

(1) C is defined, as an algebraic curve, over the field $\overline{\mathbb{Q}}$;
(2) there is a non-constant meromorphic function $\beta : C \to \mathbb{P}^1(\mathbb{C})$, branched over at most three points;
(3) C is uniformised by a subgroup M of finite index in a triangle group Δ;
(4) C is isomorphic to the compactification $\overline{\mathbb{H}/N}$, where $\mathbb{H}$ is the hyperbolic plane and N is a subgroup of finite index in the congruence subgroup $\Gamma(2)$ of level 2 in the modular group $PSL_2(\mathbb{Z})$.

The function β in (2), called a *Belyĭ function*, corresponds to the surface coverings induced by the inclusions $M \leq \Delta$ and $N \leq \Gamma(2)$ in (3) and (4); by applying a Möbius transformation, the ramification points of β can, without loss of generality, be taken to be contained in $\{0, 1, \infty\}$. The group $\Gamma(2)$ is freely generated by the parabolic Möbius transformations

$$X : z \mapsto \frac{z}{-2z+1} \qquad \text{and} \qquad Y : z \mapsto \frac{z-2}{2z-3}$$

fixing 0 and 1, and there is an epimorphism $\Gamma(2) \to \Delta$, with N the inverse image of M. The compactification in (4) is obtained by adding points to fill the punctures in $\mathbb{H}/N$ corresponding to the cusps of N, that is, its orbits on $\mathbb{P}^1(\mathbb{Q})$. Further background is given by Girondo and González-Diez in [**15**], and by Lando and Zvonkin in [**32**].

Under the above conditions, C is called a *Belyĭ curve*. It carries a *dessin*, that is, a bipartite map in which the black and white vertices are the fibres of β over 0 and 1, and the edges are the inverse images of the unit interval $I = [0, 1]$. The *monodromy group* G of D is the monodromy group of β, restricted to a smooth covering of $\mathbb{P}^1(\mathbb{C}) \setminus \{0, 1, \infty\}$; this is the transitive group of permutations of the sheets of this covering obtained by lifting closed paths in $\mathbb{P}^1(\mathbb{C}) \setminus \{0,1,\infty\}$ to C. Equivalently, G is the group of permutations of the edges of D generated by the monodromy permutations x and y about 0 and 1, obtained by following the orientation of C around the black and white vertices; this is isomorphic to the group induced by Δ on the cosets of M, or equivalently by $\Gamma(2)$ on the cosets of N, with X and Y inducing x and y.

The automorphism group $\operatorname{Aut} D$ of a dessin D (preserving orientation and vertex colours) is the group of permutations of the edges commuting with x and y, or equivalently with the monodromy group G. It can be identified with the group of covering transformations of β, permuting the fibre over a base-point. It acts semiregularly on the edges, and we say that D is a *regular dessin* if it acts transitively (and hence regularly). This is equivalent to M and N being normal subgroups of Δ and $\Gamma(2)$, in which case

$$\operatorname{Aut} D \cong G \cong \Delta/M \cong \Gamma(2)/N,$$

with $\operatorname{Aut} D$ and G acting on edges as the left and right regular representations of the same group. From now on we will always assume that D is regular.

To maintain symmetry between the three ramification points of β, it is sometimes useful to extend D to a tripartite triangulation $\beta^{-1}(\mathbb{P}^1(\mathbb{R}))$ of C by adding a generator Z to $\Gamma(2)$ satisfying $XYZ = 1$, so that the corresponding element $z = (xy)^{-1}$ of G represents the monodromy permutation of β around ∞. We say that D has *type* (l, m, n) where l, m and n are the orders of x, y and z. Then the Riemann-Hurwitz formula implies that D has genus

$$g = 1 + \frac{1}{2}\left(1 - \frac{1}{l} - \frac{1}{m} - \frac{1}{n}\right)|G|.$$

3. Counting regular dessins

The correspondences in Section 2 provide a purely group-theoretic approach to dessins. In particular, they give a bijection between isomorphism classes of regular dessins D and normal subgroups N of finite index in a free group

$$F = F_2 = \langle X, Y \mid - \rangle = \langle X, Y, Z \mid XYZ = 1\rangle$$

of rank 2, with $\operatorname{Aut} D \cong F/N$. For each 2-generator finite group G, let $\mathcal{R}(G)$ denote the set of (isomorphism classes of) regular dessins D with $\operatorname{Aut} D \cong G$, and let

$$\mathcal{N}(G) = \{N \lhd F \mid F/N \cong G\},$$

so that the above bijection restricts to a bijection between the finite sets $\mathcal{N}(G)$ and $\mathcal{R}(G)$, with each $N \in \mathcal{N}(G)$ corresponding to a dessin $D = D(N) \in \mathcal{R}(G)$. (The set corresponding to $\mathcal{N}(G)$, with F a free group of arbitrary rank, was studied by Dunwoody in **[12]**.)

There is also a natural bijection between $\mathcal{N}(G)$ and the set of orbits of $\operatorname{Aut} G$ on generating pairs (x, y) for G, or equivalently on generating triples (x, y, z) for G satisfying $xyz = 1$. This is because epimorphisms $F \to G$ correspond to choices of generating pairs (x, y) for G, and two such epimorphisms have the same kernel N if and only if they differ by an automorphism of G. The number

$$r = r(G) = |\mathcal{R}(G)|$$

of regular dessins associated with G is therefore equal to the number

$$d_2(G) = \frac{\phi_2(G)}{|\mathrm{Aut}\, G|}$$

of such orbits, where $\phi_2(G)$ is the number of generating pairs (x, y) for G (since these are permuted semiregularly by $\mathrm{Aut}\, G$).

Hall [**20**] developed a very useful method for evaluating functions such as $\phi_2(G)$. Since every pair of elements of G generate a unique subgroup H we have

$$|G|^2 = \sum_{H \leq G} \phi_2(H).$$

From this, Möbius inversion in the subgroup lattice of G shows that

$$\phi_2(G) = \sum_{H \leq G} \mu_G(H)|H|^2,$$

where μ_G is the Möbius function for G, defined recursively by

$$\sum_{K \geq H} \mu_G(K) = \delta_{H,G},$$

and $\delta_{H,G}$ denotes the Kronecker delta, equal to 1 if $H = G$ and 0 otherwise. It follows that

$$r(G) = \frac{1}{|\mathrm{Aut}\, G|} \sum_{H \leq G} \mu_G(H)|H|^2.$$

(This method can be extended to count normal subgroups of *any* finitely generated group F with a given finite quotient group G: see, for instance [**24, 28**] where F is a triangle group, and [**25**] where F is a surface group.)

Example 3.1 It is easily seen that $r(C_n)$ is a multiplicative function of n, that is, if m and n are coprime then $r(C_{mn}) = r(C_m)r(C_n)$. It follows, by considering the case where n is a prime power, that

$$r(C_n) = n \prod_{p|n} \left(1 + \frac{1}{p}\right),$$

where p ranges over the distinct primes dividing n.

Example 3.2 If G is the dihedral group D_n of order $2n$, with $n > 2$, then there are, up to automorphisms, just three generating triples (x, y, z) for G with $xyz = 1$: in each case, two of x, y and z are reflections, and the third is a rotation of order n. Thus $r(D_n) = 3$. However, if $n = 2$ then G is a Klein four-group $V_4 = C_2 \times C_2$, with $r(D_2) = 1$.

Example 3.3 In [**20**], Hall computed the Möbius function μ_G for a number of groups G, including the simple groups $L_2(p) = PSL_2(p)$ for primes $p \geq 5$. He showed that $r(A_4) = 4$, $r(S_4) = 9$, $r(A_5) = 19$, $r(\hat{A}_5) = 76$ and $r(A_6) = 53$, where $\hat{A}_5$ is the binary icosahedral group, isomorphic to $SL_2(5)$, and that

$$r(L_2(p)) = \frac{1}{4}(p+1)(p^2 - 2p - 1) - \epsilon,$$

where $\epsilon = 49, 40, 11$ or 2 as $p \equiv \pm 1$ and ± 1, or $p \equiv \pm 1$ and ± 3, or $p \equiv \pm 2$ and ± 1, or $p \equiv \pm 2$ and ± 3 mod (5) and mod (8). Thus $r(L_2(p)) = 19$, 57, 254 and 495 for $p = 5, 7, 11$ and 13 respectively. (Note that $L_2(5) \cong A_5$ and $L_2(9) \cong A_6$, so for these groups $r(G) = 19$ and 53 respectively.)

Example 3.4 Downs [8] extended Hall's computation of μ_G to $L_2(q)$ and $PGL_2(q)$ for all prime powers q; see [9] for a proof for $L_2(2^e)$ and a statement of results for $L_2(q)$ where q is odd, and [10] for some combinatorial applications by Downs and the author. The general results are rather complicated, but when $G = L_2(2^e)$ with $e > 1$ we find that

$$r(G) = \frac{1}{e}\sum_{f|e}\mu\left(\frac{e}{f}\right)2^f(2^{2f}-2^f-3)$$

where μ is the classical Möbius function on $\mathbb{N}$ (see [10]). Thus for $e = 2$ and 3 we have $r(G) = 19$ (as expected, since $L_2(4) \cong A_5$) and 142. Downs and the author [11] have recently obtained the similar formula

$$r(G) = \frac{1}{e}\sum_{f|e}\mu\left(\frac{e}{f}\right)2^f(2^{4f}-2^{3f}-9)$$

where G is the simple Suzuki group $Sz(2^e)$ [4, 43, 45] for some odd $e > 1$.

For any finite group G we have $\phi_2(G) \leq |G|^2$, so

$$r(G) \leq \frac{|G|^2}{|\mathrm{Aut}\, G|}.$$

In particular, if G has trivial centre, so that $\mathrm{Inn}\, G \cong G$, then

$$r(G) \leq \frac{|G|}{|\mathrm{Out}\, G|}$$

where $\mathrm{Out}\, G := \mathrm{Aut}\, G/\mathrm{Inn}\, G$ is the outer automorphism group of G. As a consequence of the classification of finite simple groups, it is known that any such group can be generated by two elements. In fact, results of Dixon [6], of Kantor and Lubotzky [31], and of Liebeck and Shalev [33] show that if G is a non-abelian finite simple group, then a randomly-chosen pair of elements generate G with probability approaching 1 as $|G| \to \infty$, so for such groups this upper bound is asymptotically sharp, that is,

$$r(G) \sim \frac{|G|}{|\mathrm{Out}\, G|} \quad \text{as} \quad |G| \to \infty.$$

The values of $|G|$ and $|\mathrm{Out}\, G|$ given in [4, 45] show that for each of the infinite families of non-abelian finite simple groups, $|\mathrm{Out}\, G|$ grows much more slowly than $|G|$ (indeed, in many cases it is bounded), so that $r(G)$ grows almost as quickly as $|G|$. For instance, $r(A_n) \sim n!/4$ as $n \to \infty$. (See [7, 37] for more precise results concerning generating pairs for A_n.)

One can illustrate how close this asymptotic estimate is as follows. For any finite group G we have

$$\phi_2(G) = |G^2 \setminus \bigcup_M M^2| \geq |G|^2 - \sum_M |M|^2,$$

where M ranges over the maximal subgroups of G. If G is perfect then each such M has $|G : M|$ conjugates, so

$$\phi_2(G) \geq |G|^2\left(1 - \sum_{i=1}^{k}\frac{1}{|G : M_i|}\right),$$

where M_i ranges over a set of representatives of the k conjugacy classes of maximal subgroups of G. It follows that if G also has trivial centre (and in particular if G is a non-abelian finite simple group) then

$$1 \geq \frac{r(G)}{|G|/|\mathrm{Out}\, G|} \geq 1 - \sum_{i=1}^{k} \frac{1}{|G:M_i|}.$$

When G is large, the sum on the right is typically very small, so that

$$r(G) \approx \frac{|G|}{|\mathrm{Out}\, G|}.$$

Thus, for the simple groups $G = L_2(q)$, $Sz(q)$, $R(q)$ and $U_3(q)$ (see [**4**]) we have

$$\sum_{i=1}^{k} \frac{1}{|G:M_i|} = \frac{1}{q^m} + O\left(\frac{1}{q^{m+1}}\right) \quad \text{as} \quad q \to \infty,$$

where $m = 1, 2, 3, 3$ respectively, with the sum dominated by the term corresponding to the doubly transitive permutation representation of degree $q^m + 1$. The situation is similar for the 26 sporadic simple groups: for instance, if G is O'Nan's group $O'N$ [**40**] (the 13th largest of these groups), with $k = 13$ conjugacy classes of maximal subgroups [**4, 44, 48**], we have

$$\frac{|G|}{|\mathrm{Out}\, G|} = \frac{460815505920}{2} = 230407752960$$

and

$$\sum_{i=1}^{k} \frac{1}{|G:M_i|} = 0.00001726863378\ldots,$$

so

$$230403774132 \leq r(G) \leq 230407752960.$$

If G is the Monster, the largest sporadic simple group, then $|\mathrm{Out}\, G| = 1$ and so

$$r(G) \approx |G| \approx 8.08017 \times 10^{53}.$$

4. Classifications

For some groups G, the dessins in $\mathcal{R}(G)$ have been classified.

Example 4.1 If $G = C_n$ then three of the dessins in $\mathcal{R}(G)$ have genus 0, obtained from epimorphisms $F \to G$ sending one of X, Y and Z to the identity; these correspond to Belyĭ functions $C = \mathbb{P}^1(\mathbb{C}) \to \mathbb{P}^1(\mathbb{C})$ given by $\beta : t \mapsto t^n$, $1/(1-\beta)$ and $(\beta - 1)/\beta$, resulting in dessins of types $(n, 1, n)$, $(n, n, 1)$ and $(1, n, n)$ respectively; they have vertices, of three different colours, at 0, at ∞, and at the nth roots of 1. The remaining dessins depend on the arithmetic nature of n. In the simplest case, when n is an odd prime p, they all have type (p, p, p), they lie on the Lefschetz curves of genus $(p-1)/2$, with affine models $u^p = v^m(v-1)$ for $m = 1, \ldots, p-2$, and they have Belyĭ functions $\beta : (u, v) \mapsto v$.

Example 4.2 If $G = D_n$ with $n > 2$ then the three dessins in $\mathcal{R}(G)$ all have genus 0, and their types are permutations of $(2, 2, n)$. They have vertices in $\mathbb{P}^1(\mathbb{C})$, of three different colours, at 0 and ∞, at the nth roots of 1, and at the nth roots of -1.

Example 4.3 For $G = A_5$, the $r(G) = 19$ dessins in $\mathcal{R}(G)$ have been described (as oriented hypermaps) by Breda and the author in [**3**] (see [**38**] for representative generating pairs). There are six of genus 0, each with type a permutation of $(2,3,5)$; there are three each of genera 4 and 5, with permutations of $(2,5,5)$ and $(3,3,5)$ as types; there are six of genus 9, with permutations of $(3,5,5)$ as types (three having the generators of order 5 conjugate in G, and three with them not conjugate), and there is one of genus 13 and type $(5,5,5)$.

Example 4.4 Let G be the non-abelian group of order pq, where p and q are primes with $p \equiv 1 \bmod (q)$. This is a semidirect product of a normal subgroup $P \cong C_p$ by a complement $Q \cong C_q$, and it has automorphism group $\operatorname{Aut} G \cong AGL_1(p)$ of order $p(p-1)$. Hall's method gives $r(G) = d_2(G) = q^2-1$. The dessins in $\mathcal{R}(G)$ have been investigated by Streit and Wolfart in [**42**] and (together with the present author) in [**30**]. There are $3(q-1)$ of genus $(pq-3p+2)/2$, whose types are permutations of (q,q,p), and $(q-1)(q-2)$ of genus $(p-1)(q-2)/2$ and type (q,q,q). These dessins are defined over the cyclotomic field $\mathbb{Q}(\zeta_q)$, where $\zeta_q := \exp(2\pi i/q)$, with the orbits of the absolute Galois group $\Gamma = \operatorname{Gal} \overline{\mathbb{Q}}/\mathbb{Q}$ on them all having length $q-1$.

5. Universal covers

For any 2-generator finite group G, let

$$K(G) = \bigcap_{N \in \mathcal{N}(G)} N,$$

the intersection of all $N \triangleleft G$ with $F/N \cong G$. As the intersection of finitely many normal subgroups of finite index in F, $K(G)$ is also a normal subgroup of finite index in F. It corresponds to a regular dessin

$$U(G) = D(K(G)) = \bigvee_{D \in \mathcal{R}(G)} D$$

called the *universal cover* of G, the smallest dessin covering each $D \in \mathcal{R}(G)$, with automorphism group

$$\overline{G} := \operatorname{Aut} U(G) \cong F/K(G).$$

The dessin $U(G)$ has some special properties. For instance, by its construction, $K(G)$ is the only normal subgroup of F with quotient isomorphic to $\overline{G}$. This uniquenessss implies that $U(G)$ is invariant under the action of the absolute Galois group $\Gamma = \operatorname{Gal} \overline{\mathbb{Q}}/\mathbb{Q}$, and is therefore defined over $\mathbb{Q}$.

Example 5.1 If $G = C_n$ then

$$K(G) = F'F^n,$$

the group generated by the commutators and nth powers of elements of F, so

$$\overline{G} = F/F'F^n \cong C_n \times C_n,$$

and $U(G)$ is the nth degree Fermat dessin, on the Fermat curve

$$x_0^n + x_1^n + x_2^n = 0$$

in $\mathbb{P}^2(\mathbb{C})$. Here $\overline{G}$ acts by multiplying homogeneous coordinates x_i by nth roots of 1. (In fact, we obtain the same universal cover $U(G)$ and group $\overline{G}$ whenever G is a 2-generator abelian group of exponent n.)

Example 5.2 If $G = D_n$ with $n > 1$ odd then $\overline{G}$ is an extension of a normal subgroup C_n^3 by a Klein four-group D_2, each of its three involutions centralising one of the three direct factors and inverting the other two. (Herrlich [**21**, Prop. 4.5] has also constructed this covering group in the context of origamis.) The dessin $U(G)$ is an n^3-sheeted regular covering of $U(C_2)$, branched over its vertices $\pm 1, \pm i, 0$ and ∞. It has type $(2n, 2n, 2n)$, and its genus is $2n^3 - 3n^2 + 1$. For even n we have $|\overline{G}| = n^3/2$, the type of $U(G)$ is (n, n, n), and its genus is $(n^3 - 3n^2 + 4)/4$.

Example 5.3 If G is the non-abelian group of order pq in Example 4.4, then $\overline{G}$ is a semidirect product of a normal subgroup $\tilde{P} \cong C_p^{q^2-1}$ by a complement $\overline{Q} \cong C_q^2$. As an $\mathbb{F}_p\overline{Q}$-module, $\tilde{P}$ is the quotient of the regular module $\mathbb{F}_p\overline{Q}$ by the principal submodule, i.e. the direct sum of all the (1-dimensional) non-principal irreducible $\mathbb{F}_p\overline{Q}$-modules. Factoring out all but one of these $q^2 - 1$ submodules gives a quotient of $\overline{G}$ isomorphic to $G \times C_q$, and factoring out the second direct factor of this yields G. The $q^2 - 1$ equivalence classes of epimorphisms $F \to G$ all arise in this way. The dessin $U(G)$ has type (pq, pq, pq) and genus

$$g = 1 + \frac{1}{2}p^{q^2-2}q(pq - 3).$$

It is a regular covering of the Fermat dessin $U(Q)$ of degree q, branched over its $3q$ black, white and red vertices.

The group $\overline{G} = F/K(G)$ is naturally embedded in the cartesian product

$$\prod_{N \in \mathcal{N}(G)} F/N \cong G^r$$

of $r = r(G)$ copies of G as the subgroup of G^r generated by (x_i) and (y_i) where x_i and y_i are the monodromy generators of G corresponding to the ith dessin in some numbering of $\mathcal{R}(G)$. It follows that $\overline{G}$ satisfies all the identical relations satisfied by G. For instance, if G is nilpotent of class c, is solvable of derived length d, or has exponent e, then $\overline{G}$ also has the same property. Since $\overline{G}$ is also a 2-generator group, this might suggest that it is not much larger than G. The examples in the next section show that this is not always so.

6. Non-abelian finite simple groups

The examples $G = C_n$ and D_n show that when $\overline{G}$ is embedded in G^r, it may be a proper subgroup, even though it projects onto each of the r direct factors. However, if G is a non-abelian finite simple group then $\overline{G} = G^r$. (This result and its proof are probably well-known to most group-theorists, but they are included here for the benefit of non-specialists.) For background on the finite simple groups, see [**4, 45**].

LEMMA 6.1. *If $H = H_1 \times \cdots \times H_r$ where $H_1, \ldots, H_r$ are non-abelian simple groups, then the maximal normal subgroups of H are those of the form $\prod_{i \neq j} H_i$ where $j = 1, \ldots, r$.*

Proof. Clearly these are maximal normal subgroups of H. Conversely, let N be a maximal normal subgroup, so $H_j \not\leq N$ for some $j = 1, \ldots, r$. Then $H_j \cap N$ is a proper normal subgroup of H_j, so $H_j \cap N = 1$ by the simplicity of H_j. Since H_j and N are both normal in H, their commutator $[H_j, N]$ is contained in $H_j \cap N$. Thus $[H_j, N] = 1$, so N is contained in the centraliser $C_H(H_j)$ of H_j in H, which is $\prod_{i \neq j} H_i$. Hence $N = \prod_{i \neq j} H_i$ by the maximality of N. □

Indeed, it follows that the only normal subgroups of H have the form $\prod_{i \in I} H_i$ where $I \subseteq \{1, \ldots, r\}$. Note that finiteness is not assumed in this proof.

COROLLARY 6.2. *Let $N_1, \dots, N_r$ be distinct normal subgroups of a group Φ, with each $H_i := \Phi/N_i$ non-abelian and simple. If $K = N_1 \cap \dots \cap N_r$ then*

$$\Phi/K \cong H_1 \times \dots \times H_r.$$

Proof. We use induction on r. The case $r = 1$ is trivial, so let $r > 1$, assume the result for intersections of $r-1$ normal subgroups, and define $L = N_1 \cap \dots \cap N_{r-1}$. If $L \leq N_r$ then N_r/L is a maximal normal subgroup of the group $\Phi/L \cong H_1 \times \dots \times H_{r-1}$, so Lemma 6.1 implies that $N_r = N_i$ for some $i < r$, against our hypothesis. Thus $L \not\leq N_r$, so LN_r is a normal subgroup of Φ properly containing N_r, and hence $LN_r = \Phi$ by the maximality of N_r. Since $L \cap N_r = K$ we therefore have

$$\Phi/K = (N_r/K) \times (L/K) \cong (\Phi/L) \times (\Phi/N_r) \cong (H_1 \times \dots \times H_{r-1}) \times H_r,$$

as required. □

COROLLARY 6.3. *If G is a non-abelian finite simple group then $\overline{G} \cong G^r$ where $r = r(G)$.*

Proof. In Corollary 6.2 take $\Phi = F$ and $\{N_1, \dots, N_r\} = \mathcal{N}(G)$, so that $K = K(G)$ and each $H_i \cong G$; then $\overline{G}$ can be identified with G^r, as claimed. □

Guralnick and Kantor [**18**] have shown that if G is a non-abelian finite simple group then every non-identity element is a member of a generating pair. It follows that if such a group G has exponent e then $U(G)$ has type (e, e, e), so by the Riemann-Hurwitz formula [**32**, Remark 1.2.21] it has genus

$$g = 1 + \frac{e-3}{2e}|G|^r.$$

In this case, the numerical estimates given earlier for $r(G)$ show that $\overline{G}$, and hence g, can be rather large.

Example 6.1 In the case of the smallest non-abelian finite simple group, namely $G = A_5$, we have $r(G) = 19$, so $\overline{G} = G^{19}$, of order

$$60^{19} = 609359740010496 \times 10^{17} \approx 6.1 \times 10^{31}.$$

Since G has exponent 30, it follows that $U(G)$ has type $(30, 30, 30)$ and genus

$$1 + \frac{9}{20} \times 60^{19} = 27421188300472320000000000000000001 \approx 2.742 \times 10^{31}.$$

Example 6.2 For the second smallest non-abelian finite simple group, $G = L_2(7)$, we have $r(G) = 57$, so $\overline{G} = G^{57}$, of order $168^{57} \approx 7.035 \times 10^{126}$. The universal cover $U(G)$ has type $(84, 84, 84)$ and genus

$$1 + \frac{27}{56} \times 168^{57} \approx 3.392 \times 10^{126}.$$

Example 6.3 The Monster simple group G has order

$$|G| = 2^{46}.3^{20}.5^9.7^6.11^2.13^3.17.19.23.29.31.41.47.59.71$$
$$= 808017424794512875886459904961710757005754368000000000$$
$$\approx 8.080 \times 10^{53}$$

and exponent

$$e = 2^5.3^3.5^2.7.11.13.17.19.23.29.31.41.47.59.71$$
$$= 1165654792878376600800 \approx 1.166 \times 10^{21}.$$

Since $|\mathrm{Out}\,G| = 1$ we have $r(G) \approx |G|$, so

$$|\overline{G}| = |G|^r \approx |G|^{|G|} \approx (8.080 \times 10^{53})^{8.080\times 10^{53}} \approx 10^{10^{55.639}}.$$

The universal cover $U(G)$ has type (e, e, e) and genus approximately $|\overline{G}|/2$.

Example 6.4 For comparison, we consider a quasisimple example. Let G be the binary icosahedral group $\hat{A}_5 \cong SL_2(5)$, with $\mathrm{Aut}\,G \cong S_5 \cong PGL_2(5) \cong \mathrm{Aut}\,A_5$. Then $r(G) = 76$, with each generating triple for the central quotient $G/Z(G) \cong A_5$ lifting to four for G: the members of each such quadruple are related to each other by multiplying two members of a triple by the central involution in G. Each generating triple for G arises in this way, so the $r(A_5) = 19$ orbits of S_5 on generating pairs for A_5 yield $r(G) = 4 \times 19 = 76$ orbits on those for G. Each $N \in \mathcal{N}(A_5)$ contains a unique normal subgroup $N^* = N \cap F'F^2$ of F with $F/N^* \cong G \times V_4$; this quotient has an elementary abelian centre E of order 8, and the four maximal subgroups of E complementing $Z(G)$ lift to the four subgroups of F in $\mathcal{N}(G)$ corresponding to N. Using this, one can show that $K(G)$ is a subgroup of index 2 in $K(A_5) \cap F'F^2$, and that $\overline{G} = C \times V_4$ where C is the central product of 19 copies of G (the quotient of G^{19} formed by identifying the centres of the direct factors). The dessin $U(G)$, an 8-sheeted branched covering of $U(A_5)$, has type $(60, 60, 60)$ and genus

$$231556701203988480000000000000001 \approx 2.316 \times 10^{32}.$$

7. Operations on dessins and T_2-systems

The automorphism group of the free group F permutes the subgroups of F. Since inner automorphisms leave invariant each conjugacy class of subgroups, this induces an action of the outer automorphism group

$$\Omega := \mathrm{Out}\,F = \mathrm{Aut}\,F/\mathrm{Inn}\,F$$

of F on isomorphism classes of dessins. In **[23]**, James interpreted Ω as the group of all operations on oriented hypermaps (which, in the finite case, are equivalent to dessins); see **[29, 30]** for operations on various other categories of maps, hypermaps and dessins. Here we will consider, for each finite group G, the isomorphic actions of Ω on $\mathcal{N}(G)$ and on $\mathcal{R}(G)$.

For any integer $n \geq 1$, the automorphism group of the free group F_n of rank n is generated by the elementary Nielsen transformations: permuting the free generators, inverting one of them, and multiplying one of them by another [**36**, Theorem 3.2]. When $n = 2$, with $F_2 = F$, one can identify $\Omega = \mathrm{Out}\,F$ with $GL_2(\mathbb{Z})$ through its faithful induced action on the abelianisation $F^{\mathrm{ab}} = F/F' \cong \mathbb{Z}^2$ of F [**34**, Ch. I, Prop. 4.5]. If we take the images of X and Y in F^{ab} as a basis, the elementary Nielsen transformations are represented by the matrices

$$E_i = \begin{pmatrix} 0 & 1 \\ 1 & 0 \end{pmatrix}, \quad \begin{pmatrix} -1 & 0 \\ 0 & 1 \end{pmatrix}, \quad \begin{pmatrix} 1 & 0 \\ 0 & -1 \end{pmatrix}, \quad \begin{pmatrix} 1 & 1 \\ 0 & 1 \end{pmatrix} \quad \text{and} \quad \begin{pmatrix} 1 & 0 \\ 1 & 1 \end{pmatrix}$$

for $i = 1, \ldots, 5$. This is not a minimal set of generators: for instance, E_1 and E_4 suffice to generate Ω, since $\langle E_4, E_5 \rangle = SL_2(\mathbb{Z})$ and $E_5 = E_1E_4E_1$.)

The element E_1 of Ω is represented by the automorphism of F transposing X and Y. This corresponds to the operation of interchanging the colours of the black and white vertices of a dessin, and is equivalent to replacing the Belyĭ function β with $1 - \beta$, or to transposing the monodromy generators x and y of G. This sends

$Z = (XY)^{-1}$ to Z^X, so it sends a dessin of type (l, m, n) to one of type (m, l, n) on the same surface. The element

$$S := E_4^{-1}E_5 = \begin{pmatrix} 0 & -1 \\ 1 & 1 \end{pmatrix}$$

has order 6; this lifts only to automorphisms of infinite order [**27**], but

$$T := S^2 = (E_4^{-1}E_5)^2 = \begin{pmatrix} -1 & -1 \\ 1 & 0 \end{pmatrix}$$

lifts to the automorphism $X \mapsto Z \mapsto Y \mapsto X$ which permutes the vertex colours black, white and red in a 3-cycle. The elements T and E_1 generate a subgroup Σ of Ω isomorphic to S_3, described by Machì in [**35**], inducing all permutations of the three colours and preserving the underlying surface. The element S^3 is the central involution $-I$ of Ω, induced by the automorphism inverting X and Y and reversing the orientation of each dessin. (Factoring out $\langle -I \rangle$ gives the central quotient $\Omega/Z(\Omega) \cong PGL_2(\mathbb{Z})$, the group of operations on all hypermaps, ignoring orientation [**23, 29**].) The subgroup $\Omega_1 := \langle E_1, S \rangle = \Sigma \times \langle -I \rangle \cong S_3 \times C_2 \cong D_6$ of Ω preserves the genus of each dessin, but its elements may reverse the orientation and permute the vertex colours.

The element

$$U := SE_4^{-1} = \begin{pmatrix} 0 & -1 \\ 1 & 0 \end{pmatrix}$$

corresponds to the automorphism $X \mapsto Y^{-1}$, $Y \mapsto X$ of order 4. This and its inverse transpose the colours of the black and white vertices, while reversing the cyclic permutations of edges around the vertices of one colour. Unlike the operations in Ω_1, these can (and usually do) change the genus of a dessin, even though the monodromy and automorphism groups are preserved. The elements U and E_1 generate a subgroup $\Omega_2 \cong D_4$; in addition to $U^2 = -I$, this contains the elements

$$E_2 = UE_1 = \begin{pmatrix} -1 & 0 \\ 0 & 1 \end{pmatrix} \quad \text{and} \quad E_3 = E_1U = \begin{pmatrix} 1 & 0 \\ 0 & -1 \end{pmatrix}$$

represented by the automorphisms of F inverting X or Y respectively. The operations corresponding to E_2 and E_3 reverse the cyclic permutations of edges around the black or white vertices of a dessin; they are sometimes called Petrie operations, since they transpose faces and Petrie polygons (closed zig-zag paths). These are special cases of operations $H_{i,j} : x \mapsto x^i, y \mapsto y^j$ on dessins considered by Streit, Wolfart and the author in [**30**], and also of certain operations on regular maps introduced earlier by Wilson in [**46**].

The subgroups Ω_1 and Ω_2 generate Ω. In fact, Ω is the free product

$$\Omega = \Omega_1 *_{\Omega_0} \Omega_2 \cong D_6 *_{D_2} D_4$$

of these two subgroups, amalgamating their common subgroup

$$\Omega_0 := \langle E_1, -I \rangle \cong D_2 \cong V_4.$$

Thus Ω is generated by operations of finite order; these have been classified by Pinto and the author in [**27**].

The orbits of Ω on $\mathcal{R}(G)$, or equivalently on $\mathcal{N}(G)$, correspond to the T_2-systems in G, that is, the orbits of $\mathrm{Aut}\,F \times \mathrm{Aut}\,G$ acting by composition on epimorphisms $F \to G$ and thus on generating pairs (x, y) for G. (These are called T-systems when the common size of the generating sets and the rank of the free group is arbitrary.)

Let $\nu = \nu(G)$ denote the number of orbits of Ω in each of these actions. Then $\mathcal{N}(G)$ and $\mathcal{R}(G)$ each split into $\nu(G)$ orbits $\mathcal{N}_i(G)$ and $\mathcal{R}_i(G)$ of length r_i for $i = 1, \ldots, \nu$, with $r_1 + \cdots + r_\nu = r(G)$. Each orbit determines a characteristic subgroup

$$K_i(G) = \bigcap_{N \in \mathcal{N}_i(G)} N$$

of F, corresponding to an Ω-invariant regular dessin

$$U_i(G) = D(K_i(G)) = \bigvee_{D \in \mathcal{R}_i(G)} D$$

with automorphism group $\operatorname{Aut} U_i(G) \cong F/K_i(G)$. We have

$$K(G) = \bigcap_{i=1}^{\nu(G)} K_i(G) \quad \text{and} \quad U(G) = \bigvee_{i=1}^{\nu(G)} U_i(G).$$

Example 7.1 It is known [**13, 38**] that if G is abelian then $\nu(G) = 1$, so all dessins in $\mathcal{R}(G)$ are equivalent under Ω. The same applies if G is a dihedral group. On the other hand, Dunwoody [**13**] has shown that for each prime p there are 2-generator p-groups G of nilpotence class 2 with $\nu(G)$ arbitrarily large.

An observation of Graham Higman, based on a theorem of Nielsen [**39**] (see also [**36**, Theorem 3.9]), states that if $G = \langle x, y \rangle$ then the union of the orbits of $\operatorname{Aut} G$ containing $c := [x, y]$ and $c^{-1} = [y, x]$ is invariant under Ω, and hence so is the order of c. In many cases this can be used to give lower bounds for $\nu(G)$.

Example 7.2 Neumann and Neumann [**38**] showed that if $G = A_5$ then the 19 elements of $\mathcal{N}(G)$ form two orbits of Ω, of lengths 9 and 10, so the same happens to the corresponding dessins. In fact, those of type a permutation of $(2, 5, 5)$, $(3, 3, 5)$ or $(3, 5, 5)^-$ form an orbit $\mathcal{R}_1(G)$ of length $r_1 = 3 + 3 + 3 = 9$, while those of type a permutation of $(2, 3, 5)$, $(3, 5, 5)^+$ or $(5, 5, 5)$ form an orbit $\mathcal{R}_2(G)$ of length $r_2 = 6 + 3 + 1 = 10$; here the superscript $+$ or $-$ indicates that the generators of order 5 are or are not conjugate in G. These orbits are distinguished by the property that, for each dessin, the commutator $[x, y]$ has order 3 or 5 respectively. They correspond to characteristic subgroups $K_i(G)$ of F with quotients G^{r_i}, such that $K(G) = K_1(G) \cap K_2(G)$, and to Ω-invariant regular dessins $U_i(G)$ of type $(30, 30, 30)$ and genera

$$1 + \frac{9}{20} \times 60^{r_i} = 4934963200000001 \quad \text{and} \quad 296097792000000001,$$

with $U(G) = U_1(G) \vee U_2(G)$.

Garion and Shalev [**14**] have shown that if G is a non-abelian finite simple group, then $\nu(G) \to \infty$ as $|G| \to \infty$, as conjectured by Guralnick and Pak [**19**].

Example 7.3 Let $G = L_2(p)$ for some prime $p \geq 5$. Then it follows from Dickson's description [**5**, Ch. XII] of the subgroups of G that the elements

$$x = \pm \begin{pmatrix} 1 & a \\ 0 & 1 \end{pmatrix} \quad \text{and} \quad y = \pm \begin{pmatrix} 1 & 0 \\ 1 & 1 \end{pmatrix}$$

generate G provided $a \neq 0$. This generating pair corresponds to a regular dessin $D_a \in \mathcal{R}(G)$ of type (p, p, k), where k is the order in G of an element with trace

$\pm(a+2)$. These generators have commutator

$$c = [x, y] = \pm \begin{pmatrix} a^2+a+1 & a^2 \\ -a & 1-a \end{pmatrix},$$

so c and c^{-1} have trace $\pm(a^2+2)$. Higman's commutator criterion therefore implies that if $\pm(a^2+2) \neq \pm(b^2+2)$ then D_a and D_b lie in different orbits of Ω, so $\nu(G) \geq (p-1)/4$.

An interesting invariant of a regular dessin D is the permutation group induced by Ω on the orbit containing D.

Example 7.4 According to Neumann and Neumann [**38**], Ω acts on $\mathcal{R}(A_5)$ as $S_9 \times S_{10}$. This is the largest possible group with orbits of lengths 9 and 10, so neither of the two orbits $\mathcal{R}_i(A_5)$ admits any non-trivial Ω-invariant structure.

Example 7.5 If $G = C_n$ then Ω acts on $F/K(G) = F/F'F^n \cong \mathbb{Z}_n^2$ as the subgroup $GL_2^{\pm 1}(\mathbb{Z}_n)$ of $GL_2(\mathbb{Z}_n)$ consisting of the matrices of determinant ± 1. (To see this, simply reduce the matrices E_i modulo (n), and note that the images of E_4 and E_5 generate $SL_2(\mathbb{Z}_n)$.) It follows that Ω acts on $\mathcal{R}(C_n)$ as $GL_2^{\pm 1}(\mathbb{Z}_n)$ acts on the subgroups of $\mathbb{Z}^2$ with quotient $\mathbb{Z}_n \cong C_n$. For instance, if n is a prime p then $\mathcal{R}(G)$ can be identified with the projective line $\mathbb{P}^1(\mathbb{F}_p)$, and Ω acts on it as $L_2(p)$ if $p = 2$ or $p \equiv 1 \bmod (4)$ (so that -1 is a square in $\mathbb{F}_p$), but as $PGL_2(p)$ otherwise. In the latter case, the triple transitivity of $PGL_2(p)$ ensures that there are no Ω-invariant binary or ternary relations on $\mathcal{R}(C_n)$; however, the quaternary relation induced by evaluating the cross-ratio is invariant in both cases.

Example 7.6 Let G be the non-abelian group of order pq in Examples 4.4 and 5.3. Since $xyz = 1$, the images $(\overline{x}, \overline{y}, \overline{z})$ in $(G/P)^3 \cong C_q^3 \cong \mathbb{F}_q^3$ of the generating triples (x, y, z) for G are the $q^2 - 1$ non-identity elements of a subgroup $S \cong F/F'F^q \cong C_q^2 \cong \mathbb{F}_q^2$. These are fixed by $\operatorname{Aut} G$, whereas Ω acts on them as $GL_2^{\pm 1}(q)$ (see Example 7.5), so Ω acts transitively on $\mathcal{R}(G)$.

If $q = 2$ then Ω acts 3-transitively on $\mathcal{R}(G)$ as S_3, but if $q > 2$ then its action is imprimitive, with the non-zero elements of the 1-dimensional linear subspaces of S forming $q+1$ blocks of size $q-1$; these are permuted transitively by Ω, acting as $L_2(q)$ or $PGL_2(q)$ on $\mathbb{P}^1(\mathbb{F}_q)$ as $q \equiv 1$ or $-1 \bmod (4)$ (again, see Example 7.5). These blocks are the orbits of $\Gamma = \operatorname{Gal} \overline{\mathbb{Q}}/\mathbb{Q}$, which acts on $\mathcal{R}(G)$ as $\operatorname{Gal} \mathbb{Q}(\zeta_q)/\mathbb{Q} \cong \mathbb{F}_q^*$, inducing the group $\mathbb{F}_q^*.I \cong C_{q-1}$ of scalar matrices on S. Thus the actions of Ω and Γ on $\mathcal{R}(G)$ commute, generating a group $\langle GL_1^{\pm 1}(q), \mathbb{F}_q^*.I\rangle$ which has index 2 or 1 in $GL_2(q)$ as $q \equiv 1$ or $-1 \bmod (4)$.

The actions of Γ and Ω on each of the sets $\mathcal{R}(G)$ considered above commute with each other (in some cases, though not Example 7.6, because one group acts trivially). In fact, the action of Γ always commutes with that of the Machì subgroup Σ of Ω, which simply permutes the three vertex colours, but the following example shows that on some sets $\mathcal{R}(G)$ it does not commute with Ω.

Example 7.7 There are three non-regular dessins of genus 0 and type $(6, 2, 6)$ consisting of plane trees with black and white vertices of valencies $3, 2, 1$ and $2, 2, 1, 1$: one is reflexible and the others form a chiral pair (see Fig. 1). Defined over the splitting field of the polynomial $25t^3 - 12t^2 - 24t - 16$, they form an orbit of Γ, which acts on them as the Galois group S_3 of this polynomial (see [**15**, Example 4.58] or [**32**, §2.2.2.3], for instance). They each have monodromy group $G \cong S_6$, so their minimal regular covers form an orbit of Γ consisting of three dessins of type $(6,2,6)$ and genus 61 in $\mathcal{R}(S_6)$; again, one is reflexible and the others form a chiral pair. There is an element of Γ permuting them in a 3-cycle, whereas the element $-I$ of Ω induces a transposition on them, so the actions of Γ and Ω on $\mathcal{R}(S_6)$ do not commute. (In fact, $\langle\Gamma,\Omega_1\rangle$ acts on its orbit containing these three regular dessins as $S_3 \times S_3$ on the 18 cosets of a subgroup $C_2 \times 1$, with Γ inducing $S_3 \times 1$ and $\Omega_1 = \langle -I\rangle \times \Sigma$ inducing $C_2 \times S_3$.)

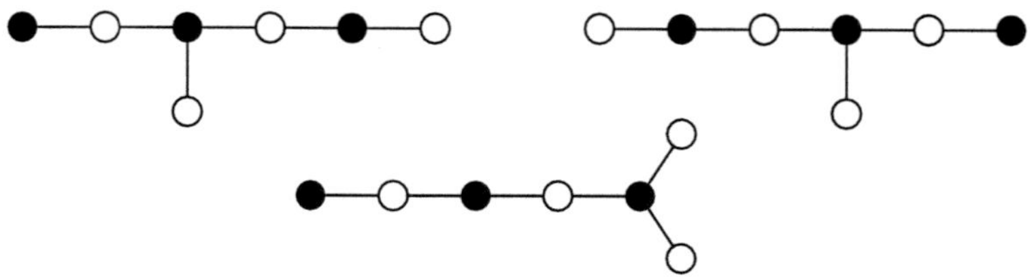

FIGURE 1. Three dessins of genus 0

8. Concluding remarks

Dessins can be regarded as unbranched finite coverings of $\mathbb{P}^1(\mathbb{C}) \setminus \{0,1,\infty\}$, or equivalently as finite permutation representations of its fundamental group $F = F_2$, so the results obtained here can also be applied to coverings of other spaces with the same fundamental group. These include a bouquet of two circles, with the regular objects the Cayley graphs of 2-generator groups [**41**], and a punctured torus, leading to the theory of origamis: see the survey [**22**] by Herrlich and Schmithüsen for parallels between origamis and dessins. In certain other categories, the objects can also be identified with the permutation representations of a particular group: thus one can use the free products $V_4 * C_2$ and $C_2 * C_2 * C_2$ for maps and hypermaps, string Coxeter groups for polytopes, and fundamental groups for covering spaces, etc. The enumerative techniques described here, along with the coverings and operations, can all be applied in similar ways in these contexts: see [**26**] for this more general approach.

References

[1] Paula Beazley Cohen, Claude Itzykson, and Jürgen Wolfart, *Fuchsian triangle groups and Grothendieck dessins. Variations on a theme of Belyĭ*, Comm. Math. Phys. **163** (1994), no. 3, 605–627. MR1284798 (96a:11056)

[2] G. V. Belyĭ, *Galois extensions of a maximal cyclotomic field* (Russian), Izv. Akad. Nauk SSSR Ser. Mat. **43** (1979), no. 2, 267–276, 479. MR534593 (80f:12008)

[3] Antonio J. Breda d'Azevedo and Gareth A. Jones, *Platonic hypermaps*, Beiträge Algebra Geom. **42** (2001), no. 1, 1–37. MR1824748 (2003c:05062)

[4] J. H. Conway, R. T. Curtis, S. P. Norton, R. A. Parker, and R. A. Wilson, *Atlas of finite groups*, Oxford University Press, Eynsham, 1985. Maximal subgroups and ordinary characters for simple groups; With computational assistance from J. G. Thackray. MR827219 (88g:20025)

[5] Leonard Eugene Dickson, *Linear groups: With an exposition of the Galois field theory*, with an introduction by W. Magnus, Dover Publications, Inc., New York, 1958. MR0104735 (21 #3488)

[6] John D. Dixon, *The probability of generating the symmetric group*, Math. Z. **110** (1969), 199–205. MR0251758 (40 #4985)

[7] John D. Dixon, *Asymptotics of generating the symmetric and alternating groups*, Electron. J. Combin. **12** (2005), Research Paper 56, 5 pp. (electronic). MR2180793 (2006m:05014)

[8] M. L. N. Downs, PhD thesis, University of Southampton, Southampton, UK, 1988.

[9] Martin Downs, *The Möbius function of* $\mathrm{PSL}_2(q)$*, with application to the maximal normal subgroups of the modular group*, J. London Math. Soc. (2) **43** (1991), no. 1, 61–75, DOI 10.1112/jlms/s2-43.1.61. MR1099086 (92d:20071)

[10] M. L. N. Downs and G. A. Jones, *Enumerating regular objects with a given automorphism group*, Discrete Math. **64** (1987), no. 2-3, 299–302, DOI 10.1016/0012-365X(87)90199-3. MR887369 (88g:20102)

[11] M. L. N. Downs and G. A. Jones, Enumerating regular objects associated with Suzuki groups, arXiv:1309.5215 [math.GR].

[12] M. J. Dunwoody, *On relation groups*, Math. Z. **81** (1963), 180–186. MR0153744 (27 #3705)

[13] M. J. Dunwoody, *On T-systems of groups*, J. Austral. Math. Soc. **3** (1963), 172–179. MR0153745 (27 #3706)

[14] Shelly Garion and Aner Shalev, *Commutator maps, measure preservation, and T-systems*, Trans. Amer. Math. Soc. **361** (2009), no. 9, 4631–4651, DOI 10.1090/S0002-9947-09-04575-9. MR2506422 (2010f:20019)

[15] E. Girondo and G. González-Diez, *Introduction to Compact Riemann Surfaces and Dessins d'Enfants*, London Math. Soc. Student Texts 79, Cambridge University Press, Cambridge, 2011.

[16] G. González-Diez and A. Jaikin-Zapirain, The absolute Galois group acts faithfully on regular dessins and on Beauville surfaces, preprint, 2013.

[17] Alexandre Grothendieck, *Esquisse d'un programme* (French, with French summary), Geometric Galois actions, 1, London Math. Soc. Lecture Note Ser., vol. 242, Cambridge Univ. Press, Cambridge, 1997, pp. 5–48. With an English translation on pp. 243–283. MR1483107 (99c:14034)

[18] Robert M. Guralnick and William M. Kantor, *Probabilistic generation of finite simple groups*, J. Algebra **234** (2000), no. 2, 743–792, DOI 10.1006/jabr.2000.8357. Special issue in honor of Helmut Wielandt. MR1800754 (2002f:20038)

[19] Robert Guralnick and Igor Pak, *On a question of B. H. Neumann*, Proc. Amer. Math. Soc. **131** (2003), no. 7, 2021–2025, DOI 10.1090/S0002-9939-02-06752-7. MR1963745 (2004a:20029)

[20] P. Hall, The Eulerian functions of a group, *Q. J. Math.* 7 (1936), 134–151.

[21] Frank Herrlich, *Teichmüller curves defined by characteristic origamis*, The geometry of Riemann surfaces and abelian varieties, Contemp. Math., vol. 397, Amer. Math. Soc., Providence, RI, 2006, pp. 133–144, DOI 10.1090/conm/397/07468. MR2218004 (2007d:32012)

[22] Frank Herrlich and Gabriela Schmithüsen, *Dessins d'enfants and origami curves*, Handbook of Teichmüller theory. Vol. II, IRMA Lect. Math. Theor. Phys., vol. 13, Eur. Math. Soc., Zürich, 2009, pp. 767–809, DOI 10.4171/055-1/19. MR2516744 (2010f:14036)

[23] Lynne D. James, *Operations on hypermaps, and outer automorphisms*, European J. Combin. **9** (1988), no. 6, 551–560, DOI 10.1016/S0195-6698(88)80052-0. MR970391 (90a:05069)

[24] Gareth A. Jones, *Ree groups and Riemann surfaces*, J. Algebra **165** (1994), no. 1, 41–62, DOI 10.1006/jabr.1994.1097. MR1272578 (95d:20030)

[25] Gareth A. Jones, *Enumeration of homomorphisms and surface-coverings*, Q. J. Math. Oxford Ser. (2) **46** (1995), no. 184, 485–507, DOI 10.1093/qmath/46.4.485. MR1366619 (96m:20059)

[26] G. A. Jones, Combinatorial categories and permutation groups, arXiv:1309.6119 [math.CO].

[27] Gareth A. Jones and Daniel Pinto, *Hypermap operations of finite order*, Discrete Math. **310** (2010), no. 12, 1820–1827, DOI 10.1016/j.disc.2009.12.019. MR2610286 (2011k:05105)

[28] Gareth A. Jones and Stephen A. Silver, *Suzuki groups and surfaces*, J. London Math. Soc. (2) **48** (1993), no. 1, 117–125, DOI 10.1112/jlms/s2-48.1.117. MR1223897 (94i:30037)

[29] Gareth Jones and David Singerman, *Maps, hypermaps and triangle groups*, The Grothendieck theory of dessins d'enfants (Luminy, 1993), London Math. Soc. Lecture Note Ser., vol. 200, Cambridge Univ. Press, Cambridge, 1994, pp. 115–145. MR1305395 (95m:20055)

[30] G. A. Jones, M. Streit, and J. Wolfart, *Wilson's map operations on regular dessins and cyclotomic fields of definition*, Proc. Lond. Math. Soc. (3) **100** (2010), no. 2, 510–532, DOI 10.1112/plms/pdp033. MR2595748 (2011e:14063)

[31] William M. Kantor and Alexander Lubotzky, *The probability of generating a finite classical group*, Geom. Dedicata **36** (1990), no. 1, 67–87, DOI 10.1007/BF00181465. MR1065213 (91j:20041)

[32] Sergei K. Lando and Alexander K. Zvonkin, *Graphs on surfaces and their applications*, Encyclopaedia of Mathematical Sciences, vol. 141, Springer-Verlag, Berlin, 2004. With an appendix by Don B. Zagier; Low-Dimensional Topology, II. MR2036721 (2005b:14068)

[33] Martin W. Liebeck and Aner Shalev, *The probability of generating a finite simple group*, Geom. Dedicata **56** (1995), no. 1, 103–113, DOI 10.1007/BF01263616. MR1338320 (96h:20116)

[34] Roger C. Lyndon and Paul E. Schupp, *Combinatorial group theory*, Springer-Verlag, Berlin-New York, 1977. Ergebnisse der Mathematik und ihrer Grenzgebiete, Band 89. MR0577064 (58 #28182)

[35] A. Machì, *On the complexity of a hypermap*, Discrete Math. **42** (1982), no. 2-3, 221–226, DOI 10.1016/0012-365X(82)90219-9. MR677055 (84a:05024)

[36] Wilhelm Magnus, Abraham Karrass, and Donald Solitar, *Combinatorial group theory*, Second revised edition, Dover Publications, Inc., New York, 1976. Presentations of groups in terms of generators and relations. MR0422434 (54 #10423)

[37] Attila Maróti and M. Chiara Tamburini, *Bounds for the probability of generating the symmetric and alternating groups*, Arch. Math. (Basel) **96** (2011), no. 2, 115–121, DOI 10.1007/s00013-010-0216-z. MR2773211 (2012c:20200)

[38] Bernhard H. Neumann and Hanna Neumann, *Zwei Klassen charakteristischer Untergruppen und ihre Faktorgruppen* (German), Math. Nachr. **4** (1951), 106–125. MR0040297 (12,671d)

[39] J. Nielsen, *Die Isomorphismen der allgemeinen, unendlichen Gruppe mit zwei Erzeugenden* (German), Math. Ann. **78** (1917), no. 1, 385–397, DOI 10.1007/BF01457113. MR1511907

[40] Michael E. O'Nan, *Some evidence for the existence of a new simple group*, Proc. London Math. Soc. (3) **32** (1976), no. 3, 421–479. MR0401905 (53 #5731)

[41] David Singerman and Jürgen Wolfart, *Cayley graphs, Cori hypermaps, and dessins d'enfants*, Ars Math. Contemp. **1** (2008), no. 2, 144–153. MR2453475 (2009i:05108)

[42] Manfred Streit and Jürgen Wolfart, *Characters and Galois invariants of regular dessins*, Rev. Mat. Complut. **13** (2000), no. 1, 49–81. MR1794903 (2001h:14026)

[43] Michio Suzuki, *On a class of doubly transitive groups*, Ann. of Math. (2) **75** (1962), 105–145. MR0136646 (25 #112)

[44] Robert A. Wilson, *The maximal subgroups of the O'Nan group*, J. Algebra **97** (1985), no. 2, 467–473, DOI 10.1016/0021-8693(85)90059-6. MR812997 (87d:20022)

[45] Robert A. Wilson, *The finite simple groups*, Graduate Texts in Mathematics, vol. 251, Springer-Verlag London, Ltd., London, 2009. MR2562037 (2011e:20018)

[46] Stephen E. Wilson, *Operators over regular maps*, Pacific J. Math. **81** (1979), no. 2, 559–568. MR547621 (81a:57004)

[47] J. Wolfart, *ABC for polynomials, dessins d'enfants and uniformization—a survey*, Elementare und analytische Zahlentheorie, Schr. Wiss. Ges. Johann Wolfgang Goethe Univ. Frankfurt am Main, 20, Franz Steiner Verlag Stuttgart, Stuttgart, 2006, pp. 313–345. MR2310190 (2009d:14031)

[48] Satoshi Yoshiara, *The maximal subgroups of the sporadic simple group of O'Nan*, J. Fac. Sci. Univ. Tokyo Sect. IA Math. **32** (1985), no. 1, 105–141. MR783183 (86m:20025)

School of Mathematics, University of Southampton, Southampton SO17 1BJ, United Kingdom

E-mail address: `G.A.Jones@maths.soton.ac.uk`

Growth in Hecke groups

Martin Kreuzer and Gerhard Rosenberger

This paper is dedicated to Professor Emilio Bujalance on occasion of his sixtieth birthday.

ABSTRACT. For the Hecke groups $\mathcal{H}(\lambda_q) = \langle s,t : s^2 = (st)^q = 1\rangle$, we determine and collect a number of growth functions, their generating series and growth rates: numbers of elements of a given length, numbers of subgroups of a certain index, numbers of normal subgroups, numbers of normal subgroups of genus one, and numbers of free subgroups. Depending on the type of these growth functions, we calculate the corresponding generating series or the generating Dirichlet series. A particular emphasis is put on explicit formulas representing these series as rational functions or via Euler product decompositions.

In [**16**], while studying Dirichlet series satisfying certain functional equations, E. Hecke introduced an infinite family of groups $\mathcal{H}(\lambda_q)$, where $q \geq 3$. Geometrically, they are discrete subgroups of $\mathrm{PSL}_2(\mathbb{R})$ consisting of linear fractional transformations preserving the upper half plane $\mathbb{H} = \{z \in \mathbb{C} \mid \Im(z) > 0\}$. For every $q \geq 3$, the group $\mathcal{H}(\lambda_q)$ is generated by the transformations $S: z \mapsto -1/z$ and $T: z \mapsto z+\lambda_q$ where $\lambda_q = 2\cos(\pi/q)$. Algebraically, the Hecke group $\mathcal{H}(\lambda_q)$ is given by the group presentation

$$\mathcal{H}(\lambda_q) = \langle s,t : s^2 = (st)^q = 1\rangle.$$

and can also be viewed as the free product $\mathcal{H}(\lambda_q) = C_2 * C_q$ of two cyclic groups of orders 2 and q, respectively. Hecke groups are a natural generalization of the modular group which is nothing but $\mathcal{H}(\lambda_3)$.

In the first part of this paper, we describe the Hilbert-Dehn functions for the two main presentations of a Hecke group, i.e. the growth functions of the number of shortest words representing group elements. Rather than referring to the abstract theory of Fuchsian groups, we provide explicit recursive formulas based on the computation of the Gröbner bases of the defining ideals of the corresponding group rings. As a consequence, we show that the generating series of these functions are rational power series, and we exhibit explicit formulas for them.

The next topic is the growth of the number of subgroups of a given (finite) index of a **group** G. A compact way of representing this subgroup counting function $a_n(G)$ is its generating Dirichlet series, called the zeta function of the group G. We

2010 *Mathematics Subject Classification.* Primary 20H10; Secondary 16Z05, 20E07, 11M41.

Key words and phrases. Hecke group, Hilbert function, Dehn function, subgroup growth, zeta function.

The authors thank the referee for his careful reading of the paper and his useful suggestions.

recall the result of [8] (generalizing the special case in [13]) that the zeta function of a finitely generated, nilpotent group has a natural Euler product expansion. In the case of the Hecke groups, we have to be content with recursive equations for $a_n(\mathcal{H}(\lambda_q))$ and with the computation of the initial values of $a_n(\mathcal{H}(\lambda_p))$ for a prime p.

More interesting results, in particular surprising connections with special functions in analytic number theory, are found if we restrict ourselves to counting the numbers of special subgroups of index n of a fixed Hecke group. One such setting is the Dirichlet series whose coefficients are the numbers of free normal subgroups of genus one and index n. For several groups, including $\mathcal{H}(\lambda_3)$, $\mathcal{H}(\lambda_4)$, and $\mathcal{H}(\lambda_6)$, we obtain zeta functions satisfying functional equations and defining automorphic forms. For instance, for the groups $\mathcal{H}(\lambda_3)$ and $\mathcal{H}(\lambda_6)$ we get an automorphic $(\sqrt{3}, 1, 1)$-form, and $\mathcal{H}(\lambda_4)$ yields an automorphic $(2, 1, 1)$-form.

Another case of interest is the counting function $f_n(\mathcal{H}(\lambda_q))$ for free subgroups of index n. Here we have an explicit, general recursive formula, and in several important special cases the counting function satisfies a homogeneous linear differential equation of Riccati type with integral coefficients.

Clearly, even in the relatively easy setting of Hecke groups, there are comparatively few general growth results and many open questions. Are there connections between the Hilbert-Dehn function and the various subgroup counting functions? Are there more explicit formulas for the subgroup counting function, the normal subgroup counting function, or the free subgroup counting function? Are there special properties of the corresponding generating series or Dirichlet series? With this paper we hope to provide a clear account of the state-of-the-art and a solid basis and reference for further research.

Unless explicitly stated otherwise, we shall adhere to the notation and definitions given in [3], [22], and [23].

1. Hilbert-Dehn Functions of Hecke Groups

Algebraically, the Hecke group $\mathcal{H}(\lambda_q)$ is defined by the group presentation

$$\mathcal{H}(\lambda_q) \;=\; \langle s,t:\; s^2 = (st)^q = 1\rangle.$$

In the following, we shall use methods from computer algebra to study the growth of the length of the elements in this group. Therefore it will be convenient to let $u = t^{-1}$ and to use the monoid presentation

$$H_q' \;=\; \langle s,t,u:\; s^2 = (st)^q = tu = ut = 1\rangle.$$

Furthermore, the set $\{a, b\}$ consisting of the elements $a = s$ and $b = st$ is also a system of generators of the group $\mathcal{H}(\lambda_q)$, and it has the advantage that both generators have a finite order. Consequently, letting $c = b^{-1}$, we shall also consider the monoid presentation

$$H_q \;=\; \langle a,b,c:\; a^2 = b^q = bc = cb = 1\rangle.$$

Our first goal is to study the Hilbert-Dehn function of $\mathcal{H}(\lambda_q)$ with respect to these two presentations. The Hilbert-Dehn function of a finitely presented monoid counts the number of elements whose shortest representation (in terms of the given generators) has a given length. More generally, let K be a field and let $P = K\langle x_1, \dots, x_n\rangle$ be the non-commutative polynomial ring (i.e. the free associative algebra) in the indeterminates $x_1, \dots, x_n$. Then we can define the Hilbert-Dehn function of any residue class ring of P as follows.

DEFINITION 1.1. Let I be a two-sided ideal in P, and for every $i \in \mathbb{Z}$ let $P_{\leq i}$ be the K-vector subspace of P generated by all words of length $\leq i$ in the letters $x_1, \ldots, x_n$. Then the **Hilbert-Dehn function** of the ring $R = P/I$ is defined by

$$\mathrm{HD}_R(i) = \dim_K(P_{\leq i}/(I \cap P_{\leq i})) - \dim_K(P_{\leq i-1}/(I \cap P_{\leq i-1}))$$

Given a monoid presentation $M = \langle x_1, \ldots, x_n : w_1 = \cdots = w_s = 1 \rangle$, we choose a field K and consider the **monoid ring**

$$K\langle M \rangle = P/\langle w_1 - 1, \ldots, w_s - 1 \rangle.$$

Then the Hilbert-Dehn function of M, denoted by $\mathrm{HD}_M(i) = \mathrm{HD}_{K\langle M \rangle}(i)$, measures the number of elements of M whose shortest representation as a word in the residue classes $\bar{x}_1, \ldots, \bar{x}_n$ has length i. It is also called the **growth function** of M. If we want to compute this function via computer algebra, we can use the following result which is a non-commutative version of Macaulay's Basis Theorem (see [**22**], Thm. 1.5.7 and [**23**], Thm. 5.1.8).

PROPOSITION 1.2. *In the setting above, let I be a two-sided ideal in P, and let $R = P/I$. For every $i \geq 0$, the value $\mathrm{HD}_R(i)$ of the Hilbert-Dehn function of R can be computed as follows.*

(1) *Choose a length compatible word ordering σ and determine a σ-Gröbner basis G of I.*
(2) *Let $\mathcal{O}_\sigma(I)_i$ be the set of words of length i which are not subwords of a leading word of one of the elements of G.*
(3) *Return* $\mathrm{HD}_R(i) = \#\mathcal{O}_\sigma(I)_i$.

The problem with this approach is that, in general, the Gröbner basis G need not be finite. Hence this proposition my lead to an infinite computation. However, for the presentations of $\mathcal{H}(\lambda_q)$ considered in this paper, we shall show that there exist finite Gröbner bases and the proposition leads to an effective result.

Before dealing with the general situation, let us have a look at a couple of simple cases.

EXAMPLE 1.3. **(The Modular Group)**
Let $H_3 = \langle a, b, c : a^2 = b^3 = bc = cb = 1 \rangle$ be the modular group, and let `llex` be the length-lexicographic word ordering (i.e. we first compare the length of two words and then we break ties using the usual lexicographic ordering). Then the group ring of H_3 over a field K satisfies

$$K\langle H_3 \rangle = K\langle a, b, c \rangle / \langle a^2 - 1,\ b^3 - 1,\ bc - 1,\ cb - 1 \rangle$$

and has the `llex`-Gröbner basis

$$G_{H_3} = \{a^2 - 1,\ bc - 1,\ cb - 1,\ b^2 - c,\ c^2 - b\}.$$

The first 10 values of $\mathrm{HD}_{H_3}(i)$ are

$$\mathrm{HD}_{H_3} : 1, 3, 4, 6, 8, 12, 16, 24, 32, 48.$$

It is evident (and will be shown later) that this Hilbert-Dehn function satisfies the recursive equation $\mathrm{HD}_{H_3}(i) = 2 \cdot \mathrm{HD}_{H_3}(i-2)$ for $i \geq 3$ with initial values $\mathrm{HD}_{H_3}(0) = 1$, $\mathrm{HD}_{H_3}(1) = 3$, and $\mathrm{HD}_{H_3}(2) = 4$.

The next case yields a pleasant surprise.

EXAMPLE 1.4. Let us consider the Hecke group for $q = 4$ and its presentation $H_4 = \langle a, b, c : \ a^2 = b^4 = bc = cb = 1 \rangle$. Then the llex-Gröbner basis of the ideal $\langle a^2 - 1,\ b^4 - 1,\ bc - 1,\ cb - 1 \rangle$ is given by

$$G_{H_4} = \{a^2 - 1,\ bc - 1,\ cb - 1,\ b^2 - c^2,\ c^3 - b\}$$

and the first 10 values of the Hilbert-Dehn function are

$$\mathrm{HD}_{H_4} :\ 1,\ 3,\ 5,\ 8,\ 13,\ 21,\ 34,\ 55,\ 89,\ 144.$$

Of course, we recognize the well-known Fibonacci numbers $F(i)$ given by $F(1) = F(2) = 1$ and $F(i) = F(i-1) + F(i-2)$ for $i \geq 3$. The recursive equation $\mathrm{HD}_{H_4}(i) = \mathrm{HD}_{H_4}(i-1) + \mathrm{HD}_{H_4}(i-2)$ for $i \geq 3$ and the initial values $\mathrm{HD}_{H_4}(0) = 1$, $\mathrm{HD}_{H_4}(1) = 3$, and $\mathrm{HD}_{H_4}(2) = 5$ show that we have $\mathrm{HD}_{H_4}(i) = F(i+3)$ for $i \geq 1$.

Our next two propositions provide the llex-Gröbner bases for the two monoid presentations of $\mathcal{H}(\lambda_q)$ introduced above. As before, we let llex denote the length-lexicographic term ordering.

PROPOSITION 1.5. *Let $q \geq 3$, let $\mathcal{H}(\lambda_q)$ be the q-th Hecke group, and let K be a field.*

(1) *With respect to the presentation*

$$H'_q = \langle s, t, u : \ s^2 = (st)^q = tu = ut = 1 \rangle$$

an llex-*Gröbner basis of the two-sided ideal $\langle s^2 - 1,\ (st)^q - 1,\ tu - 1,\ ut - 1 \rangle$ in the non-commutative polynomial ring $K\langle s, t, u \rangle$ is given by*

$$G_{H'_q} = \begin{cases} \{s^2 - 1,\ tu - 1,\ ut - 1,\ f_1,\ f_2\} & \text{if } q = 2m \text{ is even,} \\ \{s^2 - 1,\ tu - 1,\ ut - 1,\ g_1,\ g_2,\ g_3,\ g_4\} & \text{if } q = 2m + 1 \text{ is odd,} \end{cases}$$

where $f_1 = (st)^m - (us)^m$, $f_2 = (su)^m - (ts)^m$, $g_1 = (st)^m s - (us)^m u$, $g_2 = (su)^m s - (ts)^m t$, $g_3 = (st)^m t (st)^m - (us)^m u (us)^m$, and finally $g_4 = (su)^m u (su)^m - (ts)^m t (ts)^m$.

(2) *With respect to the presentation*

$$H_q = \langle a, b, c : \ a^2 = b^q = bc = cb = 1 \rangle$$

an llex-*Gröbner basis of the two-sided ideal $\langle a^2 - 1,\ b^q - 1,\ bc - 1,\ cb - 1 \rangle$ in the non-commutative polynomial ring $K\langle a, b, c \rangle$ is given by*

$$G_{H_q} = \{a^2 - 1,\ bc - 1,\ cb - 1,\ f_1, f_2\}$$

where $f_1 = b^m - c^m$, $f_2 = c^{m+1} - b^{m-1}$ if $q = 2m$ is even, and where $f_1 = b^{m+1} - c^m$, $f_2 = c^{m+1} - b^m$ if $q = 2m + 1$ is odd.

PROOF. The fact that these sets generate the correct ideals is straightforward to verify. In order to prove that they are llex-Gröbner bases, we use the non-commutative version of the Buchberger Criterion (see for instance [**24**], Prop. 2.10).

For example, let us check that the set G_{H_q} is an llex-Gröbner basis in the case that $q = 2m$ is even. The first polynomial has a leading word which has no non-trivial obstructions with the other leading words. Between the second and third polynomials there are two obstructions, leading to the S-polynomials $S_{23} = (cbc - c) - (cbc - c) = 0$ and $S'_{23} = (bcb - b) - (bcb - b) = 0$, both of which are zero. Between the leading words of the second and the fourth polynomial there is one obstruction, leading to $S_{24} = (b^m c - b^{m-1}) - (b^m c - c^{m+1} = c^{m+1} - b^{m-1} \longrightarrow 0$ by reduction via f_2. Similarly, we have non-trivial obstructions and reductions

$S_{25} = (bc^{m+1} - c^m) - (bc^{m+1} - b^m) = b^m - c^m \longrightarrow 0$ via f_1, $S_{34} = (cb^m - b^{m-1}) - (cb^m - c^{m+1}) = c^{m+1} - b^{m-1} \longrightarrow 0$ via f_2, and $S_{35} = (c^{m+1}b - c^m) - (c^{m+1}b - b^m) = b^m - c^m \longrightarrow 0$ via f_1. Between the last two polynomials there is only a trivial obstruction.

Altogether, it follows that G_{H_q} is an `llex`-Gröbner basis for even $q = 2m$. The other cases can be dealt with in the very same fashion. □

Knowing a Gröbner basis for the defining ideal I enables us to describe the set of irreducible words $\mathcal{O}_\sigma(I)_i$ for each degree i. Thus we can find and prove recursive equations satisfied by the Hilbert-Dehn functions of the group rings $K\langle G\rangle = P/I$. Our next two propositions achieve this task for the two presentations of the Hecke groups $\mathcal{H}(\lambda_q)$ introduced above.

PROPOSITION 1.6. *Let $q \geq 3$, and let $H'_q = \langle s,t,u :\ s^2 = (st)^q = tu = ut = 1\rangle$ be the presentation of the q-th Hecke group $H'_q = \mathcal{H}(\lambda_q)$ introduced above. Then the Hilbert-Dehn function of H'_q satisfies $\mathrm{HD}_{H'_q}(0) = 1$, $\mathrm{HD}_{H'_q}(i) = 3\cdot 2^{i-1}$ for $1 \leq i \leq q-1$, $\mathrm{HD}_{H'_q}(q) = 3\cdot 2^{q-1} - 2$, and $\mathrm{HD}_{H'_q}(q+1) = 3\cdot 2^q - 8$, as well as the recursive equation*

$$\mathrm{HD}_{H'_q}(i) \;=\; 2\cdot \mathrm{HD}_{H'_q}(i-1) - \mathrm{HD}_{H'_q}(i-q)$$

for every $i \geq q+2$.

PROOF. Using the method of Proposition 1.2, we choose a field K, let $\sigma = \mathtt{llex}$ and apply the Gröbner basis determined in the preceding proposition. Thus the leading word ideal is generated by the leading words of the elements of this Gröbner basis and the irreducible words are the words which do not have any of these leading words as subwords.

Let us start with the case of an even number $q = 2m$. The leading words of the Gröbner basis given in Proposition 1.5 are s^2, tu, ut, $(st)^m$, and $(su)^m$. The last two words have length q. So, to prove the claim for $1 \leq i \leq q$, it suffices to show that there are $3\cdot 2^{i-1}$ words of length i in the letters s,t,u which do not contain s^2, tu, or ut as a subword.

To prove this, we argue by induction on i. For $i = 1$, there are obviously three words s, t, and u. For $i \geq 1$, let $g(i)$ be the number of words ending in t or u, and let $h(i)$ be the number of words ending in s. Then we have $h(1) = 1$, $g(1) = 2$, $g(2) = 4$, and the recursive fomulas $h(i) = g(i-1)$ as well as $g(i) = g(i-1) + 2g(i-2)$. This leads to $g(i) = 2^i$ and $\mathrm{HD}_{H'_q}(i) = g(i) + g(i-1) = 3\cdot 2^{i-1}$.

For $i = q$, the value of $\mathrm{HD}_{H'_q}(i)$ has to be decreased by two, since the words $(st)^m$ and $(su)^m$ have to be skipped in this count. For $i = q+1$, there are four further words that have to be skipped, namely the multiples $(st)^m t$, $(su)^m)u$, $u(st)^m$, and $t(su)^m$. Notice that the other four multiples of $(st)^m$ and $(su)^m$ are not irreducible, e.g. $(su)^m s = s(us)^m$ is equivalent to $s(st)^m = (ts)^{m-1}t$ modulo f_1. Hence we get $\mathrm{HD}_{H'_q}(q+1) = 2\cdot \mathrm{HD}_{H'_q}(q) - 4 = 3\cdot 2^q - 8$.

Next we prove the recursive equation satisfied by $\mathrm{HD}_{H'_q}(i)$ for $i \geq q+2$. For extending an irreducible word w of length $i-1$ to an irreducible word of length i, we have usually two choices. If w ends in s, we can append t or u. If w ends in t, we can append t or s. If w ends in u, we can append u or s. This would indicate that the number of irreducible words doubles at each step. However, there is an exceptional situation when the given word ends in $(st)^{m-1}s$ or in $(su)^{m-1}s$. In this cases we may append only u to $(st)^{m-1}s$ and only v to $(su)^{m-1}s$, as we would otherwise produce a reducible word. To count the irreducible words of length $i-1$ of the form $w = w'(st)^{m-1}s$, we note that w' has length $i-q$ and has to end with t^2 or u. Similarly, if $w = w''(su)^{m-1}s$, then w'' has to end with u^2 or t. By changing the last letter, we see that the irreducible words w' ending with t^2 are in 1-1 correspondence with the irreducible words of length $i-q$ ending with ts. Similarly, the words w'' ending with u^2 correspond 1-1 to the irreducible words ending with us. Altogether, the words w' and w'' are in 1-1 correspondence with all the irreducible words of length $i-q$, and their number is $\mathrm{HD}_{H'_q}(i-q)$.

The proof that the Hilbert-Dehn function is the stated one in the case of odd $q = 2m+1$ proceeds analogously and is left to the interested reader. □

The next proposition deals with the computation of the Hilbert-Dehn function for the second presentation of $\mathcal{H}(\lambda_q)$ given above.

PROPOSITION 1.7. *Let $q \geq 3$, and let $H_q = \langle a, b, c : a^2 = b^q = bc = cb = 1 \rangle$ be the presentation of the q-th Hecke group $H_q = \mathcal{H}(\lambda_q)$ introduced above.*

(1) *If $q = 2m$ is an even number, then the Hilbert-Dehn function of H_q satisfies $\mathrm{HD}_{H_q}(0) = 1$, $\mathrm{HD}_{H_q}(i) = 3 \cdot 2^{i-1}$ for $1 \leq i \leq m-1$, $\mathrm{HD}_{H_q}(m) = 3 \cdot 2^{m-1} - 1$, and $\mathrm{HD}_{H_q}(m+1) = 3 \cdot 2^m - 4$, as well as the recursive equation*

$$\mathrm{HD}_{H_q}(i) = 2 \cdot \mathrm{HD}_{H_q}(i-1) - \mathrm{HD}_{H_q}(i-m-1)$$

for $i \geq m+2$.

(2) *If $q = 2m+1$ is an odd number, then the Hilbert-Dehn function of H_q satisfies $\mathrm{HD}_{H_q}(0) = 1$, $\mathrm{HD}_{H_q}(i) = 3 \cdot 2^{i-1}$ for $1 \leq i \leq m$, $\mathrm{HD}_{H_q}(m+1) = 3 \cdot 2^m - 2$, $\mathrm{HD}_{H_q}(m+2) = 3 \cdot 2^{m+1} - 6$, and $\mathrm{HD}_{H_q}(m+3) = 3 \cdot 2^{m+2} - 16$, as well as the recursive equation*

$$\mathrm{HD}_{H_q}(i) = \cdot \mathrm{HD}_{H_q}(i-1) + 2 \cdot \mathrm{HD}_{H_q}(i-2) - 2 \cdot \mathrm{HD}_{H_q}(i-m-2)$$

for $i \geq m+4$.

PROOF. To prove (1), let $q = 2m$ be even. By Proposition 1.5, the leading words of the elements of an `llex`-Gröbner basis of the two-sided ideal $\langle a^2-1, bc-1, cb-1, b^q-1 \rangle$ are a^2, bc, cb, b^m, and c^{m+1} in this case. Therefore the proof of the preceding proposition shows the claim for $0 \leq i \leq m$. To check that the given value of $\mathrm{HD}_{H_q}(m+1)$ is correct, we note that four further words of length $m+1$ turn out to be reducible, namely the three extensions b^{m+1}, cb^m, and $b^m c$ of b^m as well as the new leading word c^{m+1}.

Thus we prove the recursive equation for $i \geq m+2$ next. We argue again by induction on i. If we extend an irreducible word w of length $i-1$ to an irreducible word of length i we have usually two letters available: the letters b and c if w ends in a, the letters a and b if w ends in b, and the letters a and c if w ends in c. There are two exceptional cases when we have only one choice (namely a) available, namely if w ends in b^{m-1} or in c^m. In the first case, the word w is of the form $w = w'ab^{m-1}$ with an irreducible word w' of length $i-m-1$ which ends in b or c. In the second case, the word w is of the form $w = w''c^m$ with an irreducible word w'' of length $i-m-1$ which ends in a. Altogether, the number of exceptional cases is the number of irreducible words of length $i-m-1$, i.e. it is $\mathrm{HD}_{H_q}(i-m-1)$.

Now we prove (2). Let $q = 2m+1$ be an odd number. The leading word ideal is generated by a^2, bc, cb, b^{m+1}, and c^{m+1} in this case. For $0 \leq i \leq m$, the proof of the preceding proposition shows the claim, and for $i = m + 1$ it suffices to note that the two new leading terms b^{m+1} and c^{m+1} reduce the number of irreducible words by two. For $i = m + 2$, we have to reduce the number $3 \cdot 2^{m+1}$ by six, since the words b^{m+1}, cb^{m+1}, $b^{m+1}c$, c^{m+2}, bc^{m+1}, and $c^{m+1}b$ are reducible. A similar count shows that 16 out of the $3 \cdot 2^{m+2}$ words of length $m + 3$ become reducible by considering multiples of b^{m+1} and c^{m+1}.

Thus it remains to prove the recursive formula for $i \geq m + 2$. As before, we argue by induction on i. We let $A(i)$ be the number of irreducible words of length i ending in the letter a, $B(i)$ the number of irreducible words ending in b, and $C(i)$ the number of irreducible words ending in c. Since the set of leading words generating the leading word ideal is symmetric with respect to interchanging b and c, we have $B(i) = C(i)$, and therefore $\mathrm{HD}_{H_q}(i) = A(i) + 2B(i)$. When we want to extend an irreducible word w of length $i - 1$ to one of length i, we can use the letter a for all words ending in b or c. This shows $A(i) = 2B(i - 1)$. If w ends in b, we can use a or c, unless w ends in b^m. In the latter case w is of the form $w = w'b^m$ with an irreducible word w' of length $i - m$ ending in a. The analogous claim holds if w ends in c. Thus we get $C(i) = A(i - 1) + B(i - 1) - A(i - m - 1)$ and $B(i) = A(i - 1) + C(i - 1) - A(i - m - 1)$. Altogether, this yields

$$\begin{aligned}
\mathrm{HD}_{H_q}(i) &= A(i) + B(i) + C(i) \\
&= 2B(i-1) + 2A(i-1) + 2B(i-1) - 2A(i-m-1) \\
&= 2\,\mathrm{HD}_{H_q}(i-1) - 2A(i-m-1) \\
&= \mathrm{HD}_{H_q}(i-1) - 2A(i-m-1) + 2\,\mathrm{HD}_{H_q}(i-2) - 2A(i-m-2) \\
&= \mathrm{HD}_{H_q}(i-1) + 2\,\mathrm{HD}_{H_q}(i-2) - 2A(i-m-2) - 4B(i-m-2) \\
&= \mathrm{HD}_{H_q}(i-1) + 2\,\mathrm{HD}_{H_q}(i-2) - 2\,\mathrm{HD}_{H_q}(i-m-2)
\end{aligned}$$

as was to be shown. □

REMARK 1.8. The generating series for the case of an odd number $q = 2m + 1$ has the denominator $1 - z - 2z^2 + 2z^{m+2}$, as explained in the next section. This denominator factors into $1 - z - 2z^2 + 2z^{m+2} = (1 - z)(1 - 2z^2 - 2z^3 - \cdots - 2z^{m+1})$. This shows that $\mathrm{HD}_{H_q}(i)$ satisfies also the recursion

$$\mathrm{HD}_{H_q}(i) = 2 \cdot \mathrm{HD}_{H_q}(i-2) + 2 \cdot \mathrm{HD}_{H_q}(i-3) + \cdots + 2 \cdot \mathrm{HD}_{H_q}(i-m-1)$$

which holds for $i \geq m + 2$, as one can check immediately.

2. Hilbert-Dehn Series of Hecke Groups

A more convenient way of administering the information contained in the Hilbert-Dehn function of a finitely presented group consists in coding it into its generating function, the Hilbert-Dehn series. This series is defined as follows.

DEFINITION 2.1. Let $M = \langle x_1, \ldots, x_n : w_1 = \cdots = w_s = 1 \rangle$ be a finitely presented monoid, and let z denote a new indeterminate. Then the power series

$$\mathrm{HDS}_M(z) = \sum_{i=0}^{\infty} \mathrm{HD}_M(i)\, z^i \in \mathbb{Z}[[z]]$$

is called the **Hilbert-Dehn series** or the **(spherical) growth series** of M (or, more precisely, of the given presentation of M).

It is well-known that HDS(z) is a rational power series if $\mathrm{HD}_M(i)$ satisfies a linear recurrence relation for $i \gg 0$ (see for instance [**23**], Thm. 5.2.6). Thus, by the results of Section 1, the Hilbert-Dehn series of the presentations of Hecke groups introduced there are rational power series. Our first task in this section is to determine them. To this end, we shall follow the procedure outlined in the proof of [**23**], Thm. 5.2.6.

PROPOSITION 2.2. *Let $q \geq 3$, and let $H'_q = \langle s,t,u : s^2 = (st)^q = tu = ut = 1\rangle$ be the presentation of the q-th Hecke group $H'_q = \mathcal{H}(\lambda_q)$ introduced in Section 1. Then the Hilbert-Dehn series of H'_q is given by*

$$\mathrm{HDS}_{H'_q}(z) = \frac{1+2z+2z^2+\cdots+2z^{q-1}+z^q}{1-z-z^2-\cdots-z^{q-1}}$$

PROOF. By Proposition 1.6, the Hilbert-Dehn function $\mathrm{HD}_{H'_q}(i)$ satisfies the recursive equation $\mathrm{HD}_{H'_q}(i) = 2\cdot\mathrm{HD}_{H'_q}(i-1) - \mathrm{HD}_{H'_q}(i-q)$ for $i \geq q+2$. Therefore the denominator of one representation of $\mathrm{HDS}_{H'q}(i)$ is $(1-2z+z^q)$. The coefficients $d_i \in \mathbb{Z}$ of the corresponding numerator $g(z) = d_0 + d_1 z + \cdots + d_{q+1}z^{q+1}$ satisfy $d_0 = \mathrm{HD}_{H'_q}(0) = 1$, $d_1 = \mathrm{HD}_{H'_q}(1) - 2 = 1$, $d_2 = \mathrm{HD}_{H'_q}(2) - 2\,\mathrm{HD}_{H'_q}(1) = 0, \ldots,$ $d_{q-1} = \mathrm{HD}_{H'_q}(q-1) - 2\,\mathrm{HD}_{H'_q}(q-2) = 0$, $d_q = \mathrm{HD}_{H'_q}(q) - 2\,\mathrm{HD}_{H'_q}(q-1) + 1 = -1$, and $d_{q+1} = \mathrm{HD}_{H'_q}(q+1) - 2\,\mathrm{HD}_{H'_q}(q) + 3 = -1$. This yields $g(z) = 1+z-z^q-z^{q+1} = (1-z)(1+z)(1-z-z^2-\cdots-z^{q-1})$. Combining this with the factorization $1-2z+z^q = (1-z)(1-z-z^2-\cdots-z^{q-1})$, we find

$$\mathrm{HDS}_{H'_q}(z) = \frac{(1-z)(1+z)(1-z-z^2-\cdots-z^{q-1})}{(1-z)(1-z-z^2-\cdots-z^{q-1})} = \frac{1+2z+2z^2+\cdots+2z^{q-1}+z^q}{1-z-z^2-\cdots-z^{q-1}}$$

as claimed. □

In a similar way, we can determine the Hilbert-Dehn series of the second Hecke group presentation given in Section 1.

PROPOSITION 2.3. *Let $q \geq 3$, and let $H_q = \langle a,b,c : a^2 = b^q = bc = cb = 1\rangle$ be the presentation of the q-th Hecke group $H_q = \mathcal{H}(\lambda_q)$ introduced in Section 1.*

(1) *If $q = 2m$ is an even number, then the Hilbert-Dehn series of H_q is given by*

$$\mathrm{HDS}_{H_q}(z) = \frac{1+2z+2z^2+\cdots+2z^m}{1-z-z^2-\cdots-z^m}$$

(2) *If $q = 2m+1$ is an odd integer, then the Hilbert-Dehn series of H_q is given by*

$$\mathrm{HDS}_{H_q}(z) = \frac{1+3z+4z^2+4z^3+\cdots+4z^m+2z^{m+1}}{1-2z^2-2z^3-\cdots-2z^{m+1}}$$

PROOF. First we prove (1). Let $q = 2m$ be even. By Proposition 1.7.1, the Hilbert-Dehn function $\mathrm{HD}_{H_q}(i)$ satisfies the recursive equation $\mathrm{HD}_{H_q}(i) = 2\cdot\mathrm{HD}_{H_q}(i-1) - \mathrm{HD}_{H_q}(i-m-1)$ for $i \geq m+2$. Therefore the denominator of one representation of $\mathrm{HDS}_{H_q}(z)$ is given by $1-2z+z^{m+1} = (1-z)(1-z-z^2-\cdots-z^m)$. When we compute the coefficients $d_i \in \mathbb{Z}$ of the corresponding numerator $g(z) = d_0 + d_1z + \cdots + d_{m+1}z^{m+1}$, the same calculations as in the proof of the preceding propostion show $g(z) = 1+z-z^m-z^{m+1} = (1-z)(1+z)(1-z-z^2-\cdots-z^{m-1})$. Thus the Hilbert-Dehn series is in the case of an even number q given by

$$\mathrm{HDS}_{H_q}(z) = \frac{1+z-z^m-z^{m+1}}{1-2z+z^{m+1}} = \frac{1+2z+2z^2+\cdots+2z^m}{1-z-z^2-\cdots-z^m}.$$

To prove (2), we let $q = 2m+1$ are argue analogously. The recursive equation $\mathrm{HD}_{H_q}(i) = 2 \cdot \mathrm{HD}_{H_q}(i-1) + 2 \cdot \mathrm{HD}_{H_q}(i-2) - 2 \cdot \mathrm{HD}_{H_q}(i-m-2)$ gives rise to the denominator $1-z-2z^2+2z^{m+2} = (1-z)(1-2z^2-2z^3-\cdots-2z^{m+1})$ of $\mathrm{HDS}_{H_q}(z)$. Letting the corresponding numerator be $g(z) = d_0 + d_1 z + \cdots + d_{m+3}z^{m+3}$, we calculate $d_0 = \mathrm{HD}_{H_q}(0) = 1$, $d_1 = \mathrm{HD}_{H_q}(1) - 1 = 2$, $d_2 = \mathrm{HD}_{H_q}(2) - \mathrm{HD}_{H_q}(1) - 2 = 1$, $d_i = \mathrm{HD}_{H_q}(i) - \mathrm{HD}_{H_q}(i-1) - 2\,\mathrm{HD}_{H_q}(i-2) = 0$ for $3 \le i \le m$, $d_{m+1} = -2$, $d_{m+2} = -2$, and $d_{m+3} = 0$. Hence we have $g(z) = 1 + 2z + z^2 - 2z^{m+1} - 2z^{m+2} = (1-z)(1+z)(1+2z+2z^2+\cdots+2z^m)$, and thus

$$\mathrm{HDS}_{H_q}(z) = \frac{1+2z+z^2-2z^{m+1}-2z^{m+2}}{1-z-2z^2+2z^{m+2}} = \frac{1+3z+4z^2+4z^3+\cdots+4z^m+2z^{m+1}}{1-2z^2-2z^3-\cdots-2z^{m+1}}.$$

□

From the recursive formulas for the Hilbert-Dehn functions of Hecke groups derived in Section 1 it is already clear that these groups have exponential growth. A more precise information about this growth is provided by the following notion.

DEFINITION 2.4. Let $M = \langle x_1, \ldots, x_n : w_1 = \cdots = w_s = 1 \rangle$ be a finitely presented monoid. The limit $\omega(M) = \limsup_{n\to\infty} \sqrt[n]{\mathrm{HD}_M(n)}$ is called the **(exponential) growth rate** or the **entropy** of M with respect to the given presentation.

It is well-known that $\omega(M) = 1/R$ where R is the radius of convergence of the generating series of $\mathrm{HD}_M(n)$. In other words, the number R is the smallest absolute value of a zero of the denominator of $\mathrm{HDS}_M(z)$. The above results allow us to determine the growth rates of the Hecke group presentations under consideration.

REMARK 2.5. Let $q \ge 3$. For the presentation $H'_q : \langle s,t,u : s^2 = (st)^q = tu = ut = 1 \rangle$ of the q-th Hecke group $H'_q = \mathcal{H}(\lambda_q)$, it can be shown that the denominator $1 - z - z^2 - \cdots - z^{q-1}$ has a unique real root in the interval $]0.5, 1[$. It is the root of smallest absolute value in $\mathbb{C}$. The same is true for the denominator of $\mathrm{HDS}_{H_q}(z)$. For the corresponding growth rates, we calculate the approximate values

q	3	4	5	6	7	8	9	10
$\omega(H'_q)$	τ	1.8393	1.9276	1.9659	1.9836	1.9920	1.9960	1.9980
$\omega(H_q)$	$\sqrt{2}$	τ	1.7693	1.8393	1.8993	1.9276	1.9535	1.9659

where $\tau = (1+\sqrt{5})/2$ is the golden ratio. By writing the denominator in the form $2 - (z^q - 1)/(z-1)$, one sees that the growth rate of $\mathcal{H}(\lambda_q)$ approaches 2 as $q \to \infty$.

3. Subgroup Growth and Zeta Functions of Finitely Generated Groups

Counting subgroups of Fuchsian groups, in particular of Hecke groups, is a useful and important topic because it is equivalent to counting covering surfaces of Riemann surfaces $\mathbb{H}/G$, where $\mathbb{H}$ is the upper half plane and G a Fuchsian group acting on $\mathbb{H}$. Furthermore, if U is a normal subgroup of G then $\mathbb{H}/U$ is a regular covering of $\mathbb{H}/G$.

In the following we let G be a finitely generated group. For every $n \ge 1$, there exist only finitely many subgroups of index n in G. Thus it makes sense to consider the following Dirichlet series.

DEFINITION 3.1. For every $n \ge 1$, let $a_n(G)$ be the number of subgroups of G of index n. Moreover, let $b_n(G)$ be the number of normal subgroups of G of index n.

(1) The formal Dirichlet series

$$\zeta_G(s) = \sum_{n=1}^{\infty} a_n(G)\, n^{-s}$$

is called the **subgroup zeta function** of the group G.

(2) The formal Dirichlet series

$$\zeta_G^N(s) = \sum_{n=1}^{\infty} b_n(G)\, n^{-s}$$

is called the **normal subgroup zeta function** of G.

In order to be able to view these series as complex analytic functions on a right half plane in $\mathbb{C}$, we need to assume that the numbers $a_n(G)$ and $b_n(G)$ grow at most polynomially in n. By [**25**], it is known that this holds if G is residually finite and has a subgroup of finite index which is soluble and of finite rank. In this case, we denote the **abscissa of convergence** of $\zeta_G(s)$ by σ_G and that of $\zeta_G^N(s)$ by σ_G^N. For a suitable version of an Euler product decomposition, we need local versions of the above Dirichlet series. They are defined as follows.

DEFINITION 3.2. Given a finitely generated group G, a prime number p, and $n \geq 0$, we let $a_{p^n}(G)$ be the number of subgroups of G of index p^n and $b_{p^n}(G)$ the number of normal subgroups of G of index p^n.

(1) The formal p-Dirichlet series

$$\zeta_G^p(s) = \sum_{n=0}^{\infty} a_{p^n}(G)\, p^{-ns}$$

is called the **subgroup Euler factor** of G at the prime p.

(2) The formal p-Dirichlet series

$$\zeta_G^{N,p}(s) = \sum_{n=0}^{\infty} b_{p^n}(G)\, p^{-ns}$$

is called the **normal subgroup Euler factor** of G at the prime p.

If the zeta function of a finitely generated group converges in some half plane, it is an important question whether there exists an Euler product expansion. The first important case in which this was shown is the case of finitely generated nilpotent groups in which the hypotheses for the existence of σ_G and σ_G^N hold.

THEOREM 3.3. *Let G be a finitely generated, nilpotent group, and let $\mathcal{P}$ be the set of prime numbers.*

(1) *For all $s \in \mathbb{C}$ such that* $\mathrm{Re}(s) > \sigma_G$, *we have* $\zeta_G(s) = \prod_{p\in\mathcal{P}} \zeta_G^p(s)$.

(2) *For all $s \in \mathbb{C}$ such that* $\mathrm{Re}(s) > \sigma_G^N$, *we have* $\zeta_G^N(s) = \prod_{p\in\mathcal{P}} \zeta_G^{N,p}(s)$.

For the case of torsion-free nilpotent groups, this was shown in [**13**]. A straightforward, purely group-theoretic proof in the general case is given in [**8**]. To get a better feeling for these subgroup zeta functions and Euler product decompositions, we consider some important examples.

EXAMPLE 3.4. Let $r \geq 1$, and let $G = \mathbb{Z}^r$ be the free abelian group of rank r. In this case, the zeta function of G satisfies

$$\zeta_{\mathbb{Z}^r}(s) = \zeta^N_{\mathbb{Z}^r}(s) = \zeta(s) \cdot \zeta(s-1) \cdots \zeta(s-r+1)$$

where $\zeta(s) = \sum_{n \geq 1} n^{-s}$ is the Riemann zeta function. Clearly, we have $\sigma_{\mathbb{Z}^r} = r$ here. (For five (!) different proofs of this result, see **[26]**, Ch. 15.)

In the case $r = 2$, i.e. for the group $G = \mathbb{Z}^2$, one can go into considerably more detail. (For further explanations and proofs, see for instance **[17]** and **[30]**.) The subgroup zeta function of G satisfies $\zeta_G(s) = \zeta(s)\zeta(s-1) = \sum_{n=1}^{\infty} \sigma_1(n) n^{-s}$ where $\sigma_1(n)$ is the sum of the divisors of n. Letting $\Gamma(s) = \int_0^\infty t^{s-1}e^{-t}dt$ be the Gamma function and $R(s) = (2\pi)^{-s}\Gamma(s)\zeta_G(s)$, we have the functional equation $R(2-s) = -R(s)$ for all $s \in \mathbb{C}$ such that $\mathrm{Re}(s) > \sigma_G = 2$. (Here $R(s)$ has a meromorphic continuation to all of $\mathbb{C}$.) Finally, the function $f(\tau) = -1/24 + \sum_{n=1}^{\infty} \sigma_1(n)\, e^{2\pi i n \tau}$ defines a modular integral of weight 2 with rational period function $q(\tau) = -1/(4\pi i \tau)$.

This case suggests that we may look for further finitely generated groups whose subgroup zeta function satisfies a functional equation and gives rise to an automorphic integral. The following example was initially discussed in **[31]**.

EXAMPLE 3.5. Let H_3 be the discrete Heisenberg group, i.e. the group consisting of all upper triangular matrices $\begin{pmatrix} 1 & a & b \\ 0 & 1 & c \\ 0 & 0 & 1 \end{pmatrix}$ such that $a, b, c \in \mathbb{Z}$. Then the subgroup zeta function of H_3 is given by

$$\zeta_{H_3}(s) = \frac{\zeta(s)\,\zeta(s-1)\,\zeta(2s-2)\,\zeta(2s-3)}{\zeta(3s-3)}$$

and satisfies $\sigma_{H_3} = 2$. Furthermore, the normal subgroup zeta function of H_3 is given by $\zeta^N_{H_3}(s) = \zeta(s)\zeta(s-1)\zeta(3s-2)$ and satisfies $\sigma^N_{H_3} = 2$.

Notice that the group H_3 could also have been introduced by its presentation

$$H_3 = \langle x, y, z \mid [x, y] = z,\ [x, z] = [y, z] = 1 \rangle$$

and that it is a subgroup of index 12 in the generalized triangle group $\langle a, b \mid a^2 = b^6 = (ababab^{-1})^2 = 1 \rangle$.

Our third example contains a direct product of groups.

EXAMPLE 3.6. Let $G = \mathbb{Z} \times C_p$, where p is a prime number and C_p the cyclic group of order p. Then the subgroup zeta function of G is given by

$$\zeta_G(s) = \zeta^N_G(s) = (1 + p^{-s+1}) \cdot \zeta(s)$$

and satisfies $\sigma_G = 1$.

In general, for an arbitrary finitely generated group G, it is not clear how to calculate the subgroup zeta function of G. In general, there is no Euler product decomposition and subgroup zeta functions do not have functional equations. However, there are additional possibilities if we concentrate on certain classes of groups (such as free products of cyclic groups) or if we consider only subgroups having further properties (such as normal subgroups or free subgroups). Thus we start our investigation of the case of Hecke groups in the next section.

4. Subgroup Growth of Hecke Groups

In 1949, M. Hall discovered a fundamental enumeration relation between subgroups of finite index in free groups and permutation representations of the free group (see [**14**]). Later this method was generalized to the case of free products by I.M.S. Dey (cf. [**7**]). Since we are interested in $\mathcal{H}(\lambda_q) = C_2 * C_q$, we describe this method explicitly.

Let G be a finitely generated group, let $n \geq 1$, and let $h_n(G)$ be the number of homomorphisms of G into the symmetric group S_n. As previously, we let $a_n(G)$ be the number of subgroups of G of index n. Then we have the recursive equation

$$a_n(G) = \frac{h_n(G)}{(n-1)!} - \frac{h_{n-1}(G)\, a_1(G)}{(n-1)!} - \frac{h_{n-2}(G)\, a_2(G)}{(n-2)!} - \cdots - \frac{h_1(G)\, a_{n-1}(G)}{1!}$$

for every $n \geq 1$ (see [**14**]). Next we introduce a new indeterminate z and consider the generating series $f(z) = \sum_{n\geq 1} \frac{h_n(G)}{n!} z^n$. Then the above equalities can be combined to a formal power series identity

$$\sum_{n=1}^{\infty} a_n(G)\, z^n = \frac{z\, f'(z)}{f(z)}$$

For an arbitrary finitely generated group G, the numbers $h_n(G)$ are about as hard to come by as the desired numbers $a_n(G)$. However, if G is a free product $G = A_1 * \cdots * A_r$ of finitely generated groups, then we have $h_n(G) = h_n(A_1) \cdots h_n(A_r)$ by [**7**]. For instance, if G is a free product of cyclic groups, this offers a good way to calculate the numbers $a_n(G)$. Let us see a case in point.

EXAMPLE 4.1. In [**10**], Hall's method was used to determine the subgroup counting function for the group $G = C_2 * C_2$ as follows. The recursive equation $h_{n+1}(C_2) = h_n(C_2) + n\, h_{n-1}(C_2)$ together with the initial values $h_0(C_2) = h_1(C_2) = h_2(C_2) = 1$ is used to derive the recursive equation

$$a_n(G) = a_{n-1}(G) + a_{n-2}(G) - a_{n-3}(G)$$

for $n > 3$ with initial values $a_1(G) = 1$ and $a_2(G) = a_3(G) = 3$. This recursion is solved by $a_{2k}(G) = a_{2k+1}(G) = 2k+1$ for $k \geq 0$. Hence the subgroup zeta function of $G = C_2 * C_2$ is

$$\zeta_G(s) = 2^{-s}\zeta(s) + \zeta(s-1).$$

The abscissa of convergence is $\sigma_G = 2$.

In [**9**], and independently in [**12**], Hall's method was used to determine the subgroup zeta function of some Hecke groups. Based on the results of Chowla et al. about the number of elements in S_n whose order divides a given integer (see [**6**]), we present a more complete answer using recursive formulas.

PROPOSITION 4.2. *Let $q \geq 3$, and let $\mathcal{H}(\lambda_q)$ be the q-th Hecke group.*

(1) *The numbers $a_n(\mathcal{H}(\lambda_q))$ satisfy the recursive equations*

$$a_n(G) = \frac{A_n(2)A_n(q)}{(n-1)!} - \frac{A_{n-1}(2)A_{n-1}(q)a_1(G)}{(n-1)!} - \cdots - \frac{A_1(2)A_1(q)a_{n-1}(G)}{1!}$$

where $A_n(d)$ denotes the number of elements of S_n whose order divides d.

(2) *The numbers $A_n(d)$ satisfy the recursive equations*

$$A_n(d) = A_{n-1}(d) + \sum_{1<k\le n;\; k|d} (n-1)(n-2)\cdots(n-k+1)\, A_{n-k}(d).$$

PROOF. In view of the isomorphism $\mathcal{H}(\lambda_q) \cong C_2 * C_q$ and the equalities $h_n(G) = h_n(C_2)\, h_n(C_q) = A_n(2)\, A_n(q)$, Claim (1) is a special case of [**7**], Thm. 6.10. The second claim is shown in [**6**]. □

A particularly simple formula results from the second part of this proposition if d is a prime number p. In this case, we have $A_n(p) = \sum_{i+jp=n} \frac{n!}{i!\,j!\,p^j}$. In particular, for the group $\mathcal{H}(\lambda_p)$ we obtain

$$h_n(\mathcal{H}(\lambda_p)) = \left(\sum_{i+2j=n} \frac{1}{i!\,j!\,2^{j-1}}\right)\left(\sum_{i+pj=n} \frac{n!}{i!\,j!\,p^j}\right)$$

Let us compute the subgroup growth of the Hecke groups $\mathcal{H}(\lambda_q)$ in some important cases.

EXAMPLE 4.3. Let us consider the modular group $G = \mathcal{H}(\lambda_3)$. Proposition 4.2 yields the recursive equations

$$a_n(G) = \frac{A_n(2)A_n(3)}{(n-1)!} - \sum_{k=1}^{n-1} \frac{1}{n!}\, A_{n-k}(2)\, A_{n-k}(3)\, a_k(G)$$

where $A_n(2)$ and $A_n(3)$ are given by the explicit formulas above and $a_1(G) = 1$. Thus it is possible to calculate the first values of $a_n(G)$. We find

n	1	2	3	4	5	6	7	8	9
$a_n(\mathcal{H}(\lambda_3))$	1	1	4	8	5	22	42	40	120

In [**10**], the authors combined the recurrence relations for the numbers $a_n(G)$ and $A_n(G)$. They showed that for $n \geq 10$ we have the recurrence relation

$$\begin{aligned} a_n(G) &= 4a_{n-3}(G) + 2a_{n-4}(G) + (n-3)a_{n-6}(G) + 2a_{n-7}(G) - (n-6)a_{n-9}(G) \\ &\quad + \sum_{i=1}^{n-7} a_{n-6-i}(G)a_i(G) - \sum_{i=1}^{n-10} a_{n-9-i}(G)a_i(G). \end{aligned}$$

If $q = p \geq 3$ is a prime number, it is possible to determine the first $2p$ values of $a_n(\mathcal{H}(\lambda_p))$ explicitly, as the next example shows. (For a proof, see [**9**], Thm. 1.)

EXAMPLE 4.4. Let $p \geq 3$ be a prime number, and let $G = \mathcal{H}(\lambda_p)$. Then we have $a_1(G) = a_2(G) = 1$, $a_3(G) = \cdots = a_{p-1}(G) = 0$, $a_p(G) = A_p(2)$, and

$$a_{p+k}(G) = \binom{p-1}{k-1} \frac{p+k}{p}\, A_{p-k}(2)$$

for $k = 1, \ldots, p-1$, where $A_n(2) = \sum_{i+2j=n} \frac{n!}{i!\,j!\,2^j}$. In particular, we have $a_{2p-1}(G) = 2p-1$. For $a_{2p}(G)$, an explicit (but more complicated) formula can be given as well (see [**9**], Thm. 1).

5. Normal Subgroup Growth of Hecke Groups

To count normal subgroups of a given index n in a finitely generated group G, we may try to use a variant of the method introduced in the preceding section. The number $b_n(G)$ of these subgroups can be calculated from the number $c_n(G)$ of group homomorphisms $\phi : G \longrightarrow S_n$ for which the centralizer of $\phi(G)$ operates transitively on $\{1, \dots, n\}$. Using combinatorial arguments, it is possible to derive the equalites

$$c_n(G) \;=\; n! \cdot \sum_{m|n} \frac{b_n(G)}{\binom{n}{m}! \; m^{n/m}}$$

for every $n \geq 1$ (see [**19**]). Unfortunately, it appears difficult to compute the numbers $c_n(G)$ in concrete cases such as the Hecke groups.

For the modular group, M. Newman determined the first values of $b_n(\mathcal{H}(\lambda_3))$ in [**29**] as follows.

EXAMPLE 5.1. Let $\mathcal{H}(\lambda_3) = C_2 * C_3 = \langle a, b \mid a^2 = b^3 = 1 \rangle$ be the modular group. For a subgroup G of $\mathcal{H}(\lambda_3)$, we call $\ell(G) = \min\{i \geq 1 \mid (ab)^i \in G\}$ the **level** of G and note that a normal subgroup G of finite index in $\mathcal{H}(\lambda_3)$ has genus one if and only if $\ell(G) = 6$ (see [**11**]).

Except for the three groups $\mathcal{H}(\lambda_3)$, $\mathcal{H}(\lambda_3)^2$, and $\mathcal{H}(\lambda_3)^3$, every normal subgroup of the modular group has an index which is a multiple of 6. The first numbers $b_n(\mathcal{H}(\lambda_3))$ are given by the following table:

n	1	2	3	6	12	18	24	30	36	42	48	54	60	66
$b_n(\mathcal{H}(\lambda_3))$	1	1	1	2	1	1	2	0	0	2	2	1	1	0

The normal subgroups of $\mathcal{H}(\lambda_3)$ of genus one have been completely described (see [**28**]) and for every genus $g(G) \geq 2$ there are only finitely many normal subgroups of $\mathcal{H}(\lambda_3)$. Many further restrictions for normal subgroups of $\mathcal{H}(\lambda_3)$ are given in [**11**].

Also for the Hecke group $\mathcal{H}(\lambda_4)$ and other Hecke groups $\mathcal{H}(\lambda_q)$ with indices $q \geq 5$, a number of individual facts are known about their normal subgroups (see for instance [**4**] and [**5**]), but a complete description for the normal subgroup counting function seems to be unknown.

In the remainder of this section we therefore restrict our attention to a special class of normal subgroups, namely normal subgroups of genus one. Recall that the genus of a free normal subgroup G of index n in $\mathcal{H}(\lambda_q)$ is given by $g = 1 - \frac{t}{2} + \frac{n(q-2)}{4q}$ where t is the number of conjugacy classes of maximal cyclic parabolic subgroups in G. For general normal subgroups of $\mathcal{H}(\lambda_q)$, the genus can be computed via the Riemann-Hurwitz formula (see [**5**]). We note that the genus of a subgroup G of $\mathcal{H}(\lambda_q)$ is exactly the genus of the quotient space $\mathbb{H}/G$, where $\mathbb{H}$ denotes the upper half plane.

In free products of finite cyclic groups $C_{r_1} * \cdots * C_{r_s}$, there are either no or infinitely many genus one normal subgroups of finite index, and in [**21**] the authors give necessary and sufficient conditions for their existence in terms of the numbers r_i. For the Hecke groups $\mathcal{H}(\lambda_q) = C_2 * C_q$, these conditions amount to $\gcd(q, 12) \geq 3$ (see also [**5**]). Only in the four cases $C_2 * C_2 * C_2$, $\mathcal{H}(\lambda_3) = C_2 * C_3$, $\mathcal{H}(\lambda_4) = C_2 * C_4$, and $C_3 * C_3$, all normal subgroups of genus one and finite index are free. In all other cases where genus one normal subgroups of finite index exist, at least some of them are not free (for details, see [**21**]).

The counting function for normal subgroups of genus one has sometimes interesting number-theoretic properties. Therefore we introduce the following notion.

DEFINITION 5.2. Let $G = C_{r_1} * \cdots * C_{r_s}$ be a finite free product of finite cyclic groups. For every $n \geq 1$, let $b_n^{(1)}(G)$ be the number of normal subgroups of genus one and index n in G. Then the formal Dirichlet series

$$\zeta_G^{N,1}(s) = \sum_{n=1}^{\infty} b_n^{(1)}(G)\, n^{-s}$$

is called the **normal genus one subgroup zeta function** of G.

Now we consider some cases in point, starting with the modular group.

EXAMPLE 5.3. Let $\mathcal{H}(\lambda_3) = C_2 * C_3$ be the modular group. Its normal subgroups of genus one were classified in [**28**]. All of them are free groups. Since the index of these subgroups is always divisible by 6, we let $a_1(n) = b_{6n}^{(1)}(\mathcal{H}(\lambda_3))$ be the number of normal subgroups of genus one and index $6n$ for every $n \geq 1$. Then it turns out that

$$a_1(n) \;=\; \tfrac{1}{6} \cdot \#\{(x,y) \in \mathbb{Z}^2 \mid x^2 + xy + y^2 = n\}$$

is a multiplicative number-theoretic function (see [**28**] and [**21**]). In particular, if $n = p^k$ is a prime power, then

$$a_1(p^k) \;=\; \begin{cases} 1 & \text{if } p = 3, \\ (1 + (-1)^k)/2 & \text{if } p = 2 \text{ or } p \neq 3,\ \left(\frac{p}{3}\right) = -1, \\ k+1 & \text{if } p \neq 3,\ \left(\frac{p}{3}\right) = 1. \end{cases}$$

Here $\left(\frac{p}{3}\right)$ denotes the Legendre symbol.

Let us define the character $\chi_{-3} : \mathbb{N} \longrightarrow \mathbb{Z}$ by $\chi_{-3}(n) = 1$ if $n \equiv 1 \pmod 3$, by $\chi_{-3}(n) = -1$ if $n \equiv -1 \pmod 3$, and by $\chi_{-3}(n) = 0$ if n is a multiple of 3. Then the function $a_1(n)$ satisfies $a_1(n) = \sum_{d \geq 1;\, d|n} \chi_{-3}(d)$. This description can be used to derive a recursive equation for $a_1(n)$.

Recall that the Möbius function is given by $\mu(n) = (-1)^k$ if n is the product of k distinct primes and $\mu(n) = 0$ otherwise. By the Möbius inversion formula, we get

$$\chi_{-3}(n) \;=\; \sum_{d \geq 1;\, d|n} \mu(\frac{n}{d})\, a_1(n).$$

For the function $a_1(n)$, this leads to the recursive equation

$$a_1(n) \;=\; \chi_{-3}(n) \;-\; \sum_{1 \leq d < n;\, d|n} \mu(\frac{n}{d})\, a_1(d).$$

Based on the explicit description of $a_1(n)$, we can also study the normal genus one zeta function of $\mathcal{H}(\lambda_3)$. We have

$$\begin{aligned} \zeta_{\mathcal{H}(\lambda_3)}^{N,1}(s) \;&=\; \sum_{n=1}^{\infty} b_n^{(1)}(\mathcal{H}(\lambda_3)) \cdot n^{-s} \;=\; \sum_{k=1}^{\infty} b_{6k}^{(1)}(\mathcal{H}(\lambda_3)) \cdot (6k)^{-s} \\ &=\; \sum_{k=1}^{\infty} a_1(k) \cdot (6k)^{-s} \;=\; 6^{-s} \cdot \varphi_3(s) \end{aligned}$$

where we let $\varphi_3(s)$ be the formal Dirichlet series $\varphi_3(s) = \sum_{k=1}^{\infty} a_1(k) \cdot k^{-s}$. The abscissa of convergence is $\sigma_{\varphi_3} = 1$.

By introducing the Dirichlet series $L_{-3}(s) = \sum_{n=1}^{\infty} \chi_{-3}(n)\, n^{-s}$ of the character χ_{-3}, we obtain a factorization $\varphi_3(s) = \zeta(s) \cdot L_{-3}(s)$ where $\zeta(s)$ is the Riemann zeta function. Notice that this is exactly the zeta function of the number field $\mathbb{Q}(\sqrt{-3})$.

Moreover, the Dirichlet series $\varphi_3(s)$ satisfies a functional equation. Let $R(s) = (2\pi/\sqrt{3})^{-s}\, \Gamma(s)\, \varphi_3(s)$. Then we have $R(s) = R(1-s)$ (see **[15]**). Hence, if we let $a_1(0) = 1/6$, then the function

$$f(\tau) \;=\; \sum_{n=0}^{\infty} a_1(n) \exp(2\pi i\, n\, \tau/\sqrt{3})$$

defines an automorphic $(\sqrt{3}, 1, 1)$-form, i.e. an entire automorphic form for the Hecke group $\mathcal{H}(\lambda_6) = G(\sqrt{3})$ of weight one and multiplier one. The $\mathbb{C}$-vector space of all automorphic $(\sqrt{3}, 1, 1)$-forms is 1-dimensional. Thus the function $f(\tau)$ is a basis for this vector space.

Following the same procedure for the group $\mathcal{H}(\lambda_6)$ does not yield new insights.

EXAMPLE 5.4. The group $\mathcal{H}(\lambda_6) = C_2 * C_6$ contains also non-free normal subgroups of genus one, but the total number of free normal subgroups of genus one is, for every index n, the same number as for the modular group (see **[21]**).

Similarly, the group $C_3 * C_3$ does not produce new numbers of normal subgroups of genus one and finite index.

EXAMPLE 5.5. The group $G = C_3 * C_3$ is a subgroup of the modular group of index 2. Hence the number $b_n^{(1)}(G)$ of normal subgroups of genus one and index n is zero if n is not a multiple of 3, and we have $b_{3k}^{(1)}(G) = a_1(k)$ for every $k \geq 1$ with the numbers $a_1(k)$ studied in Example 5.3.

The next group has a rather simple normal genus one subgroup zeta function.

EXAMPLE 5.6. Let G be the free product $G = C_2 * C_2 * C_2$. Here all normal subgroups of genus one are free and have an even index in G. Denoting the number of normal subgroups of genus one and index $2n$ by $c_1(n) = b_{2n}^{(1)}(G)$, we find that $c_1(n)$ is the sum of positive divisors of n, i.e. $c_1(n) = \sigma_1(n)$ (see **[21]**). Notice that $c_1(n)$ is a multiplicative function. For the normal genus one subgroup zeta function we obtain

$$\zeta_G^{N,1}(s) \;=\; \sum_{n=1}^{\infty} b_n^{(1)}(G)\, n^{-s} \;=\; 2^{-s} \cdot \sum_{k=1}^{\infty} c_1(k)\, k^{-s} \;=\; 2^{-s} \zeta(s)\, \zeta(s-1)$$

Our final example leads again to an interesting normal genus one zeta function.

EXAMPLE 5.7. Consider the Hecke group $\mathcal{H}(\lambda_4) = C_2 * C_4$. Every normal subgroup H of finite index and genus one is free and its index in $\mathcal{H}(\lambda_4)$ is divisible by 4. Moreover, the group H has level $\ell(H) = \min\{i > 0 \mid (ab)^i \in H\} = 4$. In fact, a free normal subgroup H of $\mathcal{H}(\lambda_4)$ has genus one if and only if $\ell(H) = 4$ (see **[21]**). Let $b_1(n)$ be the number of such subgroups of index $4n$. Then we get

$$b_1(n) \;=\; \tfrac{1}{4} \cdot \#\{(x, y) \in \mathbb{Z}^2 \mid x^2 + y^2 = n\},$$

a multiplicative number-theoretic function again (see **[21]**).

Using the character $\chi_{-4} : \mathbb{N} \longrightarrow \mathbb{Z}$ defined by $\chi_{-4}(n) = 1$ if $n \equiv 1 \pmod 4$, by $\chi_{-4}(n) = -1$ if $n \equiv -1 \pmod 4$, and by $\chi_{-4}(n) = 0$ for even n, we have the formula $b_1(n) = \sum_{d\geq 1;\, d|n} \chi_{-4}(d)$. Now we use the Möbius reverse formula again and get

$$\chi_{-4}(n) = \sum_{d\geq 1;\, d|n} \mu(\frac{n}{d})\, b_1(n)$$

This yields the recurrence relation

$$b_1(n) = \chi_{-4}(n) - \sum_{0<d<n;\, d|n} \mu(\frac{n}{d})\, b_1(d).$$

Next we study the normal genus one zeta function of $\mathcal{H}(\lambda_4)$. We introduce the Dirichlet series $L_{-4}(s) = \sum_{n=1}^{\infty} \chi_{-4}(n) n^{-s}$ and get

$$\zeta^{N,1}_{\mathcal{H}(\lambda_4)}(s) = \sum_{n=1}^{\infty} b_n^{(1)}(\mathcal{H}(\lambda_4))\, n^{-s} = 4^{-s} \sum_{k=1}^{\infty} b_1(k)\, k^{-s} = 4^{-s}\, \varphi_4(s) = 4^{-s}\, \zeta(s)\, L_{-4}(s)$$

where $\varphi_4(s) = \sum_{k=1}^{\infty} b_1(k)\, k^{-s}$. The abscissa of convergence is $\sigma_{\varphi_4} = 1$, and $\varphi_4(s)$ is exactly the zeta function of the field of Gaußian numbers $\mathbb{Q}(i)$.

Introducing $R(s) = \pi^{-s}\,\Gamma(s)\,\varphi_4(s)$, we find that the functional equation $R(s) = R(1-s)$ holds. Now we let $b_1(0) = 1/4$ and get an automorphic $(2,1,1)$-form

$$f(\tau) = \sum_{n=0}^{\infty} b_1(n)\, \exp(2\pi i\, n\, \tau)$$

i.e. an entire automorphic form of weight one and multiplier one for the theta group $G(2)$ which is generated by the linear fractional transformations $z \mapsto -\frac{1}{z}$ and $z \mapsto z+2$ (see [**17**]). Group theoretically, the theta group is a subgroup if index 3 in the modular group and isomorphic to $C_2 * C_\infty$. The $\mathbb{C}$-vector space of all automorphic $(2,1,1)$-forms for $G(2)$ is 1-dimensional. Consequently, $f(\tau)$ is a basis of this vector space.

6. Free Subgroup Growth of Hecke Groups

In the preceding section we saw that many normal subgroups of genus one in Hecke groups are free. This leads us to the idea to study the growth of the numbers of free subgroups of a given finite index in a finitely generated group G.

DEFINITION 6.1. For every $n \geq 1$, let $f_n(G)$ be the number of free subgroups of index n in G. Then the formal power series

$$F_G(z) = \sum_{n=1}^{\infty} f_n(G)\, z^n$$

is called the **free subgroup counting series** of G.

The most obvious example is the case of a free group, since all of its subgroups are free.

EXAMPLE 6.2. For the free group F_r of rank r, the number of group homomorphisms $F_r \longrightarrow S_n$ is clearly $(n!)^r$. As in [**14**], it follows that the numbers $f_n(F_r)$ satisfy $f_1(F_r) = 1$ and

$$f_n(F_r) = n\,(n!)^{r-1} - \sum_{i=1}^{n-1} ((n-i)!)^{r-1} \cdot f_i(F_r).$$

For arbitrary finitely generated groups, not much is known about their free subgroup counting series. One approach to compute the numbers $f_n(G)$ is to use the following variant of the method of Hall (see Section 5) which was first suggested in [**18**]. It is based on the observation that the number $f_n(G)$ is related to the number of homomorphisms from G to S_n with the property that the preimages of all stabilizers are free or trivial. Assume that G is non-trivial, and let $k_n(G)$ be the number of these homomorphisms. Then we have the recursive equations

$$f_n(G) = \frac{k_n(G)}{(n-1)!} - \frac{k_{n-1}(G)\,f_1(G)}{(n-1)!} - \frac{k_{n-2}(G)\,f_2(G)}{(n-2)!} - \cdots - \frac{k_1(G)\,f_{n-1}(G)}{1!}$$

for all $n \geq 1$ where $f_1(G) = 1$ if G is free and $f_1(G) = 0$ otherwise. As for the analogous formulas for the numbers $a_n(G)$, these recursive equations are, in general, only of limited help, since it is not easy to determine the numbers $k_n(G)$.

The situation improves if we restrict our attention to finite free products of cyclic groups, a case including the Hecke groups. In this setting, the following explicit formulas for the numbers $k_n(G)$ were worked out in [**32**], Thm. 1.4.

PROPOSITION 6.3. *Let $G = C_{r_1} * \cdots * C_{r_d} * (C_\infty)^{*u}$ and $t = \operatorname{lcm}(r_1, \ldots, r_d)$ if $d \geq 1$ and $t = 1$ if $d = 0$.*

(1) *If t does not divide n, then we have $k_n(G) = 0$.*
(2) *For $n = kt$, we have $k_n(G) = (n!)^u \cdot \prod_{i=1}^d \frac{n!}{(n/r_i)!\, r_i^{n/r_i}}$.*

Explicit, but very complicated formulas for $f_n(G)$ were derived in [**2**] from these values of $k_G(n)$. In the following we give the results only for some basic cases, including certain Hecke groups.

EXAMPLE 6.4. For the group $G = C_2 * C_2$, we have $f_{2k}(G) = 1$ and $f_{2k-1}(G) = 0$ for all $k \geq 1$ (see [**32**]). In particular, the free subgroup counting series is the rational power series $F_{C_2 * C_2}(z) = z^2/(1-z^2)$.

EXAMPLE 6.5. For the free group $F_2 = C_\infty * C_\infty$, we get $f_1(F_2) = 1$ and for $n \geq 1$ the recursive equation

$$f_{n+1}(F_2) = (n+2)\, f_n(F_2) + \sum_{k=1}^{n-1} f_k(F_2)\, f_{n-k}(F_2).$$

Notice that we found linear recursive equations in Example 6.2 above.

EXAMPLE 6.6. Let $G(2) = C_2 * C_\infty$ be the theta group. Then we have $f_n(G(2)) = 0$ if n is odd, $f_2(G(2)) = 1$, and for $k \geq 1$ the recursive equation

$$f_{2(k+1)}(G(2)) = (2k+3)\, f_{2k}(G(2)) + \sum_{i=1}^{n-1} f_{2i}(G(2))\, f_{2(k-i)}(G(2)).$$

EXAMPLE 6.7. Consider the modular group $\mathcal{H}(\lambda_3) = C_2 * C_3$. Then the numbers $f_n(\mathcal{H}(\lambda_3))$ are zero if n is not a multiple of 6. We have $f_6(\mathcal{H}(\lambda_3)) = 5$ and the recursive equation

$$f_{6(k+1)}(\mathcal{H}(\lambda_3)) = 6\,(k+1)\, f_{6k}(\mathcal{H}(\lambda_3)) + \sum_{i=1}^{k-1} f_{6i}(\mathcal{H}(\lambda_3))\, f_{6(k-i)}(\mathcal{H}(\lambda_3))$$

for $k \geq 1$. (This result was first given in [**33**].) This function grows very fast, as its initial values show:

n	6	12	18	24	30	36
$f_n(\mathcal{H}(\lambda_3))$	5	60	1105	27120	828250	30220800

Writing the free subgroup counting series in the Form $F(z) = \widehat{F}(u)$ with $u = z^6$, it is known that $\widehat{F}(u)$ satisfies a homogeneous linear differential equation of Riccati type with integral coefficients (see [**27**]) from which the numbers $f_n(\mathcal{H}(\lambda_3))$ can also be computed recursively.

EXAMPLE 6.8. Consider the modular group $\mathcal{H}(\lambda_4) = C_2 * C_4$. Then the numbers $f_n(\mathcal{H}(\lambda_4))$ are zero if n is not a multiple of 4. We have $f_4(\mathcal{H}(\lambda_4)) = 3$ and for $k \geq 1$ the recursive equation

$$f_{4(k+1)}(\mathcal{H}(\lambda_4)) \;=\; 4\,(k+1)\,f_{4k}(\mathcal{H}(\lambda_4)) \;+\; \sum_{i=1}^{k-1} f_{4i}(\mathcal{H}(\lambda_4))\,f_{4(k-i)}(\mathcal{H}(\lambda_4)).$$

EXAMPLE 6.9. Consider the modular group $\mathcal{H}(\lambda_6) = C_2 * C_6$. Then the numbers $f_n(\mathcal{H}(\lambda_6))$ are zero if n is not a multiple of 6. Letting $f_n = f_n(\mathcal{H}(\lambda_6))$, we have $f_6 = 15$, $f_{12} = 1695$, and for $k \geq 2$ the recursive equation

$$\begin{aligned} f_{6(k+1)} &= (36k^2 + 54k + 23)\,f_{6k} \;+\; 405\,f_{6(k-1)} \\ &\quad + \sum_{i=1}^{k-2}\left[\left(\sum_{j=0}^{i-1} f_{6(j+1)} f_{6(i-j)}\right) f_{6(k-i-1)} \;+\; 9(2k-2i+1)\,f_{6i}\,f_{6(k-i)}\right] \end{aligned}$$

Again it is known that, if we write $F(z) = \widehat{F}(u)$ with $u = z^4$, then $\widehat{F}(u)$ satisfies a homogeneous linear differential equation of Riccati type with integral coefficients (see [**27**] and [**20**]).

It is important to note that the given examples are commensurable with the modular group, and hence arithmetic Fuchsian groups. It would be interesting to find the free subgroup counting series in these cases and for other Hecke groups.

References

[1] B. Buchberger, Ein Algorithmus zum Auffinden der Basiselemente des Restklassenrings nach einem nulldimensionalen Polynomideal, PhD Thesis, Leopold-Franzens-Universität Innsbruck, Innsbruck 1965.

[2] T. Camps, M. Dörfer, and G. Rosenberger, *A recurrence relation for the number of free subgroups in free products of cyclic groups*, Aspects of infinite groups, Algebra Discrete Math., vol. 1, World Sci. Publ., Hackensack, NJ, 2008, pp. 54–74, DOI 10.1142/9789812793416_0005. MR2571511 (2010k:20043)

[3] Thorsten Camps, Volkmar große Rebel, and Gerhard Rosenberger, *Einführung in die kombinatorische und die geometrische Gruppentheorie* (German), Berliner Studienreihe zur Mathematik [Berlin Study Series on Mathematics], vol. 19, Heldermann Verlag, Lemgo, 2008. MR2378619 (2009g:20094)

[4] İ. N. Cangül, *Normal subgroups of the Hecke group $H(\sqrt{2})$*, Comm. Fac. Sci. Univ. Ankara Ser. A_1 Math. Statist. **43** (1994), no. 1-2, 129–135 (1996). MR1404795 (98b:20076)

[5] İ. N. Cangül and D. Singerman, *Normal subgroups of Hecke groups and regular maps*, Math. Proc. Cambridge Philos. Soc. **123** (1998), no. 1, 59–74, DOI 10.1017/S0305004197002004. MR1474865 (98j:20071)

[6] S. Chowla, I. N. Herstein, and W. R. Scott, *The solutions of $x^d = 1$ in symmetric groups*, Norske Vid. Selsk. Forh., Trondheim **25** (1952), 29–31 (1953). MR0054605 (14,947c)

[7] I. M. S. Dey, *Schreier systems in free products*, Proc. Glasgow Math. Assoc. **7** (1965), 61–79 (1965). MR0188279 (32 #5718)

[8] Martin Dörfer and Gerhard Rosenberger, *Zeta function of finitely generated nilpotent groups*, Groups—Korea '94 (Pusan), de Gruyter, Berlin, 1995, pp. 35–46. MR1476947 (99b:20060)

[9] Benjamin Fine and Dennis Spellman, *Counting subgroups of the Hecke groups*, Internat. J. Algebra Comput. **3** (1993), no. 1, 43–49, DOI 10.1142/S0218196793000044. MR1214004 (94b:20049)

[10] C. Godsil, W. Imrich, and R. Razen, *On the number of subgroups of given index in the modular group*, Monatsh. Math. **87** (1979), no. 4, 273–280, DOI 10.1007/BF01637030. MR538760 (80k:10019)

[11] Leon Greenberg and Morris Newman, *Normal subgroups of the modular group*, J. Res. Nat. Bur. Standards Sect. B **74B** (1970), 121–123. MR0265477 (42 #386)

[12] Michael Grady and Morris Newman, *Counting subgroups of given index in Hecke groups*, A tribute to Emil Grosswald: number theory and related analysis, Contemp. Math., vol. 143, Amer. Math. Soc., Providence, RI, 1993, pp. 431–436, DOI 10.1090/conm/143/01009. MR1210530 (94h:11040)

[13] F. J. Grunewald, D. Segal, and G. C. Smith, *Subgroups of finite index in nilpotent groups*, Invent. Math. **93** (1988), no. 1, 185–223, DOI 10.1007/BF01393692. MR943928 (89m:11084)

[14] Marshall Hall Jr., *Subgroups of finite index in free groups*, Canadian J. Math. **1** (1949), 187–190. MR0028836 (10,506a)

[15] E. Hecke, Über die Zetafunktionen beliebiger algebraischer Zahlkörper, Nachrichten von der Gesellschaft der Wissenschaften zu Göttingen, Mathematisch-Physikalische Klasse 1917 (1917), 77-89.

[16] E. Hecke, *Über die Bestimmung Dirichletscher Reihen durch ihre Funktionalgleichung* (German), Math. Ann. **112** (1936), no. 1, 664–699, DOI 10.1007/BF01565437. MR1513069

[17] Erich Hecke, *Lectures on Dirichlet series, modular functions and quadratic forms*, Vandenhoeck & Ruprecht, Göttingen, 1983. Edited by Bruno Schoeneberg; With the collaboration of Wilhelm Maak. MR693092 (85c:11042)

[18] Wilfried Imrich, *On the number of subgroups of given index in* $\mathrm{SL}_2(Z)$, Arch. Math. (Basel) **31** (1978/79), no. 3, 224–231, DOI 10.1007/BF01226441. MR521474 (80c:20062)

[19] G. Kern-Isberner, Rekursionsformeln für die Anzahl von Normalteilern in freien Produkten zyklischer Gruppen, dissertation, Universität Dortmund 1985.

[20] C. Krattenthaler and T. W. Müller, *A Riccati differential equation and free subgroup numbers for lifts of* $\mathrm{PSL}_2(\mathbb{Z})$ *modulo prime powers*, J. Combin. Theory Ser. A **120** (2013), no. 8, 2039–2063, DOI 10.1016/j.jcta.2013.08.002. MR3102174

[21] Gabriele Kern-Isberner and Gerhard Rosenberger, *Normalteiler vom Geschlecht eins in freien Produkten endlicher zyklischer Gruppen* (German, with English summary), Results Math. **11** (1987), no. 3-4, 272–288, DOI 10.1007/BF03323275. MR897303 (88i:20045)

[22] Martin Kreuzer and Lorenzo Robbiano, *Computational commutative algebra. 1*, Springer-Verlag, Berlin, 2000. MR1790326 (2001j:13027)

[23] Martin Kreuzer and Lorenzo Robbiano, *Computational commutative algebra. 2*, Springer-Verlag, Berlin, 2005. MR2159476 (2006h:13036)

[24] M. Kreuzer and X. Xiu, Non-commutative Gebauer-Möller criteria, preprint 2013, available at `http://arxiv.org/abs/1302.3805`.

[25] Alexander Lubotzky, Avinoam Mann, and Dan Segal, *Finitely generated groups of polynomial subgroup growth*, Israel J. Math. **82** (1993), no. 1-3, 363–371, DOI 10.1007/BF02808118. MR1239055 (95b:20051)

[26] Alexander Lubotzky and Dan Segal, *Subgroup growth*, Progress in Mathematics, vol. 212, Birkhäuser Verlag, Basel, 2003. MR1978431 (2004k:20055)

[27] T. W. Müller, *Parity patterns in Hecke groups and Fermat primes*, Groups: topological, combinatorial and arithmetic aspects, London Math. Soc. Lecture Note Ser., vol. 311, Cambridge Univ. Press, Cambridge, 2004, pp. 327–374, DOI 10.1017/CBO9780511550706.013. MR2073354 (2005h:20055)

[28] Morris Newman, *A complete description of the normal subgroups of genus one of the modular group*, Amer. J. Math. **86** (1964), 17–24. MR0163966 (29 #1265)

[29] Morris Newman, *Classification of normal subgroups of the modular group*, Trans. Amer. Math. Soc. **126** (1967), 267–277. MR0204375 (34 #4217)

[30] Bruno Schoeneberg, *Elliptic modular functions: an introduction*, Springer-Verlag, New York-Heidelberg, 1974. Translated from the German by J. R. Smart and E. A. Schwandt; Die Grundlehren der mathematischen Wissenschaften, Band 203. MR0412107 (54 #236)

[31] G.C. Smith, Zeta functions of torsion-free finitely generated nilpotent groups, Ph.D. Thesis, Univ. of Manchester, 1983.

[32] W. W. Stothers, *Free subgroups of the free product of cyclic groups*, Math. Comp. **32** (1978), no. 144, 1274–1280, DOI 10.2307/2006352. MR502015 (80c:20037)

[33] K. Wohlfahrt, *Über einen Satz von Dey und die Modulgruppe* (German), Arch. Math. (Basel) **29** (1977), no. 5, 455–457. MR0507036 (58 #22326)

Fakultät für Informatik und Mathematik, Universität Passau, D-94030 Passau, Germany

E-mail address: martin.kreuzer@uni-passau.de

Fakultät für Informatik und Mathematik, Universität Passau, D-94030 Passau, Germany

E-mail address: rosenber@fim.uni-passau.de

Some results on abelian groups of automorphisms of compact Riemann surfaces

Jesús Rodríguez

ABSTRACT. Breuer established necessary and sufficient conditions for a finite abelian group acting as a group of conformal automorphisms of a compact Riemann surface of given genus $g > 1$. We propose a refinement of this result in which every condition is entirely expressed in terms of the invariant factors of the abelian group and the signature of the Fuchsian group. This alternative statement results in a more concise proof of Maclachlan's solution to the minimum genus problem; in many cases, we can fix an explicit expression using some results concerning the invariant factors of the abelian group. Finally, we find an explicit solution to the minimum genus problem for the family of abelian groups, cyclic or not, of the same given order, as well as the unique abelian group attaining the minimum genus.

1. Introduction

It is known that any finite group is isomorphic to a group of conformal automorphisms[1] of some compact Riemann surface of genus $g > 1$ [**5**]. The question of whether a given finite group G can act as such a group of automorphisms for a given genus $g > 1$ or not has a well-known answer as well: this is the case if and only if the finite group is the quotient Γ/Λ of a cocompact Fuchsian group Γ by a surface Fuchsian group $\Lambda \trianglelefteq \Gamma$ of orbit genus g.

Hence, it is worth finding conditions to establish the existence of a surface kernel —i.e. order preserving— homomorphism $\Gamma \to G$ from a cocompact Fuchsian group onto the finite group. For a finite abelian group A, these conditions were first posed by Breuer [**1**]. One of them claims the existence of an epimorphism $\Gamma \to A$, the other conditions stated on the structure of Γ and A (the signature of Γ and the invariant factors that define the abelian group A). In Section 3, we show how the former condition can be replaced by another one entirely stated on the structure of Γ and A.

The same abstract finite group G, or a family of groups, can act as a group of automorphisms on Riemann surfaces of different genera. The so called *minimum genus problem* consists in finding the minimum of such genera. Closely related, the *maximum order problem* deals with the largest possible order of a group, or family of groups, of automorphisms of surfaces of a given genus g. A thorough survey on these and other matters can be found in [**3**]. Tucker [**9**] introduced the term *strong*

2010 *Mathematics Subject Classification.* Primary 30F99; Secondary 20H10, 57M99.

[1]Hereinafter, any automorphism is to be understood conformal.

symmetric genus for the concept of minimum genus, although the condition $g > 1$ is not imposed in this modern terminology. The first results were given by Harvey [**6**]. He found necessary and sufficient conditions for the existence of a surface kernel epimorphism $\Gamma \to \mathbb{Z}_N$ and then applied them to solve the minimum genus and maximum order problems for $\mathbb{Z}_N$. This method has also been used for a few other families of finite groups [**2**, **4**, **8**]. For other families, different methods have been posed. This is the case for abelian groups, whose minimum genus problem was first solved by Maclachlan [**7**]. In Section 4, we take advantage of the alternative conditions stated in Section 3 for the existence of a surface kernel epimorphism to give another shorter proof of Maclachlan's result.

The expression set up by Maclachlan provides the minimum genus of a non-cyclic abelian group in terms of a minimum to be found among several quantities defined by the structure of the abelian group; see Theorem 4.1. In many cases it is possible to determine that minimal quantity and fix an explicit solution for the minimum genus looking over the invariant factors of the abelian group. We study some of these issues in Section 4.

In Section 5, we collect both results on minimum genus of Harvey and Maclachlan to establish the minimum genus attained among abelian groups, cyclic or not, of the same finite order. This result also provides a new proof of the solution to the maximum order problem given by Breuer on the basis of Maclachlan's result on minimum genus [**1**].

2. Preliminaries

The purpose of this section is to fix some notation and terminology, and to emphasize inequality (2.1) and conditions (2.2) for the existence of epimorphisms between abelian groups for later reference.

Let $(\gamma; m_1, \dots, m_r)$ be the signature of a cocompact Fuchsian group, with integers $\gamma \geqslant 0$ and $m_i > 1$. Throughout this work, both $\mathbb{Z}_n$ and $\mathbb{Z}/n\mathbb{Z}$ will denote the cyclic group of order n. Also, we assume factorizations $m_i = p_1^{\mu_i(p_1)} \cdots p_s^{\mu_i(p_s)}$ with primes $p_1 < \cdots < p_s$ and integers $\mu_i(p_j) \geqslant 0$ such that $\mu_i(p_1) + \cdots + \mu_i(p_s) > 0$; we also write $\mu_i(p) = 0$ for a prime $p \notin \{p_1, \dots, p_s\}$. In the following, only cocompact Fuchsian groups are considered. A presentation for such a Fuchsian group Γ is

$$\Gamma \approx \langle\, a_1, b_1, \dots, a_\gamma, b_\gamma, x_1, \dots, x_r \mid x_1^{m_1},\ \dots,\ x_r^{m_r},\ x_1 \cdots x_r \textstyle\prod_{i=1}^{\gamma}[a_i, b_i] \,\rangle.$$

Let $S(\Gamma) = 2(\gamma - 1) + \sum_{i=1}^{r}(1 - 1/m_i)$. We consider integers $\hat{\mu}_i(p_j) \geqslant 0$ and permutations τ_j such that $\hat{\mu}_{\tau_j(i)}(p_j) = \mu_i(p_j)$ and $\hat{\mu}_1(p_j) \leqslant \hat{\mu}_2(p_j) \leqslant \cdots \leqslant \hat{\mu}_r(p_j)$ for each $j \in \{1, \dots, s\}$, and let $\hat{m}_i = p_1^{\hat{\mu}_i(p_1)} \cdots p_s^{\hat{\mu}_i(p_s)}$. Then, there is an integer $\hat{r}$ such that $\hat{m}_1 = \cdots = \hat{m}_{r-\hat{r}} = 1$ and $1 \neq \hat{m}_{r-\hat{r}+1} | \cdots | \hat{m}_r$, so $\hat{m}_i > 1$ when i is one of the last $\hat{r}$ integers in $\{1, \dots, r\}$. If $2(\gamma - 1) + \sum_{i=1}^{\hat{r}}(1 - 1/\hat{m}_{r-\hat{r}+i}) > 0$ then the signature $(\gamma; \hat{m}_{r-\hat{r}+1}, \dots, \hat{m}_r)$ defines a Fuchsian group $\hat{\Gamma}$. Both Fuchsian groups Γ and $\hat{\Gamma}$ have the same abelianization,

$$\Gamma_{ab} \approx \hat{\Gamma}_{ab} \approx \mathbb{Z}^{2\gamma} \oplus \mathcal{T}(\Gamma_{ab}) \approx \mathbb{Z}^{2\gamma} \oplus \mathbb{Z}/\hat{m}_{r-\hat{r}+1}\mathbb{Z} \oplus \cdots \oplus \mathbb{Z}/\hat{m}_{r-1}\mathbb{Z},$$

where

$$\mathcal{T}(\Gamma_{ab}) \approx \bigoplus_{p \in \{p_1, \dots, p_s\}} Syl_p(\mathcal{T}(\Gamma_{ab})) \approx \bigoplus_{p \in \{p_1, \dots, p_s\}} \bigoplus_{i=1}^{r-1} \mathbb{Z}/p^{\hat{\mu}_i(p)}\mathbb{Z}$$

is the torsion subgroup of Γ_{ab} [**1**, Lemma A.3]. We notice that $Syl_p(\mathcal{T}(\Gamma_{ab}))$ has at most $r-1$ factors. Also, we can check that

$$\sum_{i=1}^{r}\left(1-\frac{1}{\hat{m}_i}\right) \leqslant \sum_{i=1}^{r}\left(1-\frac{1}{m_i}\right). \tag{2.1}$$

For, consider the following matrices with the factors of periods $m_i, \hat{m}_i$ in their rows:

$$\mathcal{M} = \begin{pmatrix} p_1^{\mu_1(p_1)} & \cdots & p_s^{\mu_1(p_s)} \\ \vdots & & \vdots \\ p_1^{\mu_r(p_1)} & \cdots & p_s^{\mu_r(p_s)} \end{pmatrix} \qquad \widehat{\mathcal{M}} = \begin{pmatrix} p_1^{\hat{\mu}_1(p_1)} & \cdots & p_s^{\hat{\mu}_1(p_s)} \\ \vdots & & \vdots \\ p_1^{\hat{\mu}_r(p_1)} & \cdots & p_s^{\hat{\mu}_r(p_s)} \end{pmatrix}.$$

We can get $\widehat{\mathcal{M}}$ from $\mathcal{M}$ by an interchange of entries in pairs of consecutive rows of $\mathcal{M}$. Let $(i,j) = p_j^{\mu_i(p_j)}$ be the entry of $\mathcal{M}$ in row i and column j. First, take the first and second rows and, for each $j \in \{1,\ldots,s\}$, interchange $(1,j)$ and $(2,j)$ if $(2,j) < (1,j)$. Then we proceed with the second and third rows, and so on for the remaining rows on the produced matrices. We obtain $\widehat{\mathcal{M}}$ repeating the whole process if necessary. It is sufficient to consider only one of these steps, say, for two successive rows

$$\begin{pmatrix} p_1^{a_1} & \cdots & p_{j_1}^{b_{j_1}} & \cdots & p_{j_q}^{b_{j_q}} & \cdots & p_s^{a_s} \\ p_1^{b_1} & \cdots & p_{j_1}^{a_{j_1}} & \cdots & p_{j_q}^{a_{j_q}} & \cdots & p_s^{b_s} \end{pmatrix},$$

with integers a_j, b_j such that $0 \leqslant a_j \leqslant b_j$; we take consecutive unordered columns for readability. If we let

$$\begin{aligned} m_1 &= p_1^{a_1}\cdots p_{j_1}^{b_{j_1}}\cdots p_{j_q}^{b_{j_q}}\cdots p_s^{a_s} & \qquad m_1' &= p_1^{a_1}\cdots p_s^{a_s} \\ m_2 &= p_1^{b_1}\cdots p_{j_1}^{a_{j_1}}\cdots p_{j_q}^{a_{j_q}}\cdots p_s^{b_s} & m_2' &= p_1^{b_1}\cdots p_s^{b_s}, \end{aligned}$$

it follows that $\frac{1}{m_1}+\frac{1}{m_2} \leqslant \frac{1}{m_1'}+\frac{1}{m_2'}$, since $m_1m_2 = m_1'm_2' > 0$ and $m_1' \leqslant m_1 \leqslant m_2'$.

A finitely generated abelian group A has a unique invariant factor decomposition, $A \approx \mathbb{Z}^l \oplus \mathbb{Z}_{c_1} \oplus \cdots \oplus \mathbb{Z}_{c_t}$, with rank l and invariant factors $c_1|\cdots|c_t$. Also, A has a unique primary decomposition $A \approx \mathbb{Z}^l \oplus A_{q_1} \oplus \cdots \oplus A_{q_\lambda}$ with primes $q_1 < \cdots < q_\lambda$, where $A_q = \{x \in A | q^n x = 0 \text{ for some } n \geqslant 0\}$ is the q-primary component of A —the q-Sylow subgroup $Syl_q(\mathcal{T}(A))$ of the torsion subgroup $\mathcal{T}(A)$ of A— for each prime q dividing $|A|$. We also write $c_k = q_1^{\alpha_k(q_1)}\cdots q_\lambda^{\alpha_k(q_\lambda)}$ for $k = 1,\ldots,t$, so $0 \leqslant \alpha_1(q) \leqslant \cdots \leqslant \alpha_t(q)$ and $A_q \approx \mathbb{Z}/q^{\alpha_1(q)}\mathbb{Z} \oplus \cdots \oplus \mathbb{Z}/q^{\alpha_t(q)}\mathbb{Z}$. The integers $q_j^{\alpha_k(q_j)}$ are the elementary divisors of A.

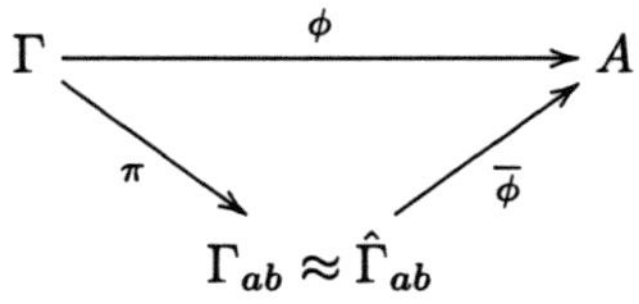

Here we are concerned with homomorphisms $\phi : \Gamma \to A$ onto a finite abelian group A. Let $\pi : \Gamma \to \Gamma_{ab}$ be the canonical homomorphism. Then, ϕ factors (uniquely, since $\Gamma' \leqslant \ker\phi$ for the commutator subgroup Γ' of Γ) through π, i.e., there is a (unique) epimorphism $\overline{\phi} : \Gamma_{ab} \to A$ such that $\phi = \overline{\phi}\circ\pi$. Conversely, given an epimorphism $\overline{\phi} : \Gamma_{ab} \to A$, then the homomorphism $\phi = \overline{\phi}\circ\pi : \Gamma \to A$ is onto.

We define $n_k(q)$ as the number of cyclic factors of A_q of order greater than or equal to $q^{\alpha_k(q)}$, and $N_k(q)$ as the number of cyclic factors of $Syl_q(\mathcal{T}(\Gamma_{ab})) \approx$

$\oplus_{i=1}^{r-1}\mathbb{Z}/q^{\hat{\mu}_i(q)}\mathbb{Z}$ of order greater than or equal to $q^{\alpha_k(q)}$:

$$n_k(q) = \#\{i \in \{1,\dots,t\} \mid \alpha_i(q) \geqslant \alpha_k(q)\},$$
$$N_k(q) = \#\{i \in \{1,\dots,r-1\} \mid \hat{\mu}_i(q) \geqslant \alpha_k(q)\}.$$

Hence, $n_k(q) \leqslant t$, $N_k(q) \leqslant r-1$, $n_k(q) \geqslant n_{k+1}(q)$ and $N_k(q) \geqslant N_{k+1}(q)$. If $\alpha_k(q) = \alpha_{k'}(q)$ then $n_k(q) = n_{k'}(q)$ and $N_k(q) = N_{k'}(q)$. Also, $N_k(q) = 0$ if $q \nmid m_i$ for all $i \in \{1,\dots,r\}$. Breuer shows in Lemmas A.1 and A.2 [**1**] necessary and sufficient conditions for an epimorphism $\Gamma_{ab} \to A$ to exist. Indeed, these conditions are expressed in terms of the number of cyclic factors of $Syl_q(\Gamma_{ab})$ and A_q for each prime q dividing $|A|$:

$$2\gamma + N_k(q) \geqslant n_k(q) \quad \text{for } k = 1,\dots,t. \tag{2.2}$$

Thus, the existence of an epimorphism $\Gamma \to A$ is equivalent to the fulfillment of inequalities (2.2) for each prime q dividing $|A|$.

It can be helpful to represent graphically the inequalities (2.2) superposing the graph of the number of factors of $Syl_q(\Gamma_{ab})$ to that of A_q for a given q. Conditions are fulfilled if and only if the second line never places above the first one —see equivalence of (2.2) and (3.1) in the proof of Theorem 3.2.

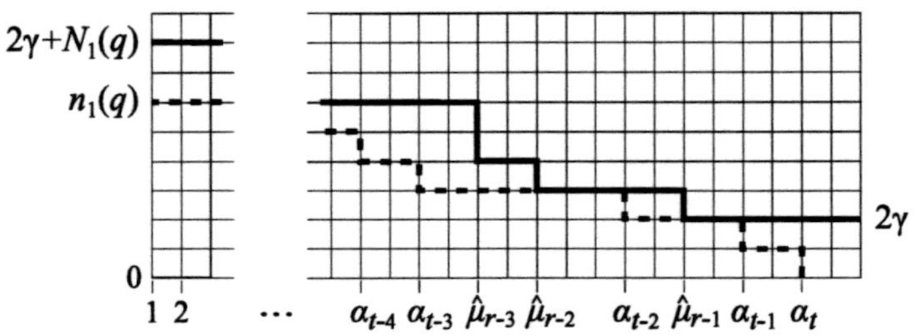

FIGURE 1. Example fulfilling conditions (2.2) for a prime q dividing $|A|$. We put the values $\alpha_1(q),\dots,\alpha_t(q)$ and $\hat{\mu}_1(q),\dots,\hat{\mu}_{r-1}(q)$ on the horizontal axis. As we move from right to left along each integer value x on the horizontal axis, the dotted line cumulatively adds up the number of factors of order q^x in A_q. Likewise, the solid line adds up the factors of $Syl_q(\mathcal{T}(\Gamma_{ab}))$ starting from the right with value 2γ. Here, $2\gamma = 2$ and $\hat{\mu}_{r-4} = \hat{\mu}_{r-3}$.

3. Existence of surface kernel homomorphisms onto an abelian group

Conditions for the existence of a surface kernel epimorphism from a Fuchsian group Γ onto a finite abelian group A were first stated by Breuer [**1**, Thm. 9.1]. Unlike for other families of groups, one of these conditions simply claims the existence of an epimorphism $\Gamma \to A$, relying this issue upon Lemmas A.1 and A.2 in [**1**], collected here in the inequalities (2.2) for each prime q dividing $|A|$. In Theorem 3.2, we embed conditions (2.2) for the existence of an epimorphism $\Gamma_{ab} \to A$ into Breuer's theorem. This way we achieve conditions only in terms of the signature of the Fuchsian group and the invariant factors defining the abelian group.

THEOREM 3.1. (Breuer) *Let A be a finite abelian group, Γ a Fuchsian group with signature $(\gamma; m_1,\dots,m_r)$ and $M = \mathrm{lcm}(m_1,\dots,m_r)$. There exists a surface kernel epimorphism $\psi : \Gamma \to A$ if and only if the following conditions are satisfied:*

(o) There exists an epimorphism $\Gamma \to A$.
(i) $\operatorname{lcm}(m_1, \dots, m_{i-1}, m_{i+1}, \dots, m_r) = M$ *for all* i.
(ii) $M \mid \exp A$; *if* $\gamma = 0$, *then* $M = \exp A$.
(iii) $r \neq 1$; *if* $\gamma = 0$, *then* $r \geqslant 3$.
(iv) If M *is even and only one of the elementary divisors of* A *is divisible by the maximum power of* 2 *dividing* M, *then the number of periods* m_i *divisible by such power of* 2 *is even.*

Now, we make use of conditions (2.2) for the existence of an epimorphism $\Gamma_{ab} \to A$ to replace condition (o) in Breuer's theorem:

THEOREM 3.2. *Let* Γ *be a Fuchsian group with signature* $(\gamma; m_1, \dots, m_r)$, $M = \operatorname{lcm}(m_1, \dots, m_r)$ *and integers* $t \geqslant 1$ *and* $c_1, c_2, \dots, c_t$ *with* $1 < c_1 | c_2 | \cdots | c_t$. *There exists a surface kernel epimorphism* $\Gamma \to \mathbb{Z}/c_1\mathbb{Z} \oplus \cdots \oplus \mathbb{Z}/c_t\mathbb{Z}$ *if and only if the following conditions are satisfied:*

(i) $\operatorname{lcm}(m_1, \dots, m_{i-1}, m_{i+1}, \dots, m_r) = M$ *for all* i.
(ii) $M \mid c_t$; *if* $\gamma = 0$, *then* $M = c_t$.
(iii) $r \neq 1$; *if* $\gamma = 0$, *then* $r \geqslant 3$.
(iv) If M *is even and only one of the elementary divisors of* $\mathbb{Z}/c_1\mathbb{Z} \oplus \cdots \oplus \mathbb{Z}/c_t\mathbb{Z}$ *is divisible by the maximum power of* 2 *dividing* M, *then the number of periods* m_i *divisible by such power of* 2 *is even.*
(v) If $t > 2\gamma$, *then* $r \geqslant t - 2\gamma + 1$ *and every elementary divisor of* $\mathbb{Z}/c_k\mathbb{Z}$ *divides, at least,* $t - 2\gamma - k + 2$ *periods* m_i *for* $k = 1, \dots, t - 2\gamma$.

PROOF. Condition (v) replaces condition (o) in Theorem 3.1. Below we prove that both conditions are equivalent. Hence, Theorems 3.1 and 3.2 are equivalent.

If there exists a surface kernel epimorphism $\Gamma \to A$, then we know, by Theorem 3.1, that conditions (i)-(iv) are satisfied. Conditions (i) and (ii) imply that $\hat{\mu}_{r-1}(q) = \hat{\mu}_r(q) \leqslant \alpha_t(q)$ for each prime q dividing $|A|$, and so dividing c_t.

We have seen before that an epimorphism $\Gamma \to A$ exists if and only if the inequalities (2.2) hold for each prime q dividing c_t.

Since $c_1 \neq 1$, there exists some prime q dividing c_t for which $0 < \alpha_1(q) \leqslant \cdots \leqslant \alpha_t(q)$ and $n_1(q) = t$. Hence, the first inequality $2\gamma + N_1(q) \geqslant n_1(q) = t$ implies $2\gamma + r - 1 \geqslant t$, since $r - 1 \geqslant N_1(q)$. So $r \geqslant t - 2\gamma + 1$, which is the first part of (v).

If $t \leqslant 2\gamma$, the inequalities (2.2) are always fulfilled for every prime q dividing c_t, since $t \geqslant n_1(q) \geqslant n_2(q) \geqslant \cdots \geqslant n_t(q)$. However, if $t > 2\gamma$, that need not necessarily be the case. Now we show that, if $t > 2\gamma$, conditions (2.2) hold if and only if

$$r \geqslant t - 2\gamma + 1 \quad \text{and} \quad \alpha_k(q) \leqslant \hat{\mu}_{r-1-t+2\gamma+k}(q) \quad \text{for} \quad k = 1, \dots, t - 2\gamma, \tag{3.1}$$

or, explicitly,

$$\begin{gathered} \alpha_{t-2\gamma}(q) \leqslant \hat{\mu}_{r-1}(q), \\ \alpha_{t-2\gamma-1}(q) \leqslant \hat{\mu}_{r-2}(q), \\ \vdots \\ \alpha_1(q) \leqslant \hat{\mu}_{r-t+2\gamma}(q), \\ \text{and } r \geqslant t - 2\gamma + 1. \end{gathered}$$

If (3.1) holds, let $k \in \{1, \dots, t - 2\gamma\}$ and l be the smallest integer in $\{1, \dots, k\}$ such that $\alpha_l(q) = \alpha_k(q)$. Then $n_k(q) = n_l(q) = t - l + 1$ and $N_k(q) = N_l(q)$. By (3.1),

$\alpha_l(q) \leqslant \hat{\mu}_{r-1-t+2\gamma+l}(q)$, so

$$N_l(q) \geqslant r-1-(r-1-t+2\gamma+l)+1 = t-2\gamma-l+1.$$

Hence, $2\gamma + N_k(q) \geqslant n_k(q)$. This proves that (3.1) implies (2.2).

On the other hand, if we assume that conditions (2.2) hold, then the last 2γ values $\alpha_{t-2\gamma+1}(q), \dots, \alpha_t(q)$ can be smaller than (except $\alpha_t(q)$, by condition (ii)), equal to or greater than $\hat{\mu}_{r-1}(q)$. However, it is always $\alpha_{t-2\gamma}(q) \leqslant \hat{\mu}_{r-1}(q)$: otherwise, the last $2\gamma+1$ values $\alpha_{t-2\gamma}(q), \dots, \alpha_t(q)$, at least, would be greater than $\hat{\mu}_{r-1}(q)$, thus $N_{t-2\gamma}(q) = 0$ and $n_{t-2\gamma}(q) \geqslant 2\gamma+1$, so the inequality $2\gamma + N_{t-2\gamma}(q) \geqslant n_{t-2\gamma}(q)$ would not be fulfilled and an epimorphism would not exist.

Also, the next value $\alpha_{t-2\gamma-1}(q)$ smaller than or equal to $\alpha_{t-2\gamma}(q)$ must satisfy $\alpha_{t-2\gamma-1}(q) \leqslant \hat{\mu}_{r-2}(q)$; otherwise, some inequality in (2.2) would not be fulfilled. Likewise, it must be $\alpha_{t-2\gamma-2}(q) \leqslant \hat{\mu}_{r-3}(q)$, $\alpha_{t-2\gamma-3}(q) \leqslant \hat{\mu}_{r-4}(q)$, and so on. Hence, (2.2) implies (3.1).

Now, since $\hat{\mu}_i(q) \leqslant \hat{\mu}_{i+1}(q)$, it is clear that conditions (3.1) hold if and only if $\alpha_k(q) \leqslant \hat{\mu}_i(q)$ for each $k \in \{1, \dots, t-2\gamma\}$ and $i = r-1-t+2\gamma+k, \dots, r-1$. Since $\hat{\mu}_r(q) = \max\{\mu_i(q)\}_{i=1,\dots,r}$, this means that $q^{\alpha_k(q)}$ divides $q^{\hat{\mu}_i}$ for $i = r-1-t+2\gamma+k, \dots, r$ and thus also divides, at least, $t-2\gamma-k+2$ periods m_i. This is condition (v). □

4. Explicit expression for the minimum genus

Conditions of Theorem 3.2 allow us to obtain a new shorter proof —see Theorem 4.1— of Maclachlan's solution of the minimum genus problem for a finite non-cyclic abelian group [**7**, Thm. 4]. Furthermore, upon closer study of this solution, we find an explicit expression for the minimum genus in many cases, as is shown in Remarks 4.4, 4.5, 4.6 and subsequent comments.

In particular, condition 3.2.(v) makes it possible to use the invariant factors $c_1, \dots, c_{t-2\gamma}$ as periods of a Fuchsian group, and this group is a candidate to minimize $S(\Gamma) = 2(\gamma-1) + \sum_{i=1}^{r}(1-1/m_i)$: elementary divisors of $\mathbb{Z}_{c_k}$ always divide certain periods m_i of any Fuchsian group satisfying conditions of Theorem 3.2, and this suggests to compose a signature with $c_1, \dots, c_{t-2\gamma}$ as periods.

THEOREM 4.1. (Maclachlan) *Let $A \approx \mathbb{Z}_{c_1} \oplus \cdots \oplus \mathbb{Z}_{c_t}$, with $t > 1$, be a finite abelian group of order $|A| > 9$ with invariant factors $c_1, c_2, \dots, c_t$ such that $1 < c_1|c_2|\cdots|c_t$. The minimum genus $g^* \geqslant 2$ of a compact Riemann surface which admits A as a group of automorphisms satisfies*

$$\frac{2(g^*-1)}{|A|} = \min_{0 \leqslant 2\gamma < t'} \left\{ 2(\gamma-1) + \sum_{i=1}^{t-2\gamma}\left(1-\frac{1}{c_i}\right) + 1 - \frac{1}{c_{t-2\gamma}} \right\}$$

where $t' = t$ if $t = 2$ or odd, and $t' = t+1$ if $t > 2$ is even (interpreting c_0 as 1).

REMARK 4.2. [**7**, Thm. 4] If $t = 2$ and $|A| > 9$, then $g^* = \frac{1}{2}(c_1c_2 - c_2) - c_1 + 1$.

REMARK 4.3. [**7**, p. 711] When A is non-cyclic ($t > 1$) of order $|A| \leqslant 9$ the minimum genus is the following:

$A \approx \mathbb{Z}_2 \oplus \mathbb{Z}_2$	$g^* = 2$
$A \approx \mathbb{Z}_2 \oplus \mathbb{Z}_4$	$g^* = 3$
$A \approx \mathbb{Z}_2 \oplus \mathbb{Z}_2 \oplus \mathbb{Z}_2$	$g^* = 3$
$A \approx \mathbb{Z}_3 \oplus \mathbb{Z}_3$	$g^* = 4$

PROOF OF THEOREM 4.1. Let $\mathcal{F}$ be the family of Fuchsian groups for which there exists a surface kernel homomorphism onto A, and let Γ_γ be any Fuchsian group in $\mathcal{F}$ with orbit genus γ, say, with signature $(\gamma;\ m_1, \dots, m_r)$

For any integer γ such that $0 \leqslant 2\gamma < t$, the signature

$$(4.1) \qquad (\gamma\,;\, c_1\,,\, c_2\,,\, \dots\,,\, c_{t-2\gamma-1}\,,\, c_{t-2\gamma}\,,\, c_{t-2\gamma})$$

defines a Fuchsian group if and only if $|A| = c_1 \cdots c_t > 9$. We denote it by $\widetilde{\Gamma}_\gamma$. That signature fulfills conditions of Theorem 3.2, so $\widetilde{\Gamma}_\gamma \in \mathcal{F}$. We notice that, by condition (v) of Theorem 3.2 and since $2\gamma < t$, it must be $r \geqslant 2$. By conditions (i) and (v) of Theorem 3.2, we have $c_{t-2\gamma}|\hat{m}_r$, $c_{t-2\gamma-1}|\hat{m}_{r-1}$, $c_{t-2\gamma-2}|\hat{m}_{r-2}$, ..., $c_1|\hat{m}_{r-t+2\gamma}$. It follows that

$$\left(1-\frac{1}{c_1}\right)+\cdots+\left(1-\frac{1}{c_{t-2\gamma}}\right)+\left(1-\frac{1}{c_{t-2\gamma}}\right) \leqslant \left(1-\frac{1}{\hat{m}_1}\right)+\cdots+\left(1-\frac{1}{\hat{m}_r}\right),$$

so the signature $(\gamma;\ \hat{m}_{r-\hat{r}+1}, \dots, \hat{m}_r)$ defines a Fuchsian group, $\hat{\Gamma}_\gamma$, since $S(\widetilde{\Gamma}_\gamma) > 0$, and $S(\widetilde{\Gamma}_\gamma) \leqslant S(\hat{\Gamma}_\gamma)$ (recall $\hat{r}$ is such that $\hat{m}_{r-\hat{r}} = 1$ and $\hat{m}_{r-\hat{r}+1} > 1$). Then, by (2.1), $S(\widetilde{\Gamma}_\gamma) \leqslant S(\hat{\Gamma}_\gamma) \leqslant S(\Gamma_\gamma)$.

For any γ with $2\gamma \geqslant t > 1$, the signature $(\gamma;\ \text{—})$ defines a Fuchsian group Λ_γ and fulfills conditions of Theorem 3.2, so $\Lambda_\gamma \in \mathcal{F}$. Then $S(\Lambda_\gamma) = 2(\gamma-1) < S(\Gamma_\gamma)$ for any $\Gamma_\gamma \in \mathcal{F}$. Also, $S(\Lambda_{\gamma'}) < S(\Lambda_\gamma)$ if $\gamma' < \gamma$. Let $\overline{\gamma} = \min\{\gamma \in \mathbb{Z} \mid 2\gamma \geqslant t,\ \gamma > 1\}$. Then, $\overline{\gamma} = 2$ if $t = 2$, $\overline{\gamma} = (t+1)/2$ if t is odd, and $\overline{\gamma} = t/2$ if $t > 2$ is even; the corresponding values of $S(\Lambda_{\overline{\gamma}})$ are 2, $t-1$ and $t-2$, respectively.

Therefore, we have to compare $S(\widetilde{\Gamma}_\gamma)$ and $S(\Lambda_{\overline{\gamma}})$ for any γ satisfying $0 \leqslant 2\gamma < t$. If $\gamma = 0$, then $S(\widetilde{\Gamma}_\gamma) < S(\Lambda_{\overline{\gamma}})$ for $t = 2$ or t odd (recall that $S(\widetilde{\Gamma}_\gamma) > 0$ when $|A| > 9$ and $0 \leqslant \gamma < t$). But, if t is even, it could be $S(\widetilde{\Gamma}_\gamma) > S(\Lambda_{\overline{\gamma}})$ for every γ such that $2\gamma < t$. To take this possibility into account, we define $c_0 = 1$, $t' = t$ if $t = 2$ or odd, and $t' = t+1$ if $t > 2$ is even. □

Now we get some insight into the expression for g^* in Theorem 4.1. In many cases, we can avoid calculating the minimum, since it is possible to find out in advance which γ satisfying $0 \leqslant \gamma < t$ provides it, by simple inspection of the invariant factors of the abelian group. In the rest of this section we assume $t \geqslant 3$.

We can write $g^* = 1 + \frac{|A|}{2}\ \min\{S_0, S_1, \dots, S_{\gamma'}\}$, where

$$S_\gamma = 2(\gamma-1) + \sum_{i=1}^{t-2\gamma}\left(1-\frac{1}{c_i}\right) + \left(1-\frac{1}{c_{t-2\gamma}}\right)$$

and $\gamma' = \lfloor t/2 \rfloor$ is the integer part of $t/2$. Let also

$$q_i = \frac{c_{i+1}}{c_i} \quad \text{for } i = 1, \dots, t-1.$$

If t is odd, we put the elements $\{q_1, \dots, q_{t-1}\}$ into pairs, reversing the order of subindices:

$$(q_{t-1}, q_{t-2}), \dots, (q_4, q_3), (q_2, q_1).$$

If t is even, we can consider A as an abelian group with $t+1$ invariant factors, $A \approx \{0\} \oplus \mathbb{Z}_{c_1} \oplus \cdots \oplus \mathbb{Z}_{c_t}$, without changing the expressions for S_γ and g^*; now the $(t+1)-1 = t$ quotients of two consecutive invariant factors become $c_1/1 = c_1$, $c_2/c_1 = q_1$, ..., $c_t/c_{t-1} = q_{t-1}$, resulting in the sequence of pairs

$$(q_{t-1}, q_{t-2}), \dots, (q_3, q_2), (q_1, c_1).$$

The expression of $S_{\gamma+1}$ in terms of the foregoing S_γ is

$$S_{\gamma+1} = S_\gamma + \frac{2}{c_{t-2\gamma}} + \frac{1}{c_{t-2\gamma-1}} - \frac{1}{c_{t-2\gamma-2}} = S_\gamma + \frac{2-(q_{t-2\gamma-2}-1)q_{t-2\gamma-1}}{c_{t-2\gamma}}$$

for $\gamma = 0, \dots, \gamma' - 1$ (interpreting $c_0 = 1$ and $q_0 = c_1$ if t is even). Writing them explicitly,

$$S_1 = S_0 + \frac{2-(q_{t-2}-1)q_{t-1}}{c_t},$$

$$S_2 = S_1 + \frac{2-(q_{t-4}-1)q_{t-3}}{c_{t-2}}, \quad \text{etc.},$$

we observe that the difference $S_1 - S_0$ depends on q_{t-1}, q_{t-2}, c_t, which are the largest subindices for q and c. As γ increases, these subindices reduce by 2 each step until the largest value $\gamma = \gamma' - 1$. The last difference $S_{\gamma'} - S_{\gamma'-1}$ depends on q_2, q_1, c_3 (or on $q_0 = c_1$ and q_1 if t is even). It follows that:

$$\text{(4.2)}\quad \begin{aligned}
&a)\quad S_0 < S_1 \iff \begin{cases} q_{t-2} = 1 \\ \quad\text{or} \\ (q_{t-1}, q_{t-2}) = (1,2) \end{cases} \iff \begin{cases} c_{t-2} = c_{t-1} \\ \quad\text{or} \\ 2c_{t-2} = c_{t-1} = c_t \end{cases} \\
&b)\quad S_0 = S_1 \iff (q_{t-1}, q_{t-2}) = \begin{cases} (2,2) \\ \text{or} \\ (1,3) \end{cases} \iff \begin{cases} 4c_{t-2} = 2c_{t-1} = c_t \\ \quad\text{or} \\ 3c_{t-2} = c_{t-1} = c_t \end{cases} \\
&c)\quad S_0 > S_1 \quad \text{otherwise.}
\end{aligned}$$

The same happens for each pair $S_\gamma, S_{\gamma+1}$ and $(q_{t-2\gamma-1}, q_{t-2\gamma-2})$. So, in general, S_γ is smaller as γ increases, and g^* is mostly given by $S_{\gamma'}$:

REMARK 4.4. If $c_1 \neq c_2 \neq \cdots \neq c_t$, then

$$g^* = 1 + \frac{|A|}{2}\, S_{\gamma'}$$

since, in this case, $q_i > 1$ for all i, so $S_\gamma \geqslant S_{\gamma+1}$ for all $\gamma \in \{0, \dots, \gamma' - 1\}$.

On the other hand, we can have $S_\gamma \leqslant S_{\gamma+1}$ for all γ:

REMARK 4.5. If either $c_{i+1} = c_i$ or $c_{i+1} = 2c_i$ for each $i \in \{1, \dots, t-1\}$, then

$$g^* = 1 + \frac{|A|}{2}\, S_0.$$

In both remarks, we have only made use of the case $(q_{t-1}, q_{t-2}) = (2,2)$ in (4.2.b). It is straightforward to include the other case $(q_{t-2\gamma-1}, q_{t-2\gamma-2}) = (1,3)$, for values of γ in $\{1, \dots, \gamma'\}$, to enlarge the set of cases with the same result $g^* = 1 + |A|S_0/2$ or $1 + |A|S_{\gamma'}/2$.

Also, we can get $g^* = 1 + |A|S_\gamma/2$ for each $\gamma \in \{0, \dots, \gamma'\}$ combining conditions in both remarks to get $S_0 \geqslant S_1 \geqslant \dots \geqslant S_{\gamma-1} \geqslant S_\gamma \leqslant S_{\gamma+1} \leqslant \dots \leqslant S_{\gamma'}$:

REMARK 4.6. Let $\gamma \in \{0, \dots, \gamma'\}$. If $c_{t-2\gamma-1} \neq c_{t-2\gamma} \neq \cdots \neq c_t$ and $c_{i+1} = c_i$ or $c_{i+1} = 2c_i$ for all $i \in \{1, \dots, t-2\gamma-2\}$, then

$$g^* = 1 + \frac{|A|}{2}\, S_\gamma.$$

Conditions (4.2) make it difficult to study which S_γ provides g^* in general. But we can still give some hints that ease the situation by reducing the number of candidates among $\{S_0, \ldots, S_{\gamma'}\}$ to obtain g^*.

It can help to assign a symbol to each pair $(q_{t-2\gamma-1}, q_{t-2\gamma-2})$ for $\gamma = 0, \ldots, \gamma'-1$ as follows:

$$
\begin{array}{ll}
{}_{\gamma}\nearrow^{\gamma+1} & \text{if } (q_{t-2\gamma-1}, q_{t-2\gamma-2}) = (\ldots, 1) \text{ or } (1,2),\\
{}_{\gamma}\rightarrow_{\gamma+1} & \text{if } (q_{t-2\gamma-1}, q_{t-2\gamma-2}) = (2,2) \text{ or } (1,3),\\
{}^{\gamma}\searrow_{\gamma+1} & \text{otherwise.}
\end{array}
$$

So, for example, the sequence of pairs $(q_{t-1}, q_{t-2}), \ldots, (q_4, q_3), (q_2, q_1)$ —ending with (q_1, c_1) if t is even— is represented by

$$
{}^{0}\searrow_{1}\nearrow^{2}\searrow_{3}\searrow_{4}\longrightarrow_{5}\;\cdots\;{}^{\gamma'-2}\searrow^{\gamma'-1}\searrow_{\gamma'}
$$

if and only if

$$
S_0 > S_1 < S_2 > S_3 > S_4 = S_5 \;\cdots\; S_{\gamma'-2} > S_{\gamma'-1} > S_{\gamma'}.
$$

Obviously, the case ${}^{\gamma-1}\searrow_{\gamma}\nearrow^{\gamma+1}$ indicates that S_γ is a candidate to obtain g^*, but not $S_{\gamma-1}$ or $S_{\gamma+1}$. Also, S_0 or $S_{\gamma'}$ are candidates if ${}_{0}\nearrow$ or $\searrow_{\gamma'}$ appear, respectively, but not if we have ${}^{0}\searrow$ or $\nearrow^{\gamma'}$.

When ${}_{\gamma}\nearrow^{\gamma+1}$ occurs, then $S_{\gamma+1} = S_\gamma + 1/c_{t-2\gamma}$ or $S_{\gamma+1} = S_\gamma + 2/c_{t-2\gamma}$ corresponding to each possibility in (4.2.a).

It is easy to see that, in general, $S_{\gamma+1} > S_\gamma \geqslant S_{\gamma+2}$ in the case ${}_{\gamma}\nearrow^{\gamma+1}\searrow_{\gamma+2}$, with the only exception when $(q_{t-2\gamma-1}, q_{t-2\gamma-2}) = (1,1)$ and $(q_{t-2\gamma-3}, q_{t-2\gamma-4}) = (3,2)$, in which case $S_{\gamma+1} > S_{\gamma+2} > S_\gamma$, since $S_{\gamma+1} = S_\gamma + 2/c_{t-2\gamma}$ and $S_{\gamma+2} = S_\gamma + 1/c_{t-2\gamma}$. We introduce another symbol for this last situation:

$$
\begin{aligned}
{}_{\gamma}\nearrow\!\!\!\searrow^{\gamma+2} &\iff S_\gamma < S_{\gamma+2} < S_{\gamma+1}\\
&\iff (q_{t-2\gamma-1}, q_{t-2\gamma-2}) = (1,1) \text{ and } (q_{t-2\gamma-3}, q_{t-2\gamma-4}) = (3,2),
\end{aligned}
$$

so S_γ is a candidate to obtain g^*, but not $S_{\gamma+1}$ or $S_{\gamma+2}$. In the following examples, we assume $S_{\gamma+2} < S_\gamma < S_{\gamma+1}$ whenever we write ${}_{\gamma}\nearrow^{\gamma+1}\searrow_{\gamma+2}$ (the case $\nearrow\!\!\!\searrow$ is excluded from that sequence and we split it apart as a separate case).

EXAMPLE 4.7. $t = 3$. Here $\gamma = 0, 1$, $S_0 = 2 - \dfrac{1}{c_1} - \dfrac{1}{c_2} - \dfrac{2}{c_3}$, $S_1 = 2 - \dfrac{2}{c_1}$ and, considering the pair (q_2, q_1), there are only three cases:

$$
{}^{0}\searrow_{1}\quad S_1 \qquad\qquad {}_{0}\rightarrow_{1}\quad S_1 = S_0 \qquad\qquad {}_{0}\nearrow^{1}\quad S_0.
$$

EXAMPLE 4.8. $t = 4$. We have $\gamma = 0, 1, 2$ and

$$
S_0 = 3 - \frac{1}{c_1} - \frac{1}{c_2} - \frac{1}{c_3} - \frac{2}{c_4} \qquad S_1 = 3 - \frac{1}{c_1} - \frac{2}{c_2} \qquad S_2 = 2.
$$

We consider the sequence $(q_3, q_2), (q_1, c_1)$; there are ten cases:

${}^{0}\searrow^{1}\searrow_{2}$	S_2	${}_{0}\nearrow^{1}\nearrow^{2}$	S_0	${}^{0}\rightarrow^{1}\searrow_{2}$	S_2
${}^{0}\searrow_{1}\rightarrow_{2}$	$S_2 = S_1$	${}_{0}\nearrow^{1}\rightarrow^{2}$	S_0	${}_{0}\rightarrow_{1}\rightarrow_{2}$	$S_2 = S_1 = S_0$
${}^{0}\searrow_{1}\nearrow^{2}$	S_1	${}_{0}\nearrow^{1}\searrow_{2}$	S_2	${}_{0}\rightarrow_{1}\nearrow^{2}$	$S_1 = S_0$
		${}_{0}\nearrow\!\!\searrow{}^{2}$	S_0		

EXAMPLE 4.9. $t = 5$. Here, the sequence is $(q_4, q_3), (q_2, q_1)$, $\gamma = 0, 1, 2$ and

$$S_0 = 4 - \frac{1}{c_1} - \frac{1}{c_2} - \frac{1}{c_3} - \frac{1}{c_4} - \frac{2}{c_5} \qquad S_1 = 4 - \frac{1}{c_1} - \frac{1}{c_2} - \frac{2}{c_3} \qquad S_2 = 4 - \frac{2}{c_1}.$$

The cases are the same as in the previous example.

EXAMPLE 4.10. The group $A \approx \mathbb{Z}_7 \oplus \mathbb{Z}_{14} \oplus \mathbb{Z}_{14} \oplus \mathbb{Z}_{84} \oplus \mathbb{Z}_{336}$ attains the minimum genus $g^* = 1 + |A|\, S_2/2 = 1 + |A|\,(t-1-2/c_1)/2 = 71914753$, since the corresponding sequence is $(4, 6), (1, 2)$, so the diagram is ${}_{0}\nearrow^{1}\searrow_{2}$.

EXAMPLE 4.11. Consider the groups

$$\begin{aligned} &A \approx \mathbb{Z}_{15} \oplus \mathbb{Z}_{90} \oplus \mathbb{Z}_{450} \oplus \mathbb{Z}_{900} \oplus \mathbb{Z}_{900}, && \text{with sequence } (1,2),(5,6), \\ &B \approx \mathbb{Z}_{3} \oplus \mathbb{Z}_{450} \oplus \mathbb{Z}_{450} \oplus \mathbb{Z}_{900} \oplus \mathbb{Z}_{900}, && \text{with sequence } (1,2),(1,150), \text{ and} \\ &C \approx \mathbb{Z}_{3} \oplus \mathbb{Z}_{450} \oplus \mathbb{Z}_{450} \oplus \mathbb{Z}_{810000}, && \text{with sequence } (1800,1),(150,3). \end{aligned}$$

The three groups have order $N = 492075000000$ and diagram ${}_{0}\nearrow^{1}\searrow_{2}$. Hence

$$\begin{aligned} g_A^* &= 1 + NS_2/2 = 1 + N(t - 1 - 2/c_1)/2 = 951345000001, \\ g_B^* &= 1 + NS_2/2 = 1 + N(t - 1 - 2/c_1)/2 = 820125000001, \\ g_C^* &= 1 + NS_2/2 = 1 + N(t - 2)/2 = 492075000001, \end{aligned}$$

for the corresponding value of S_2, t and c_1 in each case. We notice that $g_C^* < g_B^* < g_A^*$. The following Section deals with this issue.

5. Minimum genus and maximum order for abelian groups of the same given order

Harvey solved the minimum genus and maximum order problems for cyclic groups [**6**, Thm. 6]. In this section, by means of this result and Theorem 4.1, we solve the minimum genus problem for the family of abelian groups, cyclic or not, of the same given order $N > 1$. From this result, we also obtain a new proof of the solution to the maximum order problem stated by Breuer [**1**, Cor. 9.6].

We first consider, in Lemma 5.1, abelian groups having order N and a fixed number t of non trivial invariant factors. Applying this result, we obtain, in Theorem 5.2, the abelian group of order N that provides the minimum genus, by comparing the resulting genera for admissible values of t.

Let $N = p_1^{\alpha_1} \cdots p_s^{\alpha_s}$ be the prime factorization of N, where $p_i < p_{i+1}$. The number t of non trivial invariant factors ($c_i > 1$) of an abelian group $\mathbb{Z}_{c_1} \oplus \cdots \oplus \mathbb{Z}_{c_t}$ of order $c_1 \cdots c_t = N$, with $c_1 | \cdots | c_t$, is bounded by the greatest exponent in the prime factorization of N, that is, $1 \leqslant t \leqslant \max_{j=1,\dots,s}\{\alpha_j\}$.

LEMMA 5.1. *Let $N = p_1^{\alpha_1} \cdots p_s^{\alpha_s} > 1$, with $s \geqslant 1$, $\alpha_j > 0$ and primes $p_1, \ldots, p_s$ such that $p_j < p_{j+1}$, and an integer $t > 1$ such that $t \leqslant \max_{j=1,\ldots,s}\{\alpha_j\}$. Then the abelian group*

$$\mathbb{Z}_p \oplus \overset{t-1}{\cdots} \oplus \mathbb{Z}_p \oplus \mathbb{Z}_{N/p^{t-1}}$$

attains the minimum genus

$$g^* = \begin{cases} (t-1)(p-1)\frac{N}{2p} - p^{t-1} + 1 & \text{if } t = 2,\ t \text{ is odd or } p - t + 1 \leqslant \frac{2p^t}{N}, \\ (t-2)\frac{N}{2} + 1 & \text{otherwise,} \end{cases}$$

among all abelian groups of order N with t non trivial invariant factors, where p is the smallest prime such that $p^t | N$.

PROOF. Let $A \approx \mathbb{Z}_{c_1} \oplus \cdots \oplus \mathbb{Z}_{c_t}$ with $t > 1$, $|A| = c_1 \cdots c_t = N$ and $1 < c_1 | \cdots | c_t$. Hence, there must be, at least, a prime $p_j \in \{p_1, \ldots, p_s\}$ such that $p_j | c_i$ for all $i \in \{1, \ldots, t\}$, so $p_j^t | N$. Let p be the smallest such prime,

$$p = \min_{j=1,\ldots,s} \left\{ p_j \mid p_j^t \text{ divides } N \right\}.$$

By Theorem 4.1, we obtain the minimum genus of A from the smallest S_γ for the admissible values $\gamma = 0, \ldots, \gamma'$. If we let A vary with t fixed, then the abelian group with invariant factors

$$c_1 = \cdots = c_{t-1} = p \qquad c_t = \frac{N}{p^{t-1}}$$

gives the smallest value of S_γ for each $\gamma = 0, \ldots, \gamma'$: this is straightforward to check when $\gamma > 0$, since such values of $c_1, \ldots, c_{t-1}$ are greater than or equal to any other value with $c_1 \cdots c_t = N$. For S_0, we observe that

$$-\frac{1}{n} - \frac{2}{qm} \leqslant -\frac{1}{qn} - \frac{2}{m}$$

for any prime $q \geqslant 2$ and integers $m \geqslant 2$, $n \geqslant 1$ with $qn | m$, since such inequality is equivalent to $0 \leqslant (\frac{m}{n} - 2)(q - 1)$.

For such invariant factors, we have $S_0 = (t-1)(p-1)/p - 2p^{t-1}/N$, and $g^* = 1 + \frac{N}{2} S_0$ if $t = 2$. When $t > 2$, and following Section 4, the corresponding sequences $(N/p^t, 1), (1,1), \ldots, (1,1)$ if t is odd, and $(N/p^t, 1), (1,1), \ldots, (1,1), (1,p)$ if t is even, lead to the cases

${}_0\nearrow^1 \nearrow^2 \cdots \nearrow^{(t-1)/2}$	if t is odd,
${}_0\nearrow^1 \nearrow^2 \cdots \nearrow^{t/2}$	if t is even and $p = 2$,
${}_0\nearrow^1 \nearrow^2 \cdots \nearrow^{t/2-1} \rightarrow^{t/2}$	if t is even and $p = 3$,
${}_0\nearrow^1 \nearrow^2 \cdots \nearrow^{t/2-1} \searrow_{t/2}$	if t is even and $p \geqslant 5$.

Hence, $S_0 < S_i$ when $i > 0$ and $g^* = 1 + \frac{N}{2} S_0$, unless $t > 2$ is even and $p \geqslant 5$, in which case $g^* = 1 + \frac{N}{2} \min\{S_0, S_{t/2}\}$. If $t > 2$ is even, then $S_{t/2} = t - 2$, so $S_0 \leqslant S_{t/2}$ if and only if $p - t + 1 \leqslant 2p^t/N$ (this includes $p \in \{2, 3\}$). □

Now, we consider different values of $t \in \{2, \ldots, \max_{j=1,\ldots,s}\{\alpha_j\}\}$. Let

$$S_0(t) = (t-1)\left(1 - \frac{1}{p(t)}\right) - \frac{2}{N}\, p(t)^{t-1}, \tag{5.1}$$

where $p(t)$ is the smallest prime q such that $q^t|N$. By comparing the values g_t^* in Lemma 5.1 with the minimum genus g_1^* corresponding to the cyclic group $\mathbb{Z}_N$, we obtain the following theorem, that can be seen as a generalization of Harvey's solution of the minimum genus problem for cyclic groups.

THEOREM 5.2. *The minimum genus $g^* > 1$ of a compact Riemann surface which admits an abelian group of automorphisms of order N is*

$$g^* = \begin{cases} 2 & \text{if } N = 2,3,4,5,6 \text{ or } 8, \\ 3 & \text{if } N = 7 \text{ or } 9, \\ \frac{1}{2}(N-1) & \text{if } N > 9 \text{ is prime}, \\ \frac{1}{2}(p_1 - 1)\left(\frac{N}{p_1} - 1\right) & \text{if } N > 9 \text{ is not prime and } \alpha_1 = 1, \\ (p_1 - 1)\left(\frac{N}{2p_1} - 1\right) & \text{if } N > 9 \text{ and } \alpha_1 > 1, \end{cases}$$

where $N = p_1^{\alpha_1} \cdots p_s^{\alpha_s}$ is the prime factorization of N, with $\alpha_j > 0$ and $p_j < p_{j+1}$. The abelian group

i) $\mathbb{Z}_N$ *if* $N \leqslant 9$ *or* $\alpha_1 = 1$,
ii) $\mathbb{Z}_{p_1} \oplus \mathbb{Z}_{N/p_1}$ *if* $N > 9$ *and* $\alpha_1 > 1$.

attains the minimum genus.

PROOF. If $N \leqslant 9$, by [**6**, Thm. 6] and Theorem 4.1 and subsequent remarks, g^* is given by the cyclic group $\mathbb{Z}_N$. For $N > 9$ we distinguish two cases: a) $\alpha_1 = 1$; b) $\alpha_1 > 1$. Let $\tau = \max_{j=1,\dots,s}\{\alpha_j\}$.

a) $\alpha_1 = 1$. If $N = p_1 \cdots p_s$, there is only one abelian group of order N, the cyclic group $\mathbb{Z}_N$. Otherwise, and also for non-cyclic groups to take place, we consider $N = p_1 p(t)^2 m(t)$ for primes $p_1 < p(t)$, and $m(t) \geqslant 1$ an integer such that $p_1 < q$ if q is a prime dividing $m(t)$ —$p(t)$ and $m(t)$ may be different for each $t \in \{2, \dots, \tau\}$.

Let g_1^* be the minimum genus for $\mathbb{Z}_N$ [**6**, Thm. 6],

$$g_1^* = \frac{1}{2}(p_1 - 1)\left(\frac{N}{p_1} - 1\right) = \underbrace{\frac{1}{2}(p_1 - 1)\frac{N}{p_1}}_{A} \underbrace{-\frac{p_1}{2}}_{B} + \frac{1}{2},$$

and, in case that $g_t^* = 1 + \frac{N}{2}S_0(t)$ by Lemma 5.1, let

$$\begin{aligned} g_t^* &= 1 + \frac{1}{2}(t-1)(p(t)-1)\frac{N}{p(t)} - p(t)^{t-1} \\ &= \underbrace{\frac{1}{2}(p(t)-1)\frac{N}{p(t)}}_{A'} + \underbrace{\frac{1}{2}(t-2)(p(t)-1)\frac{N}{p(t)} - p(t)^{t-1} + 1}_{C} \end{aligned}$$

be the minimum genus for each $t \in \{2, \dots, \tau\}$ obtained from (5.1). We notice that

$$A < A', \quad B < 0, \quad C = -p(t) \text{ if } t = 2, \quad \text{and } C \geqslant 0 \text{ if } t > 2.$$

The last inequality holds since $p(t)^t|N$ and, therefore, $N/p(t) \geqslant p(t)^{t-1}$. If $g_t^* = (t-2)N/2+1$ by Lemma 5.1, then $g_1^* < g_t^*$ as well, since $(p_1-1)/p_1 < 1 < t-2$ for $t > 2$ even, as is the case when $g_t^* = (t-2)N/2+1$. Hence, $g_1^* < g_t^*$ if $t \in \{3, \dots, \tau\}$.

We address the case $t = 2$ as follows: if $N = 2 \cdot 3^2$, then $g_1^* = g_2^* = 4$. Any other non-cyclic abelian group with $\alpha_1 = 1$ has order $N > 18$. In this case,

(a) $p_1 > 2$, or
(b) $p_1 = 2$ and $p(2) > 3$, or
(c) $p_1 = 2$, $p(2) = 3$ and $m > 1$.

It follows that

$$(p(2) - p_1)\, N \;\geqslant\; 2p_1 p(2)^2,$$

since $p(2) - p_1 \geqslant 2$ and $N \geqslant p_1 p(2)^2$ in (a) and (b), and, in (c), $p(2) - p_1 = 1$ and $N > 2p_1 p(2)^2$. Hence,

$$(p(2) - p_1)\, N - 2p_1 p(2)^2 + p_1 p(2)(1 + p_1) \;>\; (p(2) - p_1)\, N - 2p_1 p(2)^2 \;\geqslant\; 0,$$

thus

$$p(2)N + p_1^2 p(2) \;>\; p_1 N + 2p_1 p(2)^2 - p_1 p(2).$$

Dividing by $-2p_1 p(2)$ and adding $N/2 + 1/2$ to both sides, we get $g_1^* < g_2^*$.

Then, $g_1^* \leqslant g_t^*$ for any $t \in \{2, \dots, \tau\}$ —equality holds only when $N = 18$— and the cyclic group $\mathbb{Z}_N$ attains the minimum genus when $\alpha_1 = 1$.

b) If $\alpha_1 > 1$ then, by [**6**, Thm. 6],

$$g_1^* = \frac{1}{2}(p_1 - 1)\frac{N}{p_1}.$$

For non-cyclic groups of order N and $t \in \{2, \dots, \tau\}$ invariant factors, either

$$g_t^* = 1 + \frac{N}{2} S_0(t) = 1 + (t-1)\left(1 - \frac{1}{p(t)}\right) - \frac{2}{N}\, p(t)^{t-1}$$

or $g_t^* = (t-2)N/2 + 1$ by Lemma 5.1. In particular, if $t = 2$,

$$g_2^* = \frac{1}{2}(p(2) - 1)\frac{N}{p(2)} - (p(2) - 1).$$

Since $1 < p_1 \leqslant p(2)$, it follows that $g_2^* < g_1^*$. Also $g_2^* < g_t^*$ if $g_t^* = (t-2)N/2 + 1$, since $(p(2) - 1)/p(2) < 1 < t - 2$ for $t > 2$ even.

Now it remains to check that $g_2^* < g_t^*$ for all $t \in \{3, \dots, \tau\}$ whenever $g_t^* = 1 + \frac{N}{2} S_0(t)$. For, we first change slightly the notation and define

$$S_0(p, t) = (t-1)\left(1 - \frac{1}{p}\right) - \frac{2}{N}\, p^{t-1},$$

If we prove that

i) $S_0(p, t-1) < S_0(p, t)$ for all $t \in \{3, \dots, \alpha\}$ if $N = p^\alpha m$ with p prime and integers $\alpha \geqslant 3$ and $m \geqslant 1$ such that $p \nmid m$, and
ii) $S_0(p, t) < S_0(q, t)$ for an integer $t \geqslant 2$ and $p < q$ primes such that $p^t|N$ and $q^t|N$,

then it follows that $S_0(p_1, 2)$ minimizes $S_0(p, t)$ for $N = p_1^{\alpha_1} \cdots p_s^{\alpha_s}$ with primes $p_j < p_{j+1}$, $p \in \{p_1, \dots, p_s\}$ and $t \in \{2, \dots, \tau\}$, thus $g_2^* \leqslant g_t^*$ for all $t \in \{2, \dots, \tau\}$.

i) We notice that

$$S_0(p, t) - S_0(p, t-1) = 1 - \frac{p^{\alpha-1}m - 2(p^{t-1} - p^{t-2})}{N} \;>\; 0$$

for all $t \in \{3, \ldots, \alpha\}$, since $p^{\alpha-1}m - 2(p^{t-1} - p^{t-2}) < N$ if $t \geqslant 3$, $p > 1$, $\alpha \geqslant 3$ and $m \geqslant 1$.

ii) Since $p < q$, we can write $N = p^{\alpha_1} q^{\alpha_2} m$, where $\alpha_1 > 1$, $\alpha_2 > 1$ and $m \geqslant 1$ is an integer such that $p \nmid m$ and $q \nmid m$. If $t \in \{2, \ldots, \min\{\alpha_1, \alpha_2\}\}$, then

$$\begin{aligned} S_0(q,t) - S_0(p,t) &= (t-1)\left(\frac{1}{p} - \frac{1}{q}\right) - \frac{2q^{t-1} - 2p^{t-1}}{N} \\ &= \frac{1}{pqm}\left((t-1)(q-p)m - A\right) > 0, \end{aligned}$$

with

$$A = \frac{2}{p^{\alpha_1 - 1} q^{\alpha_2 - t}} - \frac{2}{p^{\alpha_1 - t} q^{\alpha_2 - 1}},$$

since $t \geqslant 2$, $q > p \geqslant 2$ and $m \geqslant 1$; therefore $(t-1)(q-p)m \geqslant 1$ and $0 < A < 1$. Then $S_0(p,t) < S_0(q,t)$ whenever $p < q$ and $t \in \{2, \ldots, \min\{\alpha_1, \alpha_2\}\}$. □

The group $\mathbb{Z}_2 \oplus \mathbb{Z}_{2g+2}$ has order $4g+4$ and acts as a group of automorphisms of a compact Riemann surface of genus g: for, consider the triangle group with signature $(0;\ 2, 2g+2, 2g+2)$ in Theorem 3.2 —see also [**1**, Ex. 9.9]. In fact, this is the maximum order for a finite abelian group acting on genus g, as Breuer [**1**, Cor. 9.6] proved from Maclachlan's result [**7**, Thm. 4] —see Theorem 4.1— for the minimum genus. This result follows easily from Theorem 5.2 as well.

COROLLARY 5.3. *The maximum order for an abelian group of automorphisms of a compact Riemann surface of genus $g > 1$ is $4g+4$.*

PROOF. Let A be any abelian group of automorphisms of a compact Riemann surface of genus g of order $N = p_1^{\alpha_1} \cdots p_s^{\alpha_s}$, with integers $s \geqslant 1$, $\alpha_j > 0$ and primes $p_j < p_{j+1}$. The genus g must be greater than or equal to the minimum genus g^* provided by Theorem 5.2. Hence, $N \leqslant 4g+4$ is trivially satisfied when $N \leqslant 9$: $N < 12 = 4g^* + 4 \leqslant 4g+4$ if $N \in \{2,3,4,5,6,8\}$, and $N < 16 = 4g^* + 4$ if $N \in \{7, 9\}$.

Now let $N > 9$. Then $N = 2g^* + 1 \leqslant 2g + 1 < 4g + 4$ if N is prime. If $N > 9$ is not prime and $\alpha_1 = 1$, let $N = p_1 q$ for an integer $q > 1$, $p_1 \nmid q$ and $p_1 < p$ for every prime $p|q$. Then $4g^* + 4 = N(2(p_1 - 1)/p_1 - 2/q) + 6$, so $4g^* + 4 = N + 2$ if $p_1 = 2$, $4g^* + 4 = N + q$ if $p_1 = 3$. Since $2(p_1-1)/p_1$ grows and $2/q$ decreases with increasing p_1, and $q \geqslant 7$ if $p_1 = 5$, then $4g^* + 4 \geqslant \frac{46}{35}N + 4$ if $p_1 \geqslant 5$. Hence, $4g + 4 \geqslant 4g^* + 4 > N$ provided that $N > 9$ is not prime and $\alpha_1 = 1$.

Finally, we prove that $4g^* + 4 \geqslant N$ if $N > 9$ and $\alpha_1 > 1$. Indeed, let us write $N = p_1^2 q$ for an integer $q \geqslant 3$ if $p_1 \in \{2, 3\}$ and $q = 1$ or $q \geqslant p_1$ otherwise. Then $4g^* + 4 = N(2(p_1-1)/p_1 - 4/p_1 q) + 8$, so $4g^* + 4 = N$ if $p_1 = 2$, and it is also straightforward to check that $4g^* + 4 \geqslant N$ for the cases $\{p_1 = 3, q = 3\}$ and $\{p_1 = 5, q = 1\}$. We have $4g^* + 4 \geqslant N$ also for greater values of p_1 and q, since $2(p_1-1)/p_1$ grows and $4/p_1 q$ decreases with increasing values of p_1 and q. □

Acknowledgements

The author wishes to express his gratitude to Emilio Bujalance and Javier Cirre for their advice and encouragement throughout the preparation of this paper. The author would also like to thank the referee for his comments and suggestions.

References

[1] Thomas Breuer, *Characters and automorphism groups of compact Riemann surfaces*, London Mathematical Society Lecture Note Series, vol. 280, Cambridge University Press, Cambridge, 2000. MR1796706 (2002i:14034)

[2] E. Bujalance, F. J. Cirre, J. M. Gamboa, and G. Gromadzki, *On compact Riemann surfaces with dihedral groups of automorphisms*, Math. Proc. Cambridge Philos. Soc. **134** (2003), no. 3, 465–477, DOI 10.1017/S030500410200662X. MR1981212 (2004c:20086)

[3] Emilio Bujalance, Francisco-Javier Cirre, and Grzegorz Gromadzki, *A survey of research inspired by Harvey's theorem on cyclic groups of automorphisms*, Geometry of Riemann surfaces, London Math. Soc. Lecture Note Ser., vol. 368, Cambridge Univ. Press, Cambridge, 2010, pp. 15–37, DOI 10.1007/978-3-642-14828-6. MR2665004 (2011g:14070)

[4] Budha Prasad Chetiya and Kuntala Patra, *K-metacyclic groups of automorphisms of compact Riemann surfaces*, Far East J. Math. Sci. **2** (1994), no. 2, 127–136. MR1385096 (97d:20064)

[5] Leon Greenberg, *Maximal Fuchsian groups*, Bull. Amer. Math. Soc. **69** (1963), 569–573. MR0148620 (26 #6127)

[6] W. J. Harvey, *Cyclic groups of automorphisms of a compact Riemann surface*, Quart. J. Math. Oxford Ser. (2) **17** (1966), 86–97. MR0201629 (34 #1511)

[7] C. Maclachlan, *Abelian groups of automorphisms of compact Riemann surfaces*, Proc. London Math. Soc. (3) **15** (1965), 699–712. MR0179348 (31 #3596)

[8] Murad Özaydin, Charlotte Simmons, and Jennifer Taback, *Surface symmetries and* $\mathrm{PSL}_2(p)$, Trans. Amer. Math. Soc. **359** (2007), no. 5, 2243–2268, DOI 10.1090/S0002-9947-06-04011-6. MR2276619 (2007j:57018)

[9] Thomas W. Tucker, *Finite groups acting on surfaces and the genus of a group*, J. Combin. Theory Ser. B **34** (1983), no. 1, 82–98, DOI 10.1016/0095-8956(83)90009-6. MR701174 (85b:20055)

Departamento de Matemáticas Fundamentales, Universidad Nacional de Educación a Distancia, Madrid 28040, Spain

E-mail address: jesus.rodriguez.ms@gmail.com

Abelian varieties and group actions

Rubí E. Rodríguez

To Emilio Bujalance: feliz cumpleaños!

ABSTRACT. In this survey we review part of the theory of abelian varieties with group actions.

As an application, we study the group algebra decomposition for principally polarized abelian varieties admitting an involution. We give the dimensions and polarizations of the factors and the kernel of the corresponding isogenies, for each conjugacy type of symplectic involution, and describe the associated period matrices.

1. Introduction

Writing a survey for a volume honoring Emilio Bujalance on a topic related to, but not exactly, his own is challenging. In doing it, I tried to convey the passion he displays when writing mathematics; of course any failure to do so is my responsibility.

This survey is on abelian varieties and group actions; it contains part of the work done on the subject by the Chilean Complex Geometry group. The connection to Emilio's work is in the group action part, where he has made a tremendous contribution to the study of symmetries. Our hope is that he and his group will become interested in abelian varieties, and will extend to them the study of anticonformal actions.

The contents of this work are as follows. Section 2 recalls the basic results on complex tori; Section 3 specializes them to abelian varieties. The fundamental references for both sections are [**31**], [**4**], [**3**] and [**33**]. In Section 4 we describe group actions on abelian varieties: the beautiful analog to Poincaré's Complete Reducibility Theorem for an abelian variety with group action is given, in both the group algebra and the isotypical decomposition. In Section 5 we give a new application: a characterization of abelian varieties admitting an involution. In particular, we give their isotypical decomposition and describe their period matrices; we also study the subfamilies of Jacobians of compact Riemann surfaces admitting an involution.

2. Complex tori

In this section we study complex tori, a straightforward generalization to higher dimensions of compact Riemann surfaces of genus one.

2010 *Mathematics Subject Classification.* Primary 14H40, 14K10.
The author was partially supported by Fondecyt grant 1141099.

A *complex torus* $T = V/L$ of dimension g is the quotient of a complex vector space V of dimension g by a lattice L (a discrete subgroup of maximal rank $2g$) in V. Thus T is a compact complex manifold (of dimension g) and a commutative complex Lie group, and the natural quotient map $p : V \to T$ is holomorphic. Conversely, any connected compact complex Lie group of dimension g is a complex torus of dimension g.

In a very concrete way, a torus $T = V/L$ may be described by its period matrix: choose bases $\{e_1, \dots, e_g\}$ for V and $\{\lambda_1, \dots, \lambda_{2g}\}$ for L, and write each λ_j in terms of the e_i: $\lambda_j = \sum_{i=1}^{g} \lambda_{ij} e_i$.

Then the matrix

$$\Pi = (\lambda_{ij}) \tag{2.1}$$

in $M(g \times 2g, \mathbb{C})$ is called a *period matrix* for T.

Furthermore, since L is a lattice the matrix Π has rank g, and we can normalize to see that there are bases for V and L with respect to which the period matrix for T has the form

$$\Pi = (I_g \ Z)$$

with I_g the $g \times g$ identity matrix, and Z a complex $g \times g$ matrix with $\det(\operatorname{Im} Z) \neq 0$.

A *homomorphism* $f : T_1 \to T_2$ between two tori T_1 and T_2 is an analytic map that is also a group homomorphism. For fixed tori T_1 and T_2, the set $\operatorname{Hom}(T_1, T_2)$ of all homomorphisms between them is an abelian group with the addition operation. Two tori T_1 and T_2 are called *isomorphic* if there is a bijective homomorphism between them.

In particular, $\operatorname{End}(T) := \operatorname{Hom}(T, T)$ is the *ring of endomorphisms* of a torus T, with the operations addition and composition, and its units form the *automorphism group* of T, denoted by $\operatorname{Aut}(T)$.

Note that every homomorphism $f : T_1 \to T_2$ between two tori $T_j = V_j/L_j$, $1 \le j \le 2$, gives rise to two maps:

$$\rho_a(f) : V_1 \to V_2 \quad \text{and} \quad \rho_r(f) : L_1 \to L_2,$$

where $\rho_a(f)$, the *analytic representation* of f, is the unique complex-linear map from V_1 to V_2 lifting f, and $\rho_r(f)$, the *rational representation* of f, is the $\mathbb{Z}$-linear map from L_1 to L_2 given by the restriction of $\rho_a(f)$ to the lattice L_1.

In other words, we have two faithful representations of $\operatorname{Hom}(T_1, T_2)$, given by

$$\begin{aligned} \rho_a &: \operatorname{Hom}(T_1, T_2) \longrightarrow \operatorname{Hom}_{\mathbb{C}}(V_1, V_2), \text{ and} \\ \rho_r &: \operatorname{Hom}(T_1, T_2) \longrightarrow \operatorname{Hom}_{\mathbb{Z}}(L_1, L_2), \end{aligned}$$

showing that $\operatorname{Hom}(T_1, T_2)$ is finitely generated.

The extensions of ρ_a and ρ_r to $\operatorname{Hom}_{\mathbb{Q}}(T_1, T_2) := \operatorname{Hom}(T_1, T_2) \otimes_{\mathbb{Z}} \mathbb{Q}$ will be denoted by the same letters. In particular, when $T_1 = T_2 = T$, ρ_a and ρ_r are both representations of the ring $\operatorname{End}(T)$ and of the algebra $\operatorname{End}_{\mathbb{Q}}(T)$.

Given two tori T_1 and T_2, where bases have been chosen for the corresponding V_1, L_1, V_2 and L_2, consider the respective period matrices Π_1 and Π_2 as in (2.1). If f is in $\operatorname{Hom}_{\mathbb{Q}}(T_1, T_2)$, then, with respect to the same bases, $\rho_a(f)$ is given by a matrix M in $M(g \times g, \mathbb{C})$ and $\rho_r(f)$ is given by a matrix R in $M(2g \times 2g, \mathbb{Q})$; moreover, they satisfy

$$M\Pi_1 = \Pi_2 R. \tag{2.2}$$

Conversely, given M and R satisfying (2.2), there is a corresponding torus homomorphism between T_1 and T_2. In particular, f is an isomorphism if and only if R is in $\mathrm{GL}(2g, \mathbb{Z})$ and M is in $\mathrm{GL}(g, \mathbb{C})$.

If $T = V/L$ is a torus and W is a subspace of V such that $W \cap L$ is a lattice in W, we call $S = W/(W \cap L)$ a *subtorus* of T. If $f : T_1 \to T_2$ is a homomorphism, then $\mathrm{Im}(f)$ is a subtorus of T_2 and $\ker(f)$ is a closed subgroup of T_1. The connected component of $\ker(f)$ containing 0 is a subtorus of T_1, of finite index in $\ker(f)$.

A (nonzero) complex torus is called *simple* if its only subtori are itself and the trivial one $\{0\}$. It is easy to see that for a simple torus T its endomorphism algebra $\mathrm{End}_{\mathbb{Q}}(T)$ is a skew-field of finite dimension over $\mathbb{Q}$. In general, $\mathrm{End}_{\mathbb{Q}}(T)$ is completely arbitrary; according to a theorem by Oort and Zarhin [**34**], for any finite dimensional algebra A there is a complex torus T with $A = \mathrm{End}_{\mathbb{Q}}(T)$; see also [**21**] and [**44**]. As we will see in the next section, this is not the case for abelian varieties; as a consequence of Poincaré's Reducibility Theorem, the endomorphism algebra of an abelian variety is semisimple.

The following special homomorphisms will be of interest in the rest of this work. An *isogeny* between two tori is a surjective homomorphism with finite kernel; equivalently, a homomorphism $f : T_1 \to T_2$ is an isogeny if and only if it is surjective and $\dim T_1 = \dim T_2$.

Examples of isogenies are provided by taking a finite subgroup H of a torus T and considering the natural projection from T to the torus T/H; of course all isogenies may be thought of in this way. The *degree* and the *exponent* of an isogeny f are defined as the cardinality and the exponent of the (finite) subgroup $\ker(f)$.

In particular, for a positive integer n and a torus T, consider the endomorphism n_T of T given by multiplication by n:

$$\begin{aligned} n_T : T &\longrightarrow T \\ x &\dashrightarrow nx. \end{aligned}$$

The kernel of n_T is a finite subgroup of T, denoted by $T[n]$, and its elements are called the *points of order n* in T; note that $T[n]$ is isomorphic to $(\mathbb{Z}/n\mathbb{Z})^{2g}$, and hence the order of $T[n]$ is n^{2g}. Of course $T/T[n]$ is a torus isogenous to T. If an endomorphism f of T is an isogeny of exponent n, then $\ker(f) \subseteq T[n]$.

Isogenies are especially interesting because they are the units of $\mathrm{End}_{\mathbb{Q}}(T)$; in fact, for each isogeny $f : T_1 \to T_2$ with exponent e, there exists an isogeny $g : T_2 \to T_1$, unique up to isomorphism, such that $g \circ f = e_{T_1}$ and $f \circ g = e_{T_2}$.

In particular, the existence of an isogeny between two tori is an equivalence relation among complex tori; we will write

$$T_1 \sim T_2$$

and say that T_1 and T_2 are *isogenous* to each other in this case. Most classification theorems about complex tori are given up to isogeny, as we will see in the next section.

3. Abelian varieties

We will be mainly interested in complex tori that are also projective varieties; these correspond to complex tori that possess sufficiently many meromorphic functions, the so-called abelian varieties. We will use the classical definition for a polarization, as it is the most useful one for our purposes here. More modern definitions, in terms of line bundles, may be found for instance in [**4**].

A *polarization* (or a Riemann form) on a torus $T = V/L$ is a non-degenerate real alternating form E on V such that

$$E(\imath u, \imath v) = E(u, v) \text{ for all } u, v \text{ in } V, \quad \text{and} \quad E(L \times L) \subseteq \mathbb{Z}; \tag{3.1}$$

here $\imath$ denotes a complex number with $\imath^2 = -1$.

A *polarized abelian variety* $\mathcal{A} = (T, E)$ of dimension g is a pair consisting of a complex torus $T = V/L$ of dimension g and a polarization E on T. An *abelian variety* is a complex torus that admits a polarization.

A polarization E is said to be of *type* $(d_1, d_2, \ldots, d_g)$ if the elementary divisors for the alternating form E are $\{d_1, \ldots, d_g\}$; that is, if the d_j are natural numbers such that d_i divides d_{i+1}, for all i in $\{1, \ldots, g-1\}$, and if there exists a basis for L such that the matrix for E with respect to this basis has the form

$$E = \begin{pmatrix} 0 & D \\ -D & 0 \end{pmatrix},$$

where D is the diagonal matrix $D = (d_1, d_2, \ldots, d_g)$; such a basis for L is called *symplectic*. If the type is $(1, 1, \ldots, 1)$, the abelian variety is called *a principally polarized abelian variety*, abbreviated *ppav*.

In terms of a period matrix for the torus T, the *Riemann relations* given next provide necessary and sufficient conditions for T to be an abelian variety; that is, to admit a polarization.

THEOREM 3.1 (Riemann Relations). *Let T be a complex torus and let Π denote a period matrix for T. Then T is an abelian variety if and only if there exists a non-degenerate alternating $2g \times 2g$ integral matrix E such that*

a) $\Pi E^{-1} \Pi^t = 0$, *and*

b) $-\imath \Pi E^{-1} \overline{\Pi}^t$ *is positive definite,*

where ^{t} *represents the transposed matrix.*

The Riemann Relations immediately tell us that not every complex torus is an abelian variety, since for instance a) implies that in an abelian variety a set of generators for the lattice elements has to satisfy certain algebraic relations. In fact, abelian varieties correspond precisely to those complex tori that may be embedded in projective space, or, equivalently, to those complex tori that are also algebraic varieties (see [**4**]).

It is easy to see from the Riemann relations that if $\mathcal{A} = (T, E)$ is a polarized abelian variety of type $(d_1, \ldots, d_g)$, then there exist a basis for V and a symplectic basis for L with respect to which a period matrix for T has the form

$$\Pi = (D\ Z), \tag{3.2}$$

where $D = (d_1, d_2, \ldots, d_g)$ is the diagonal matrix reflecting the type of E, and Z belongs to the Siegel space $\mathbb{H}_g$ of complex $g \times g$ *symmetric* matrices Z such that $\mathrm{Im}(Z)$ is positive definite.

It is not hard to see that every compact Riemann surface X of genus one (or, equivalently, every complex torus of dimension one) is a ppav in a natural way: if $X = \mathbb{C}/L$, we can always normalize L to be generated by 1 and a complex number

τ with $\operatorname{Im}\tau > 0$. Then we set

$$E(a_1 + b_1\tau, a_2 + b_2\tau) = \frac{-\operatorname{Im}\left((a_1 + b_1\tau)(a_2 + b_2\overline{\tau})\right)}{\operatorname{Im}\tau} = a_1b_2 - a_2b_1, \tag{3.3}$$

for all real values of a_1, a_2, b_1, b_2. Note that by thinking of 1 and τ as representing 1-cycles that generate the lattice $H_1(X, \mathbb{Z})$, the formula in (3.3) gives exactly the (geometric) intersection number of the 1-cycles $a_1 + b_1\tau$ and $a_2 + b_2\tau$ for all integral values of a_1, a_2, b_1, b_2.

More generally, if X is any compact Riemann surface of genus $g \geq 1$, then its *Jacobian variety*

$$JX = H^{1,0}(X, \mathbb{C})^*/H_1(X, \mathbb{Z})$$

is a ppav, since it admits a canonical principal polarization given by extending the (geometric) intersection number in the lattice $L = H_1(X, \mathbb{Z})$ to the complex vector space $V = H^{1,0}(X, \mathbb{C})^*$, the dual to the space of analytic differentials on X, where the injection of L in V is given by

$$\begin{aligned} H_1(X, \mathbb{Z}) &\longrightarrow H^{1,0}(X, \mathbb{C})^* \\ \alpha &\dashrightarrow (\omega \to \int_\alpha w). \end{aligned}$$

A *homomorphism* $f : \mathcal{A}_1 = (T_1, E_1) \to \mathcal{A}_2 = (T_2, E_2)$ *between polarized abelian varieties* is a homomorphism $f : T_1 \to T_2$ between the two tori with finite kernel and such that $E_1(v, u) = E_2(\rho_a(f)(v), \rho_a(f)(u))$ for all v, u in V_1. In this case we say that f *preserves the polarizations.* Of course, there may exist homomorphisms between the underlying tori T_1 and T_2 that do not preserve the corresponding polarizations in the corresponding abelian varieties.

Any subtorus $S \subseteq T$ of an abelian variety is an abelian variety: if E denotes a polarization of T then its restriction $E_{|S}$ to S is a polarization on S. We will call $(S, E_{|S})$ an *abelian subvariety* of the polarized abelian variety (T, E).

An important property of abelian varieties (that does not hold for complex tori in general) is that every abelian subvariety has a *complementary* one: If $\mathcal{B}$ is an abelian subvariety of the abelian variety $\mathcal{A}$, there exists an abelian subvariety $\mathcal{C}$ of $\mathcal{A}$ such that $\mathcal{B} \cap \mathcal{C}$ is finite and the addition map

$$\begin{aligned} \mathcal{B} \times \mathcal{C} &\longrightarrow \mathcal{A} \\ (b, c) &\dashrightarrow b + c \end{aligned}$$

is an isogeny. This important result is called *Poincaré's Reducibility Theorem.*

An abelian variety is called *simple* if the only abelian subvarieties (or, equivalently, the only subtori) it contains are itself and $\{0\}$. Applying Poincaré's Reducibility Theorem, the following results follow.

THEOREM 3.2 (Poincaré's Complete Reducibility Theorem). *Given an abelian variety $\mathcal{A}$, there exist simple abelian subvarieties $\mathcal{X}_1, \ldots, \mathcal{X}_r$ of $\mathcal{A}$, not isogenous to each other, and positive integers $k_1, \ldots, k_r$ such that*

$$\mathcal{A} \sim \mathcal{X}_1^{k_1} \times \ldots \times \mathcal{X}_r^{k_r}. \tag{3.4}$$

The $\mathcal{X}_j$ and k_j are unique up to isogeny and permutation of the factors.

COROLLARY 3.3. *Given an abelian variety $\mathcal{A}$, $\operatorname{End}_{\mathbb{Q}}(\mathcal{A})$ is a semisimple algebra. If (3.4) is an isogeny decomposition for $\mathcal{A}$ into simple factors, then*

$$\operatorname{End}_{\mathbb{Q}}(\mathcal{A}) \simeq M(k_1 \times k_1, F_1) \times \ldots \times M(k_r \times k_r, F_r),$$

where

$$F_j = \mathrm{End}_{\mathbb{Q}}(\mathcal{X}_j)$$

are skew-fields of finite dimension over $\mathbb{Q}$.

REMARK 3.4. The skew-fields of finite dimension over $\mathbb{Q}$ together with a positive involution were classified by Albert in [**1**] and [**2**]. Those that appear as $\mathrm{End}_{\mathbb{Q}}(\mathcal{A})$ for a simple abelian variety $\mathcal{A}$ have this property, they all carry the Rosati involution.

4. Group actions on abelian varieties

In this section we will discuss automorphisms of ppavs; that is, bijective homomorphisms from a ppav $\mathcal{A}$ to itself that preserve the polarization of $\mathcal{A}$. This includes giving the corresponding analog to Poincaré's Complete Reducibility Theorem for an abelian variety admitting the action of a group G: a G-equivariant decomposition into non-isogenous abelian subvarieties with G-action.

It follows from (2.2) and (3.2) that an isomorphism between ppavs with period matrices $(I_g\ Z_1)$ and $(I_g\ Z_2)$ is given by a matrix M in $\mathrm{GL}(g,\mathbb{C})$ and a matrix R in $\mathrm{GL}(2g,\mathbb{Z})$ such that

$$M(I_g\ Z_1) = (I_g\ Z_2)R. \tag{4.1}$$

Since R preserves the polarization $E = \begin{pmatrix} 0 & I_g \\ -I_g & 0 \end{pmatrix}$, R belongs to the symplectic group

$$\mathrm{Sp}(2g,\mathbb{Z}) = \{R \in \mathrm{GL}(2g,\mathbb{Z}) : R^tER = E\}.$$

It follows from (4.1) that $\mathrm{Sp}(2g,\mathbb{Z})$ acts on the Siegel space $\mathbb{H}_g$ as follows

$$R * Z = \begin{pmatrix} A & B \\ C & D \end{pmatrix} * Z = (A+ZC)^{-1}(B+ZD),$$

and that this action precisely identifies two period matrices corresponding to isomorphic ppavs. In other words, the quotient

$$\mathbb{H}_g/\,\mathrm{Sp}(2g,\mathbb{Z}) := \mathfrak{A}_g$$

is the moduli space of (isomorphism classes of) ppavs of dimension g.

In particular, an automorphism of a ppav given by a matrix Z in $\mathbb{H}_g$ is given by an element of the symplectic group fixing Z.

In the next section we give the group algebra decomposition of a ppav with group action. This is analogous to Poincaré's Complete Irreducibility Theorem, in that it decomposes a ppav with action of a group G isogenously and G-equivariantly into a product of subvarieties with G-action.

4.1. The isotypical decomposition. (see [**4**, Section 13.6]). Let $\mathcal{A}$ be an abelian variety of dimension g with a faithful action by a finite group G; that is, there is a monomorphism from G to $\mathrm{Aut}(\mathcal{A})$; we will say that $\mathcal{A}$ is a *G-abelian variety.* This action induces a homomorphism of semisimple $\mathbb{Q}$-algebras

$$\rho : \mathbb{Q}[G] \to \mathrm{End}_{\mathbb{Q}}(\mathcal{A}).$$

We denote an element of the rational group algebra and its image in $\mathrm{End}_{\mathbb{Q}}(\mathcal{A})$ by the same letter.

Any element $\alpha \in \mathbb{Q}[G]$ defines an abelian subvariety

$$\mathcal{A}^{\alpha} := \mathrm{Im}(m\alpha) \subseteq \mathcal{A},$$

where m is some positive integer such that $m\alpha \in \mathrm{End}(\mathcal{A})$. This definition does not depend on the chosen integer m. Let

$$\mathbb{Q}[G] = Q_1 \times \cdots \times Q_r$$

denote the decomposition of $\mathbb{Q}[G]$ as a product of simple $\mathbb{Q}$-algebras Q_i. The factors Q_i correspond canonically to the irreducible rational representations W_i of the group G. The corresponding decomposition of $1 \in \mathbb{Q}[G]$, given by

$$1 = e_1 + \cdots + e_r$$

with central idempotents e_i in Q_i, induces an isogeny

$$\mathcal{A}^{e_1} \times \cdots \times \mathcal{A}^{e_r} \to \mathcal{A} \tag{4.2}$$

given by addition; this is the *isotypical decomposition* of $\mathcal{A}$. Note that the components $\mathcal{A}^{e_i}$ are G-stable complex subtori of $\mathcal{A}$ with

$$\mathrm{Hom}_G(\mathcal{A}^{e_i}, \mathcal{A}^{e_j}) = 0$$

for $i \neq j$.

If W_i is the irreducible rational representation of G corresponding to e_i, we also denote

$$\mathcal{A}_{W_i} := \mathcal{A}^{e_i}.$$

The idempotents e_i are determined as follows: Let χ_i be a character of one of the irreducible $\mathbb{C}$-representations associated to W_i and let K_i be the character field

$$K_i = \mathbb{Q}(\chi_i(g), g \in G).$$

Then

$$e_i = \frac{\deg \chi_i}{|G|} \sum_{g \in G} \mathrm{tr}_{K_i|\mathbb{Q}}(\chi_i(g^{-1}))g. \tag{4.3}$$

The isotypical components $\mathcal{A}_{W_i}$ can be decomposed further. According to Schur's Lemma,

$$D_i := \mathrm{End}_G(W_i)$$

is a skew-field of finite dimension

$$n_i = \frac{\deg \chi_i}{m_i} \tag{4.4}$$

over $\mathbb{Q}$, where m_i denotes the Schur index of χ_i (see **[9]**). It follows that there is a set of primitive idempotents $\{q_{i1}, \cdots, q_{in_i}\}$ in $Q_i \subset \mathbb{Q}[G]$ such that

$$e_i = q_{i1} + \cdots + q_{in_i}.$$

Moreover, the abelian subvarieties $\mathcal{A}^{q_{ij}}$ are mutually isogenous for fixed i and for all $j = 1, \ldots, n_i$. If $\mathcal{B}_{W_i}$ denotes one of them, we obtain an isogeny

$$\mathcal{B}_{W_i}^{n_i} \to \mathcal{A}_{W_i}$$

for every $i = 1, \ldots, r$. Combining with (4.2) we obtain an isogeny

$$\mathcal{B}_{W_1}^{n_1} \times \cdots \times \mathcal{B}_{W_r}^{n_r} \to \mathcal{A}, \tag{4.5}$$

called the *group algebra decomposition* of the G-abelian variety $\mathcal{A}$. Note that, whereas (4.2) is uniquely determined, (4.5) is not. It depends on the choice of the q_{ij} as well as on the choice of the $\mathcal{B}_{W_i}$.

As we have seen, the group algebra and the isotypical decomposition of a G-abelian variety come from algebraic data of the group G, its irreducible representations; further information about the decomposition, such as the dimensions and the polarizations of the factors $\mathcal{A}_{W_i}$ and B_{W_i} depend on the *geometry* of the action. For instance, in the case of the Jacobian $\mathcal{A} = JX$ of a compact Riemann surface X, the dimensions of the factors are explicitly given in terms of the monodromy of the action of the group on X, see [**42**]. In the general case of an abelian variety, the dimensions depend on the analytic representation, see [**10**].

The group algebra decomposition of a G-abelian variety has been proven to be useful in several directions: among them, finding families of Jacobians with complex multiplication for infinitely many values of the genus (see [**7**]), many new examples of Prym-Tyurin varieties (see [**8**]), and families of Jacobians isogenous to the product of elliptic curves (see [**12**]).

In Section 3 we saw that the Jacobian varieties JX of compact Riemann surfaces X of genus g are ppavs of dimension g. It follows from Torelli's Theorem that the Abel period map that associates to each (isomorphism class of) such X (the isomorphism class of) its Jacobian variety JX is an injective map from the moduli space $\mathcal{M}_g$ of curves of genus g to the moduli space $\mathfrak{A}_g$ of ppavs of dimension g.

Let $g \geq 2$. Since $\dim \mathcal{M}_g = 3g - 3$ and $\dim \mathfrak{A}_g = g(g+1)/2$, the period map is not surjective for $g \geq 4$; its image is called the *Jacobian* locus.

Another well known locus in $\mathfrak{A}_g$ is that of *classical Prym varieties*, which we describe next. Consider a compact Riemann surface X of genus $2g+1$, with $g \geq 1$, admitting a fixed point free involution j. Then $Y = X/\langle j \rangle$ is a compact Riemann surface of genus g, and the quotient map $p : X \to Y$ induces a natural homomorphism $p^* : JY \to JX$. Consider the subvariety $P = P(X, j)$ of JX defined to be the complementary subvariety to $p^*(JY)$. It has dimension $g + 1$, and it may be shown (see Section 6.1) that the restriction of the canonical polarization E of JX to P has type $(2, \dots, 2)$; that is, P carries a *principal* polarization (namely, $\frac{1}{2}E_{|P}$). A ppav obtained in this way is called a classical Prym variety. They were originally studied by Wirtinger in [**51**] and later by Mumford in [**31**] and Welters in [**50**], among many others.

Abelian varieties with nontrivial automorphism group have been studied extensively, since they compose the singular locus of $\mathcal{A}_g$; see for instance [**4**], [**10**], [**6**], [**13**], [**14**], [**19**], [**22**], [**25**], [**26**], [**28**], [**29**], [**38**], [**39**], [**45**], [**47**], [**49**], [**50**], [**52**], [**53**] for abelian varieties, [**9**], [**7**], [**11**], [**17**], [**18**], [**23**], [**30**], [**36**], [**42**], [**54**] for Jacobians, and [**8**], [**15**], [**24**], [**26**], [**32**], [**35**], [**40**], [**41**], [**43**], [**46**], [**48**], [**51**] for Pryms.

5. Application

In the previous section we saw that to every compact Riemann surface X admitting a fixed point free involution j we can associate a ppav, the classical Prym variety $P(X, j)$. In this section we will study the more general case of a ppav $\mathcal{A}$ admitting an involution j. Applying the theory developed in previous sections, we will completely classify such pairs $(\mathcal{A}, j)$, including their group algebra decomposition, period matrices, and dimensions and polarizations of the factors.

Consider a principally polarized abelian variety $\mathcal{A}$ admitting an involution j preserving the polarization on $\mathcal{A}$.

Since the group $G = \langle j \rangle$ is isomorphic to the cyclic group of order two, it has two irreducible representations: W_0, the trivial one, and W_1, given by multiplication by -1. The group algebra decomposition (4.5) for $\mathcal{A}$ in this case gives that there exist two complementary abelian subvarieties $\mathcal{A}_0 = \mathcal{B}_0$ and $\mathcal{A}_1 = \mathcal{B}_1$ of $\mathcal{A}$, where the involution acts trivially on $\mathcal{A}_0$ and as multiplication by (-1) on $\mathcal{A}_1$, and such that the addition map

$$\mathcal{A}_0 \times \mathcal{A}_1 \xrightarrow{a} \mathcal{A} \tag{5.1}$$

is a G-equivariant isogeny.

We are interested in the dimensions of $\mathcal{A}_0$ and $\mathcal{A}_1$, in their induced polarizations, and in the kernel of the isogeny a.

Observe that, in the notation of the previous section,

$$e_{W_0} = \frac{1}{2}(1_{\mathcal{A}} + j) \text{ and } e_{W_1} = \frac{1}{2}(1_{\mathcal{A}} - j),$$

and

$$\mathcal{A}_0 = \operatorname{Im}(1_{\mathcal{A}} + j) \text{ and } \mathcal{A}_1 = \operatorname{Im}(1_{\mathcal{A}} - j).$$

Choosing a symplectic basis for L, the rational representation $\rho_r(j)$ for j corresponds to a symplectic involution.

For x, y and z nonnegative integers such that $2x + y + z = \dim \mathcal{A}$, consider the block matrix

$$W(x, y, z) = (J_1, \ldots, J_1, -I_y, I_z) \tag{5.2}$$

where the blocks are

$$J_1 = \begin{pmatrix} 1 & 0 \\ 1 & -1 \end{pmatrix}$$

repeated x times, and I_n is the $n \times n$ identity matrix, with $n = y$ first and then $n = z$.

Reiner classified all symplectic involutions in [**37**]: up to conjugation in the symplectic group, $\rho_r(j)$ is a block matrix of the form

$$\rho_r(j) = (W(x, y, z), W(x, y, z)^t), \tag{5.3}$$

where $W(x, y, z)$ is as in (5.2), for unique x, y and z nonnegative integers such that $2x + y + z = \dim \mathcal{A}$. We will say that the involution j is of *type* (x, y, z).

For each fixed type (x, y, z) we will denote by

$$\mathcal{F}_{(x,y,z)}$$

the family of all pairs $(\mathcal{A}, j)$, where $\mathcal{A}$ is a ppav and j is an involution of type (x, y, z) acting on $\mathcal{A}$.

Given $(\mathcal{A}, j)$ in $\mathcal{F}_{(x,y,z)}$, there exists a symplectic basis

$$B = \{\alpha_1, \ldots, \alpha_{2x}, \gamma_1, \ldots, \gamma_y, \varepsilon_1, \ldots, \varepsilon_z, \beta_1, \ldots, \beta_{2x}, \delta_1 \ldots, \delta_y, \eta_1, \ldots, \eta_z\} \tag{5.4}$$

for the lattice L such that the matrix for j with respect to this basis has the form (5.3); we will call B *a basis adapted to* j.

The following two technical results will be useful in the proof of Theorem 5.3.

LEMMA 5.1. *Let* $(\mathcal{A}, j)$ *be in* $\mathcal{F}_{(x,y,z)}$*, and consider the corresponding group algebra decomposition of* $\mathcal{A}$ *given in* (5.1).*Let* B *be a basis adapted to* j*, as in* (5.4).

Then bases for the corresponding sublattices of L, L_0 for $\mathcal{A}_0$ and L_1 for $\mathcal{A}_1$, are given respectively as follows.

$$L_0 = \langle 2\alpha_1 + \alpha_2, \dots, 2\alpha_{2x-1} + \alpha_{2x}, \varepsilon_1, \dots, \varepsilon_z, \beta_1, \beta_3, \dots, \beta_{2x-1}, \eta_1, \dots, \eta_z \rangle_{\mathbb{Z}}$$
$$L_1 = \langle \alpha_2, \alpha_4, \dots, \alpha_{2x}, \gamma_1, \dots, \gamma_y, -\beta_1 + 2\beta_2, \dots, -\beta_{2x-1} + 2\beta_{2x}, \delta_1, \dots, \delta_y \rangle_{\mathbb{Z}}$$

PROOF. Observe that according to our description of $\mathcal{A}_0$ and $\mathcal{A}_1$, the sublattices L_0 and L_1 correspond to the maximal sublattices of L respectively pointwise fixed and multiplied by (-1) by the action of $\rho_r(j)$.

Since the involution $\rho_r(j)$ is given by (5.3) with respect to the basis B, the result follows. □

LEMMA 5.2. *Consider $W(x,y,z)$ as in (5.3) and let*

$$S(x,y,z) = \{Z \in M(g \times g, \mathbb{C}) : Z^t = Z \text{ and } W(x,y,z)Z = ZW(x,y,z)^t\}.$$

If $Z \in S(x,y,z)$, then

$$Z = \begin{pmatrix} X & Y & U \\ Y^t & Z_y & 0_{y,z} \\ U^t & 0_{z,y} & Z_z \end{pmatrix},$$

where X is a $2x \times 2x$ complex symmetric matrix of the form

$$X = \begin{pmatrix} X_1 & X_2 & \dots & X_x \\ & X_{x+1} & \dots & X_{2x-1} \\ & & \ddots & \\ & & & X_{\frac{x(x+1)}{2}} \end{pmatrix}, \text{ with } X_j = \begin{pmatrix} 2a_j & a_j \\ a_j & b_j \end{pmatrix},$$

Y is a $y \times 2x$ complex matrix of the form

$$Y = \begin{pmatrix} Y_1 & Y_2 & \dots & Y_y \\ Y_{y+1} & & \dots & Y_{2y} \\ & & \vdots & \\ & & & Y_{xy} \end{pmatrix}, \text{ with } Y_j = \begin{pmatrix} 0 \\ c_j \end{pmatrix},$$

U is a $z \times 2x$ complex matrix of the form

$$U = \begin{pmatrix} U_1 & U_2 & \dots & U_z \\ U_{z+1} & & \dots & U_{2z} \\ & & \vdots & \\ & & & U_{xz} \end{pmatrix}, \text{ with } U_j = \begin{pmatrix} 2d_j \\ d_j \end{pmatrix},$$

Z_n is a complex symmetric $n \times n$ matrix, with $n = y$ first and then $n = z$, and $0_{y,z}$ is the zero $y \times z$ matrix.

PROOF. This is an easy computation. □

We can now give our main result, with the dimensions and induced polarizations on the factors in the group algebra decomposition of a ppav with involution. It also includes a description of the kernel of the decomposition, and a concrete expression for the period matrix of the ppav.

THEOREM 5.3. *Let $(\mathcal{A}, j)$ be in $\mathcal{F}_{(x,y,z)}$.*
Consider the corresponding group algebra decomposition of $\mathcal{A}$ given in (5.1). Then

(1)

$$\dim \mathcal{A}_0 = x + z, \quad \textit{and the polarization type of } \mathcal{A}_0 \textit{ is } (1^z, 2^x),$$
$$\dim \mathcal{A}_1 = x + y, \quad \textit{and the polarization type of } \mathcal{A}_1 \textit{ is } (1^y, 2^x),$$

where n^a means $n, n, \dots, n$ repeated a times.

(2) *Let B be a basis adapted to j, as in* (5.4). *The kernel of the isogeny a of* (5.1) *is the subgroup of order 2^{2x} of the points of order two in $\mathcal{A}_0 \times \mathcal{A}_1$, generated by the $2x$ points*

$$\left\{ \left(\frac{1}{2}\alpha_{2k}, -\frac{1}{2}\alpha_{2k}\right), \left(\frac{1}{2}\beta_{2k-1}, -\frac{1}{2}\beta_{2k-1}\right); 1 \le k \le x \right\}.$$

(3) *The period matrix of $\mathcal{A}$ with respect to the basis B has the form $(I_g\, Z)$, with Z in $\mathbb{H}_g \cap S(x,y,z)$, where $S(x,y,z)$ is as in Lemma* 5.2.

PROOF. First observe that Lemma 5.1 gives the dimensions for $\mathcal{A}_0$ and $\mathcal{A}_1$. Then note that the positive geometric intersection numbers for the 1-cycles generating L_0 given in Lemma 5.1 are

$$(2\alpha_{2j-1} + \alpha_{2j}, \beta_{2j-1}) = 2, (\varepsilon_k, \eta_k) = 1, 1 \le j \le x, 1 \le k \le z,$$

from where the type of the induced polarization on $\mathcal{A}_0$ follows. The polarization on $\mathcal{A}_1$ may be computed similarly, thus proving 1).

Note that for u in $\mathcal{A}_0$ and v in $\mathcal{A}_1$ the following holds:

$$(1+j)(u) = 2u, (1+j)(v) = 0, (1-j)(u) = 0, (1-j) = 2v.$$

Therefore the isogeny

$$\begin{aligned} b : &\mathcal{A} \longrightarrow \mathcal{A}_0 \times \mathcal{A}_1 \\ &w \dashrightarrow ((1+j)(w), (1-j)(w)) \end{aligned}$$

makes the following diagram commute:

$$\begin{array}{ccc} \mathcal{A}_0 \times \mathcal{A}_1 & \xrightarrow{\;2_{\mathcal{A}_0 \times \mathcal{A}_1}\;} & \mathcal{A}_0 \times \mathcal{A}_1, \\ & {\scriptstyle a}\searrow \quad \nearrow{\scriptstyle b} & \\ & \mathcal{A} & \end{array}$$

and it follows that $\ker(a) \subseteq (\mathcal{A}_0 \times \mathcal{A}_1)[2]$.

But

$$\ker(a) = \{(u, -u) : u \in \mathcal{A}_0 \cap \mathcal{A}_1\},$$

and now it follows from Lemma 5.1 that indeed

$$\mathcal{A}_1 \ni \frac{1}{2}\alpha_{2j} \equiv_L \frac{1}{2}(2\alpha_{2j-1} + \alpha_{2j}) \in \mathcal{A}_0 \,, \ 1 \le j \le x$$

and similarly for $\frac{1}{2}\beta_{2j-1}$, thus proving 2).

3) follows from Lemma 5.2. □

COROLLARY 5.4. *The family $\mathcal{F}_{(x,y,z)}$ depends on*

$$x^2 + x + xy + xz + \frac{1}{2}(y^2 + y + z^2 + z)$$

moduli.

PROOF. Let $m = x^2 + x + xy + xz + \frac{1}{2}(y^2 + y + z^2 + z)$. With the notation of Lemma 5.2, the period matrices of the form $(I_g\, Z)$ with Z in $\mathbb{H}_g \cap S(x,y,z)$ form an open set in $\mathbb{C}^m$, with coordinate functions given by a_j, b_j for each X_j, $1 \le j \le \frac{x(x+1)}{2}$, c_k for Y_k, $1 \le k \le xy$, d_s for W_s, $1 \le s \le xz$, the usual $\frac{y(y+1)}{2}$ coordinates for Z_y and the usual $\frac{z(z+1)}{2}$ coordinates for Z_z. □

6. The case of Jacobian varieties

In this section we consider a compact Riemann surface X of genus g at least equal to two, with an involution j acting on it. Then the Jacobian variety JX of X is a principally polarized abelian variety of dimension g admitting an involution, and we can apply the results of the previous section. We will denote by $2n$ the number of fixed points of j on X, with $n \ge 0$, and by γ the genus of the quotient surface $X/\langle j \rangle$; by the Riemann Hurwitz formula,

$$g = 2\gamma + n - 1.$$

In this case we can further identify the two subvarieties $\mathcal{A}_0$ and $\mathcal{A}_1$ appearing in the group algebra decomposition (5.1) for $\mathcal{A} = JX$: if we denote the canonical projection by $p : X \to Y = X/\langle j \rangle$, then

$$\mathcal{A}_0 = p^*(JY) \quad \text{and } \mathcal{A}_1 = P(X/Y)$$

where $p^* : JY \to JX$ is the pullback of p, and $P(X/Y)$ is the generalized Prym variety of the covering $p : X \to Y$.

A nice description of symplectic bases adapted to the action of an involution on a compact Riemann surface is given in Fay [**20**]. Comparing this description with the general case of an involution on a principally polarized abelian variety as given in Section 2, we can prove the following result.

THEOREM 6.1. *Let X be a compact Riemann surface of genus g at least equal to two, with an involution j acting on X with $2n$ fixed points. Denote by γ the genus of the quotient surface $X/\langle j \rangle$.*

Then the type of the involution j is $(\gamma - 1, 0, 1)$ if $n = 0$, and $(\gamma, n-1, 0)$ if $n \ge 1$.

We will now prove Theorem 6.1 by analyzing three cases separately: $n = 0$, $n = 1$ and $n > 1$.

6.1. The classical case $n = 0$. In this case $g = 2\gamma - 1$, and it is shown in [**20**] that there exists a symplectic basis $\widetilde{\alpha}_0, \widetilde{\alpha}_1, \ldots, \widetilde{\alpha}_{g-1}, \widetilde{\beta}_0, \widetilde{\beta}_1, \ldots, \widetilde{\beta}_{g-1}$ for X such that

$$j(\widetilde{\alpha}_0) = \widetilde{\alpha}_0, j(\widetilde{\beta}_0) = \widetilde{\beta}_0, \widetilde{\alpha}_{k+\gamma-1} = j(\widetilde{\alpha}_k) \;\;,\;\; \widetilde{\beta}_{k+\gamma-1} = j(\widetilde{\beta}_k)\;,\; 1 \le k \le \gamma - 1.$$

It follows that

$$L_0 = \langle \widetilde{\alpha}_0, \widetilde{\alpha}_k + \widetilde{\alpha}_{k+\gamma-1}, \widetilde{\beta}_0, \widetilde{\beta}_k + \widetilde{\beta}_{k+\gamma-1} : 1 \le k \le \gamma - 1 \rangle$$

with $\dim \mathcal{A}_0 = \gamma$ and polarization type $(1, 2^{\gamma-1})$, and

$$L_1 = \langle \widetilde{\alpha}_k - \widetilde{\alpha}_{k+\gamma-1}, \widetilde{\beta}_k - \widetilde{\beta}_{k+\gamma-1} : 1 \le k \le \gamma - 1 \rangle$$

with $\dim \mathcal{A}_1 = \gamma - 1$ and polarization type $(2^{\gamma-1})$.

We conclude that the type of j is $(\gamma-1,0,1)$ in this case.

Observe that $\mathcal{A}_1$ is a classical Prym variety, as it has a principal polarization.

6.2. The case $n=1$. In this case $g=2\gamma$, and it is shown in [**20**] that there exists a symplectic basis $\widetilde{\alpha}_1,\widetilde{\beta}_1,\ldots,\widetilde{\alpha}_\gamma,\widetilde{\beta}_\gamma,\ldots,\widetilde{\alpha}_g,\widetilde{\beta}_g$ such that

$$\widetilde{\alpha}_{k+\gamma}=-j(\widetilde{\alpha}_k)\ ,\ \widetilde{\beta}_{k+\gamma}=-j(\widetilde{\beta}_k)\ ,\ 1\leq k\leq\gamma$$

It follows that

$$L_0=\langle\widetilde{\alpha}_k-\widetilde{\alpha}_{k+\gamma},\widetilde{\beta}_k-\widetilde{\beta}_{k+\gamma}\rangle$$

with $\dim\mathcal{A}_0=\gamma$ and polarization type (2^γ), and

$$L_1=\langle\widetilde{\alpha}_k+\widetilde{\alpha}_{k+\gamma},\widetilde{\beta}_k+\widetilde{\beta}_{k+\gamma}\rangle$$

with $\dim\mathcal{A}_1=\gamma$ and polarization type (2^γ).

We conclude that the type of j is $(\gamma,0,0)$ in this case.

Observe that $\mathcal{A}_1$ is again a classical Prym variety, as it has a principal polarization, and so is $\mathcal{A}_0$.

6.3. The case $n>1$. In this case $g=2\gamma+n-1$, and it is shown in [**20**] that there exists a symplectic basis $\widetilde{\alpha}_1,\widetilde{\beta}_1,\ldots,\widetilde{\alpha}_\gamma,\widetilde{\beta}_\gamma,\ldots,\widetilde{\alpha}_g,\widetilde{\beta}_g$ such that

$$\widetilde{\alpha}_{k+\gamma+n-1}=-j(\widetilde{\alpha}_k)\ ,\ \widetilde{\beta}_{k+\gamma+n-1}=-j(\widetilde{\beta}_k)\ ,\ 1\leq k\leq\gamma$$

$$\widetilde{\alpha}_i=-j(\widetilde{\alpha}_i),\widetilde{\beta}_i=-j(\widetilde{\beta}_i),\ \gamma+1\leq i\leq\gamma+n-1$$

It follows that

$$L_0=\langle\widetilde{\alpha}_k-\widetilde{\alpha}_{k+\gamma+n-1},\widetilde{\beta}_k-\widetilde{\beta}_{k+\gamma+n-1};1\leq k\leq\gamma\rangle$$

with $\dim\mathcal{A}_0=\gamma$ and polarization type (2^γ), and

$$L_1=\langle\widetilde{\alpha}_k+\widetilde{\alpha}_{k+\gamma+n-1},\widetilde{\beta}_k+\widetilde{\beta}_{k+\gamma+n-1},\widetilde{\alpha}_i,\widetilde{\beta}_i;1\leq k\leq\gamma,\gamma+1\leq i\leq\gamma+n-1\rangle$$

with $\dim\mathcal{A}_1=\gamma+n-1$ and polarization type $(1^{n-1},2^\gamma)$.

We conclude that the type of j is $(\gamma,n-1,0)$ in this case, and we have thus completed the classification of ppavs admitting an involution.

COROLLARY 6.2. *For a given type (x,y,z), there are no Jacobian varieties in the corresponding family $\mathcal{F}_{(x,y,z)}$ if (x,y,z) is different from $(x,0,1)$ and different from $(x,y,0)$.*

If the type is $(x,0,1)$, then the dimension of the intersection of the Jacobian locus and $\mathcal{F}_{(x,0,1)}$ is $3x$.

If the type is $(x,y,0)$, then the dimension of the intersection of the Jacobian locus and $\mathcal{F}_{(x,y,0)}$ is $3x+y-1$.

PROOF OF COROLLARY. Observe that the dimension of the family of compact Riemann surfaces X of genus $g=2\gamma-1+n$ admitting an involution with $2n$ fixed points is $3\gamma-3+2n$ (see [**16**] and [**5**]), and that an involution on X induces an involution on JX. □

REMARK 6.3. It follows from Corollary 6.2 and Theorem 6.1 that the families $\mathcal{F}_{(x,0,1)}$ and $\mathcal{F}_{(x,y,0)}$ contain Jacobian varieties of Riemann surfaces and also ppavs that are not Jacobians. It would be interesting to solve the Schottky problem in this case: find a characterization of the Jacobian varieties in the respective families.

References

[1] Albert, A. A. *A solution of the principal problem in the theory of Riemann matrices*, Ann. of Math. **35**, 500–515, 1934.

[2] A. Adrian Albert, *On the construction of Riemann matrices. II*, Ann. of Math. (2) **36** (1935), no. 2, 376–394, DOI 10.2307/1968578. MR1503230

[3] Christina Birkenhake and Herbert Lange, *Complex tori*, Progress in Mathematics, vol. 177, Birkhäuser Boston, Inc., Boston, MA, 1999. MR1713785 (2001k:14085)

[4] Christina Birkenhake and Herbert Lange, *Complex abelian varieties*, 2nd ed., Grundlehren der Mathematischen Wissenschaften [Fundamental Principles of Mathematical Sciences], vol. 302, Springer-Verlag, Berlin, 2004. MR2062673 (2005c:14001)

[5] Emilio Bujalance, Francisco Javier Cirre, José Manuel Gamboa, and Grzegorz Gromadzki, *Symmetries of compact Riemann surfaces*, Lecture Notes in Mathematics, vol. 2007, Springer-Verlag, Berlin, 2010. MR2683160 (2011h:30062)

[6] Angel Carocca, Víctor González-Aguilera, and Rubí E. Rodríguez, *Weyl groups and abelian varieties*, J. Group Theory **9** (2006), no. 2, 265–282, DOI 10.1515/JGT.2006.018. MR2220419 (2007b:14097)

[7] Angel Carocca, Herbert Lange, and Rubí E. Rodríguez, *Jacobians with complex multiplication*, Trans. Amer. Math. Soc. **363** (2011), no. 12, 6159–6175, DOI 10.1090/S0002-9947-2011-05560-1. MR2833548 (2012h:11091)

[8] Angel Carocca, Herbert Lange, Rubí E. Rodríguez, and Anita M. Rojas, *Prym-Tyurin varieties via Hecke algebras*, J. Reine Angew. Math. **634** (2009), 209–234, DOI 10.1515/CRELLE.2009.073. MR2560410 (2010k:14047)

[9] Angel Carocca and Rubí E. Rodríguez, *Jacobians with group actions and rational idempotents*, J. Algebra **306** (2006), no. 2, 322–343, DOI 10.1016/j.jalgebra.2006.07.027. MR2271338 (2007h:14038)

[10] Angel Carocca and Rubí E. Rodríguez *Abelian varieties with Hecke algebra action.* Preprint (2013).

[11] Angel Carocca, Sevín Recillas, and Rubí E. Rodríguez, *Dihedral groups acting on Jacobians*, Complex manifolds and hyperbolic geometry (Guanajuato, 2001), Contemp. Math., vol. 311, Amer. Math. Soc., Providence, RI, 2002, pp. 41–77, DOI 10.1090/conm/311/05446. MR1940163 (2004b:14044)

[12] Carocca, A., Rodríguez, R. E. and Rojas, A. M. *Symmetric group actions on Jacobian varieties*, this volume.

[13] C. Chevalley, A. Weil, and E. Hecke, *Über das verhalten der integrale 1. gattung bei automorphismen des funktionenkörpers* (German), Abh. Math. Sem. Univ. Hamburg **10** (1934), no. 1, 358–361, DOI 10.1007/BF02940687. MR3069638

[14] Olivier Debarre, *Complex tori and abelian varieties*, SMF/AMS Texts and Monographs, vol. 11, American Mathematical Society, Providence, RI; Société Mathématique de France, Paris, 2005. Translated from the 1999 French edition by Philippe Mazaud. MR2158864 (2006e:14001)

[15] Ron Donagi, *Decomposition of spectral covers*, Astérisque **218** (1993), 145–175. Journées de Géométrie Algébrique d'Orsay (Orsay, 1992). MR1265312 (95f:14065)

[16] Clifford J. Earle, *H. E. Rauch, function theorist*, Differential geometry and complex analysis, Springer, Berlin, 1985, pp. 15–31. MR780033 (86h:01063)

[17] Torsten Ekedahl and Jean-Pierre Serre, *Exemples de courbes algébriques à jacobienne complètement décomposable* (French, with English and French summaries), C. R. Acad. Sci. Paris Sér. I Math. **317** (1993), no. 5, 509–513. MR1239039 (94j:14029)

[18] Jordan S. Ellenberg, *Endomorphism algebras of Jacobians*, Adv. Math. **162** (2001), no. 2, 243–271, DOI 10.1006/aima.2001.1994. MR1859248 (2003c:11061)

[19] Federigo Enriques and Francesco Severi, *Mémoire sur les surfaces hyperelliptiques* (French), Acta Math. **32** (1909), no. 1, 283–392, DOI 10.1007/BF02403219. MR1555054

[20] John D. Fay, *Theta functions on Riemann surfaces*, Lecture Notes in Mathematics, Vol. 352, Springer-Verlag, Berlin-New York, 1973. MR0335789 (49 #569)

[21] Akira Fujiki, *Finite automorphism groups of complex tori of dimension two*, Publ. Res. Inst. Math. Sci. **24** (1988), no. 1, 1–97, DOI 10.2977/prims/1195175326. MR944867 (90f:32031)

[22] Víctor González-Aguilera and Rubí E. Rodríguez, *Families of irreducible principally polarized abelian varieties isomorphic to a product of elliptic curves*, Proc. Amer. Math. Soc. **128** (2000), no. 3, 629–636, DOI 10.1090/S0002-9939-99-05415-5. MR1676344 (2000e:14076)

[23] Alexander Grothendieck, *Sur quelques points d'algèbre homologique* (French), Tôhoku Math. J. (2) **9** (1957), 119–221. MR0102537 (21 #1328)

[24] Vassil Kanev, *Spectral curves, simple Lie algebras, and Prym-Tjurin varieties*, Theta functions—Bowdoin 1987, Part 1 (Brunswick, ME, 1987), Proc. Sympos. Pure Math., vol. 49, Amer. Math. Soc., Providence, RI, 1989, pp. 627–645. MR1013158 (91b:14028)

[25] H. Lange and S. Recillas, *Abelian varieties with group action*, J. Reine Angew. Math. **575** (2004), 135–155, DOI 10.1515/crll.2004.076. MR2097550 (2005h:14107)

[26] Herbert Lange and Sevin Recillas, *Prym varieties of pairs of coverings*, Adv. Geom. **4** (2004), no. 3, 373–387, DOI 10.1515/advg.2004.022. MR2071812 (2005g:14060)

[27] Herbert Lange, Rubí E. Rodríguez, and Anita M. Rojas, *Polarizations on abelian subvarieties of principally polarized abelian varieties with dihedral group actions*, Math. Z. **276** (2014), no. 1-2, 397–420, DOI 10.1007/s00209-013-1206-1. MR3150211

[28] Solomon Lefschetz, *On certain numerical invariants of algebraic varieties with application to abelian varieties (cont.)*, Trans. Amer. Math. Soc. **22** (1921), no. 4, 407–482, DOI 10.2307/1988964. MR1501180

[29] Solomon Lefschetz *Hyperelliptic surfaces and Abelian varieties*, Topics in Algebraic Geometry I, Washington, 349–395, 1928.

[30] Jean-Yves Mérindol, *Variétés de Prym d'un revêtement galoisien* (French), J. Reine Angew. Math. **461** (1995), 49–61, DOI 10.1515/crll.1995.461.49. MR1324208 (96b:14035)

[31] Mumford, D. *Abelian Varieties.* Oxford 1972.

[32] David Mumford, *Prym varieties. I*, Contributions to analysis (a collection of papers dedicated to Lipman Bers), Academic Press, New York, 1974, pp. 325–350. MR0379510 (52 #415)

[33] Frans Oort and Yuri G. Zarhin, *Complex tori*, Indag. Math. (N.S.) **7** (1996), no. 4, 473–487, DOI 10.1016/S0019-3577(97)89134-3. MR1620112 (99d:14044)

[34] Frans Oort and Yuri Zarhin, *Endomorphism algebras of complex tori*, Math. Ann. **303** (1995), no. 1, 11–29, DOI 10.1007/BF01460976. MR1348352 (97h:14063)

[35] Angela Ortega, *Prym-Tyurin varieties coming from correspondences with fixed points*, J. Algebra **311** (2007), no. 1, 268–281, DOI 10.1016/j.jalgebra.2006.10.015. MR2309888 (2008e:14042)

[36] Sevín Recillas and Rubí E. Rodríguez, *Prym varieties and fourfold covers. II. The dihedral case*, The geometry of Riemann surfaces and abelian varieties, Contemp. Math., vol. 397, Amer. Math. Soc., Providence, RI, 2006, pp. 177–191, DOI 10.1090/conm/397/07472. MR2218008 (2007h:14039)

[37] Irving Reiner, *Automorphisms of the symplectic modular group*, Trans. Amer. Math. Soc. **80** (1955), 35–50. MR0073603 (17,458a)

[38] Gonzalo Riera, *Automorphisms of abelian varieties associated with Klein surfaces*, J. London Math. Soc. (2) **51** (1995), no. 3, 442–452, DOI 10.1112/jlms/51.3.442. MR1332882 (97c:14045)

[39] Gonzalo Riera and Rubí E. Rodríguez, *Riemann surfaces and abelian varieties with an automorphism of prime order*, Duke Math. J. **69** (1993), no. 1, 199–217, DOI 10.1215/S0012-7094-93-06910-4. MR1201698 (93k:14059)

[40] Gonzalo Riera and Rubí E. Rodríguez, *The period matrix of Bring's curve*, Pacific J. Math. **154** (1992), no. 1, 179–200. MR1154738 (93e:14033)

[41] John Ries, *The Prym variety for a cyclic unramified cover of a hyperelliptic Riemann surface*, J. Reine Angew. Math. **340** (1983), 59–69, DOI 10.1515/crll.1983.340.59. MR691961 (85a:14020)

[42] Anita M. Rojas, *Group actions on Jacobian varieties*, Rev. Mat. Iberoam. **23** (2007), no. 2, 397–420, DOI 10.4171/RMI/500. MR2371432 (2009e:14047)

[43] Schottky, F. and Jung, H. *Neue Sätze über Symmetralfunctionen und die Abel'schen Functionen der Riemann'schen Theorie*, S.-B. Akad. Wiss. (Berlin), Phys. Math. Kl. **1**, 282–297, 1909.

[44] Atsushi Shimizu, *On complex tori with many endomorphisms*, Tsukuba J. Math. **8** (1984), no. 2, 297–318. MR767962 (86c:32036)

[45] Goro Shimura and Yutaka Taniyama, *Complex multiplication of abelian varieties and its applications to number theory*, Publications of the Mathematical Society of Japan, vol. 6, The Mathematical Society of Japan, Tokyo, 1961. MR0125113 (23 #A2419)

[46] Rudi Salomon, *Prym varieties associated to graphs*, J. Algebra **313** (2007), no. 2, 828–845, DOI 10.1016/j.jalgebra.2007.01.001. MR2329573 (2008e:14043)

[47] A. Silverberg and Yu. G. Zarhin, *Isogenies of abelian varieties*, J. Pure Appl. Algebra **90** (1993), no. 1, 23–37, DOI 10.1016/0022-4049(93)90133-E. MR1246271 (94j:14040)

[48] A. N. Tjurin, *Five lectures on three-dimensional varieties* (Russian), Uspehi Mat. Nauk **27** (1972), no. 5, (167), 3–50. MR0412196 (54 #323)

[49] Gerard van der Geer and Frans Oort, *Moduli of abelian varieties: a short introduction and survey*, Moduli of curves and abelian varieties, Aspects Math., E33, Vieweg, Braunschweig, 1999, pp. 1–21, DOI 10.1007/978-3-322-90172-9_1. MR1722536 (2001a:14048)

[50] G. E. Welters, *Curves of twice the minimal class on principally polarized abelian varieties*, Nederl. Akad. Wetensch. Indag. Math. **49** (1987), no. 1, 87–109. MR883371 (88c:14061)

[51] Wirtinger, W., *Untersuchungen über Theta Funktionen*, Teubner, Berlin, 1895.

[52] Yu. G. Zarkhin, *Endomorphisms of abelian varieties, cyclotomic extensions, and Lie algebras* (Russian, with Russian summary), Mat. Sb. **201** (2010), no. 12, 93–102; English transl., Sb. Math. **201** (2010), no. 11-12, 1801–1810. MR2760104 (2012d:14076)

[53] Yu. G. Zarhin, *Endomorphisms and torsion of abelian varieties*, Duke Math. J. **54** (1987), no. 1, 131–145, DOI 10.1215/S0012-7094-87-05410-X. MR885780 (88h:14046)

[54] Yuri G. Zarhin, *The endomorphism rings of Jacobians of cyclic covers of the projective line*, Math. Proc. Cambridge Philos. Soc. **136** (2004), no. 2, 257–267, DOI 10.1017/S0305004103007102. MR2040573 (2004m:11088)

Departamento de Matemáticas, Universidad de La Frontera, Temuco, Chile
E-mail address: rubi.chile@gmail.com

The remarkable Accola-Maclachlan surfaces

David Singerman

Dedicated to Emilio Bujalance on the occasion of his 60th birthday

ABSTRACT. The Accola-Maclachlan surfaces were found independently in 1968 by R. Accola and C. Maclachlan who showed that they admit $8(g+1)$ automorphisms, the least value of the size of the maximum group of automorphisms of a Riemann surface of genus g. Here we point out other interesting properties. For $g \neq 2, 5$ they have the the largest number of anticonformal automorphisms of order 2; they are the unique Platonic M-surfaces; the automorphisms group acts transitively on their Weierstrass points, a rare property for a Riemann surface.

1. Introduction

The most well-known theorem about automorphism groups of Riemann surfaces of genus $g \geq 2$ is that they are finite and their order is bounded above by $84(g-1)$. This is a famous theorem of Hurwitz from 1893 [**H**]. Hurwitz knew that this bound was attained because in 1878, Klein had already found his beautiful surface of genus 3, the Klein quartic, which has 168 automorphisms (see [**L**] for a translation of this famous paper). In 1961 Macbeath [**M**] showed that this bound is attained for infinitely many values of g. However, it turns out that there are comparatively few such values. (Conder [**C**] has shown that for $1 < g < 11905$, there are only 32 values of g for which the Hurwitz bound is attained; the order of the automorphism group is now less than one million.) If $\mu(g)$ denotes the maximum order of a group of automorphisms of a Riemann surface of genus g, then $\mu(g) \leq 84(g-1)$. An obvious question is about the lower bound for $\mu(g)$. This was found independently in 1968 by Robert Accola [**A**] and Colin Maclachlan [**M1**] who showed that for every $g \geq 2$ there is a Riemann surface of genus g with $8(g+1)$ automorphisms so that $\mu(g) \geq 8(g+1)$.

I would also like to dedicate this paper to Colin Maclachlan who died recently. He was a wonderful mathematician and friend.

We now describe the surfaces for which this lower bound is attained. The approaches of Accola and Maclachlan are different. Accola constructed a two-sheeted cover of the sphere branched over the vertices of a regular $(2g+2)$-gon. This polygon has $4g+4$ automorphisms and these lift to $8g+8$ automorphisms of

---

2010 *Mathematics Subject Classification.* Primary: 30F10, 20H10.

its two-sheeted cover. From Accola's description we can deduce that the equation of the complex algebraic curve is

$$w^2 = z^{2g+2} - 1.$$

Maclachlan constructed the surface using Fuchsian groups. Let Γ denote the triangle group with periods 2, 4, $2g+2$. This group has a presentation

$$\langle X, Y | X^4 = Y^{2g+2} = (XY)^2 = 1 \rangle.$$

Let G be the group with presentation

(1) $$\langle R, S | R^4 = S^{2g+2} = (RS)^2 = (R^{-1}S)^2 = 1 \rangle.$$

The group G has as centre the subgroup C of order 2 generated by R^2, and $G/C \cong D_{2g+2}$, the dihedral group of order $4g+4$. Thus G has order $8g+8$. The group G is also the automorphism group of a regular map of type $\{4, 2g+2\}$, which we shall call the *Accola-Maclachlan* map of genus g. This map has $N/2$ edges, $N/4$ faces and $N/(2g+2)$ vertices, where $N = |G| = 8g+8$. See section 8.8, Maps on a two-sheeted Riemann surface in [**CM**]. Figure 8.5 of this book gives a picture of this map when $g = 2$.

There is an epimorphism $\theta : \Gamma \longrightarrow G$ defined by $\theta(X) = R$, $\theta(Y) = S$. The kernel of θ is a surface group K and by the Riemann-Hurwitz formula the genus of the Riemann surface $\mathcal{U}/K$ is equal to g where $\mathcal{U}$ is the upper half-plane. We now show that this is the same surface as the one found by Accola.

We can find the signature of $\Lambda = \theta^{-1}(C)$ by standard Fuchsian group techniques [**S1**], as the permutation representation of Γ on the cosets of Λ is the same as the permutation representation of G on the cosets of C. We find that the only elliptic periods of Λ are those induced by the generator X of Γ which has $2g+2$ cycles of length 2 so it follows from the permutation theorem in [**S1**] that there are $2g+2$ periods equal to 2 in Λ and no other periods. By the Riemann-Hurwitz formula the genus of Λ is zero so that Λ has signature $(0; 2^{(2g+2)})$, with $2g+2$ periods equal to 2. It follows from a theorem of Maclachlan [**M2**] that $\mathcal{U}/K$ is a hyperelliptic surface and its automorphism group is G. Also note that R^2, being the unique central involution in G, must be the hyperelliptic involution. Now $K \triangleleft \Lambda$ with index 2 and $\Lambda \triangleleft \Gamma$ with index $4g+4$, the quotient being D_{2g+2}. Now a normal subgroup of a triangle group corresponds to a regular map on a surface, in this case the sphere as Λ has genus zero. Thus the regular map corresponding to Λ must be a dihedron (or its dual the hosohedron), so that $\mathcal{U}/K$ is a two-sheeted cover of the dihedron with $2g+2$ branch points corresponding to the $2g+2$ elliptic fixed points. These branch points are the vertices and so the surfaces found by Accola and Maclachlan are the same. These surfaces are known as *Accola-Maclachlan surfaces*. We denote the Accola-Maclachlan surface of genus g by AM_g.

Property 1 of the Accola-Maclachlan surfaces. *For every $g \geq 2$ the size of the automorphism group of AM_g is $8(g+1)$, the least value of the size of the maximum group of automorphisms of a Riemann surface of genus g.*

It was later shown that if $g \equiv 3 \bmod 4$ then there is exactly one more family of surfaces with $8(g+1)$ automorphisms, namely the Kulkarni surfaces [**K**].

2. Properties of AM_g

The two most fundamental properties of AM_g follow immediately from the above descriptions of this surface. A Riemann surface is called *hyperelliptic* if it is a two-sheeted branched cover of the sphere. By the Riemann-Hurwitz formula there must be $2g+2$ branch-points, where g is the genus of the surface. A Riemann surface is called *platonic* if it is of the form $\mathcal{U}/K$ for K a normal subgroup of a $(2,m,n)$ triangle group. Platonic surfaces are the ones that underlie regular maps. From our two descriptions of AM_g we get the following.

THEOREM 1. *AM_g is a platonic hyperelliptic surface.*

PROOF. From the Accola description, AM_g is hyperelliptic; from the Maclachlan definition, AM_g is platonic. □

3. Symmetries of AM_g

By a symmetry of a Riemann surface X, we mean an anticonformal involution $T : X \longrightarrow X$, and a symmetric surface is one that admits a symmetry. If we consider the group G above of automorphisms of AM_g we see that it admits an automorphism $\alpha : G \longrightarrow G$, defined by $\alpha(R) = R^{-1}, \alpha(S) = S^{-1}$. By the criteria in [**S3**], [**JSW**], [**BCGG**, Theorem 1.5.10], this shows that AM_g is symmetric. If J is the hyperelliptic involution then $G/\langle J\rangle$ is a group of automorphisms of the sphere and so $G/\langle J\rangle \cong C_n$, D_n, A_4, S_4 or A_5. (This remark applies to all hyperelliptic surfaces.)

We now examine these cases separately and show that we usually get the maximum number of symmetries when $G/\langle J\rangle \cong D_n$ and then G is an Accola-Maclachlan group. Let us start by examining this case. Now G has a presentation of the form (1).

If we adjoin a symmetry T we get a group $G^* = \langle G, T\rangle$ with presentation

(2) $$\langle R,S,T|T^2 = (RT)^2 = (ST)^2 = R^4 = S^{2g+2} = (RS)^2 = (R^{-1}S)^2 = 1\rangle$$

From the relations in (2), we have $TR = R^{-1}T$ and $TS = S^{-1}T$ and $SR = R^{-1}S^{-1}$, $SR^{-1} = RS^{-1}$ and thus every anticonformal automorphism has the form TR^iS^j. (Also note that as $TRT^{-1} = R^{-1}$ $TST^{-1} = S^{-1}$, conjugation by T induces the automorphism α with the properties above.) We need to find the transformations TR^iS^j of order 2. As $\{T,R\}$ and $\{T,S\}$ generate dihedral groups TR^i and TS^j all have order 2. Now we see when TRS^i has order 2. As $(RS)^2 = (R^{-1}S)^2 = 1$, $(TRS^i)^2 = TRS^iTRS^i = R^{-1}S^{-i}RS^i$.

Now by induction $RS^i = S^{-i}R$ if i is even and $RS^i = S^{-i}R^{-1}$ if i is odd.

Thus $(TRS^i)^2 = S^{2i}$ if i is even and $(TRS^i)^2 = R^2S^{2i}$ if i is odd.

As S has period $2g+2$ we see that if g is odd then TRS^{g+1} is a symmetry. If g is even then $(TRS^{g+1})^2 = R^2 = J$, the hyperelliptic involution. Anticonformal transformations whose square is the hyperelliptic involution are called *pseudosymmetries*. These transformations were first discussed by Earle [**E**]. Now if U is a symmetry of a Riemann surface X then so is $UJ = UR^2$. Thus a complete list of symmetries of X is $TS^i, TS^iR^2, i = 0,1\dots 2g+1$, TR, TR^3, and also TRS^{g+1}, TR^3S^{g+1} if g is odd.

Thus

THEOREM 2. *The number of symmetries of AM_g is equal to $4g+8$ if g is odd and $4g+6$ if g is even.*

(This is a different result than in [**S4**] where there were a number of errors.)

Alternatively, we can get this result using the Accola-Maclachlan maps, see section 1. The Accola-Maclachlan map is an example of a hyperelliptic map, one that admits an involution such that the quotient of this map by the involution is a planar map, i.e a map of genus 0. In our case the involution is R^2 and as $R^4 = I$ the quotient map is of type $\{2, 2g+2\}$ and so is a hosohedron (beach-ball) with $2g+2$ edges, $2g+2$ faces and vertices at the poles. In this way, we see that the Accola-Maclachlan map is just a two-sheeted cover of the beach-ball with the $2g+2$ branch-points lying over the $2g+2$ face-centres which all lie on the equator. If we regard this beach-ball as being the Riemann sphere then the branch-points lie over the $(2g+2)$th roots of unity. (This is like Accola's original description.) Its rotation group is D_{2g+2} generated by a rotation s about the axis joining the poles and through an angle $2\pi/(2g+2)$, and a rotation r about an axis through 1 and -1. Thus $s(z) = e^{2\pi i/(2g+2)}z$ and $r(z) = 1/z$. Now $r^2 = s^{2g+2} = (rs)^2 = 1$. To get the full automorphism group, including orientation-reversing transformations, we adjoin the reflection $t(z) = \bar{z}$. Now $t^2 = (tr)^2 = (ts)^2 = r^2 = s^{2g+2} = (rs)^2 = 1$. The antipodal automorphism which is given by $s^{g+1}tr$ commutes with all the elements of D_{2g+2} and so the full automorphism group of the hosohedron is $C_2 \times D_{2g+2}$. The symmetries of the hosohedron are elements of order two outside the rotation group and so consist of the elements $t, ts^{g+1}, ts^i r$, where $i = 0, \ldots, 2g+1$, giving $2g+4$ elements in all. Now if V is a symmetry of AM_g then so is $VJ = VR^2$. Both these symmetries project to the same symmetry of the beach-ball and so the number of symmetries of AM_g is twice the number of symmetries and pseudosymmetries of the beach-ball. This is $2 \times (2g+4) = 4g+8$ as before. Also, we showed that if g is even then there are two pseudosymmetries, those symmetries that project to J.

We now examine the other cases. We do the first of these in some detail. This is where $G/\langle J\rangle \cong S_4$ which we regard as the rotation group of the cube and which has presentation $\langle r, s | r^2 = s^3 = (rs)^4 = 1\rangle$. Now the cube has 9 planes of symmetry, one bisecting the mid-points of the top and bottom faces, one of the front and back faces and one of the left and right faces and 6 joining the pairs of opposite edges. We also have the antipodal map, giving 10 symmetries of the cube. Now consider the two-sheeted cover Y of the cube branched over the 6 face centres. As the number of branch points is $2g+2$ we get a hyperelliptic surface of genus 2 (this also follows from the Riemann-Hurwitz formula). Now the square of each symmetry of the cube is the identity and so lifts to either I or J and thus each symmetry of the cube lifts to a symmetry or pseudosymmetry of the surface. There are 20 of these. We now look for the pseudosymmetries. In [**Z**] and [**S4**] it is shown that if V is a pseudosymmetry of a surface X then $X/\langle V\rangle$ is the real projective plane and in the natural homomorphism from G^* to $G^*/\langle J\rangle$, V is mapped to the antipodal map. Thus if V_1, V_2 are two pseudosymmetries then $V_2 = V_1$ or $V_2 = V_1 J$. Thus there are at most two pseudosymmetries. (Also it was shown in [**S4**] and [**Z**] that if a pseudosymmetry exists then the genus of the surface is even.) The automorphism group of Y has the following presentation,

$$\langle R, S | R^8 = S^3 = (RS)^2 = 1, R^4 S = SR^4\rangle$$

where the last relation expresses the commutativity of R^4, the hyperelliptic involution, with S. The extended automorphism group has presentation

$$\langle R, S, T | T^2 = (RT)^2 = (ST)^2 = R^8 = S^3 = (RS)^2 = 1, R^4 S = SR^4\rangle$$

where the last relation express the fact that R^4, the hyperelliptic involution, commutes with S. Now by a calculation using GAP we can show that TR^2SR^3S and

FIGURE 1. The Bolza surface

TR^6SR^3S are the pseudosymmetries and thus Y admits 18 symmetries. The surface Y is sometimes called the *Bolza surface.* (I would like to thank Ian Leary for help with the GAP calculation). The Easter egg in Figure 1 is a picture of the Bolza surface, together with its embedded regular map of type $\{3, 8\}$. This comes from the classic book by Burnside [**B**, Section 288].

If we consider the two-sheeted covering of the cube branched over the 8 vertices we get a surface of genus 3. Again this will have 20 symmetries and as the genus is odd, there are no pseudosymmetries. This is the same number of symmetries as for the Accola-Maclachlan surface of genus 3. We also get 20 symmetries for the two-sheeted cover branched over 12 edge centres and this surface has genus 5. Now consider the two-sheeted covers of the icosahedron. Every plane of symmetry goes through a pair of opposite edges and thus there are 15 planes of symmetry and with the antipodal map this gives 16 symmetries. Thus every two-sheeted cover has 32 symmetries and pseudosymmetries. The branching is over the 12 faces and this gives a two-sheeted cover of genus 5 (again using the result that there are $2g + 2$ branch points), or over the the 20 vertices giving a surface of genus 9 or the 30 edge centres giving a surface of genus 14. (We cannot have an orbit of A_5 consisting of 6 points on the sphere because then the stabilizer of a point would be a cyclic group of order 10 which does not lie in A_5. Also, the tetrahedon has at most four symmetries and so a two-sheeted cover would have at most 8 symmetries, not exceeding the Accola-Maclachlan bound. Other orbits, such as unions of vertices and faces would have more points and thus the genera of the quotient surfaces would be bigger and again the Accola-Maclachlan bound would still hold.) Finally, we consider the case when we have a cyclic automorphism group. The largest cyclic group of automorphisms of a surface of genus g has order $4g + 2$, the Wiman bound. This also occurs for a hyperelliptic surface. This will extend to a dihedral group D_{4g+2} containing $4g + 2$ symmetries, less than the number for the Accola-Maclachlan surfaces.

We summarize with the following theorem.

THEOREM 3. *Let $M(g)$ denote the maximum number of symmetries of a hyperelliptic surface of genus g. Then*

(1) *$M(2)$=18*

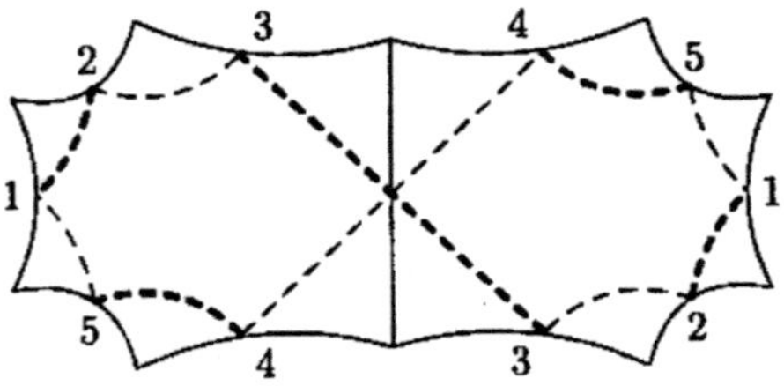

FIGURE 2. The Accola-Maclachlan surface of genus 2

(2) *M(3)=20*
(3) *M(5)=32*
(4) *M(g)=4g+6 (g even, $g \geq 4$)*
(5) *M(g)=4g+8 (g odd, $g \geq 7$).*

Note that when $g = 3$, the two-sheeted cover of the cube branched over the vertices has the same number of symmetries as AM_3. Symmetries of Accola-Maclachlan surfaces are also discussed in [**BBCGG**]. Also note that for $g = 5$, there are no pseudosymmetries, because as we remarked above, pseudosymmetric surfaces have even genus, [**S4**], [**Z**].

Property 2 of the Accola-Maclachlan surfaces. *For every $g \geq 2$ ($g \neq 2, 5$) AM_g has the largest number of symmetries of all hyperelliptic surfaces of genus g.*

4. *M*-surfaces

The fixed point set of a symmetry T, if not empty, consists of a collection of simple closed curves on the surface. Harnack's theorem tells us that the number of these curves is at most $g + 1$. A Riemann surface of genus g that admits a symmetry fixing $g+1$ simple closed curves is called an M-surface, or sometimes an M-curve if one is thinking in terms of algebraic geometry. Then the fixed curves are called *ovals* and describe the real locus of the curve. (Symmetric Riemann surfaces correspond to real algebraic curves.) Now, by results originally due to Natanzon [**N**], if an M-surface is non-hyperelliptic then it admits a unique M-symmetry, (a symmetry with $g + 1$ ovals), but if the surface is hyperelliptic it admits two M-symmetries whose product is the hyperelliptic involution. In [**MS**] the following result was proved:

THEOREM 4. *The Accola-Maclachlan surface AM_g is an M-surface and in fact the only platonic M-surface of genus g.*

This gives us

Property 3 of the Accola-Maclachlan surfaces. *For every $g \geq 2$ the Accola-Maclachlan surface of genus g is the only platonic M-surface of genus g.*

In Figure 2, we give a picture of the Accola-Maclachlan surface of genus 2, where opposite sides of the decagon are identified. The heavy dashed lines are the ovals of one symmetry and the lighter dashed lines are the ovals of that symmetry composed with J. Taken together these form the edges of a face of the associated regular map of type $\{2g+2, 4\} = \{6, 4\}$. The 6 Weierstrass points, in this case fixed points of the hyperelliptic involution, are the points labelled 1-5 together with the centre point.

5. Transitivity on Weierstrass points

Every Riemann surface of genus $g \geq 2$ has a finite set of Weierstrass points. Their definition depends on the properties of poles of meromorphic functions on the surface (see [**FK**]). These points are defined purely in terms of the complex structure of the surface and thus they are invariant under automorphisms of the surface. A natural problem is to find the Riemann surfaces whose automorphism group acts transitively on the Weierstrass points. This question was examined in [**LS**] and there the problem was solved completely for hyperelliptic surfaces. The surfaces where the transitivity property occurs are the branched covers of the regular maps on the sphere which we investigated in section 2 above. Apart from the infinite family of Accola-Maclachlan surfaces, there are only a few other examples.

Property 4 of the Accola-Maclachlan surfaces. *The automorphism group of AM_g acts transitively on the Weierstrass points of AM_g.*

In [**LS**] we found 12 examples of non-hyperelliptic surfaces where this phenomenon occurs and all these correspond to regular maps, so the surfaces are platonic. These include some very well-known Riemann surfaces, such as the Klein surface of genus 3, Bring's surface of genus 4, the Macbeath surface of genus 7 and the fourth order Fermat curve. of genus 3. We use the result originally in Paul Watson's Southampton thesis of 1995 which identifies various Weierstrass points as vertices, face-centres or edge-centres of regular maps [**W**], [**SW**]. Are there examples which do not correspond to regular maps? Note that the Accola-Maclachlan surfaces correspond to regular maps of type $\{4, 2g+2\}$. This leads to the following conjectures.

(1) All surfaces whose automorphism group acts transitively on the Weierstrass points are platonic.
(2) There are only finitely many non-hyperelliptic surfaces whose automorphism group acts transitively on the Weierstrass points.
(3) The only infinite family of surfaces whose automorphism group acts transitively on the Weierstrass points are the Accola-Maclachlan surfaces.

Acknowledgement

The author would like to thank the referee for his very careful reading of the manuscript and pointing out several stylistic errors.

References

[A] Robert D. M. Accola, *On the number of automorphisms of a closed Riemann surface*, Trans. Amer. Math. Soc. **131** (1968), 398–408. MR0222281 (36 #5333)

[BBCGG] S. A. Broughton, E. Bujalance, A. F. Costa, J. M. Gamboa, and G. Gromadzki, *Symmetries of Accola-Maclachlan and Kulkarni surfaces*, Proc. Amer. Math. Soc. **127** (1999), no. 3, 637–646, DOI 10.1090/S0002-9939-99-04534-7. MR1468184 (99f:14036)

[BCGG] Emilio Bujalance, Francisco Javier Cirre, José Manuel Gamboa, and Grzegorz Gromadzki, *Symmetries of compact Riemann surfaces*, Lecture Notes in Mathematics, vol. 2007, Springer-Verlag, Berlin, 2010. MR2683160 (2011h:30062)

[B] W. Burnside, *Theory of groups of finite order*, Dover Publications, Inc., New York, 1955. 2d ed. MR0069818 (16,1086c)

[C] Marston Conder, *The genus of compact Riemann surfaces with maximal automorphism group*, J. Algebra **108** (1987), no. 1, 204–247, DOI 10.1016/0021-8693(87)90135-9. MR887205 (88f:20063)

[CM] H. S. M. Coxeter and W. O. J. Moser, *Generators and relations for discrete groups*, 4th ed., Ergebnisse der Mathematik und ihrer Grenzgebiete [Results in Mathematics and Related Areas], vol. 14, Springer-Verlag, Berlin-New York, 1980. MR562913 (81a:20001)

[E] Clifford J. Earle, *On the moduli of closed Riemann surfaces with symmetries*, Advances in the theory of Riemann surfaces (Proc. Conf., Stony Brook, N.Y., 1969), Princeton Univ. Press, Princeton, N.J., 1971, pp. 119–130. Ann. of Math. Studies, No. 66. MR0296282 (45 #5343)

[FK] Hershel M. Farkas and Irwin Kra, *Riemann surfaces*, Graduate Texts in Mathematics, vol. 71, Springer-Verlag, New York-Berlin, 1980. MR583745 (82c:30067)

[H] A. Hurwitz, Über algebraische Gebilde mit eindeutigen Transformationen in sich. Math.Ann. 41(1893).

[JSW] G. A. Jones, D. Singerman and P. D. Watson, Symmetries of quasiplatonic Riemann surfaces, Arxiv1402.2575

[K] Ravi S. Kulkarni, *A note on Wiman and Accola-Maclachlan surfaces*, Ann. Acad. Sci. Fenn. Ser. A I Math. **16** (1991), no. 1, 83–94. MR1127698 (92j:30045)

[LS] Zoë Laing and David Singerman, *Transitivity on Weierstrass points*, Ann. Acad. Sci. Fenn. Math. **37** (2012), no. 1, 285–300, DOI 10.5186/aasfm.2012.3711. MR2920441

[L] *The eightfold way*, Mathematical Sciences Research Institute Publications, vol. 35, Cambridge University Press, Cambridge, 1999. The beauty of Klein's quartic curve; Edited by Silvio Levy. MR1722410 (2000f:14001)

[M] A. M. Macbeath, *On a theorem of Hurwitz*, Proc. Glasgow Math. Assoc. **5** (1961), 90–96. MR0146724 (26 #4244)

[M1] C. Maclachlan, *A bound for the number of automorphisms of a compact Riemann surface.*, J. London Math. Soc. **44** (1969), 265–272. MR0236378 (38 #4674)

[M2] C. Maclachlan, *Smooth coverings of hyperelliptic surfaces*, Quart. J. Math. Oxford Ser. (2) **22** (1971), 117–123. MR0283194 (44 #427)

[MS] Adnan Melekoğlu and David Singerman, *Reflections of regular maps and Riemann surfaces*, Rev. Mat. Iberoam. **24** (2008), no. 3, 921–939, DOI 10.4171/RMI/560. MR2490203 (2010e:30033)

[N] S. M. Natanzon, *Automorphisms of the Riemann surface of an M-curve* (Russian), Funktsional. Anal. i Prilozhen. **12** (1978), no. 3, 82–83. MR509395 (82b:14020)

[S1] David Singerman, *Subgroups of Fuschian groups and finite permutation groups*, Bull. London Math. Soc. **2** (1970), 319–323. MR0281805 (43 #7519)

[S2] David Singerman, *Finitely maximal Fuchsian groups*, J. London Math. Soc. (2) **6** (1972), 29–38. MR0322165 (48 #529)

[S3] David Singerman, *Symmetries of Riemann surfaces with large automorphism group*, Math. Ann. **210** (1974), 17–32. MR0361059 (50 #13505)

[S4] David Singerman, *Symmetries and pseudosymmetries of hyperelliptic surfaces*, Glasgow Math. J. **21** (1980), no. 1, 39–49, DOI 10.1017/S0017089500003967. MR558273 (81c:30080)

[SW] D. Singerman and P .D. Watson, *Weierstrass points on regular maps*, Geometriae Dedicata **66** (1997), 69–88.

[W] Paul Daniel Watson, *Symmetries and automorphisms of compact Riemann surfaces*, ProQuest LLC, Ann Arbor, MI, 1995. Thesis (Ph.D.)–University of Southampton (United Kingdom). MR2714937

[Z] Robert Zarrow, *A canonical form for symmetric and skew-symmetric extended symplectic modular matrices with applications to Riemann surface theory*, Trans. Amer. Math. Soc. **204** (1975), 207–227. MR0407324 (53 #11100)

University of Southampton

Patterns in semigroups associated with combinatorial configurations

Klara Stokes and Maria Bras-Amorós

ABSTRACT. We extend the list of known linear patterns admitted by numerical semigroups associated with combinatorial configurations. This is done through the construction of configurations from combinations of several smaller configurations. These results may be used to construct configurations with certain parameters, and therefore contribute with answers to the existence question for these configurations.

1. Introduction

A (v_r, b_k)-configuration is a set of points and a set of lines such that each point is on r lines and each line has k points. It is also assumed that there is at most one line through each pair of points. Observe that this is not an obvious assumption, for example the theory of biplanes relies on the assumption that there are exactly two lines through each pair of points.

The existence of intersecting lines on surfaces is an important question in geometry. A famous example of this, which can be found in most books on algebraic geometry, is the set of 27 lines on a cubic surface, given by the Cayley-Salmon theorem. When these lines intersect, they do so in pairs or in triples. The points in which three lines meet are called Eckardt points.

If the surface does not contain any Eckardt point, then each of the 27 lines intersect with 10 other lines in 10 distinct points in a total of $27 \cdot 10/2 = 135$ points. In this case, these 135 points and 27 lines have the property that each point is on 2 lines and each line contains 10 of the points, so it is an example of a $(135_2, 27_{10})$-configuration [5]. An equivalent description is as a 10-regular graph on 27 vertices, in which the lines are taken as the vertices and the points are the edges. This graph is then the dual configuration of the former, in which the role of the points and the lines are interchanged, so it has 135 edges (or lines) with two points each.

2010 *Mathematics Subject Classification.* Primary 05B30, 20M99.

Key words and phrases. Numerical semigroup, pattern, combinatorial configuration, partial linear space.

The first author was financed by the Swedish government through the National Graduate School in Computer Science (CUGS). Both authors received partial support from the Spanish MEC projects ARES (CONSOLIDER INGENIO 2010 CSD2007-00004) and ICWT (TIN2012-32757).

On any cubic surface, among the 27 lines one can find two sets of six lines each, such that lines in the same set are skew, and each line in one set is skew to exactly one line in the other set. Then there are 30 points of intersection between pairs of these lines, and 5 intersection points on each line. These points and lines form a $(30_2, 12_5)$-configuration, called the Schäfli double six. Again, its dual is a graph, a 5-regular graph on 12 vertices and with 30 edges. There are 36 Schäfli double sixes in a generic cubic surface. The 15 lines in the complement of a double six in the total of 27 lines, together with the 15 tangent planes through the triples of these lines, form a $(15_3, 15_3)$-configuration that is known as the Cremona-Richmond configuration. It is the smallest configuration without triangles, and is also the smallest example of a generalized quadrangle.

The study of polygons inscribed in conics is a classic subject in geometry. In 1640, Pascal proved that if a hexagon is inscribed in a conic, then the three points in which pairs of opposite sides meet will lie on a straight line. The hexagon is not required to be regular, nor convex, so if the 6 corner points are chosen in sufficiently general position they will define 60 different hexagons with 60 different Pascal lines, such that they intersect in quadruples in exactly 45 points. These 45 points and 60 lines form a $(45_4, 60_3)$-configuration, called the Pascal configuration [**9**].

Other important configurations in geometry are of course the finite projective and affine planes, the Desargues $(10_3, 10_3)$-configuration and the Pappus $(9_3, 9_3)$-configuration, describing incidences in the projective plane, as ensured by their respective theorems.

It is clear that the incidences of the configurations described above can be defined without considering the surface, the conic and the plane, respectively. In this context, the configuration is called a combinatorial configuration. It is natural to ask for which parameters combinatorial (v_r, b_k)-configuration exist. Certainly, a combinatorial (v_r, b_k)-configuration always exists whenever a geometric (v_r, b_k)-configuration exists, however the opposite is not true in general, although it depends on what is meant by "geometric".

The results in this article give affirmative answers to the existence question of combinatorial configurations with parameters (v_r, b_k) under the assumption of the existence of either

- two configurations with parameters $((v_1)_r, (b_1)_k)$ and $((v_2)_r, (b_2)_k)$ such that $v = v_1 + v_2 - nk/\gcd(r,k)$ and $n \in \{1, \dots, \gcd(r,k)\}$, or
- m configurations with parameters $((v_i)_r, (b_i)_k)$ for $i = 1, \dots, m$ and $m = rk/\gcd(r,k)$, such that $v = v_1 + \cdots + v_m + k/\gcd(r,k)$.

In the latter case, we also characterize the conditions for when it is possible to take m smaller than $rk/\gcd(r,k)$.

For proving this we provide explicit constructions of the larger configuration from the smaller configurations. Other examples of this kind of constructions can be found for example in [**9**, Chapter 6, Section 5].

2. The numerical semigroup associated with configurable parameters

Combinatorially, a necessary condition on the parameters for the existence of configurations is $vr = bk$, which is a consequence of counting the incidences in two different ways. This relation clearly implies that given r and k, not all natural numbers are valid choices for v and b. Indeed, let $r' = r/gcd(r,k)$ and $k' = k/gcd(r,k)$. Then $b = \frac{vr'}{k'}$ is an integer, but r' and k' are coprime, so v must

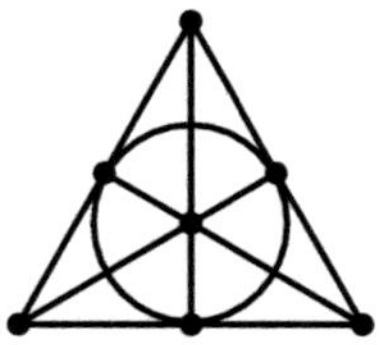

Projective plane over $\mathbb{F}_2$
$(d, r, k) = (7, 3, 3)$

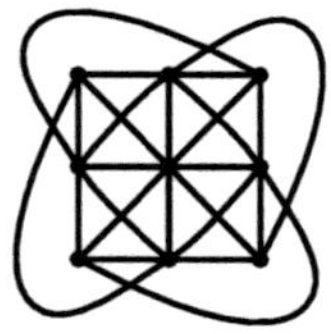

Affine plane over $\mathbb{F}_3$
$(d, r, k) = (3, 4, 3)$

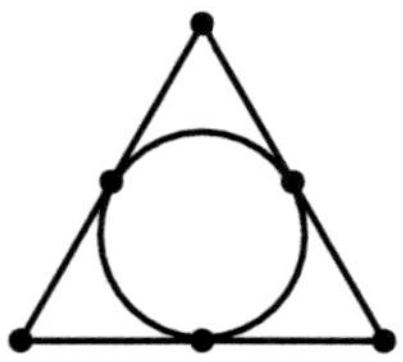

Nonbalanced configuration
$(d, r, k) = (2, 2, 3)$

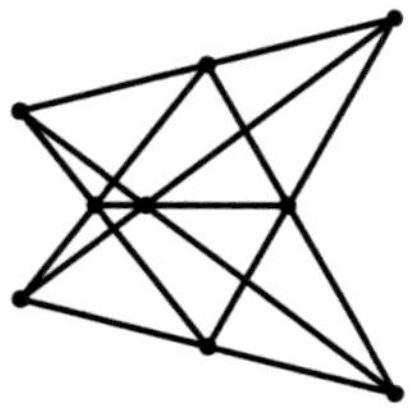

Pappus configuration
$(d, r, k) = (9, 3, 3)$

FIGURE 1. Examples of configurations

be a multiple of $k' = k/gcd(r, k)$. We denote by d the unique integer that satisfies $v = dk'$, that is, $d = v\gcd(r, k)/k$. Observe that also $d = \frac{v\gcd(r,k)}{k} = \frac{vr\gcd(r,k)}{rk} = \frac{bk\gcd(r,k)}{rk} = \frac{b\gcd(r,k)}{r}$.

In this article we will use the reduced parameter set (d, r, k), instead of the traditional (v_r, b_k). The reduced parameters have the obvious advantage of including the condition $vr = bk$, so that before considering other conditions for the existence of configurations, d could hypothetically be any natural number. Indeed, for $r = k = 1$, one could argue that there is a combinatorial configuration for each $d \in \mathbb{N}$, consisting of a set of d points and a set of d lines each passing through one point. However, if $d > 1$ this configuration is not connected. In this article we consider only connected configurations. If $r = k = 2$, then d can take any integer value greater or equal to 3, corresponding to a polygon on d vertices. If $r = 4$ and $k = 3$, then d can also take any integer value greater or equal to 3, but v is obtained from multiplying d by $k = 3$, so the possible values for v are $9, 12, 15, \ldots$.

The second necessary condition on the parameters of combinatorial configurations is the lower bound on the number of points $v = dk/\gcd(r, k) \geq r(k-1) + 1$, saying that given a point p there must be at least enough points to fill the lines through p, and $b = dr/\gcd(r, k) \geq k(r-1)+1$, saying that given a line l there must be at least r lines on each point on l. This translates in terms of d to the condition $r + k \leq 2\frac{d}{\gcd(r,k)} + 2$ for $r = k$ and $r + k \leq 2\frac{d}{\gcd(r,k)} + 1$ for $r \neq k$, see **[10]**.

These two necessary conditions for the existence of combinatorial configurations are however not sufficient. The smallest known example of this fact is for $r = k = 5$. Then $d = v \geq 5 \cdot 4 + 1 = 21$, and the finite projective plane of order 4 is a $(21, 5, 5)$-configuration meeting this lower bound. However, there exists no combinatorial configuration with parameters $(22, 5, 5)$. We say that (d, r, k) is *configurable* if there exists at least one (d, r, k)-configuration.

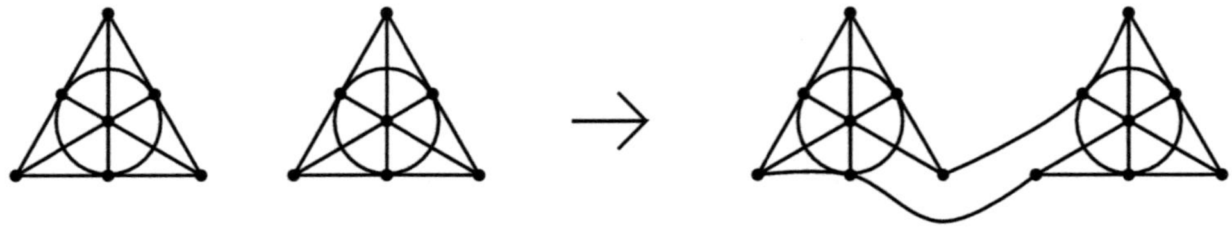

FIGURE 2. Connecting the union of two configurations

As a tool for the study of configurability, in [**3**] we associated a numerical semigroup with the combinatorial configurations. A numerical semigroup is a subset $S \subset \mathbb{N} \cup \{0\}$, such that S is closed under addition, $0 \in S$ and the complement $(\mathbb{N} \cup \{0\}) \setminus S$ is finite. The multiplicity of a numerical semigroup is its smallest non-zero element. The conductor of a numerical semigroup is the smallest element such that all subsequent natural numbers belong to the numerical semigroup. The gaps of a numerical semigroup are the elements in the complement of the numerical semigroup.

It was proven in [**3**] that for any $r, k \in \mathbb{N}$, with $r, k \geq 2$, the set $D_{(r,k)}$ of natural numbers d such that (d, r, k) is configurable forms a numerical semigroup. The proof is based on the fact that a set of positive integers generate a numerical semigroup if and only if they are coprime [**8**]. It is therefore enough to prove that

(1) $0 \in D_{(r,k)}$,
(2) $D_{(r,k)}$ is closed under addition, and
(3) at least two elements of $D_{(r,k)}$ are coprime.

For the first fact, consider the empty configuration. The two remaining facts are proven by combining several configurations into larger configurations.

Observe that for $d_1, d_2 \in D_{(r,k)}$ to imply $d_1 + d_2 \in D_{(r,k)}$, the existence of two (d_1, r, k) and (d_2, r, k) configurations should imply the existence of a $(d_1 + d_2, r, k)$-configuration. Then the two first configurations have $v_1 = d_1 k / \gcd(r, k)$ and $v_2 = d_2 k / \gcd(r, k)$ points respectively, and the third configuration has v points with $v = (d_1 + d_2)k / \gcd(r, k) = d_1 k / \gcd(r, k) + d_2 k / \gcd(r, k) = v_1 + v_2$. So the simplest way to ensure that $d_1 + d_2 \in D_{(r,k)}$ is to construct a configuration from the union of two configurations with parameters (d_1, r, k) and (d_2, r, k). Connectivity can be ensured by simply swapping two incidences between the two disjoint configurations, see Figure 2.

Finally, the proof requires two coprime elements in $D_{(r,k)}$. The first element can be obtained as the integer d associated with the configuration obtained by taking r parallel classes of lines of a finite affine plane of order $q \geq \max(r, k)$ and restricting them to the points on k of the lines in an additional parallel class of lines. This configuration has qk points and qr lines, so $d = q \gcd(r, k)$.

The second combinatorial configuration with an associated integer that is coprime with d can be constructed through a suitable combination of existing configurations. In [**3**] we showed that if $d \in D_{(r,k)}$, then also $md + 1 \in D_{(r,k)}$, with $m = rk / \gcd(r, k)$. The results in this article imply, for example, that $2d - 1 \in D_{(r,k)}$, which is significantly smaller. This is important, since one of the most interesting questions regarding the numerical semigroup $D_{(r,k)}$ is to give good upper bounds on its conductor. Such upper bounds typically come as a function of the multiplicity, when this is known, and otherwise, of other known elements in $D_{(r,k)}$.

3. Linear non-homogeneous patterns

A configuration with parameters of the form (d, r, r) is called balanced, or symmetric. Observe that the fact that $r = k$ does not necessary imply that the configuration has a lot of symmetry, so maybe the term balanced is preferable. In a balanced configuration the number of points equals the number of lines, indeed, $r = k$ and $vr = bk$ together imply that $v = b$. The integer d associated with a balanced configuration satisfies $d = v = b$.

The results in this section were motivated by two constructions of balanced configurations described in [**7**]. Given two configurations (d_1, r, r) and (d_2, r, r) these constructions give two configurations with parameters $(d_1 + d_2 - 1, r, r)$ and $(d_1 + d_2 + 1, r, r)$. In terms of the numerical semigroup $D_{(r,r)}$, this translates to $d_1, d_2 \in D_{(r,r)} \Rightarrow d_1 + d_2 - 1 \in D_{(r,r)}$ and $d_1, d_2 \in D_{(r,r)} \Rightarrow d_1 + d_2 + 1 \in D(r, r)$. In particular, given an element $d \in D_{(r,r)}$ we get $2d - 1, 2d, 2d + 1 \in D_{(r,r)}$ and since d and $2d - 1$ are always coprime, this is enough for proving the finiteness of the complement of $D_{(r,r)}$ in $\mathbb{N} \cup \{0\}$.

A pattern of length n admitted by a numerical semigroup S is a polynomial $p(X_1, \ldots, X_n)$ with non-zero integer coefficients, such that for every ordered sequence of n elements $s_1 \geq \cdots \geq s_n$ from S, we have $p(s_1, s_2, \ldots, s_n) \in S$. A pattern is called linear and homogenous if the pattern polynomial is linear and homogenous. The concept of linear homogeneous patterns is a generalization of the Arf property. A numerical semigroup has the Arf property if $s_1 + s_2 - s_3 \in S$ for every triple $s_1 \geq s_2 \geq s_3 \in S$. So, S is Arf if and only if S admits the linear and homogeneous pattern $X_1 + X_2 - X_3$. This article treats patterns admitted by the numerical semigroup $D_{(r,k)}$, like the following family of patterns, first presented in [**11**].

THEOREM 3.1. *Let $D_{(r,k)}$ be a numerical semigroup associated with the (r, k)-configurations. Then $D_{(r,k)}$ admits the pattern*

$$X_1 + X_2 - n$$

for all $n \in \{1, \ldots, \gcd(r, k)\}$.

The proof of Theorem 3.1 can be found in [**11**] and is based on a construction of a configuration with reduced parameters $(d_1 + d_2 - n, r, k)$ from two configurations with reduced parameters (d_1, r, k) and (d_2, r, k), respectively. Figure 3 illustrates this construction for $(r, k) = (3, 5)$. In this case $\gcd(r, k) = 1$, so the only possible choice of n is $n = 1$.

Here we give further examples of patterns admitted by the numerical semigroup $D_{(r,k)}$.

THEOREM 3.2. *The numerical semigroup $D_{(r,k)}$ admits the pattern $X_1 + \cdots + X_n + 1$ with $n = rk/\gcd(r, k)$.*

PROOF. Take n combinatorial configurations $C_1, \ldots, C_n$ with reduced parameter sets $(d_1, r, k), \ldots, (d_n, r, k)$. On each of these configurations C_i, consider a pair of a point and a line (p_i, l_i) such that p_i and l_i are incident and remove this incidence. Instead let the n lines l_i all meet in sets of r in $k/\gcd(r, k)$ new points $p'_1, \ldots, p'_{k/\gcd(r,k)}$, and join the n points in sets of k over $r/\gcd(r, k)$ new lines $l'_1, \ldots, l'_{r/\gcd(r,k)}$. It is clear that this construction gives r lines through all points from the original configurations, as well as through the new points. Also, there

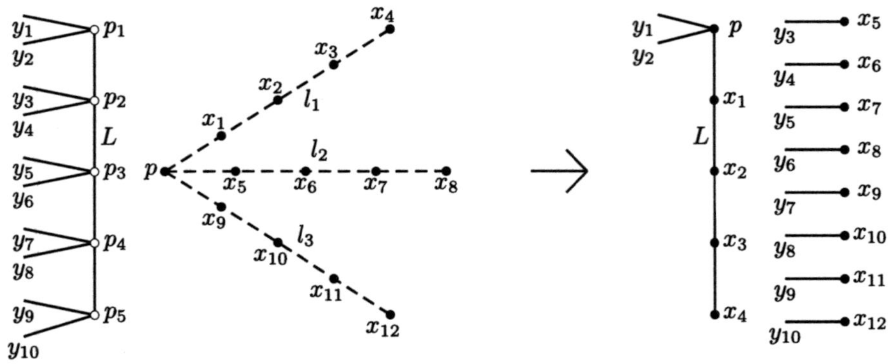

FIGURE 3. An example of the construction implying the linear pattern $X_1 + X_2 - 1$ admitted by $D_{(3,5)}$. The white points and the dashed lines in the two (partially drawn) configurations on the left are removed and the corresponding points and lines in the resulting configuration is shown on the right.

are k points on all lines from the original configurations, as well as on the new lines. Since the point and line pairs all come from different configurations, it is clear that there are at most one line through each pair of points. It is not obvious that the resulting incidence geometry is connected, however, if the result was not connected, then it can be made connected by swapping two incidences as was previously described in Section 2 and illustrated in Figure 2.

For each $i \in \{1, \ldots, n\}$, the original configuration C_i had associated integer d_i and $v_i = d_i k / \gcd(r,k)$ points and $b_i = d_i r / \gcd(r,k)$ lines, so the new incidence structure is clearly a configuration with parameters $(v_1 + \cdots + v_n + k/\gcd(r,k), b_1 + \cdots + b_n + r/\gcd(r,k), r, k)$, so that it has reduced parameters $(d_1 + \cdots + d_n + 1, r, k)$. Therefore, the numerical semigroup $D_{(r,k)}$ admits the pattern $X_1 + \cdots + X_n + 1$. $\square$

We are interested in characterizing how the points and the lines that were used in the construction of Theorem 3.2 should be chosen, if we dropped the assumption that they belong to different configurations. In this way, it would be possible to reduce the number of variables in the corresponding pattern. The following is a first step in this direction.

DEFINITION 3.3 (Point-line pair sets with Property A). Let $\mathcal{C}^n = \bigcup_{i=1}^n C_i$ be a disjoint union of n (d_i, r, k)-configurations with union of points $\mathcal{P}^n = P_1 \cup \cdots \cup P_n$ and lines $\mathcal{L}^n = L_1 \cup \cdots \cup L_n$. A pair of a subset of points $P \subseteq \mathcal{P}^n$ and a subset of lines $L \subseteq \mathcal{L}^n$ has Property A if

- there is a bijection $f : P \to L$ such that the image of a point p is a line $f(p)$ through that point,
- two points $p, q \in P$ either are not collinear, or are collinear only on the line $f(p)$ or on the line $f(q)$, and
- two lines $l, m \in L$ either do not meet, or meet only in the point $f^{-1}(l)$ or in the point $f^{-1}(m)$.

It is clear that the point-line pair sets consisting of the pairs (p_i, l_i) in Theorem 3.2 has Property A, since no pair of points nor pair of lines belong to the same original configurations. In general, ensuring Property A for all (r,k)-configurations is a sufficient requirement for the pattern $X_1 + \cdots + X_n + 1$ also when $n < rk/\gcd(r,k)$, that is, when several points and lines are taken from the same configuration.

THEOREM 3.4. *Assume that there exists a disjoint union of n configurations $\mathcal{C}^n = (\mathcal{P}^n, \mathcal{L}^n)$ with reduced parameters (d_i, r, k) for $i = 1, \ldots, n$, such that there are point-line pair sets $(P, L) \subseteq (\mathcal{P}^n, \mathcal{L}^n)$ satisfying Property A. Then there also exists a configuration with reduced parameters $(d_1 + \cdots + d_n + 1, r, k)$.*

PROOF. Replace each incidence $(p, f(p))$ with

- the incidence $(q, f(p))$, where q is one of $k/\gcd(r,k)$ new points, so that through each of these points q there pass exactly r lines, and
- the incidence (p, l), where l is one of $r/\gcd(r,k)$ new lines, so that each new line contains exactly k points from P.

Clearly, the resulting incidence geometry has $v_1 + \cdots + v_n + k/\gcd(r,k)$ points and $b_1 + \cdots + b_n + r/\gcd(r,k)$ lines, where $v_i = d_i k/\gcd(r,k)$ and $b_i = d_i r/\gcd(r,k)$ is the number of points and lines of the ith configuration in $\mathcal{C}^n$, respectively. Also, there are r lines through each point, and there are k points on every line.

It is also true that through any two points there is at most one line. Indeed, if two of the original points were on two lines, then one line would be old and the other would be new. But two points p and q can only be collinear on a new line if both were points in P, and if they are also on an old line then they were collinear in $\mathcal{C}^n$. However, if $p, q \in P$ were collinear in $\mathcal{C}^n$ then they were so either on the line $f(p)$ or on the line $f(q)$, and the incidences $(p, f(p))$ and $(q, f(q))$ were removed in the construction. Also, two new points cannot be collinear twice. Indeed, since the lines that go through the new points are all different, two new points cannot even be collinear once.

It can also be proven that any two lines meet in at most one point. Indeed, suppose that two lines meet in at least two points. If the two lines belonged to the original configurations, then one of the points in which they meet would be old, and the other new. But two lines that meet in a new point must both belong to L, and if $l, m \in L$ are also collinear in an old point, that is, if they were collinear in $\mathcal{C}^n$, then they were so either in $f^{-1}(l)$ or in $f^{-1}(m)$, but the incidences $(f^{-1}(l), l)$ and $(f^{-1}(m), m)$ were removed in the construction. Finally, all points on the new lines are different, so the new lines do not meet.

Consequently, the construction gives a combinatorial configuration with reduced parameters $(d_1 + \cdots + d_n + 1, r, k)$. □

Theorem 3.4 allows us to select some of the lines and points in Theorem 3.2 from the same configuration. In this way, it is possible to prove that the numerical semigroup $D_{(r,k)}$ admits a pattern involving fewer variables. It can be observed that Property A is actually more than is needed, at least for $r \neq k$. Indeed, two points in P that are assigned different new lines in the construction can be allowed to be collinear in $\mathcal{C}^n$, and the construction would still work without removing these incidences. Similarly, two lines in L that are assigned different new points in the construction can be allowed to meet in $\mathcal{C}^n$, and the construction would still work without removing these incidences. The exact requirements for the construction to work are described here below as Property A'.

DEFINITION 3.5 (Point-line pair sets with Property A'). Let $\mathcal{C}^n = \cup_{i=1}^n C_i$ be a disjoint union of n (d_i, r, k)-configurations with union of points $\mathcal{P}^n = P_1 \cup \cdots \cup P_n$ and lines $\mathcal{L}^n = L_1 \cup \cdots \cup L_n$. A pair of a subset of points $P \subseteq \mathcal{P}^n$ and a subset of lines $L \subseteq \mathcal{L}^n$ has property A if

- there is a bijection $f : P \to L$ such that the image of a point p is a line $f(p)$ through that point,
- P can be partitioned into $r/\gcd(r,k)$ parts, each of cardinality k, such that two points p and q in each part, either are not collinear, or are collinear only on the line $f(p)$ or on the line $f(q)$, and
- L can be partitioned into $k/\gcd(r,k)$ parts, each of cardinality r, such that two lines l and m in each part either do not meet, or meet only in the point $f^{-1}(l)$ or in the point $f^{-1}(m)$.

Observe that when $r = k$, then we have $r/\gcd(r,k) = k/\gcd(r,k) = 1$, so Property A and Property A' are equivalent.

Theorem 3.4 leads to several questions, of which we believe that the following two are of particular interest.

(1) What is the smallest n such that for any (r, k), there is always a union of configurations $\mathcal{C}^n$ such that there is a pair of subsets (P, L) with property A?
(2) When is it possible to set $n = 1$?

As an example of possible implications of the answers to these two questions, we note that Theorem 3.4 implies the well-known result stated below as Corollary 3.6 [**6**, **7**]. Several alternative proofs exist, of which perhaps the one that uses cyclic configurations is the simplest. For an illustration of the triangle construction described in Corollary 3.6 see Figure 4.

COROLLARY 3.6. There is a combinatorial $(d, 3, 3)$-configuration for all $d \geq 7$.

PROOF. For $r = k = 3$ we have $n = rk/\gcd(r,k) = 3$. Given a $(d, 3, 3)$-configuration with triangles, the three points P and the three lines L on a triangle satisfy Property A'. Given a $(d, 3, 3)$-configuration without triangles, consider a path starting at a point p of three lines L intersecting in two points p', p''. Then $P = \{p, p', p''\}$ and L satisfy Property A'. Therefore Theorem 3.4 implies that from any $(d, 3, 3)$ configuration we can construct a $(d+1, 3, 3)$-configuration. The smallest $(d, 3, 3)$-configuration is the Fano plane, with $d = 7$, and the result follows. □

Observe that the parameters $r = k = 3$ of Corollary 3.6 were not chosen arbitrarily. The construction in Theorem 3.4 can in general not be applied to a single (d, r, k)-configuration when d is small and $r, k > 3$. For given r, k, the smallest possible configurations are the linear spaces, when they exist. In a linear space all points are collinear, and this obstructs the existence of more than 3 points with Property A'. However, this does not imply that the associated numerical semigroup $D_{(r,k)}$ does not admit the pattern $X + 1$. Indeed, the numerical semigroup $D_{(4,4)}$ admits the pattern $X + 1$, but it is not possible to use Theorem 3.4 to construct a $(14, 4, 4)$-configuration from the unique configuration with parameters $(13, 4, 4)$, the projective plane over $\mathbb{F}_3$.

We may now improve the upper bound of the conductor of $D_{(r,k)}$ that was given in [**11**], using our new knowledge of the set of patterns admitted by this numerical semigroup.

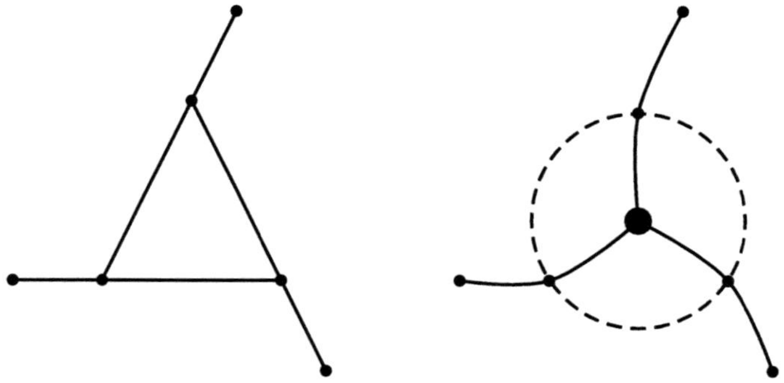

FIGURE 4. The new (thick) point and the new (dashed) line of a $(d+1,r,r)$-configuration to the right, added to a triangle in the original $(d,3,3)$-configuration to the left.

THEOREM 3.7. *Let S be a numerical semigroup with multiplicity m and conductor c. Let M be the maximum integer such that S admits the pattern x_1+x_2-a for $a \in \{1,\dots,M\}$. Let N be the minimum integer such that S admits $x_1+\ldots+x_N+1$. Then $c \le \left\lceil \frac{m-1}{1/N+M} \right\rceil (m-M)+M$.*

PROOF. The intervals

$$I_x = \left[(x+1)m - xM, (x+1)m + \left\lfloor \frac{(x+1)}{N} \right\rfloor\right]$$

belong to S. If there is a gap between I_x and I_{x-1}, then $(x+1)m - xM > xm + \lfloor \frac{x}{N} \rfloor + 1$, so that $m-1 > \lfloor \frac{x}{N} \rfloor + xM$ and $m-1 > \frac{x}{N} + xM$. So, the largest x for which I_x and I_{x-1} are separated by at least one gap is at most $\left\lceil \frac{m-1}{1/N+M} \right\rceil - 1$. Therefore, the conductor must be at least the first element of I_x with $x = \left\lceil \frac{m-1}{1/N+M} \right\rceil - 1$, that is, we have $c \le \left\lceil \frac{m-1}{1/N+M} \right\rceil (m-M)+M$. □

In particular, if $m = M$ then $c = m$, so that the numerical semigroup is ordinary, implying that $N = 1$. Note also that necessarily $M \le m$.

4. Nonlinear patterns and pattern composition

We have given patterns that are admitted by the numerical semigroups $D_{(r,k)}$ associated with combinatorial configurations. So far, these patterns have all been linear, although not always homogeneous. Curiosity may invoke the question whether there exist numerical semigroups that admit patterns of higher degrees. It can be argued that a numerical semigroup is essentially a linear structure, so that the most interesting patterns should be linear. However, it is certainly true that there exists nonlinear patterns admitted by numerical semigroups. Indeed, by definition, $X_1 + X_2$ is a pattern admitted by any numerical semigroup, implying that $X_1 \cdots X_n = \sum_1^{X_1} \cdots \sum_1^{X_{n-1}} X_n$ is also a pattern, and in extension, any polynomial $p(X_1,\dots,X_n)$ with positive integer coefficients such that $p(0,\dots,0) = n \in S$ is a pattern admitted by any numerical semigroup S. Observe that a numerical semigroup S allows the constant pattern n if and only if n is in S.

More generally, the sum of two patterns is always a pattern, and so is the product of two patterns. Indeed, if $p(X_1, \ldots, X_n) = s_1 \in S$, and $q(X_1, \ldots, X_m) = s_2 \in S$, then also $s_1 + s_2 \in S$ and also $as_1 \in S$ for any $a \in \mathbb{N}$, so in particular, $s_1 s_2 \in S$. For numerical semigroups associated with combinatorial configurations, the natural way of proving that the semigroup admits a pattern is to provide a construction that implies the existence of the pattern. The pattern $X_1 + X_2$ corresponds to the union construction described in Figure 2. As just described above, higher degree polynomial patterns can be obtained through repetition of this construction. Also other types of constructions that give nonlinear patterns can be imagined.

5. Conclusions

We have given new patterns admitted by the numerical semigroups associated with combinatorial configurations. The existence of these patterns is proven through recursive constructions of configurations. We have also identified the requirements for patterns of the form $X_1 + \cdots + X_n + 1$ in these numerical semigroups and improved the upper bound of the conductor: the integer c such that there exists a combinatorial configuration with parameters (d, r, k) for any $d \geq c$.

This article only treats combinatorial configurations. A geometric treatment of this subject would require additional tools.

References

[1] M. Bras-Amorós and P. A. García-Sánchez, *Patterns on numerical semigroups*, Linear Algebra Appl. **414** (2006), no. 2-3, 652–669, DOI 10.1016/j.laa.2005.11.007. MR2214415 (2007b:20129)

[2] M. Bras-Amorós, P. A. García-Sánchez, and A. Vico-Oton, *Nonhomogeneous patterns on numerical semigroups*, Internat. J. Algebra Comput. **23** (2013), no. 6, 1469–1483, DOI 10.1142/S0218196713500306. MR3109453

[3] M. Bras-Amorós and K. Stokes, *The semigroup of combinatorial configurations*, Semigroup Forum **84** (2012), no. 1, 91–96, DOI 10.1007/s00233-011-9343-5. MR2886000

[4] A. A. Davydov, G. Faina, M. Giulietti, S. Marcugini, F. Pambianco. *On constructions and parameters of symmetric configurations v_k*, Arxiv:1203.0709v1 (2012).

[5] I. V. Dolgachev, *Abstract configurations in algebraic geometry*, The Fano Conference, Univ. Torino, Turin, 2004, pp. 423–462. MR2112585 (2005k:14091)

[6] H. Gropp. "Configurations." In *The handbook of combinatorial designs*, 2nd. ed., Discrete Mathematics and its Applications, Chapman & Hall/CRC, Boca Raton, FL, 2007, pp. 352–355. Edited by C. J. Colburn and J. H. Dinitz. MR2246267 (2007i:05001)

[7] B. Grünbaum, *Configurations of points and lines*, Graduate Studies in Mathematics, vol. 103, American Mathematical Society, Providence, RI, 2009. MR2510707 (2011j:52001)

[8] J. C. Rosales and P. A. García-Sánchez, *Numerical semigroups*, Developments in Mathematics, vol. 20, Springer, New York, 2009. MR2549780 (2010j:20091)

[9] T. Pisanski and B. Servatius, *Configurations from a graphical viewpoint*, Birkhäuser Advanced Texts: Basler Lehrbücher. [Birkhäuser Advanced Texts: Basel Textbooks], Birkhäuser/Springer, New York, 2013. MR2978043

[10] K. Stokes. *Combinatorial structures for anonymous database search*, PhD thesis, Universitat Rovira i Virgili, 2011.

[11] K. Stokes and M. Bras-Amorós, *Linear, non-homogeneous, symmetric patterns and prime power generators in numerical semigroups associated to combinatorial configurations*, Semigroup Forum **88** (2014), no. 1, 11–20, DOI 10.1007/s00233-013-9493-8. MR3164148

LINKÖPINGS UNIVERSITET, SE-581 83 LINKÖPING, SWEDEN
Current address: Universitat Oberta de Catalunya, IN3, Barcelona, Catalonia, Spain
E-mail address: kstokes@uoc.edu

DEPARTMENT OF COMPUTER ENGINEERING AND MATHEMATICS, UNIVERSITAT ROVIRA I VIRGILI, AVINGUDA PAÏSOS CATALANS 26, 43007 TARRAGONA, CATALONIA, SPAIN
E-mail address: maria.bras@urv.cat

Computing equations, automorphisms and symmetries of Riemann surfaces

Peter Turbek

For Emilio, on his sixtieth birthday

ABSTRACT. We discuss techniques that will enable researchers who use Fuchsian and NEC groups to determine automorphisms and symmetries of Riemann surfaces, to find explicit equations for some of the surfaces they construct. We also illustrate methods for determining the number of ovals fixed by a symmetry in particular cases.

1. Introduction

Given a compact Riemann surface X of genus $g \geq 2$, there exists a surface Fuchsian group $\Gamma < \mathrm{PSL}(2, \mathbb{R})$ such $\mathcal{H}/\Gamma$, the orbit space of the upper half plane $\mathcal{H}$ under the action of Γ, is biholomorphic to X. Each such surface can also be defined by an equation $F(x, y) = 0$, where $F(x, y) \in \mathbb{C}[x][y]$, unfortunately given an arbitrary X defined as $\mathcal{H}/\Gamma$, determining a corresponding defining equation is a nearly intractable problem. The situation improves slightly if the surface has a solvable group of automorphisms which yield an orbit space of the Riemann sphere. In particular, equations have been found for a variety of surfaces that possess automorphism groups that are maximal in some sense, such as the Accola-Maclachlan or Kulkarni surfaces, and explicit equations for their automorphisms are known.

The purpose of this manuscript is to discuss techniques that will enable researchers who use Fuchsian and NEC groups to determine automorphisms and symmetries of Riemann surfaces, to find explicit equations for some of the surfaces they construct. It also illustrates methods for determining the number of ovals fixed by a symmetry in particular cases.

There are many references for the classical algebraic geometry discussed in this paper, including **[3]** and **[5]**. Examples of specific Riemann surfaces with explicit defining equations and formulas for their automorphisms and symmetries are given in **[1]**. There are also computer programs that can be used facilitate calculations with Riemann surfaces, see **[2]**.

2010 *Mathematics Subject Classification.* Primary 20H10, 30F50, Secondary 14P .

2. Results concerning defining equations

Suppose $F(x, y) = 0$ is an irreducible polynomial that is the defining equation of a Riemann surface X. There are two points of view you can take. The first is to view X as the set of all ordered pairs (a, b) with the property that $F(a, b) = 0$; however we must make adjustments for singular points of F and the point at ∞. The solutions of F, with these modifications, yield a Riemann surface. The other point of view is to think of x and y as elements of $\mathbb{C}(X)$, the field of meromorphic functions on X. If we do this, then $\mathbb{C}(X) = \mathbb{C}(x, y)$ and we can view F as a polynomial in y over the rational function field $\mathbb{C}(x)$, in other words, $F(x, y) \in \mathbb{C}(x)[y]$. In addition, if F is monic when considered as a polynomial in y with coefficients in $\mathbb{C}(x)$, then $F(x, y)$ is the minimal polynomial for y over $\mathbb{C}(x)$. If we take this view, we can use results from Galois theory concerning field extensions. Note that if we determine functions $t \in \mathbb{C}(x, y)$ and $w \in \mathbb{C}(x, y)$ with the property that $\mathbb{C}(t, w) = \mathbb{C}(x, y)$, then we can determine the minimal polynomial for w over $\mathbb{C}(t)$. If $G(w, t)$ is the minimal polynomial, then $G(w, t)$ defines the same Riemann surface, since the function fields $\mathbb{C}(x, y)$ and $\mathbb{C}(t, w)$ are the same. We can think of this new equation as giving a different view of the same Riemann surface, in particular, it may give us a better view if points which were singular for $F(x, y)$ are now nonsingular for $G(w, t)$.

Suppose the point (a, b) is a solution of $F(x, y) = 0$. Recall that a point (a, b) of X is nonsingular if one of $F_x(a, b) = \partial F/\partial x(a, b)$ or $F_y(a, b) = \partial F/\partial y(a, b)$ is nonzero, and the point is singular if $F_x(a, b) = F_y(a, b) = 0$. If $F_x(a, b) \neq 0$, then $y - b$ is a local parameter at (a, b) and if $F_y(a, b) \neq 0$ then $x - a$ is a local parameter. We explain exactly what we mean by this. For simplicity, suppose the point $(a, b) = (0, 0)$ and suppose $F_y(0, 0) \neq 0$. Let $G(x)$ consist of all the terms of F that do not contain a y and let $H(x, y)$ consist of all the terms of F that contain a power of y. Then we can rewrite $0 = F(x, y) = yH_1(x, y) + x^k G_1(x)$, where $yH_1 = H$, $x^k G_1(x) = G(x)$ and $k \geq 1$ is the highest power of x that divides $G(x)$. Since $F_y(0, 0) \neq 0$, we must have that $H_1(0, 0) \neq 0$. In addition, since k is the highest power of x that divides $G(x)$, we know that $G_1(0) \neq 0$. Therefore

$$y = -x^k \frac{G_1(x)}{H_1(x, y)}, \quad \text{and} \quad -\frac{G_1(0)}{H_1(0, 0)} \neq 0.$$

This means that y can be expressed in the form $y = x^k m$ for a meromorphic function m that does not have a zero or pole at $(0, 0)$ and that $\text{ord}(y) = k\text{ord}(x)$ at $(0, 0)$. The set of meromorphic functions that are defined at $(0, 0)$ (in other words, that do not have a pole at $(0, 0)$) forms a Discrete Valuation Ring and the ideal of functions that have a zero at $(0, 0)$ is generated by (x), since y can be expressed in terms of x. For this to occur, it means that $\text{ord}(x) = 1$ at $(0, 0)$ and we say that x is a local parameter. At the point $(0, 0)$, the Riemann surface X is locally biholomorphic to x in a neighborhood of $x = 0$, so the variable x gives a completely accurate picture of the surface X near $(0, 0)$. We chose the point $(0, 0)$ for simplicity, the same results are true for any nonsingular point (a, b) on the surface. In summary, if $F_x(a, b) \neq 0$, then $y - b$ is a local parameter at (a, b) and if $F_y(a, b) \neq 0$ then $x - a$ is a local parameter.

If (a, b) is a singular point, then there will be one or more points on the Riemann surface lying over (a, b) and possibly neither $x - a$ nor $y - b$, will be a local parameter. If the singular point is of interest to us, we can try to determine a function t with the property that $\text{ord}(t) = 1$ at (a, b). For example, if $\text{ord}(y - b) = 5$ and $\text{ord}(x - a) = 2$, then $t = (y-b)/(x-a)^2$ is a local parameter, and since $\mathbb{C}(t,x) = \mathbb{C}(x,y)$ we could write the defining equation for X in terms of t and x instead. We will see examples of this in greater detail later.

We discuss several basic results concerning defining equations and meromorphic function fields of Riemann surfaces, focusing on cyclic branched covers of the Riemann sphere.

(1) Assume $y^n - r(x)/s(x)$ is a defining equation of X, where $r(x)$ and $s(x)$ are relatively prime polynomials. Then there is a defining equation for X of the form $y_1^n - f(x)$, where $f(x)$ is a polynomial. One such equation is obtained by defining $y_1 := s(x)y$ and $f(x) = r(x)s^{n-1}(x)$.
(2) Let $y^n - (x-a)^m f(x)$ be a defining equation for X, where $(x-a)$ and $f(x)$ are relatively prime polynomials and $m \geq n$. Assume $m = nq + r$, where $0 \leq r < n$. Then X has a defining equation of the form $y_1^n - (x-a)^r f(x)$, by defining $y_1 := y/(x-a)^q$.
(3) Assume X has a defining equation of the form $y^n - (x-a)^{dm} f(x)$, where $\gcd(n, dm) = d$, $f(x)$ is a polynomial relatively prime to $(x-a)$ and assume there exist integers u and v such that $mu + vn/d = 1$ and $(u,n) = 1$. Then there is a defining equation for X of the form $y_1^n - (x-a)^d f_1(x)$. One such equation can be obtained by noting that $y^{nu} = (x-a)^{dmu} f^u(x) = (x-a)^{d-nv} f^u(x)$. Therefore we define $y_1 = (x-a)^v y^u$ and $f_1(x) = f^u(x)$. Since $(u,n) = 1$, y can be expressed in terms of y_1 and x, thus $\mathbb{C}(x, y_1) = \mathbb{C}(x,y) = \mathbb{C}(X)$.
(4) Assume $F(x,y)$ is the defining equation of a Riemann surface, in particular, assume F is irreducible. Suppose $F(x,y) = a_n(x)y^n + a_{n-1}(x)y^{n-1} + \ldots + a_1(x)y + a_0(x)$. Then for an arbitrary number a, there are n points (counting multiplicities) on the Riemann surface at which the meromorphic function $x-a$ has a zero. Therefore the degree of the zero divisor of $x-a$ is n. If we write F in the form $F(x,y) = a_m(y)x^m + a_{m-1}(y)x^{m-1} + \ldots + a_1(y)x + a_0(y)$, then the zero divisor of the meromorphic function $y-b$ has degree m.
(5) Let P be a point on the Riemann surface X. The set of meromorphic functions defined at P forms a discrete valuation ring and the ideal of functions that vanish at P is principally generated by any local parameter t. Every nonzero element of $w \in \mathbb{C}(X)$ has an order; if $\operatorname{ord}(w) = k > 0$, w has a zero of order k at P, if $\operatorname{ord}(w) = 0$, then $w(P) \neq 0$ and if $\operatorname{ord}(w) = k < 0$ then w has a pole of order k at P. For nonzero functions w and v, $\operatorname{ord}(wv) = \operatorname{ord}(w) + \operatorname{ord}(v)$ and if the orders of w and v are different at P, then $\operatorname{ord}(w+v) = \min(\operatorname{ord}(w), \operatorname{ord}(v))$. This implies that if $w^n - (x-a)^r f(x) = 0$ on X and if $x-a$ is relatively prime to $f(x)$ then $n\operatorname{ord}_P(w) = r\operatorname{ord}_P(x-a)$, at any point P that lies over $x-a$.
(6) Let $y^n - (x-a)^{dm} f(x) = 0$ be a defining equation for X where $d = \gcd(n, dm)$ and $f(x)$ is a polynomial that is relatively prime to $x-a$. We will show this implies that there exists d points of X which lie over the point $(a,0)$ and we will determine a change of coordinates which separates each of the points and yields a local parameter at each point. To accomplish this, note that $y^n/(x-a)^{dm} - f(x) = 0$ and that $f(a) \neq 0$. Define $w := y^{n/d}/(x-a)^m$, so $w^d - f(x) = 0$ and at any point P of X lying over $(a,0)$, we have $w^d(P) = f(a) \neq 0$. The function w has d different possible values when $x = a$, so there must be d different points of X at which w takes on these values, which means there are at least d points of X that lie over $(a,0)$. However, if P is any point of X lying over $x = a$, then $\operatorname{ord}_P(w) = 0$, therefore $(n/d)\operatorname{ord}_P(y) = m\operatorname{ord}_P(x-a)$.

Therefore, at each such point P, $\mathrm{ord}_P(x-a)$ is at least n/d. Since the degree of the zero divisor of $x-a$ is n, this means there must be exactly d points lying over $(a,0)$, that $x-a$ has order n/d at each of them, and y has order m at each of them. We now determine a local parameter at the point P. First note that if $m=1$, then y is a local parameter. For the general case, let u and v be integers such that $mu+vn/d=1$. Then $t:=y^u(x-a)^v$ has order $mu+vn/d=1$, so t is a local parameter. We now show that $\mathbb{C}(w,t)=\mathbb{C}(x,y)$, so X can be defined in terms of w and t. Note that $t^m w^v = y^{um+vn/d}(x-a)^{vm-vm}=y$, so $(x-a)^m$ and $(x-a)^v$ are elements of $\mathbb{C}(w,t)$. Therefore $(x-a)^{um+vn/d}=(x-a)\in\mathbb{C}(w,t)$, so $\mathbb{C}(w,t)=\mathbb{C}(x,y)$. In fact, one can check that $t^{n/d}/w^u = x-a$. Since $w^d - f(x)=0$, we obtain a defining equation of the form $w^d - f(a+t^{n/d}/w^u)$. At any of the points $(t,w)=(0,e^{2\pi ik/d})$ that lie over the point $(x,y)=(a,0)$, we note that $f(a+t^{n/d}/w^u)=f(a)\neq 0$.

(7) We continue with the previous item under the assumption that $y^n-(x-a)^{dm}f(x)$ is a real equation, so that complex conjugation, denoted by σ, is a symmetry of X. A natural question to ask is how many ovals are fixed by σ. If n is even, then for each real number r with $(r-a)^{dm}f(r)>0$, there are two solutions for y, say (r,s^+) and (r,s^-) where the value of s is positive and negative respectively. Suppose $r>a$, $(r-a)^{dm}f(r)>0$, and there is no root of $f(x)$ between r and a. In order to determine the behavior of the real solutions near $x=a$, we make the change of variables $t:=y^u(x-a)^v$ and $w:=y^{n/d}/(x-a)^m$ indicated in the previous item. Recall that $w^d-f(x)=0$, so there are d points of X lying over $x=a$. If d is odd, then there is only one real solution, however if d is even, then there are two solutions, say Q^+ and Q^- where w is positive or negative respectively. Since t is a local parameter, X is locally biholomorphic to a neighborhood of $t=0$ at a point P that lies over $x=a$. As r decreases to a, the points $(x,y)=(r,s^+)$ correspond to points (t,w) where $t>0$ and $w>0$ that must approach the point Q^+, because $w>0$. At Q^+, $t=0$, but w remains positive in a neighborhood of $t=0$. Since fixed points under a symmetry form ovals (and do not have endpoints), the oval must continue to points of the form (t,w) where $t<0$ and $w>0$. Translating this back to coordinates (x,y), we see that $t=y^u(x-a)^v<0$, and $w=y^{n/d}/(x-a)^m>0$, and combining this with $mu+vn/d=1$ we can determine the signs of $(x-a)$ and y, which will indicate whether (r,s^+) joins with (r,s^-), or for example, (r,s^+) joins with points of the form (r,s^+) or (r,s^-) where now $r<a$. The examples in the third part of the paper will illustrate this technique.

(8) Suppose $y^n-x^kf(x)=0$ is the defining equation of X where $f(x)$ has degree m, $f(x)$ is not divisible by x and let $d=\gcd(m+k,n)$. Then there are d points of X lying over $x=\infty$. To see this, define $t=1/x$ and note that when $x=\infty$, then $t=0$. We obtain

$$y^n-\frac{1}{t^k}f\left(\frac{1}{t}\right)=0,\ \text{so}\ \left(y^{\frac{n}{d}}t^{\frac{m+k}{d}}\right)^d - t^m f(1/t) = w^d - g(t) = 0,$$

where $w=y^{\frac{n}{d}}t^{\frac{m+k}{d}}$, $g(m)=t^mf(1/t)$ is a polynomial in t and $g(0)\neq 0$. Note that the function w has d possible values when $t=0$, this means there must be d points of X lying over $x=\infty$.

(9) We can always choose the coordinates of three arbitrary points on the Riemann sphere. For example, suppose the Riemann surface X is associated to a surface Fuchsian group Γ and X admits a cyclic group of automorphisms G of order n. Suppose the Fuchsian group Λ associated to X/G has signature $(0; [m_1, m_2, \ldots, m_k])$, where for each i, $m_i | n$ and there is a canonical elliptical generator e_i or order m_i. We can choose e_0, e_1, and e_2 to fix $x = 0$, $x = 1$ and $x = -1$ respectively, for example. Once this is done, we have no control over the coordinates of the remaining points, so we assume e_i fixes a_i for some $a_i \in \mathbb{C}$. For each ramification index m_i, define $d_i = n/m_i$, so there are d_i points of X lying over a_i. Therefore X has a defining equation of the form

$$y^n - x^{d_1 r_1}(x-1)^{d_2 r_2}(x+1)^{d_3 r_3}(x-a_4)^{d_4 r_4} \ldots (x-a_k)^{d_k r_k},$$

where each r_i is relatively prime to n. Using (3) above, we possibly can choose one of the r_i to equal 1, for example we possibly can rewrite the equation as

$$y^n - f(x) =$$
$$y^n - x^{d_1}(x-1)^{d_2 r_2}(x+1)^{d_3 r_3}(x-a_4)^{d_4 r_4} \ldots (x-a_k)^{d_k r_k},$$

where the values of the r_i's in the two equations are probably different. Once this is done, you have no control over the remaining r_i's; different values will most likely result in non-biholomorphic Riemann surfaces. Note that, since $x = 0, 1, -1, a_3, \ldots, a_k$ are the only points of the Riemann sphere that are ramified in X, then $x = \infty$ is not ramified. This means that the degree of $f(x)$ must be divisible by n; the choices of the r_i must ensure that this occurs.

If we want to choose the three points on the Riemann sphere that are ramified in X to be $x = 0, 1, \infty$, then we can choose $x = \infty$ to have ramification index m_1, $x = 0$ to have ramification index m_2 and $x = 1$ to have ramification index m_3. Then a defining equation for X is

$$y^n - f(x) =$$
$$y^n - x^{d_2 r_2}(x-1)^{d_3 r_3}(x-a_4)^{d_4 r_4} \ldots (x-a_k)^{d_k r_k},$$

where $d_i = n/m_i$ and the the greatest common divisor of n and the degree of f is d_1.

(10) Assume the Riemann surface Y is associated to a Fuchsian group Λ, $\Gamma \lhd \Lambda$ and X is a Riemann surface associated to Γ. Assume that $[\Lambda : \Gamma] = n$ and $e \in \Lambda$ is an elliptic element of order m that fixes the point P of Y. If e and no nonidentity powers of e are contained in Γ, then P has ramification index m in the covering of Y by X. This means that there are n/m points of X lying over P. If k is the smallest positive power of e such that $e^k \in \Gamma$, then the ramification index at P is k and there are n/k points of X lying over P.

(11) Suppose $F(x, y) = y^n - f(x)$ is a defining equation of X, where f is a polynomial. Assume that σ is an automorphism of $\mathbb{C}(x, y)$ of order k, $\sigma(x)$ is a rational function of x, and suppose we want to determine an equation $F(y, t)$ for the Riemann surface where t is fixed by σ. This is equivalent to determining a function $t \in \mathbb{C}(x, y)$ such that $\mathbb{C}(x, y) = \mathbb{C}(y, t)$ and t is fixed by σ. A simple case is when $\sigma(x) = 1/x$, which arises quite frequently. Note that y is a solution of the polynomial $T^n - f(x) \in \mathbb{C}(x)[T]$ and you can let σ act on the coefficients of this polynomial. Then

$$\hat{F} = (T - f(x))(T - f(\sigma(x)))(T - f(\sigma^2(x))) \cdots (T - f(\sigma^{k-1}(x)))$$

is an element of $\mathbb{C}(x)[T]$ and the coefficients of $\hat{F}$ are fixed by σ. Many times you can choose $t = x + \sigma(x) + \cdots + \sigma^{k-1}(x)$ and the coefficients of $\hat{F}$ can be expressed in terms of rational functions of t.

(12) If X is a Riemann surface and σ is an automorphism of X, then there is an induced automorphism σ^* of the function field $\mathbb{C}(X)$ defined by $\sigma^*(g) = g \circ \sigma$ for each $g \in \mathbb{C}(X)$. Every automorphism of $\mathbb{C}(X)$ is obtained in this way. If the Riemann surface is expressed as the points satisfying an irreducible polynomial $F(x,y) = 0$ and $\sigma(x,y) = (f_1(x,y), f_2(x,y))$, then the induced function σ^* is defined by $\sigma^*(g(x,y)) = g(f_1(x,y), f_2(x,y))$ for each meromorphic function $g(x,y) \in \mathbb{C}(X)$. In particular, we have the following. Suppose $\sigma(P) = Q$ on the Riemann surface X and suppose g is a rational function that satisfies $g(Q) = 0$. Then $\sigma^*(g)(P) = g \circ \sigma(P) = g(Q) = 0$, so $\sigma^*(g)$ has a zero at P. Therefore, if a function $g \in \mathbb{C}(X)$ has a zero at Q, then $\sigma^*(g)$ has a zero at P; in addition, the orders of the zeros are the same. Similarly, if $g \in \mathbb{C}(X)$ has a pole at Q, then $\sigma^*(g)$ has a pole at P of the same order. Suppose, instead, that we begin with an automorphism $\sigma^* \in \mathbb{C}(X)$, say $\sigma^*(x) = r_1(x,y)$ and $\sigma^*(y) = r_2(x,y)$. In this case, the corresponding automorphism σ of X is given by $\sigma(x,y) = (r_1(x,y), r_2(x,y))$.

In dealing with Klein surfaces, we can also have the symmetry σ given by complex conjugation. What is the corresponding automorphism σ^* in this case? We again use the fact that if $\sigma(P) = Q$ and $g(x,y) \in \mathbb{C}(X)$ has a zero at Q then $\sigma^*(g(x,y))$ has a zero at P. Suppose $F(x,y)$ is nonsingular at the point $P = (a+bi, c+di)$, and assume F is also nonsingular at $Q = \sigma(P) = (a-bi, c-di)$. Consider the functions $x-(a-bi)$ and $y-(c-di)$ in $\mathbb{C}(X)$ which have zeros at Q. Then $\sigma^*(x-(a-bi))$ and $\sigma^*(y-(c-di))$ must have zeros at P. Since this must happen at each nonsingular point, we conclude that $\sigma^*(x) = x$, $\sigma^*(y) = y$, and $\sigma^*(i) = -i$.

In what follows, we will not use the notation σ^*, but will refer to the automorphism on $\mathbb{C}(X)$ by σ also. It should be clear from the context whether σ is acting on the Riemann surface X or on the function field $\mathbb{C}(X)$. Our preference is to work with function fields $\mathbb{C}(X)$, since we can use Galois theory. In addition, two surfaces are the same if they have the same function field. This allows us to use several equations for the same surface, so we can use one equation which is convenient, but use a different equation to study points where the original equation has singularities.

(13) Assume $\mathbb{C}(Y)$ is a cyclic Galois extension of degree n of $\mathbb{C}(X)$. Therefore there is a meromorphic function y that satisfies an equation of the form $y^n - f = 0$, where $f \in \mathbb{C}(X)$ and $\mathbb{C}(Y) = \mathbb{C}(X)[y]$. Then $\mathbb{C}(Y)$ admits the automorphism σ such that σ is the identity on $\mathbb{C}(X)$ and $\sigma(y) = \epsilon y$, where ϵ is a primitive nth unit of unity. Assume λ is a symmetry or an automorphism of $\mathbb{C}(Y)$ of order 2, and assume $\lambda\sigma\lambda = \sigma^\alpha$. Let $\lambda(y) = a_0 + a_1 y + \ldots a_{n-1}y^{n-1}$, where each $a_i \in \mathbb{C}(X)$. Then

$$\sigma \circ \lambda(y) = a_0 + a_1(\epsilon y) + \ldots a_{n-1}(\epsilon y)^{n-1}, \text{ and}$$
$$\lambda \circ \sigma^\alpha(y) = \epsilon^{i\alpha}(a_0 + a_1 y + \ldots a_{n-1}y^{n-1}),$$

where $i = -1$ if λ is a symmetry, but $i = 1$ otherwise. We obtain that each $a_j = 0$ except for a_α, if $i = 1$, and $a_{n-\alpha}$, if $i = -1$. In particular, we have the following cases. If the automorphism λ has order two and $\lambda\sigma\lambda = \sigma^{-1}$, then $\lambda(y) = h_1/y$, for some $h_1 \in \mathbb{C}(X)$. If λ is a symmetry which commutes with σ, then $\lambda(y) = h_2/y$ for some $h_2 \in \mathbb{C}(X)$.

3. Examples

Example I: Surface with 4 symmetries of distinct topological character. In [4], a Riemann surface of genus two that possesses four symmetries yielding Klein surfaces of distinct topological type is constructed as follows. Let Λ be an NEC group with signature $(0; +; [-]; \{(2,2,2,4)\})$, and let c_0, c_1, c_2, c_3 and c_4 be canonical reflections whose successive products have the orders listed in the signature. Define the epimorphism $\theta : \Lambda \to G = \mathrm{D}_4 \oplus \mathrm{Z}_2 = \langle \hat{x}, \hat{y} \rangle \oplus \langle \hat{z} \rangle$, where $\hat{x}, \hat{y}$ and $\hat{z}$ each have order two and $\hat{x}\hat{y}$ has order four, by $\theta(c_0) = \hat{x}$, $\theta(c_1) = \hat{z}$, $\theta(c_2) = \hat{x}\hat{y}\hat{x}$, $\theta(c_3) = \hat{y}$, $\theta(c_4) = \hat{x}$. Let $\Gamma = \ker(\theta)$ and let Z be the Riemann surface corresponding to Γ. Note that $\hat{x}$, $\hat{y}$ and $\hat{z}$ induce symmetries of Z; in [4], it is shown $\hat{x}$ fixes one oval, the symmetry $\hat{z}$ is central, fixes one oval and is separating, the symmetry $\hat{y}$ fixes three ovals, and the symmetry $\hat{z}(\hat{x}\hat{y})^2$ is fixed point free and central. We will determine a defining equation for Z, determine explicit equations for the conformal automorphisms and determine the symmetries that correspond to $\hat{x}$, $\hat{y}$, $\hat{z}$, and $\hat{z}(\hat{x}\hat{y})^2$.

Note that the conformal automorphism group of Z is $G^+ = D_4$ and is generated by $\hat{x}\hat{z}$ and $\hat{y}\hat{z}$ which each have order two and their product has order four. The Fuchsian group Λ^+ corresponding to the Riemann double cover of Λ has signature $(0; +; [2,2,2,4])$ where the four points ramified by the canonical elliptical elements are on the boundary. By choosing complex conjugation to be the symmetry acting on the Riemann sphere, we can assume the four points have real coordinates.

We have the following table that lists the intermediate groups and Riemann surfaces given by the homomorphism θ. Note that $\widehat{\mathbb{C}}$ denotes the Riemann sphere and $\overline{Y}$ is the orbit space of Y under a symmetry.

Surface	Automorphism Group	Group Structure
$\overline{Y}$	$\Lambda/\Gamma = \langle \hat{x}, \hat{y}, \hat{z} \rangle$	$G \cong D_4 \oplus Z_2$
$Y = \widehat{\mathbb{C}}$	$\Lambda^+/\Gamma = \langle \hat{x}\hat{z}, \hat{y}\hat{z} \rangle$	$G^+ \cong D_4$
$X = \widehat{\mathbb{C}}$	$\Delta/\Gamma = \langle \hat{x}\hat{y} \rangle$	$\mathbb{Z}_4$
Z	Γ/Γ	id

There are two Riemann spheres to deal with, Y and X; usually it is preferable to choose coordinates on the covering Riemann sphere. In Y, there are four real points a_1, a_2, a_3 and a_4 ramified in X with ramification indices $2, 2, 2$ and 4 that are fixed points of elliptic elements that map to $\hat{x}\hat{z}$, $\hat{z}\hat{x}\hat{y}\hat{x}$, $(\hat{x}\hat{y})^2$ and $\hat{y}\hat{x}$ respectively. The group Δ does not contain $\hat{x}\hat{z}$ and $\hat{z}\hat{x}\hat{y}\hat{x}$ so the ramification index in the map $X \to Y$ is two there and there is only one point of X lying over each of the points a_1 and a_2. Since $(\hat{x}\hat{y})^2$ and $\hat{y}\hat{x}$ are contained in Δ, there is no ramification at the points a_3 and a_4, so there are two points of X lying over each of the points a_3 and a_4. We choose coordinates on the Riemann sphere X so that $x = 0$ and

$x = \infty$ are the points of X lying over a_4 and that $x = -1$ is the point lying over a_1. The meromorphic function field of X is $\mathbb{C}(x)$, the field of rational functions in x. There is an automorphism τ of X of order two which yields Y as the orbit space. By our choice of coordinates τ switches $x = 0$ and $x = \infty$ and fixes $x = -1$. This means that the automorphism must be $\tau(x) = 1/x$. Considering this as an automorphism of $\mathbb{C}(x)$, we see that the element $y = x + 1/x$ is fixed by τ. Note that $[\mathbb{C}(x) : \mathbb{C}(y)] = 2$, and x satisfies the polynomial $T^2 - yT + 1 = 0$. Therefore, $\mathbb{C}(y)$ is the function field corresponding to Y.

By our choice of coordinates, since $\tau(x) = 1/x$, $x = 1$ is also fixed by τ, so $x = 1$ must be the point lying over a_2. We have no control over the points lying over a_3, so they have coordinates $x = b$ and $x = 1/b$. In addition, since $b + 1/b$ must be real, either b is real or b lies on the unit circle, so $b = e^{i\theta}$ for some real number θ. However $y = x + 1/x$, so $x = -1$ lies over $y = -2$, $x = 1$ lies over $y = 2$ and $x = 0$ and $x = \infty$ lie over $y = \infty$. Therefore $a_1 = -2$, $a_2 = 2$ and $a_4 = \infty$. Since the points a_1, a_2, a_3 and a_4 are in order (according to the signature and the definition of the group homomorphism), we have that $a_3 > 2$. This is extremely important, since it implies that b is real, since if b were on the unit circle, then $-2 \leq b + 1/b \leq 2$. In addition, since b is real, out of the pair of numbers $\{b, 1/b\}$ we define b to be the number that is greater than one.

The points $x = 0$ and $x = \infty$ are ramified in Z with ramification index 4 and the points $x = b$ and $x = 1/b$ are ramified with ramification index 2. A defining equation for Z is

$$z^4 - f(x) = z^4 - x(x-b)^2(x-1/b)^2, \text{ so } z^4 - x(x^2 - cx + 1)^2 = 0,$$

where $b > 1$ and $c = b + 1/b > 2$. Since the degree of $f(x)$ is five, (8) yields that $z = \infty$ is a point with ramification index four. Note that Z has the automorphisms γ and τ where $\gamma(z) = iz$ and $\gamma(x) = x$. To compute the lift of $\tau(x) = 1/x$ to $\mathbb{C}(x, z)$ we use property (13) above to obtain that $\tau(z) = r/z$ for some rational function $r \in \mathbb{C}(x)$. This yields that

$$\begin{aligned} \tau(z^4) - \tau(f(x)) &= 0 \\ r(x)^4/z^4 - \frac{1}{x}\left(\frac{1}{x^2} - c\frac{1}{x} + 1\right)^2 &= 0 \\ r(x)^4 - \frac{1}{x^4}(x^2 - cx + 1)^2(x^2 - cx + 1)^2 &= 0. \end{aligned}$$

This means that $r(x) = (x^2 - cx + 1)/x$ and it is easily checked that

$$\tau(x) = 1/x \text{ and } \tau(z) = \frac{x^2 - cx + 1}{xz}$$

is an automorphism of Z. It is easy to see that γ and τ generate the conformal automorphism group $G^+ = \langle \gamma, \tau \rangle \cong D_4$. Complex conjugation σ is clearly a symmetry of Z, so Z admits the conformal and anticonformal automorphism group $G = \langle \gamma, \tau, \sigma \rangle$. Both τ and σ commute and γ^2 and σ commute. Since γ^2 is central in the conformal automorphism group G^+, this means that γ^2 is central in G. Note that $\tau\sigma$ commutes with both τ and σ and a calculation shows that $\tau\sigma$ commutes with γ, so $\tau\sigma$ is a central symmetry in G. Since γ^2 is central and conformal, $\gamma^2\tau\sigma$ is a central symmetry of G. Therefore the four distinct conjugacy classes of symmetries of Z have representatives σ, $\gamma\sigma$, $\tau\sigma$ and $\gamma^2\tau\sigma$, where the final two symmetries are central.

Since the genus of Z is two, Z is automatically hyperelliptic, in fact the orbit space of Z under the action of γ^2 is the Riemann sphere. Since our goal is to illustrate general techniques that apply to a variety of Riemann surfaces, we do not determine a defining equation for Z in the standard form $y^2 - f(x) = 0$ where $f(x)$ is squarefree, nor use it to determine properties of the symmetries of Z that would only be applicable to hyperelliptic surfaces.

We now show that σ fixes three ovals and $\tau\sigma$ is fixed point free. This means that σ corresponds to $\hat{y}$, which fixes three ovals, and $\tau\sigma$ corresponds to $\hat{z}(\hat{x}\hat{y})^2$ since they are both fixed point free. Out of the two remaining symmetries, $\gamma\sigma$ corresponds to $\hat{x}$, since neither is central, and $\gamma^2\tau\sigma$ corresponds to $\hat{z}$ because they are both central.

The symmetry σ: Since σ fixes both x and z, the action of σ on the Riemann surface Z is $\sigma(r,s) = (\overline{r}, \overline{s})$, so fixed points of σ have real coordinates. For any real value of r other than $0, b$, or $1/b$ there will be two real solutions (r, s^+) and (r, s^-) of $s^4 - r(r^2 - cr + 1)^2 = 0$ if $r > 0$ and no solution if $r < 0$. With this notation, we assume that $s^+ > 0$ and $s^- < 0$. To determine how many ovals this yields, we must determine how the branches of fixed points are joined at the points over $x = 0$, $x = b$, $x = 1/b$ and $x = \infty$. Recall we chose coordinates so that $b > 1$, $c > 2$ and $r^2 - cr + 1 = (r - b)(r - 1/b)$. Note that there is a unique point P lying over $x = 0$ and a unique point Q lying over $x = \infty$. From (6), at each point lying over $x = b$, the function $x - b$ has order two and z is a local parameter. Similarly $x - 1/b$ has order two and z is a local parameter at each point lying over $x = 1/b$. We define the function $w := z^2/((x - b)(x - 1/b))$ and note that $w^2 = x$ so $w(P) \neq 0$ for any point P lying over $(b, 0)$ or $(1/b, 0)$. In addition, $\mathbb{C}(w, z) = \mathbb{C}(x, z)$, so (z, w) gives an alternate set of coordinates of points on Z. We denote positive and negative values of a variable by s^+, s^-, t^+, t^-, and so on. There is only one point each lying over $x = 0$ and $x = \infty$; we denote them by P_0 and P_∞ respectively. The two points lying over $x = b$ have coordinates $(z, w) = (0, \sqrt{b})$ and $(z, w) = (0, -\sqrt{b})$. Similarly, the points lying over $x = 1/b$ have coordinates $(z, w) = (0, \sqrt{1/b})$ and $(z, w) = (0, -\sqrt{1/b})$

Case I, $\mathbf{x \geq b}$: Since P_∞ is the only point lying over $x = \infty$, as $r \to \infty$, the real points (r, s^+) and (r, s^-) must both approach P_∞ and the two branches must join there. If $r > b$ and $r \to b$ at each point of the form $(x, z) = (r, s^+)$, the meromorphic function w assigns the value

$$w(r, s^+) = \frac{(s^+)^2}{(r - b)(r - \frac{1}{b})} > 0,$$

because $r > b$. Therefore, as r approaches b, the points (r, s^+) must go through $(z, w) = (0, \sqrt{b})$. Since z is a local parameter and fixed points form continuous curves, after reaching $(0, \sqrt{b})$, z must become negative, so on the other side of $(0, \sqrt{b})$, z is negative but w remains positive. But at such points, $(x, z) = (r, s^-)$ and we have $w(r, s^-) = (s^-)^2/((r - b)(r - \frac{1}{b})) > 0$, which forces that $r > b$. Therefore, the branch of real points corresponding to (r, s^+) joins with the branch (r, s^-) at $(z, w) = (0, \sqrt{b})$. Since the branch also joints at ∞, we see that the points correspond to $x \geq b$ from one fixed oval of σ.

Case II, $1/b \le x \le b$: We now consider fixed points (r, s^+) and (r, s^-) where $1/b < r < b$. Note that as $r \to b$, $w(r, s^+) = (s^+)^2/((r-b)(r-\frac{1}{b})) < 0$, since $1/b < r < b$. Therefore the fixed points (r, s^+) go through $(z, w) = (0, -\sqrt{b})$. At this point, z equals zero; on the other side, z is negative and w remains negative. At such points $(x, z) = (r, s^-)$ and $w(r, s^-) = (s^-)^2/((r-b)(r-\frac{1}{b})) < 0$, which forces that $(r-b)(r-\frac{1}{b}) < 0$ so $1/b < r < b$. Therefore the two branches of points given by (r, s^+) and (r, s^-) join at $(z, w) = (0, -\sqrt{b})$. Now as $x \to 1/b$, at the points $(x, z) = (r, s^+)$ the function w has the value $w(r, s^+) = (s^+)^2/((r-b)(r-\frac{1}{b})) < 0$, since $1/b < r < b$. Therefore the fixed points (r, s^+) go through $(z, w) = (0, -\sqrt{1/b})$. At this point, z equals zero; on the other side, z is negative and w remains negative, so $w(r, s^-) = (s^-)^2/((r-b)(r-\frac{1}{b})) < 0$, which forces $(r-b)(r-\frac{1}{b}) < 0$ so $1/b < r < b$. Therefore the two branches of points given by (r, s^+) and (r, s^-) join at $(0, -\sqrt{1/b})$. Since they also joined at $(z, w) = (0, -\sqrt{b})$, this means they form one oval and go through both $(z, w) = (0, -\sqrt{1/b})$ and $(z, w) = (0, -\sqrt{b})$. So there is one fixed oval corresponding to $1/b \le x \le b$.

Case III, $0 \le x \le 1/b$: We now examine the behavior of fixed points (r, s^+) and (r, s^-) where $0 < r < 1/b$. At points of the form (r, s^+), as $r \to 1/b$, the function w satisfies $w(r, s^+) = (s^+)^2/((r-b)(r-\frac{1}{b})) > 0$, since $r < 1/b < b$. Therefore the fixed points (r, s^+) go through $(z, w) = (0, \sqrt{1/b})$. At this point, z equals zero; on the other side, z is negative and w remains positive, so points have the form (r, s^-) and $w(r, s^-) = (s^-)^2/((r-b)(r-\frac{1}{b})) > 0$, which forces $(r-b)(r-\frac{1}{b}) > 0$, so $r < 1/b$. Therefore the two branches of fixed points given by (r, s^+) and (r, s^-) where $0 < r < 1/b$ join at $(0, \sqrt{1/b})$. Since there is only one point lying over $x = 0$, they must join at that point also. This yields one fixed oval when $0 \le x \le 1/b$.

In total, there are three fixed ovals for the non-central symmetry σ. We now prove that $\tau\sigma$ is fixed point free.

The symmetry $\tau\sigma$: First note that if $(x, z) = (r, s)$ is a point on Z, then

$$\tau\sigma(r, s) = \tau(\overline{r}, \overline{s}) = \left(\frac{1}{\overline{r}}, \frac{\overline{r}^2 - c\overline{r} + 1}{\overline{rs}}\right).$$

Assume (r, s) is a fixed point of Z. This means that

$$r = \frac{1}{\overline{r}} \text{ and } s = \frac{\overline{r}^2 - c\overline{r} + 1}{\overline{rs}} = \left(\overline{r} + \frac{1}{\overline{r}} - c\right)\overline{s}^{-1}$$

which implies that $|r|^2 = 1$, so r lies on the unit circle and $|s|^2 = \overline{r} + 1/\overline{r} - c$, so $\overline{r} + 1/\overline{r} - c = 2\Re(r) - c > 0$. However, recall that $c = b + 1/b > 2$ and for a point on the unit circle $2\Re(r) < 2$, therefore this condition is never satisfied. Therefore $\tau\sigma$ is fixed point free.

This completes the proof that σ corresponds to $\hat{y}$, $\tau\sigma$ corresponds to $\hat{z}(\hat{x}\hat{y})^2$, $\gamma\sigma$ corresponds to $\hat{x}$, and $\gamma^2\tau\sigma$ corresponds to $\hat{z}$.

Example II: A surface not expressed as a plane curve. We now determine the number of ovals fixed by complex conjugation for a Riemann surface that is not expressed as a plane curve. Consider the Riemann surface W defined by the equations

$$\begin{aligned} F(x,y) =& y^6 - f(x) = y^6 - (x-a)(x-c)^3(x-1) \\ G(x,z) =& z^6 - g(x) = z^6 - x(x-b)^2(x-1) \end{aligned}$$

where a, b and c are real and $a < b < c < 0$. There are several ways to prove that F is irreducible over $\mathbb{C}(x)$, one of them is by using Eisenstein's criterion, since $x-1$, but not $(x-1)^2$ divides the constant term of F. Similarly, G is irreducible over $\mathbb{C}(x)$. Define X, Y, Z and W to be the Riemann surfaces corresponding to $\mathbb{C}(x)$, $\mathbb{C}(x,y)$, $\mathbb{C}(x,z)$ and $\mathbb{C}(x,y,z)$, respectively, and let $\epsilon = e^{\pi i/3}$. Each of the covers $Y \to X$ and $Z \to X$ has degree six. We now show that G is irreducible over the field $\mathbb{C}(x,y)$ so the cover $W \to X$ has degree 36. Letting T denote a variable, assume $\hat{G}(x,T)$ is an irreducible factor of $G(x,T)$ of degree n over $\mathbb{C}(x,y)$ and z is a root of $\hat{G}(x,T)$. Then $\mathbb{C}(x,y,z)$ is a splitting field of $G(x,T)$ over $\mathbb{C}(x,y)$, so it is a Galois field extension of $\mathbb{C}(x,y)$. In addition, $\mathbb{C}(x,y,z)$ is a splitting field for the polynomial $F(x,T)G(x,T)$ over $\mathbb{C}(x)$, so the extension $\mathbb{C}(x,y,z)$ is a Galois extension of $\mathbb{C}(x)$. However, given that z is a root of $G(x,T)$, the only automorphisms of $\mathbb{C}(x,y,z)$ over $\mathbb{C}(x,y)$ are of the form $z \mapsto \epsilon^k z$. If $k = \pm 1$ then $n = 6$, while $k = \pm 2$ implies $\hat{G}(x,T) = (T-z)(T-\epsilon^2 z)(T-\epsilon^4 z) = T^3 - h(x,y)$, for some meromorphic function $h(x,y) \in \mathbb{C}(x,y)$, so $G(x,T) = T^6 - h^2(x,y) = T^6 - x(x-b)^2(x-1)$. If $k = 3$ then $G(x,T) = T^6 - h^3(x,y) = T^6 - x(x-b)^2(x-1)$ for some meromorphic function $h(x,y) \in \mathbb{C}(x,y)$. However, if P is any point of Y lying over $x = 0$ these cases imply $2\mathrm{ord}_P(h) = \mathrm{ord}_P(x) = 1$ or $3\mathrm{ord}_P(h) = \mathrm{ord}_P(x) = 1$, which yield contradictions. Therefore $G(x,T)$ is irreducible, so $W \to X$ is a cover of degree 36 and W admits $\mathbb{Z}_6 \times \mathbb{Z}_6$ as an automorphism group.

From (6), there is one point on Y lying over $x = a$, one point lying over $x = \infty$ and three points lying over $x = c$. At any point P on Y lying over $x = c$, $z^6 - g(x) = 0$ at P, so $z^6 = g(c) \neq 0$, so there must be six points on W lying over each point P so there are 18 points of W lying over $x = c$. Similarly, there are six points of W lying over $x = a$. On Z there is one point lying over $x = 0$, two points lying over $x = b$ and two points lying over $z = \infty$. Reasoning as above, we conclude that there are six points on W lying over $x = 0$ and 12 points lying above $x = b$.

We now determine the number of points lying over $x = 1$. When $x = 1$,

$$\left(\frac{y}{z}\right)^6 = \frac{(1-a)(1-c)^3}{(1-b)^2} > 0,$$

so there must be at least six points of W that lie over $x = 1$. At each point P, $6\mathrm{ord}_P(y) = \mathrm{ord}_P(x-1)$, so $\mathrm{ord}_P(x-1) \geq 6$. This implies that there must be exactly six points on W lying over $x = 1$, $\mathrm{ord}_P(x-1) = 6$, y and z are local parameters at each point, and y/z has the values $\epsilon^k\lambda$ at these points for some positive real number λ. At $x = 1$ we will use the coordinates $(x, y, y/z)$; note that there are two points with real coordinates, namely $(x,y,y/z) = (1,0,\lambda)$ and $(x,y,y/z) = (1,0,-\lambda)$.

To examine $x = \infty$, we make the change of variables $t = 1/x$, $u = y/x$ and $\hat{z} = z/x$ we obtain

$$u^6 - t(1-at)(1-ct)^3(1-t), \text{ and } \hat{z}^6 - t^2(1-bt)^2(1-t),$$

so $\left(u^2/\hat{z}\right)^6 = 1$ when $t = 0$. Therefore there are six points of W lying over $t = 0$. At each point P, $y^6 - f(x) = 0$ implies that $6\mathrm{ord}_P(y) = 5\mathrm{ord}_P(x)$, so $\mathrm{ord}_P(x) = -6$ and $\mathrm{ord}_P(y) = -5$. An analogous argument for $z^6 - g(x) = 0$ yields that $\mathrm{ord}_P(z) = -4$.

When $x = \infty$ we make the change of variables (t, u, v) where $v = u^2/\hat{z} = y^2/(xz)$. When $t = 0$, $v^6 = 1$, so the six points of W lying over $x = \infty$ are $(t, u, v) = (0, 0, \epsilon^k)$, where $0 \leq k < 6$. At each of these points u is a local parameter, t has order six, and v is nonzero and equals a sixth root of unity. Two points have real coordinates, namely $(t, u, v) = (0, 0, 1)$ and $(t, u, v) = (0, 0, -1)$.

It may seem counterintuitive that there is one point of Y and two points of Z that lie over $x = \infty$, while six points of W lie over it. But consider the six points $(0, 0, \epsilon^k)$. Recall that $\mathbb{C}(x, y)$ is the fixed field under the automorphism that fixes x and y and maps z to ϵz. That automorphism maps $y^2/(xz) \mapsto \epsilon^{-1} y^2/(xz)$, so all of the points $(0, 0, \epsilon^k)$ are in the same orbit, and therefore get mapped to the same point of Y. So there is only one point on Y that is above $x = \infty$. On the other hand, the automorphism that maps y to ϵy but fixes x and z maps $y^2/(xz) \mapsto \epsilon^2 y^2/(xz)$, so there are two orbit spaces of points, which means there are two points of Z that lie above $x = \infty$.

The Riemann sphere X has the points $x = a$, $x = b$ $x = c$, $x = 0$, $x = 1$, and $x = \infty$ ramified with ramification indices 6, 3, 2, 6, 6, and 6 respectively. If g is the genus of W, the Riemann Hurwitz formula yields $2g - 2 = 36(-2 + 4(1 - 1/6) + 1 - 1/3 + 1 - 1/2)$ so $g = 46$.

Complex conjugation σ is an symmetry of W; we now determine the number of ovals fixed by σ. If (x, y, z) is a fixed point, then clearly $x \geq 1$ or $a \leq x \leq c$. We now show that σ fixes three ovals.

Case I, x ≥ 1: We prove that the fixed points with $x \geq 1$ from one oval. When $x > 1$ there are four branches of real points, which we denote by (x^+, y^+, z^+), (x^+, y^+, z^-), (x^+, y^-, z^+), (x^+, y^-, z^-) where each exponent refers to the sign of the variable. Near $x = \infty$, we see that (x^+, y^+, z^+) approaches $(t = 1/x, u = y/x, v = y^2/(xz)) = (0, 0, 1)$, since y^2, x and z are all positive. At that point, the local parameter u must change sign and become negative, $t = 1/x$ must remain positive (since there are no fixed points where $x < a$) and, w must remain positive. This means that y changes sign, but z does not change sign. Therefore the branches (x^+, y^+, z^+) and (x^+, y^-, z^+) join at $(t, u, v) = (0, 0, 1)$. The exact same argument shows that (x^+, y^+, z^-) and (x^+, y^-, z^-) join at $(t, u, v) = (0, 0, -1)$. We now examine the branches as x approaches 1, using the coordinates $(x, y, y/z)$, so $(x, y, y/z) = (1, 0, \lambda)$ and $(1, 0, -\lambda)$ are the points with real coordinates lying above $x = 1$. The branch (x^+, y^+, z^+) goes through the point $(1, 0, \lambda)$ at which point the local parameter y becomes negative. Since y/z remains positive, it means z also becomes negative. This means the branches (x^+, y^+, z^+) and (x^+, y^-, z^-) join at $(1, 0, \lambda)$. Similarly, (x^+, y^-, z^+) and (x^+, y^+, z^-) join at $(1, 0, -\lambda)$. Beginning with the branch (x^+, y^+, z^+), as x successively approaches $x = \infty$ and then $x = 1$, the fixed oval goes through the points and branches: (x^+, y^+, z^+), $(0, 0, 1)$, (x^+, y^-, z^+), $(1, 0, -\lambda)$, (x^+, y^+, z^-), $(0, 0, -1)$, (x^+, y^-, z^-), $(1, 0, \lambda)$ and (x^+, y^+, z^+). Therefore, there is one fixed oval when $x \geq 1$.

Case II, a ≤ x ≤ c: We prove that the fixed points with $a \leq x \leq c$ form two ovals. To do this, we need to determine coordinates at each of the points lying over $x = a$, $x = b$ and $x = c$.

(1) There are six points of W lying over $x = a$; at each point y is a local parameter and z is nonzero. Therefore we can use coordinates (x, y, z) and

note that there are two points with real coordinates $(x, y, z) = (a, 0, \gamma)$ and $(x, y, z) = (a, 0, -\gamma)$, where γ is positive and $\gamma^6 = g(a)$.

(2) There are 12 points of W lying over $x = b$, so $x - b$ has order three at each of them. From $z^6 - g(x) = 0$ we conclude z is a local parameter at each point and from $y^6 = f(x)$ we conclude that $y = \epsilon^k \delta$ at each point, where δ is a positive number with $\delta^6 = f(b)$. In addition, if $u = z^3/(x - b)$, then $u^2 = b(b - 1) \neq 0$ when $x = b$. At each point we use coordinates (z, y, u). Define the positive number ζ by $\zeta^2 = b(b-1)$. The twelve points lying over $x = b$ have the form $(z, y, u) = (0, \epsilon^k \delta, \pm\zeta)$ and the four points $(z, y, u) = (0, \pm\delta, \pm\zeta)$ are real.

(3) There are 18 points of W lying over $x = c$ so $x - c$ has order two at each of them. From $y^6 - f(x) = 0$ we conclude y is a local parameter at each point and and from $z^6 = g(x)$ we conclude that $z = \epsilon^k \eta$ at each point, where η is a positive number with $\eta^6 = g(c)$. In addition, if $v = y^2/(x-c)$, then $v^3 = (c-a)(c-1) \neq 0$ when $x = c$. At each point we use coordinates (y, z, v). Define the real number ξ by $\xi = \sqrt[3]{(c-a)(c-1)} < 0$. The 18 points lying over $x = b$ have the form $(y, z, v) = (0, \epsilon^k \eta, \epsilon^{2k} \xi)$. Note that there are only two real points: $(y, z, v) = (0, \pm\eta, \xi)$.

For each $x \neq b$ with $a < x < c$, there are four branches of fixed points under complex conjugation σ. Using the notation x_b^+ to mean $x > b$ and x_b^- to mean $x < b$, the four branches are (x_b^+, y^+, z^+), (x_b^+, y^+, z^-), (x_b^+, y^-, z^+), (x_b^+, y^-, z^-). As x approaches c, we use the coordinates $(y, z, v = y^2/(x - c))$ given in (3) and observe that the branch (x_b^+, y^+, z^+) goes through the point $(0, \eta, \xi)$ where the local parameter y must become negative and v and z must remain negative and positive, respectively, which means that $x < c$ again. Therefore the branch (x_b^+, y^+, z^+) joins the branch (x_b^+, y^-, z^+) at $(0, \eta, \xi)$. Similarly, the branches (x_b^+, y^+, z^-) and (x_b^+, y^-, z^-) join at $(y, z, v) = (0, -\eta, \xi)$. We now let x approach b from the right and switch to the coordinates $(z, y, u = z^3/(x - b))$ defined above in (2). The branches are joined in the following manner:

(1) The branch (x_b^+, y^+, z^+) goes through the point $(z, y, u) = (0, \delta, \zeta)$, where the local parameter z changes sign but y and $u = z^3/(x-b)$ do not change sign. This means that now $x < b$, so the branch (x_b^+, y^+, z^+) joins the branch (x_b^-, y^+, z^-) where now $x < b$.

(2) The branch (x_b^+, y^+, z^-) goes through the point $(z, y, u) = (0, \delta, -\zeta)$, where the local parameter z changes sign but y and $u = z^3/(x - b)$ do not change sign. This means that now $x < b$, so the branch (x_b^+, y^+, z^-) joins the branch (x_b^-, y^+, z^+) where now $x < b$.

(3) The branch (x_b^+, y^-, z^+) joins (x_b^-, y^-, z^-) at $(z, y, u) = (0, -\delta, \zeta)$.

(4) The branch (x_b^+, y^-, z^-) joins (x_b^-, y^-, z^+) at $(z, y, u) = (0, -\delta, -\zeta)$.

As x approaches $x = a$, we continue to use coordinates (x, y, z) where y is a local parameter and z is not zero. Therefore y will change sign and z will not change sign when a fixed oval goes through a fixed point with $x = a$. Therefore the branch (x_b^-, y^+, z^+) joins the branch (x_b^-, y^-, z^+) at $(x, y, z) = (a, 0, \gamma)$. Similarly the branch (x_b^-, y^+, z^-) joins the branch (x_b^-, y^-, z^-) $(x, y, z) = (a, 0, -\gamma)$.

We now show that the branches join to form two ovals when $a \leq x \leq c$. Beginning with the branch (x_b^+, y^+, z^+), and letting x approach $x = c$, then decrease to $x = a$ and then increase to $x = c$ again, the fixed oval goes through the following points and branches: (x_b^+, y^+, z^+), $(y, z, v) = (0, \eta, \xi)$, (x_b^+, y^-, z^+), $(z, y, u) = (0, -\delta, \zeta)$, (x_b^-, y^-, z^-), $(x, y, z) = (a, 0, -\gamma)$, (x_b^-, y^+, z^-), $(z, y, u) = (0, \delta, \zeta)$, and (x_b^+, y^+, z^+) which closes the loop. Similarly the fixed points that contain the branch (x_b^+, y^+, z^-) form another oval.

Therefore σ fixes a total of three ovals on W.

References

[1] Emilio Bujalance, Francisco Javier Cirre, José Manuel Gamboa, and Grzegorz Gromadzki, *Symmetries of compact Riemann surfaces*, Lecture Notes in Mathematics, vol. 2007, Springer-Verlag, Berlin, 2010. MR2683160 (2011h:30062)

[2] Computer Aided Riemann Surfaces, http://www.math.fsu.edu

[3] William Fulton, *Algebraic curves. An introduction to algebraic geometry*, W. A. Benjamin, Inc., New York-Amsterdam, 1969. Notes written with the collaboration of Richard Weiss; Mathematics Lecture Notes Series. MR0313252 (47 #1807)

[4] Grzegorz Gromadzki and Ewa Kozłowska-Walania, *On the real nerve of the moduli space of complex algebraic curves of even genus*, Illinois J. Math. **55** (2011), no. 2, 479–494 (2012). MR3020692

[5] I. R. Shafarevich, *Basic algebraic geometry*, Springer-Verlag, New York-Heidelberg, 1974. Translated from the Russian by K. A. Hirsch; Die Grundlehren der mathematischen Wissenschaften, Band 213. MR0366917 (51 #3163)

Purdue University Calumet

Current address: Hammond, IN, USA 46323

E-mail address: `turbek@purduecal.edu`